CHEMISTRY IN AMERICA, 1876–1976

CHEMISTS AND CHEMISTRY

A series of books devoted to the examination of the history and development of chemistry from its early emergence as a separate discipline to the present day. The series will describe the personalities, processes, theoretical and technical advances which have shaped our current understanding of chemical science.

ARNOLD THACKRAY, JEFFREY L. STURCHIO,
P. THOMAS CARROLL, and ROBERT BUD

CHEMISTRY IN AMERICA 1876–1976

Historical Indicators

D. REIDEL PUBLISHING COMPANY

A MEMBER OF THE KLUWER ACADEMIC PUBLISHERS GROUP

DORDRECHT / BOSTON / LANCASTER

Library of Congress Cataloging in Publication Data

Main entry under title:

Chemistry in America, 1876–1976.

(Chemists and chemistry)
Bibliography: p.
Includes index.
1. Chemistry—United States. I. Thackray, Arnold, 1939– . II. Series.
QD18.U6C47 1984 540'.973 84–13388
ISBN 978-90-277-2662-9 ISBN 978-94-015-1124-7 (eBook)
DOI 10.1007/978-94-015-1124-7

Published by D. Reidel Publishing Company,
P.O. Box 17, 3300 AA Dordrecht, Holland

Sold and distributed in the U.S.A. and Canada
by Kluwer Academic Publishers,
190 Old Derby Street, Hingham, MA 02043, U.S.A.

In all other countries, sold and distributed
by Kluwer Academic Publishers Group,
P.O. Box 322, 3300 AH Dordrecht, Holland

"That's the gist of what I want to say. Now get me some statistics to base it on."

To our wives
— who count more

CONTENTS

PREFACE

This study is an outgrowth of our interest in the history of modern chemistry. The paucity of reliable, quantitative knowledge about past science was brought home forcibly to us when we undertook a research seminar in the comparative history of modern chemistry in Britain, Germany, and the United States. That seminar, which took place at the University of Pennsylvania in the spring of 1975, was paralleled by one devoted to the work of the "Annales School". The two seminars together catalyzed the attempt to construct historical measures of change in aspects of one science, or "chemical indicators". The present volume displays our results.

Perhaps our labors may be most usefully compared with the work of those students of medieval science who devote their best efforts to the establishment of texts. Only when acceptable texts have been constructed from fragmentary and corrupt sources can scholars move on to the more satisfying business of making history. So too in the modern period, a necessary preliminary to the full history of any scientific profession is the establishing of reliable quantitative information in the form of statistical series.

This volume does not offer history. Instead it provides certain elements – indicators – that may be useful to individuals interested in the history of American chemistry and chemical industry, and suggestive for policy. Statisticians may properly be distressed by the lack of technical sophistication in the pages that follow. Sociologists will deplore our cavalier use of certain terms. Historians can rightly deprecate our failure to focus on the unique, and on the individual. Scientists should note our inability to take the measure of their ideas, and purists of all kinds will unite in abhorrence of our resort to "guesstimates", approximations, and interpolations. Even so, we trust some people will find some use in some of what is reported here.

Two technicalities deserve explanation here. First, our references follow a modified social science format. The symbol preceding the author's name designates which section of our bibliography contains the entry in question: IA, federal government sources of data (pp. 505ff); IB, other sources of data (pp. 511ff); II, bibliography, historiography, and methodology (pp. 520ff); or III, other books and articles (pp. 526ff). Second, when compiling time series, we did not necessarily stop at the terminal year in our title, but rather included the most recent data available at the time of our analyses. As time permitted, information appearing subsequently was added to the tables, although not analyzed in full. Even that level of inclusion proved impossible for some of the most recent of available compilations, including the results of the 1980 Census of Population. Readers interested in more recent data should consult the bibliographic essay (pp. 497–504) and the notes to the tables.

ACKNOWLEDGMENTS

The work of constructing even the simplest time series of chemical indicators for any period before the 1950s is tedious, complicated, and labor-intensive to a degree we would have found difficult to credit before our innocent embarkation on the task. We are correspondingly grateful to those colleagues, friends, and associates who have sustained our labors even in the darkest hour of suspect sources, dubious data, and conflicting curves.

We are indebted to the National Science Foundation for financial support, and to Murray Aborn and Ronald J. Overmann for enthusiasm beyond the call of duty. The members of the Subcommittee on Science Indicators of the Social Science Research Council have been unfailingly helpful. Several of them belong to that band of consultants we have importuned with our inquiries; our thanks go especially to Daryl Chubin, June Z. Fullmer, Karl Hufbauer, Roberta B. Miller, Robert Parke, Derek J. de Solla Price, Nathan Rosenberg, Henry Small, Spencer R. Weart, Robert Wright, and Harriet Zuckerman. Among generous commentators on a preliminary report of our findings were Richard A. Easterlin, Yehuda Elkana, Louis Galambos, Zvi Griliches, Jean-Claude Guédon, Gerald Holton, Daniel J. Kevles, Russell McCormmach, Everett Mendelsohn, John Servos, and John Ziman. Robert K. Merton and Edward Shils offered thoughtful encouragement.

Time and again, we have profited from the assistance of libraries and officials in government and scientific agencies. Our thanks go to the staffs of the U.S. National Center for Education Statistics; of the U.S. Bureau of the Census; of the U.S. Bureau of Labor Statistics (especially Maurice Moylan); of the U.S. National Archives and Records Service; and of the U.S. Patent and Trademarks Office (especially Gary Robinson of the Office of Technology Assessment and Forecast). We also depended heavily upon the patience and good will of personnel at the American Chemical Society (especially Robert Jones and David M. Kiefer); the National Opinion Research Center; the Free Library of Philadelphia; and the University of Pennsylvania libraries.

Colleagues, staff, and students in the Department of History and Sociology of Science at the University of Pennsylvania encouraged us with their interest, humor, enthusiasm, and calculated incredulity. Discussions in the Department

Workshop and in seminars helped remove some rough edges from our work. We are especially indebted to Thomas P. Hughes and Robert E. Kohler for their comments. Simon Baatz and James Capshew provided capable research assistance. Marthenia Perrin, Elizabeth Cooper, Sylvia Dreyfuss, and Judy Weglarz deserve our gratitude for unfailing secretarial support. Finally, we thank those myriad friends and relations who bore with fortitude their recruitment to the cause of chemical indicators.

Philadelphia, 4 July 1983

ILLUSTRATIONS

Readers interested in a more detailed view of the data, including estimates of its accuracy and technical limitations, should consult the appendixes and tables.

Unless the caption states otherwise, information in each figure refers only to the United States.

TEXT TABLES

CHAPTER 1

ORIENTATIONS

1.1. American Chemistry in Cultural Context

The modern natural sciences are among the most distinctive institutions of industrial society. Indeed, they may be used as definers of that society. The countries that constitute the core group of advanced industrial nations are set off from the rest of the world as much by their contributions to the specialized literature of scientific research as by their gross national products. At present, the United States dominates. According to one estimate, it is responsible for 33% of "world GNP" and 41% of all publishing scientists.[1] The fate of the natural sciences in the United States has therefore, and properly, been the subject of increasing attention in recent years. The main thrust of inquiry has been toward questions of policy. By comparison, less attention has been directed toward understanding the historical forces at work in the development of the American "scientific estate". Little is known of the dynamics by which the sciences have evolved in America over the past one hundred years, or of the cultural constellations in which those sciences are embedded.[2]

American chemistry offers a strategic site for research. For a start, chemistry occupies a central place among the natural sciences. Its methods, concepts, theories, and techniques merge imperceptibly with those of physics

[1] In Ernest Gellner's brilliant apothegm, "science is the mode of cognition of industrial society" (III, Gellner, 1964, 179). Thomas S. Kuhn has made the point that "one need not . . . inflate the importance of history of science to suppose that since 1870 science has assumed a role which no student of modern socio-economic development may responsibly ignore" (III, Kuhn, 1971, 287). The estimates of U.S. contributions to world GNP and science are taken from III, Price, 1970, 106, 109.

[2] However, the prospects for a cultural history of American science and technology now seem promising. For historiographic studies see II, Dupree, 1966; II, Hughes, 1978; and II, Rosenberg, 1983. Changes in the field are evident from a comparison of the range of issues addressed in II, Van Tassel and Hall (eds.), 1966; II, Reingold (ed.), 1979. Reassessment of the antebellum period has proceeded most rapidly, as indicated in II, Kohlstedt, 1977, but such monographs as III, Noble, 1977; III, Kevles, 1978; III, Oleson and Voss, 1979; III, Kohler, 1982; and III, Rossiter, 1982, point to the increasing attention given to modern eras. For a useful bibliographic guide, see II, Rothenberg, 1982.

on the one side and biology on the other. Chemistry plays a major role in the dynamic interchanges between science and technology which are themselves a main characteristic of modern science. There is no obvious point at which chemistry ceases and chemical engineering begins, nor at which the activities of academic researchers may be divorced from those of their industrial colleagues. Chemical knowledge, chemical activity, and chemical products are pervasive in American society. Moreover, chemistry is the largest and oldest of the natural sciences in the United States (as measured by degrees awarded in the field, and by the date of formation of its main learned society).

While American chemistry offers a strategic site for research, it also provides the investigator with special problems. The history of American chemistry has been greatly neglected, and little reliable literature exists.[3] Problems of definition are acute. The scale and complexity of the phenomena to be studied are such as to pose severe analytical problems. A full, cultural history of American chemistry remains a distant goal.

1.2. Indicators of Trends in American Chemistry

This study has the more modest objective of establishing some quantitative indicators that may provide context for any future history, and for contemporary discussion of policy. The work is limited to the past century, that is to the period from the foundation of the American Chemical Society in 1876 to its centennial in 1976 (though we have not hesitated to delve more deeply into the past, or to bring a series of indicators more nearly up-to-date whenever possible). Our aim throughout has been to establish long-run trends in statistical series -- creating what might be called "chemical indicators", by analogy with social indicators and science indicators.

In recent years considerable effort has been devoted to collecting and

[3] Section Dwb of the *Isis Cumulative Bibliography*, on chemistry in North America, lists only twenty-seven entries (II, Whitrow, 1976, 184–185). There has been no synoptic history since Edgar Fahs Smith's *Chemistry in America* (III, E. F. Smith, 1914; see also III, E. F. Smith, 1927). However, aspects of nineteenth-century American chemistry are treated in III, Beardsley, 1964, and Kenneth L. Taylor has recently provided a two-century survey (III, Taylor, 1976). The American Chemical Society has sponsored historical publications in commemoration of major anniversaries: see III, ACS, 1951a; III, Browne and Weeks, 1952; III, C&EN, 1976; IB, Miles, 1976; and IB, Skolnik and Reese, 1976. A pioneering set of essays on chemical engineering appear in III, Furter (ed.), 1980 and 1982.

refining statistics upon different aspects of contemporary scientific activity. The series of biennial reports which began with *Science Indicators 1972* gives a convenient example of the results of this concern with measuring the "health" of science. Policy-related work of this kind on indicators of American science offers one important reference point for this study.[4]

Because of the very *size* of the community, and because of the pioneering interest of the American Chemical Society, much statistical information on current American chemistry has been published within the past decade. As is the case with science indicators, this material stresses matters of policy rather than historical or cultural questions. Still, it provides another reference point.[5] The size of the chemical community has also meant that chemists are the one group of scientists for whom long-run information is contained in the reports of the U.S. Bureau of the Census. That material was drawn upon freely, together with an eclectic array of other sources offering statistical information.

The institution of chemistry consists of a complex web of political, social, intellectual, material, and emotional alliances through which many disparate groups seek to achieve a range of common purposes. To arrive

[4] The relation between science indicators and social indicators is explored in II, Duncan, 1978; see also II, Zuckerman and Miller, 1980. Among recent surveys of social indicators research, see II, Ferriss, 1979; and II, Rossi and Gilmartin, 1980. For a comprehensive bibliography of the field, see II, Wilcox *et al.*, 1972, and II, Gilmartin *et al.*, 1979.

The National Science Board has published six statistical surveys of the 'health' of American science in the *Science Indicators* series (IA, NSB, 1973, 1975, 1977, 1979, 1981, and 1983); see II, Elkana *et al.* (eds.), 1978. The science indicators movement is the most recent manifestation of a tradition of interest in the quantitative measurement of science; for discussion of earlier contributions, see II, Thackray, 1978, and II, Merton, 1977, 24–59. See also II, Leupold *et al.*, 1982.

[5] The ACS sponsors annual surveys of the economic status of chemists, and of degree conferrals in chemistry and chemical engineering; see IB, ACS, 1978a; and IB, ACS, 1978c. The Office of Manpower Studies coordinates ACS data collection and produces useful statistical compendia on *Professionals in Chemistry*, which first appeared in 1975. *Chemical and Engineering News* produces an annual survey of the chemical industry. The ACS has also sponsored a sociological survey of the chemistry profession (III, Strauss and Rainwater, 1962) and a survey of *Chemistry in the Economy* (IB, ACS, 1973) which includes useful statistical information on the chemical industry and the chemical labor force.

The Chemical Abstracts Service conducts extensive quantitative investigations of the growth and structure of chemical literature; see, for example, IB, Baker, 1976. The Westheimer Report, sponsored by the Committee on Science and Public Policy of the National Academy of Sciences, includes quantitative material on American chemical publications; see IB, NAS, 1965, 30–38.

at statistics for the whole by discovery and investigation of all constituent parts would obviously be desirable. However such a course implies research on a scale far beyond our present resources. We therefore decided that, rather than rest content with only certain subsets and an unknown bias, we would seek for an overview, using "macro-indicators". It is those indicators which have been our main concern. It would certainly have been simpler to seek data on, say, chemistry faculty at Harvard University rather than on all chemistry faculty in America, on research expenditures by Du Pont rather than by all chemical industry, or on chemists in one federal laboratory rather than on all chemists employed by the government. "Micro-indicators" of the former kind are much to be desired, and some initial use of them is made in the later stages of this study.[6]

A decision to concentrate on aggregate statistics does not eliminate problems of definition. The boundaries of American chemistry have been neither rigid nor static. Accordingly, variant definitions are used on different occasions, dependent upon the context. Usually the term "chemist" is taken to include assayers and metallurgists. From time to time, chemical engineers, biochemists, or high school chemistry teachers are included in the groupings discussed. These choices depend in part on whether chemistry is being considered as occupation, profession, or discipline. The bounds of chemistry will be set appropriately wide. For instance, discussion of the role of chemistry in industry may involve one or more of the "chemicals and allied products" industries, the more inclusive "chemical process" industries, other manufacturing industry, or non-manufacturing industry. These problems are considered at appropriate points in the body of the text.

The main focus of this study is upon explicit-discovered indicators. These indicators are measures of change that are routinely and unobtrusively produced by one or another agency – usually, but not necessarily, a government agency. Examples include statistics on PhD production, high school enrollment, or the number of patents issued. Their use as indicators awaits only their discovery, which can be no mean task. Complete dependence on such explicit-discovered indicators would imply commitment to an interim empiricism. Instead "explicit-invented" indicators have also been used wherever possible. These indicators – measures that are constructed deliberately – are usually "theory-laden" series of normative interest.[7]

[6] The issue of aggregate versus fine-structure analysis is raised in II, Holton, 1978, 45–47.

[7] This classification of science indicators is set forth in the Introduction to II, Elkana *et al.* (eds.), 1978, 3–4. Gerald Holton (II, 1978) offers astute observations on the

As with economic or social indicators, so too it would be a mistake to place undue weight on any single chemical indicator. What one seeks are clusters of indicators which, in conjunction, reveal an underlying trend that otherwise would be obscured by the noise of everyday circumstance. Again, as with other indicators, there are obvious and immediate hazards in extrapolating past and present trends into future contexts.

1.3. Indicators and History

By establishing some aggregate measures of change in American chemistry as an institution, a context is created for subsequent disaggregated studies and detailed historical analyses. Simon Kuznets and those French historians associated with *histoire sérielle* have shown two very different ways of traversing the ground that lies between statistical investigation and historical understanding. The chemical indicators contained in this report are several stages removed from that analysis of enduring trends, short-run variations, conjunctures, and visible events toward which they point.[8]

---

distinction between quantitative and qualitative indicators of science and the tension between numerical data and theory-laden indicators of science; see also II, Ziman, 1978, for an engaging analysis of the dimensions of normative interest underlying the construction of various science indicators.

[8] For a classic statement of the importance of understanding statistical trends in history, see II, Kuznets, 1951. Among French historians of the *Annales* school, Pierre Chaunu and François Furet have been persuasive advocates of *histoire sérielle*; see II, Chaunu, 1970; II, Chaunu, 1973; and II, Furet, 1971. A quotation from Furet establishes the affinity between chemical indicators and his brand of *histoire sérielle*:

> Serial history offers the conclusive advantage . . . of substituting for the elusive 'event' of positivist history the regular repetition of data selected or constructed by reason of their comparability. It does not, however, claim to give an exhaustive account of the whole body of evidence, nor to be an overall system of interpretation, nor to be a mathematical formulation. On the contrary, the division of historical reality into series leaves the historian confronted with this material broken down into different levels and subsystems, among which he is at liberty to suggest internal relationships if he chooses. . . . [To] the classic question "What is a historical fact?" quantitative and serial history both give a new answer which transforms the historian's raw material – time. (II, Furet, 1971, 153)

On the achievements and shortcomings of *Annales* history, see II, Hexter, 1972; and II, Stoianovich, 1976. On the state of quantitative history, see II, Fogel, 1975; II, Sprague, 1978; II, Kousser, 1980; and II, Bogue, 1981. An International Symposium on Quantitative Methods in the History of Science was held at the University of California, Berkeley, in August, 1976; see II, Carroll, Sturchio, and Bud, 1976. II, Hahn, 1980, is a convenient guide to quantitative studies in the history of science and related fields.

These chemical indicators are also different in purpose and possibility from three types of quantitative investigation in the history of science that have become familiar in recent years. Most familiar is bibliometric work – a term that embraces journal and author counts, citation analysis, and its recent refinement in co-citation studies. The great bulk of that work has been focused on sharply-limited specialty groups, short time intervals, and contemporary or near-contemporary periods. Policy concerns have been paramount. However, the historical possibilities inherent in bibliometric work have been repeatedly stressed, as much in the recent writings of Price and Garfield as in the pioneering investigation of Cole and Eales. When pursued over long periods of time, bibliometric studies can become a special class of historical indicators, dealing with the official, public record of science.[9]

The other two sorts of quantitative investigation are more concerned with the static portrayal of groups than with those long-run trends essential to the creation of historical indicators. Prosopography provides a way to obtain the social composition of a scientific community. The technique may enable a link to be made between the social interests of a group and its intellectual programs. Broader statistical portraits of individual scientific disciplines have also attracted increasing attention. These portraits are usually static or concerned only with particular years at close intervals.[10]

Among the reasons directing attention to these less dynamic measures are the technical difficulties inherent in the construction of extended statistical series of the kind reported in this study. No one who has not wrestled with them can truly appreciate the recalcitrance of the historical sources on which a work like this depends. However, historical indicators have particular

[9] II, Cole and Eales, 1917, is the pioneering bibliometric study. Eugene Garfield (the originator of the *Science Citation Index*) and Derek de Solla Price were early advocates of the utility of citation studies for the history of science; see II, Garfield, Sher, and Torpie, 1964; and II, Price, 1965. For a comprehensive review of the burgeoning activity in this field, consult II, Narin and Moll, 1977; II, Cozzens *et al.*, 1978; and II, Garfield, 1979, esp. Chapters 7 and 8. See also II, Edge, 1979, and II, Chubin, 1983, 91–106.

[10] An excellent review of prosopography is II, Stone, 1971. On the use of collective biography in history of science, see II, Shapin and Thackray, 1974, and II, Pyenson, 1977.

III, Forman, Heilbron, and Weart, 1975, a detailed portrait of the academic physics community in 1900, is *sui generis*. III, MacLeod and Andrews, 1967, is a compendium of science statistics for Victorian and Edwardian England. III, Pfetsch, 1974, deals quantitatively with science in Germany from 1750 to 1914. IB, Kevles and Harding, 1977, provides a comparative quantitative portrait of the chemistry, physics, and mathematics disciplines in late nineteenth- and early twentieth-century America.

promise where the science of the last one hundred years is concerned. The size of the activity, the complexity of its interrelations, the availability of at least some statistical information, and the persistence of certain enduring trends make indicators a rewarding source for the historian.

1.4. The Structure of This Study

The institution of chemistry in America may be approached from many different angles. In Chapter 2 chemistry is treated first as an occupation – "what occupies one, the means of filling one's time, temporary or regular employment". Such an approach implies a broad, inclusive view of chemists and chemistry. In the second part of the chapter, a somewhat narrower approach will be taken in terms of chemistry as a profession. A professional will be viewed as one who "is trained and skilled in the theoretic or scientific parts of a trade or occupation, as distinct from its merely mechanical parts". In contrast to the *occupation* of chemistry as measured in, say, the official Census tabulations, where self-designation results in a promiscuous mixing together of those trained in theory with those simply occupied in chemistry as mechanical activity, our accounting of the *profession* of chemistry will focus on those who possess certain intellectual skills. While membership implies possession of these skills, the activity of the profession is best understood in terms of the application of such skills "to the affairs of others or in the practice of an art".[11]

Chapters 3 and 4 consider two of the principal contexts in which chemists are employed and their skills are utilized. Both contexts offer support to the institution of chemistry in ways that transcend the question of employment. Chemical education serves social purposes beyond the employment of the present – or the training of the next – generation of chemists. Similarly, chemical industry fulfills many functions for the institution of chemistry, even as chemists play varied roles in chemical and non-chemical industries. Chapter 5 returns from these wider considerations to take a closer look at the employment of chemists. Doing so helps set the stage for a consideration of the discipline of chemistry in Chapter 6.

Chemistry is not simply an occupation, a profession, a means of education

[11] *Oxford English Dictionary*. Definitions of professionals range from Everett C. Hughes's terse 'professionals *profess*' to the list of twenty-three essential traits adduced by Geoffrey Millerson in *The Qualifying Associations* (II, 1964). The vast literature on professionalization is thoughtfully addressed in II, Freidson, 1977.

and socialization, or an industrial activity of great power and moment. It is also a scientific discipline, a part of mankind's intellectual quest. The practitioners of the discipline are trained and skilled in the theoretic parts of chemistry. They are committed to the articulation and advancement of those theoretic parts considered as an intellectual system. The emergence of the chemical discipline is a relatively modern phenomenon. That phenomenon is inextricably linked to the increased size of chemistry as an occupation, to the need to find means of symbolic exchange within the expanded profession, to the forms of socialization appropriate to an industrial world, and to the emergence of the university as the focus of research and advanced training (at least in Britain, Germany, and the United States). It is in the discipline that the institution of chemistry finds its symbols and its intellectual resources.[12] The discipline is but one facet of the institution, linked to other parts in a host of subtle relationships. A focus on the chemical discipline in America therefore provides a fitting terminus for our investigations. It places matters treated in earlier sections in perspective, and it raises questions for future research.

[12] Historians and sociologists of science have directed increasing attention in recent years to the study of scientific disciplines. Two early and influential programmatic statements are II, McCormmach, 1971, and II, Rosenberg, 1964; see also II, Rosenberg, 1970 and 1979, and III, Kohler, 1982. For a synoptic view of recent literature on scientific disciplines, see II, Edge and Mulkay, 1976, Chapter 10; II, Lemaine *et al.* (eds.), 1976; II, Weisz and Kruytbosch, 1982; and II, Graham *et al.* (eds.), 1983.

CHAPTER 2

CHEMISTRY AS OCCUPATION AND PROFESSION

2.1. Chemistry as Occupation

INDICATOR HIGHLIGHTS:

In 1870 about 800 people were employed as chemists in the United States. By 1970 the number had increased more than a hundredfold, to around 110 000. Growth at a rate equivalent to about 5% per year has been the enduring trend.

Expressed as a fraction of the work force, chemists have increased from 2 to 15 per 10 000 over the past century.

Measures of chemistry as an occupation may be derived from several sources. Among them, the reports of the U.S. Bureau of the Census offer the most carefully defined, most homogeneous, and longest time series on the aggregate level. This section will be based primarily upon Census information with discussion, when appropriate, of both variant estimates for and alternative definitions of the occupation of chemist.[1]

2.1.1. THE DIFFERENTIATION OF OCCUPATIONS

In successive Census reports one may follow the exfoliation of chemistry as a means of livelihood by watching the ever-increasing number of subcategories used to organize the known varieties of chemist. In 1880 six subcategories sufficed: assayer; chemist (practicing); metallurgist; metal worker (scientific); metal refiner; and mineralogist. A generation later, in

[1] The Bureau of the Census is recognized for the consideration its analysts give to the reliability of its reports. For example, William Kruskal singles out the demographic sections of Census reports as models of attention paid to statistical errors (II, Kruskal, 1978, 155–156). The *Technical Papers* published periodically by the Bureau are especially valuable; see II, USBC, 1960a; II, USBC, 1972; and II, Gonzalez *et al.*, 1975. The Bureau takes care to provide homogeneous time series for occupational statistics, devoting great effort to recalculations; see IA, Edwards, 1943; IA, Kaplan and Casey, 1958; and IA, USBC, 1975b, 125–126.

1910, ten subcategories were needed. By 1940 there were thirty, while by 1970 the number had risen to seventy-five (see Table 2.1 for the 1970 list). Certain of the subcategories show remarkable endurance over time. "Assayer" appeared at least as early as 1880 and was still in use in 1970, while "analytical chemist" is represented from 1910 to the present. Even so, the separately recognizable occupation of gold assayer is listed from 1910, while the 1970 classification recognizes not simply analytical chemists, but also chemical analysts, food analysts, and pharmaceutical analysts, among others.[2]

These shifts in categorization serve as reminders that, because it is an institution, a social construct, chemistry changes over time. But growth has been such a deep-rooted and enduring aspect of American chemistry in the past century that almost all change has been by way of addition and differentiation, rather than by decay and dissolution, of occupational groups. A graphic example in the case of just one small subgroup, food chemists, is provided in Figure 2.1–1. Similar trees could be constructed for each of the other subgroups in Table 2.1. This phenomenon of growth by addition – rather like the "twigging" of a tree or bush – is one enduring trend in American chemistry in the last century. To be sure, some occupational categories have died out. An example is "sampler, sugar refinery" which disappeared between 1910 and 1920. But such events have been comparatively rare.

The addition of new specialties and the differentiation of old ones is not simply a process of accumulation. As the sapling is not the mature tree in either structure or ecological function, so the 1870 occupational category of chemist is not to be equated with the 75 groupings into which it had been transformed a century later. Nonetheless, certain continuities are obvious.

In the 1870s most chemists did routine testing for industrial concerns, performing analyses based upon classical gravimetric and volumetric methods. The importance of inorganic chemistry followed naturally from the centrality

[2] The 1880 subcategories were obtained from II, USBC, 1880, and IA, USDI, 1883, II, Table 103, 1368. This breakdown was used until 1910, when a new classification was adopted (II, USBC, 1915, 396). New occupational classifications for chemists were also developed for each succeeding Census, and one may trace the differentiation of the occupation through the steady expansion of subcategories from 1920 to 1970; see II, USBC, 1921, 161; II, USBC, 1930, 186; II, USBC, 1940, 27–28; II, USBC, 1950, 2–3; II, USBC, 1960b, 3–4; II, USBC, 1971, O–5 to O–6. The 1980 Census restructured the subcategories, but details were not available at the time of our analyses.

It would be interesting to construct time series of separate subcategories of chemistry as occupation, but figures for the subcategories are not separately reported.

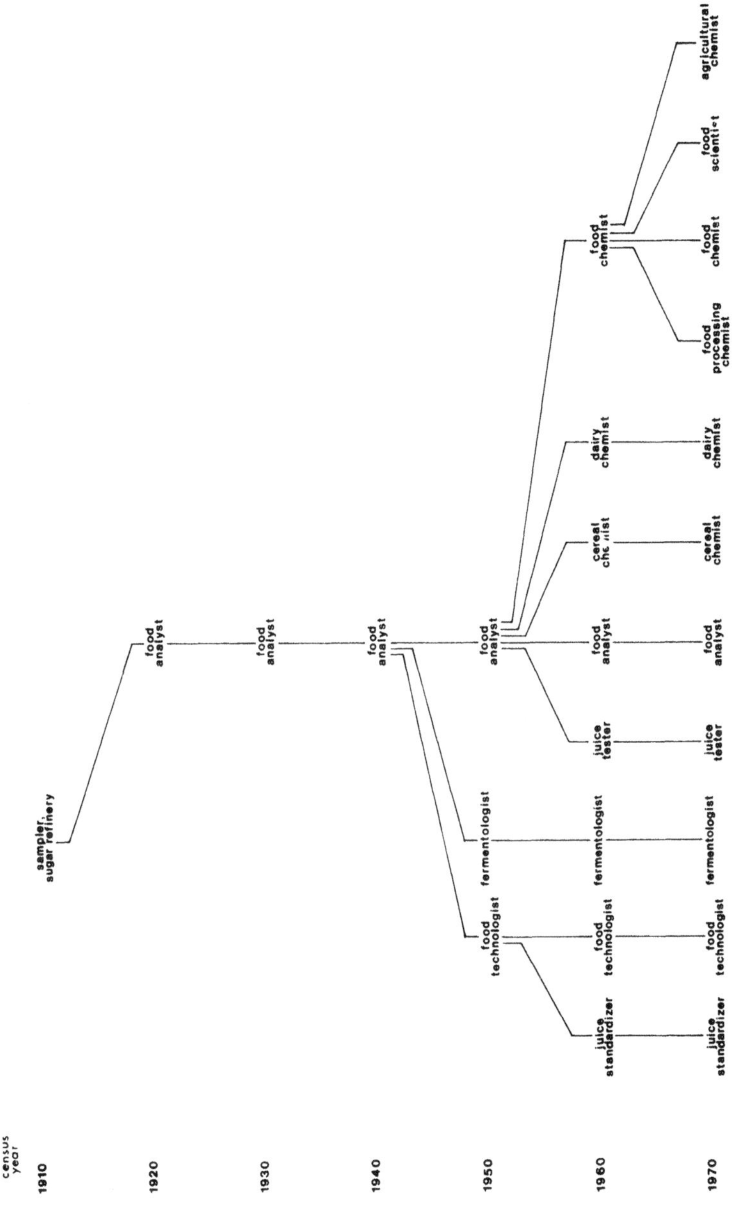

Fig. 2.1–1. Differentiation of Census occupational categories for food chemists, 1910–1970. (See Table 2.1.)

of the exploitation of mineral resources to the expansion of the American economy. Phosphates for artificial fertilizers and metallic ores for the production of iron and steel were but two examples of the materials analyzed by the bench chemist, whose contributions were vital to the control of the panoply of industrial processes. This function was supplemented after the turn of the century when the possibilities of synthetic chemistry (for instance, in the dyestuffs and pharmaceutical industries) created increased demand for the special skills of the organic chemist.[3]

The gradual displacement of the inorganic analyst from center stage is reflected in the proliferation since 1910 of jobs related to organic synthesis or natural products chemistry (for example, "oil expert, oil refinery", "pharmaceutical chemist", "dye expert", "fermentologist", "cereal chemist", or "pesticide chemist"). A more recent trend is indicated by the appearance of "chemical economist" and "chemical librarian" among the occupational titles in the 1970 Census classification; chemists are filling new service niches within the chemical community. The extent of this movement is demonstrated by a 1977 survey which found that roughly four of every ten American Chemical Society members were *not* involved in laboratory work.[4]

These two shifts in the complexion of American chemistry – the organic chemists' rise to prominence and the growing importance of nontraditional roles for certified chemists – should not obscure the enduring importance of chemical control and testing for industry. New physical methods and more sophisticated apparatus have been added to the chemist's arsenal. However, the analysis of raw materials, intermediate processes, and final products has been a major task of bench chemists throughout

[3] Occupational niches for chemists in late-19th century America are surveyed in III, Beardsley, 1964, 43–69. The significance of mineral chemistry for 19th-century American chemists is suggested by III, E. F. Smith, 1926, and III, F. W. Clarke, 1909. Chemistry's contributions to agriculture are ably assessed in III, Wiley, 1899. On the role of chemists in industry during this period, see III, Ash *et al.*, 1915; III, Haynes, 1954a, 239–399; and III, Richardson, 1908.

On the rapid development of the organic chemicals industry in America after World War I, see III, Haynes, 1945b; III, Herty, 1925; and III, Stine, 1940. For contemporary appraisals of the organic chemist's role in industrial progress, see, for example, III, Rose, 1920; III, Norris, 1932; and III, Howe, 1937, 312.

[4] III, Sanders and Seltzer, 1977, 29. Surveys of job opportunities for chemists in the past few decades yield a picture of the expanding set of service roles in industry to which the chemist's expertise has been applied. See, for example, III, ACS, 1944, 1951b; III, Ayre and Lemaire, 1977; III, Billmeyer and Kelley, 1975, 190–258; III, Carlisle, 1943; III, Coith, 1943; III, Grady and Chittum, 1940; III, Heath *et al.*, 1944; III, Sanders and Seltzer, 1977, 29–36.

the past century.[5] The prosecution of most chemical investigations – whether organic or inorganic, synthetic or analytic – depends upon the routines of the analytical chemist; at least in this respect, American chemistry today resembles its 1870s counterpart.

2.1.2. THE PROBLEMS OF MEASUREMENT

While continuities in the occupations of chemists may be traced over time, change is also apparent. Such change is easier to treat in qualitative than in quantitative ways. Two difficulties immediately appear when quantitative data are studied. The first is the conceptual problem of setting boundaries for the object under study. The problem surfaces in the choice of criteria for inclusion in the occupational and professional categories of "chemist", the disciplinary category of "chemistry", and the industrial category of "chemical industry". It is easy to distinguish, for example, between a chemist and an airplane pilot, but it is less simple to say whether a metallurgist is "really" a chemist. The decision depends upon the objectives at hand, and shifts in the character of chemistry over time mandate flexible criteria of choice. The conceptual issue overlaps with a practical problem. The available measures were generated by bodies (e.g., the U.S. Bureau of the Census) with purposes of their own in mind. Their definitions do not necessarily remain stable over time or coincide with our purposes.[6]

In the case of chemistry as occupation, the twin problems emerge with respect to the three categories of (a) chemists, assayers, and metallurgists; (b) chemists, chemical engineers, and biochemists; and (c) chemical educators. A time series on "chemistry as occupation" should include everyone who is engaged in practicing, advancing, or teaching the concepts or methods of chemistry. Over the last century, however, certain occupational fields –

[5] On the development of analytical chemistry, see III, Szabadváry, 1966; III Ewing, 1976, and III, Laitinen and Ewing, 1977. The degree to which the role of the analytical chemist has remained unchanged is suggested by comparison of III, Dudley, 1898; III, Hart, 1893, III, B. Clarke, 1937; and III, Strauss and Rainwater, 1962, 45–46, 91, 94–95.

[6] Inconsistencies in classification schemes have been a persistent problem for students of occupational change (see II, Blau and Duncan, 1967; II, Conk, 1978a; II, Hershberg and Dockhorn, 1976; II, Katz, 1972; and II, T. Smith, 1975). The most useful solution is to use a classification which identifies particular job titles with occupations of similar function over time. A related issue is the problem of comparability of categories. Where available, Census recalculations have been used to maximize the comparability of figures (cf. II, Lee *et al.*, 1957, 400).

most notably chemical engineering and biochemistry – have emerged from chemistry and assumed a life of their own. To gauge the occupation of chemistry as reflected by the use of chemical expertise within the labor force, it is necessary to *in*clude such spin-off fields. Later in our investigations, when the aim is to trace the employment of those who have graduated from academic chemistry departments, such spin-offs will be *ex*cluded.

As revealed by the discussion of the exfoliation of subcategories, the Census compilers have constantly adjusted their definitions. "Chemical engineer" made its first and only appearance as a subcategory of chemist in 1920; by 1930 it was an occupation in its own right. Biological chemists also first appeared as a subcategory in 1920, but have never split off; instead, since 1950, they have been turned into the subcategories of biological chemist and biochemist. In contrast, teachers of chemistry, including both high school and college faculty, are separately classified because of their pedagogical functions. Finally, there is the matter of chemists, assayers, and metallurgists. The three occupations have always been closely linked. The Census Bureau grouped them together until 1950. By that time the number of metallurgists had become sufficiently large to warrant a separate category. Census officials subsequently recalculated the numbers for chemists and assayers *only* for the period from 1900 to 1960. This recalculation shifted the discontinuity back to 1890, where there is minor historical and statistical distortion involved in treating metallurgists as a part of the occupational grouping of chemists.[7]

Common sense might suggest further adjustment of the Census figures to incorporate chemistry faculty and to account for chemical engineers and other spin-off groups. Other considerations militate against this. Most important is the lack of reliable long-run data with which to make such adjustments. The main aim of this report – the construction of indicators that reveal trends over long periods – requires above all else the use of homogeneous and comparable time series. The paramount criterion is that the *trends* revealed by the series be accurate. Consequently, analyses of trends presented in the main body of the text are based upon those homogeneous

[7] Metallurgists *are* included in the category for chemist in the 1970 Census. Since no separate estimate of the number of metallurgists in 1970 is available, the figures and the text discussion employ the composite total for chemists, assayers, and metallurgists for that year. In Appendix A, an attempt is made to eliminate metallurgists from the 1970 Census tabulation.

On high school and university chemistry teachers, see the note to Table 2.1.

time series which come closest to providing measures that suit our conceptual aims.[8]

To check the reliability of the findings, a variety of refinements or adjustments have been made wherever feasible. To the Census data for chemists in the labor force, for example, available estimates were added for chemistry faculty and for chemical engineers. The result of such changes is the more complex graph discussed in Appendix A.2. This Appendix reveals how efficacious the indicators approach can be, since the trends evident in the homogeneous time series are rarely altered significantly by such adjustments.

2.1.3. Indicators of the Occupation

The available data reveal the dramatic growth in chemistry as an occupation over the last century. Figure 2.1–2 indicates that chemists in the labor force have increased more than a hundred-fold over the period, starting at under 1000 in 1870 and exceeding 100 000 by 1970.[9] The graph gives a strong visual sense of the explosion of the chemical enterprise in one century's time. A shift of this magnitude cannot but transform the very nature of the undertaking, influencing such things as patterns of organization and communication within chemistry.

The transformation was gradual. As a plot of the same data on semilogarithmic axes illustrates (see Figure 2.1–3), the pattern of growth has been quite smooth. Rather than following the best-fitting exponential trend, indicated in the figure by the straight line, the time series traces a gentle convex curve, moving above the exponential trend line before World War I and then dropping below it again in recent years. As a second approximation, therefore, a classic sigmoid Gompertz curve was fitted to the same time series, as shown in Figure 2.1–4. Although one should not use the fitted curve for

[8] There are biases inherent in even the best historical time series. Measurement and sampling errors may affect reported trends, or the data collected may not measure the variables one set out to observe. Cultivating a sense of "creative self-doubt" – well developed among economic historians (see II, McCloskey, 1976, 444–445) – helps to minimize the uncertainties caused by the vagaries of historical data. In this spirit, the appendixes of this report provide documentation of our sources, assumptions, and analytical methods. We follow Simon Kuznets in his assertion that the "most important characteristic of statistical data is, of course, their *comparability* – over time and space" (II, Kuznets, 1951, 268).

[9] Documentation for all figures will be found in the tables (pp. 243–496). A note at the bottom of each figure refers readers to the appropriate tables.

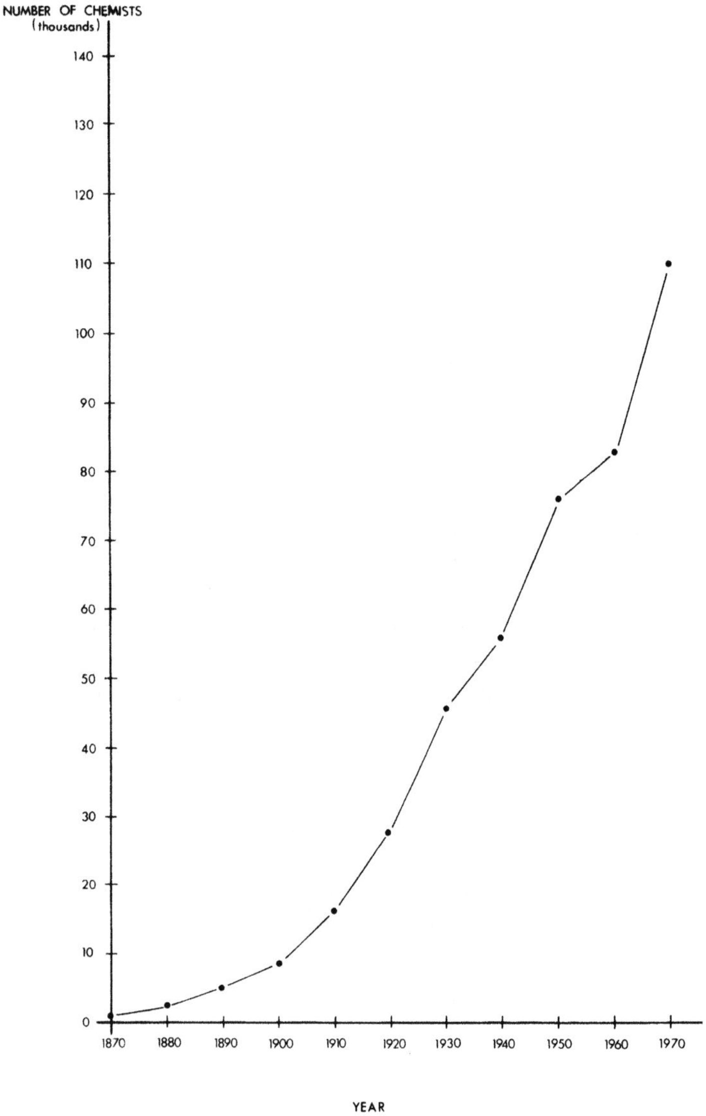

Fig. 2.1–2. The number of chemists, 1870–1970. (See Table 2.2, column 1.)

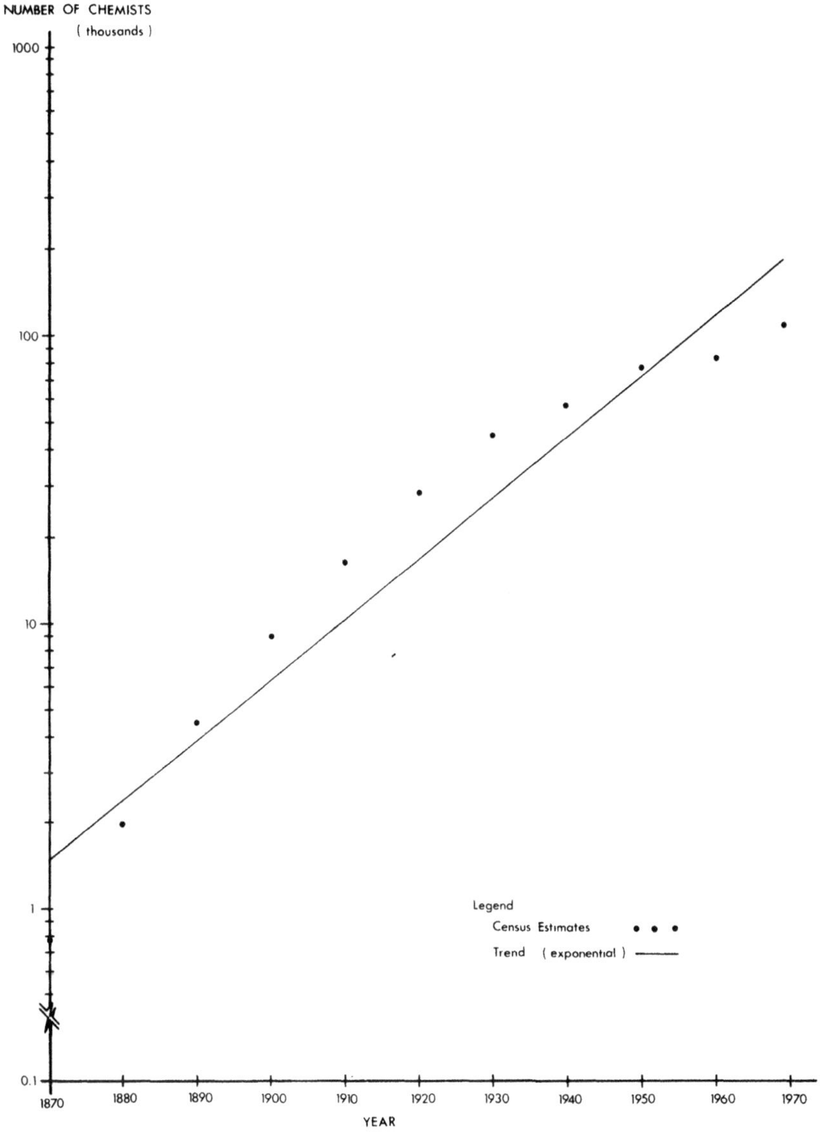

Fig. 2.1–3. Exponential trend in the number of chemists, 1870–1970. (See Table 2.2, columns 1 and 2.)

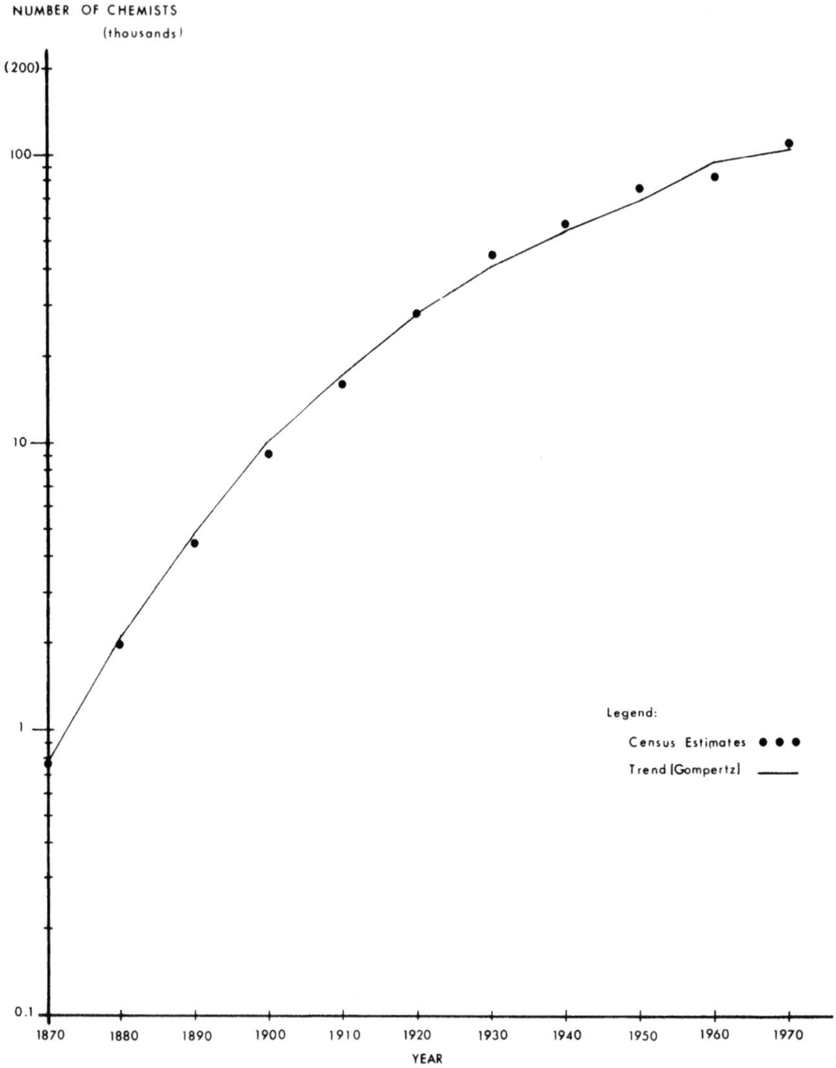

Fig. 2.1–4. Gompertz trend in the number of chemists, 1870–1970. (See Table 2.2, columns 1 and 3.)

predictive purposes – nonlinear regression techniques remaining as much art as exact science – the pattern is correct, as a plotting of the same information on arithmetic axes makes clear (see Figure 2.1–5). The implications of these curve-fitting exercises are that, while growth in the number of chemists in the labor force has been dramatic, its rate has been slowing for some years. The total acts as if it is approaching a saturation point not all that much above the current level; reasoning from the Gompertz curve displayed, the increase in the growth rate stopped just after World War II, and the saturation level is somewhere between 232 000 and 280 000.[10]

The same conclusions follow from a *comparative* look at chemistry in relation to the rest of the U.S. labor force. Figure 2.1–6 shows the time series of chemists per 10 000 workers – an indicator which a logistic growth curve fits well. The enduring trend is one in which chemists represent an increasing fraction of the labor force. But the rate of that increase is slowing, and past performance suggests a saturation around 15 chemists per 10 000 workers.

Relative size of an occupational group is only one measure of the group's power and importance in society, but it is a significant measure. The growth curve for chemistry as occupation carries important implications for the chemical profession, for the social and economic resources available to sustain the intellectual work of chemistry as a scientific discipline, and for the meaning of chemistry as an instrument of socialization into, and as a facet of, American culture.

10 Technical details concerning our trend analyses are given in Appendix E; here a note on the characteristics of particular patterns of growth must suffice. Linear growth proceeds by constant annual increments. Exponential growth exhibits a constant *rate* of change, which appears as a straight line on semi-logarithmic axes. For trends exhibiting a logistic pattern (the classic S-curve), the rate of growth increases steadily to an inflection point, after which it decreases as an upper asymptote (or saturation level) is approached. A related form of growth is described by the Gompertz curve, for which increments to the logarithms of the trend values decline by a constant percentage (see Figures 2.2–6 and 2.2–7).

II, Floud, 1973, 85–124, and II, Harnett, 1975, 439–475, offer readable introductions to time series analysis, while II, Croxton *et al.*, 1967, 214–342, is a *vade mecum* of the art. For techniques of nonlinear least-squares estimation, only highly technical texts (such as II, Bard, 1974) are available. Derek de Solla Price has discussed the implications of exponential and logistic growth for the scientific enterprise in II, Price, 1963, esp. 20–32. For a stimulating application of these concepts to the description of social change in general, see II, Hamblin *et al.*, 1973.

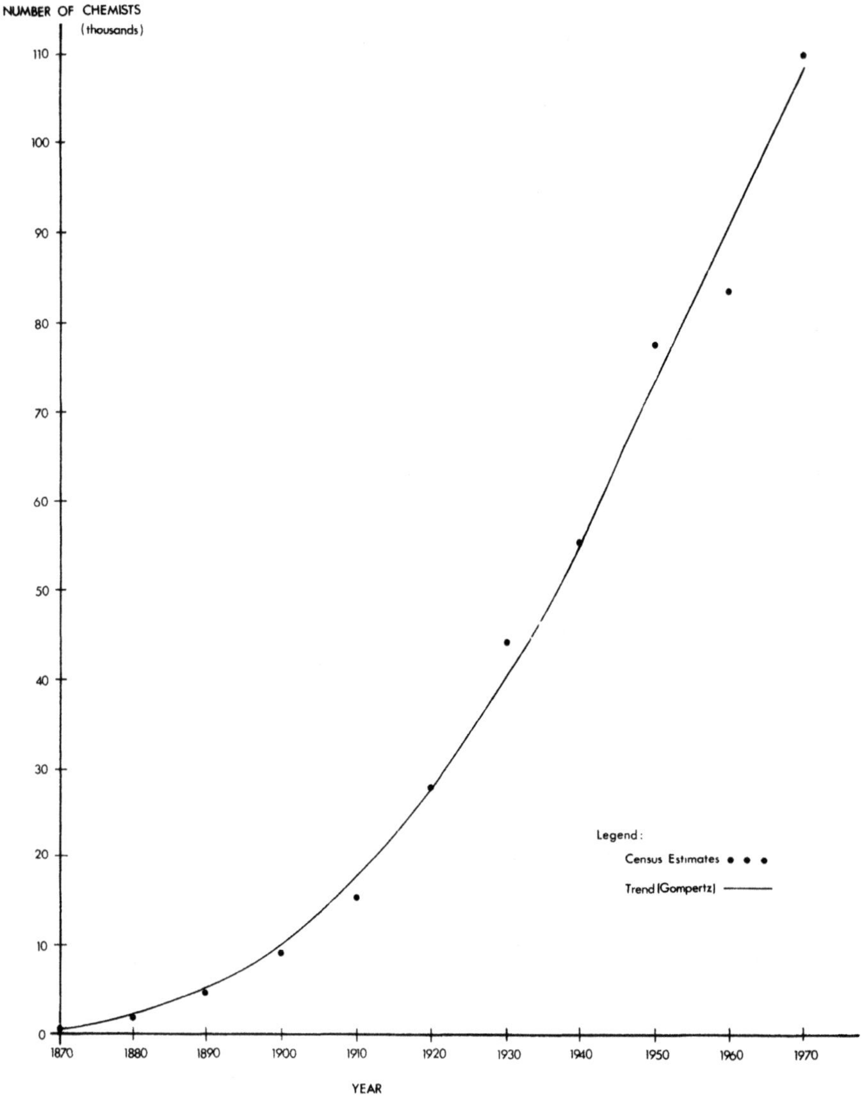

Fig. 2.1–5. Gompertz trend in the number of chemists, 1870–1970. (See Table 2.2, columns 1 and 3.)

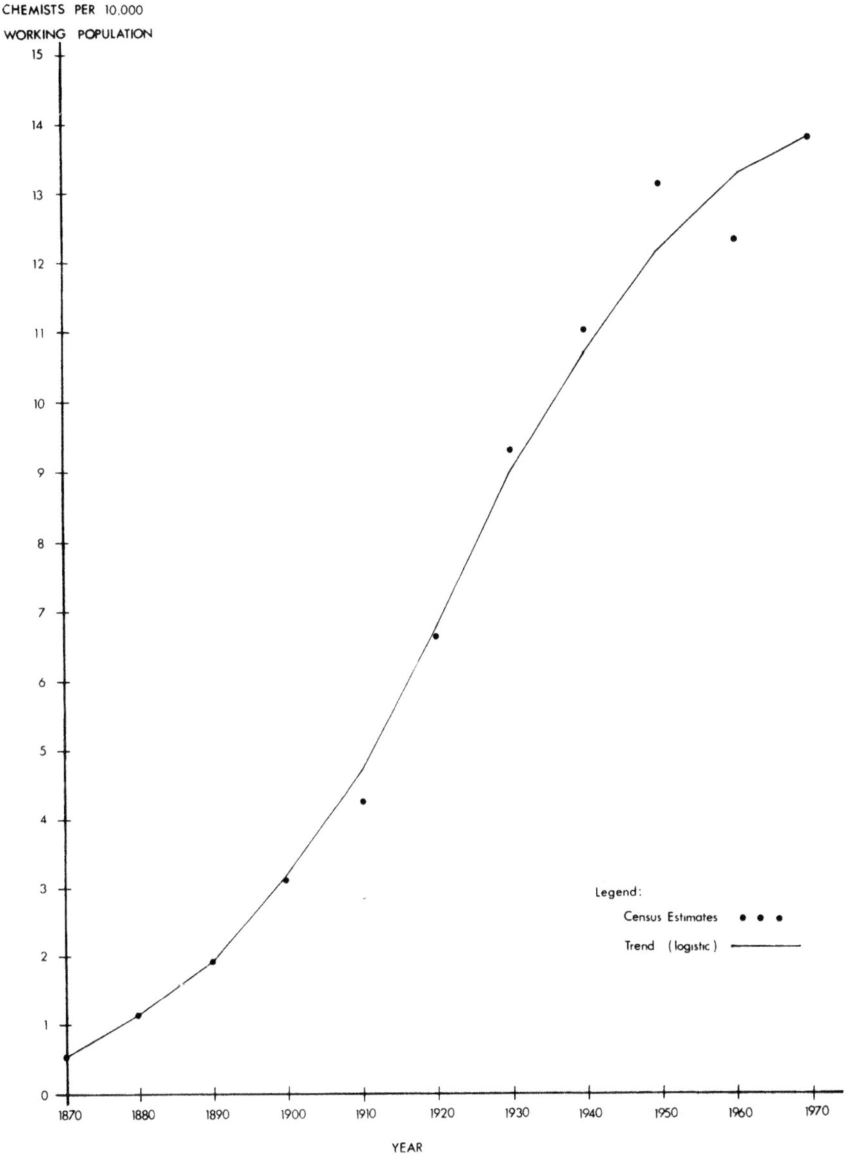

Fig. 2.1–6. Chemists per ten thousand working population, 1870–1970. (See Table 2.3, columns 3 and 4.)

2.2. Chemistry as Profession

INDICATOR HIGHLIGHTS:

American chemistry emerged as a recognizable profession in the 1890s. Its size has increased continuously since then. By the 1960s the profession and the occupation of chemistry were functionally equivalent.

Chemistry as a profession has attracted a decreasing fraction of credentialled chemists during the past half century. Chemistry has also declined in relative importance as an occupation for "professional, technical, and kindred workers" during the last quarter century.

Professionals are "trained and skilled in the theoretic or scientific parts of an occupation, as distinct from its merely mechanical parts". Training and skill imply standards of performance and evaluative mechanisms. The emergence of a new professional grouping is thus often heralded by attempts to create, and the eventual successful establishment of, a new organization or learned body.[11] This process occurred in American chemistry during the 1880s and 1890s.

2.2.1. THE AMERICAN CHEMICAL SOCIETY

The American Chemical Society, founded in 1876, was in the vanguard of American discipline-oriented societies.[12] In its early years, it was but one of several competing chemical societies and it served mainly the social interests of chemists in New York City and vicinity. Membership, which rose from 230 to 314 between 1876 and 1881, fell to a low point of 204 in 1889. A subsequent series of reforms signalled the transformation of the ACS into a national association representing the expanding professional group of American chemists.[13] Figure 2.2–1 shows this shift in the membership trend of the American Chemical Society around 1890.

[11] For a treatment of professionalization which focuses on the role of these organizations, see II, Millerson, 1964.

[12] Among the wave of discipline-oriented organizations established in the 1880s were the American Society of Naturalists (1883), Modern Language Association (1883), American Historical Association (1884), American Economic Association (1885), American Mathematical Society (1888), American Political Science Association (1888), and Geological Society of America (1888).

[13] The history of the various associations established to foster the interests of the

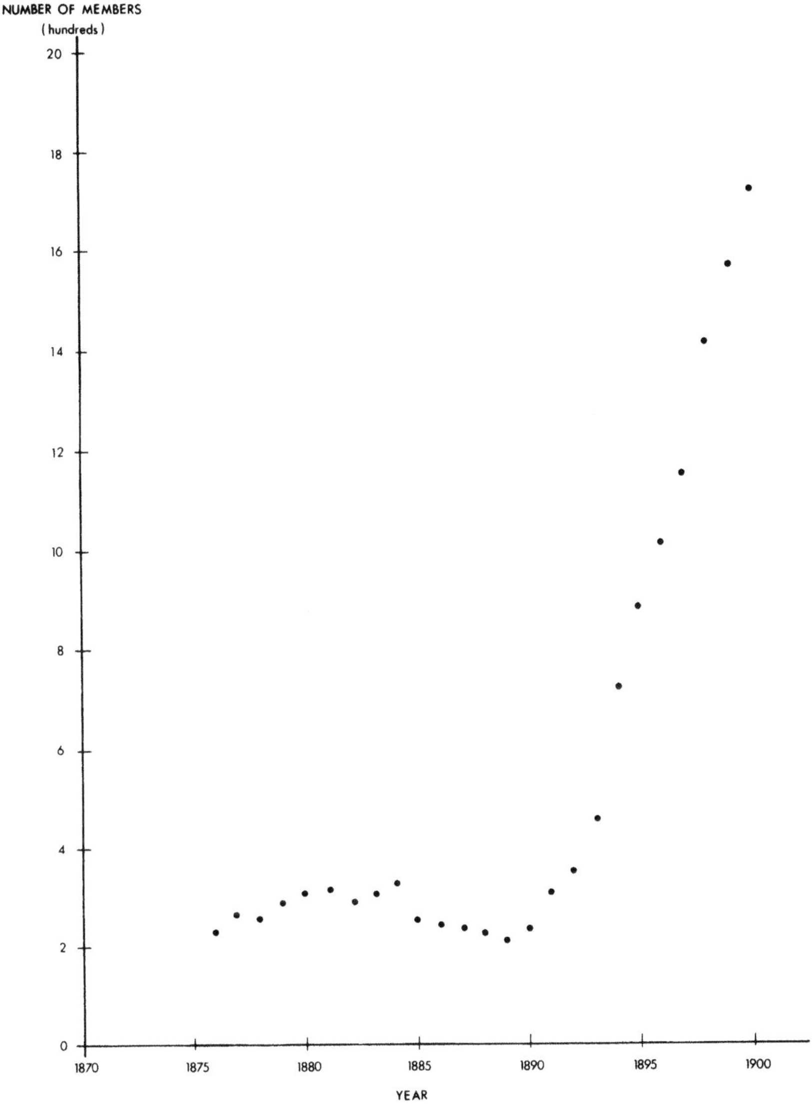

Fig. 2.2–1. Annual membership in the American Chemical Society, 1876–1900. (See Table 2.4, column 1.)

Membership in the ACS may serve as an explicit-discovered indicator of the size of the chemical profession.[14] The rapid growth which began in the 1890s continued over the next few decades, as displayed in Figure 2.2–2, which provides a century-long time series. The same data are plotted on a semilogarithmic scale in Figure 2.2–3. This latter graph juxtaposes membership in the American Chemical Society with the best-fitting exponential trend (with a growth rate of about 7% per annum), thus giving a standard of comparison for the indicator. The divergence between the exponential and the data is evident, especially in the leveling-off of ACS membership after World War II. Replacing the exponential with the best-fitting logistic curve (as displayed in Figure 2.2–4) alleviates this difficulty somewhat and provides an approximation of the size of the profession since its emergence in the late nineteenth century.

2.2.2. THE PROFESSIONALIZATION OF CHEMISTRY

The ACS and the profession it represents have infiltrated the occupation of chemistry. That this process of "professionalization" has proceeded steadily is seen from comparison of the explicit-discovered indicators of chemistry as profession (i.e., ACS membership) and chemistry as occupation or employment (i.e., Census estimates of chemists in the labor force). Their ratio over time yields an explicit-invented indicator of the degree of professionalization of chemistry. In Figure 2.2–5, the number of chemists

---

emerging body of chemists possessing advanced skills and certified knowledge is complex and untold. For a summary of the reform movement in the 1880s and early 1890s, see III, Browne and Weeks, 1952, 26–54.

[14] Specifying the "true" size of the chemical profession is a problem of greater complexity than might appear. The difficulty arises in deciding *who* to count as a chemist. Should one enumerate only those employed as chemists, only those who identify themselves as chemists, only those with chemical training, or those persons with some specified combination of these attributes? Available estimates indicate that some 200 000 persons are today employed as chemists in the United States, but this figure is distorted by the inclusion of chemical engineers. If one takes a chemistry degree as the criterion for deciding who is a chemist, then tens of thousands of persons are included who have pursued other careers, such as BS chemists who become physicians. For these reasons, ACS membership is the best *single* measure of the size of the chemistry profession, since that membership includes the maximum number of people who are both trained and employed as chemists. This problem of measurement is discussed further in IB, ACS, 1973, 423–425. See also pp. 54–64 below.

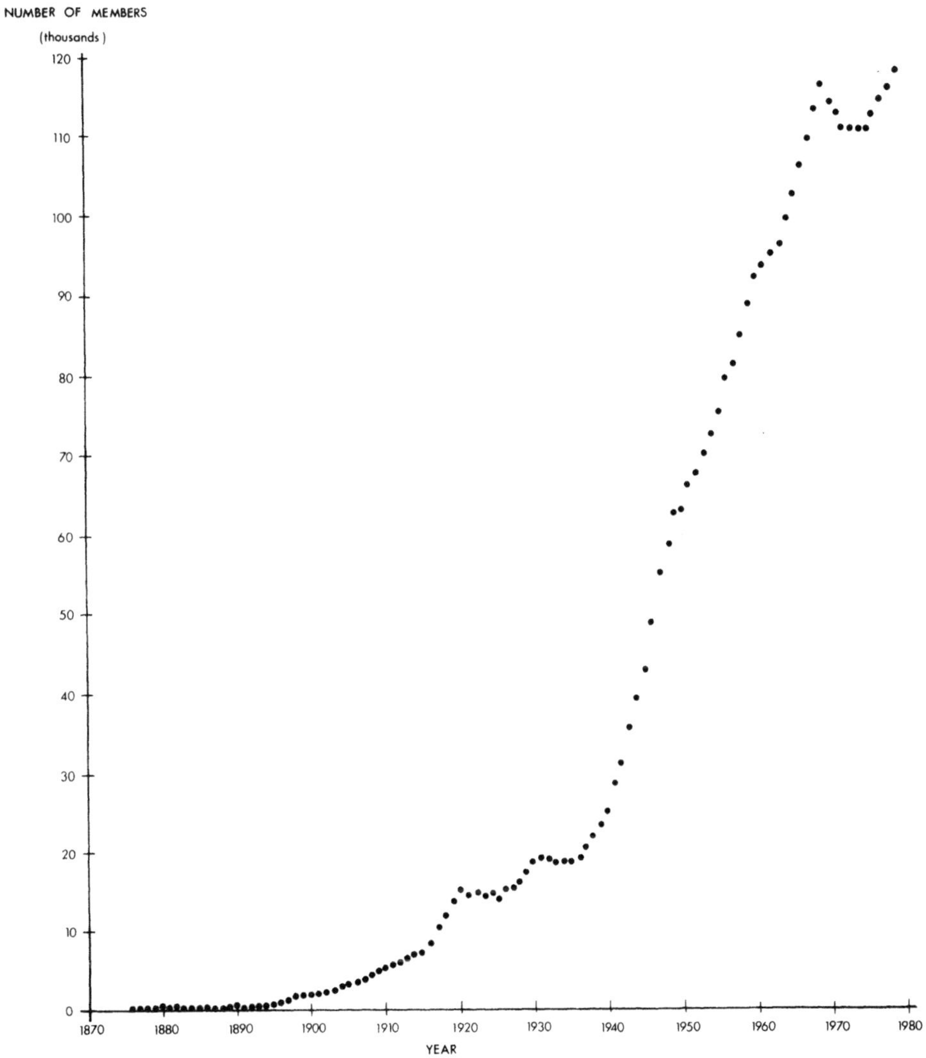

Fig. 2.2–2. Annual membership in the American Chemical Society, 1876–1979. (See Table 2.4, column 1.)

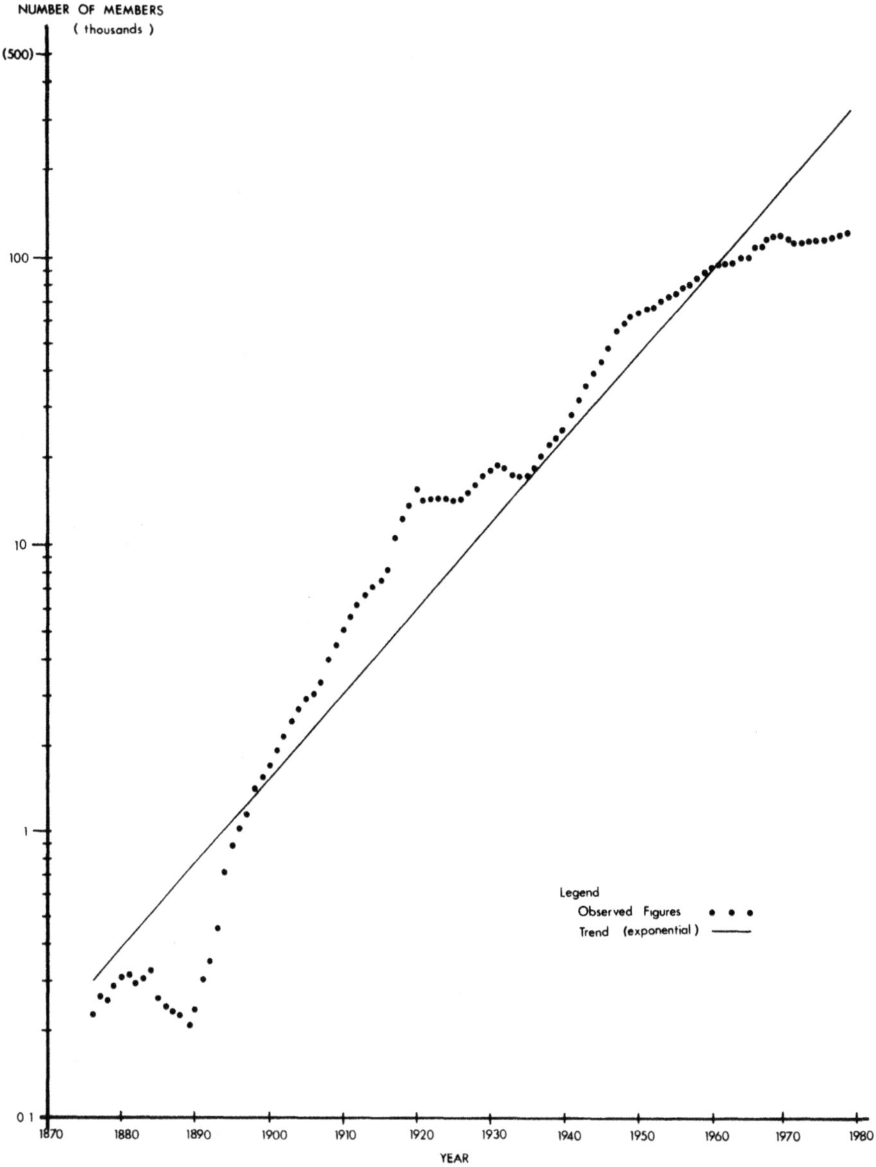

Fig. 2.2–3. Exponential trend in annual membership in the American Chemical Society, 1876–1979. (See Table 2.4, columns 1 and 2.)

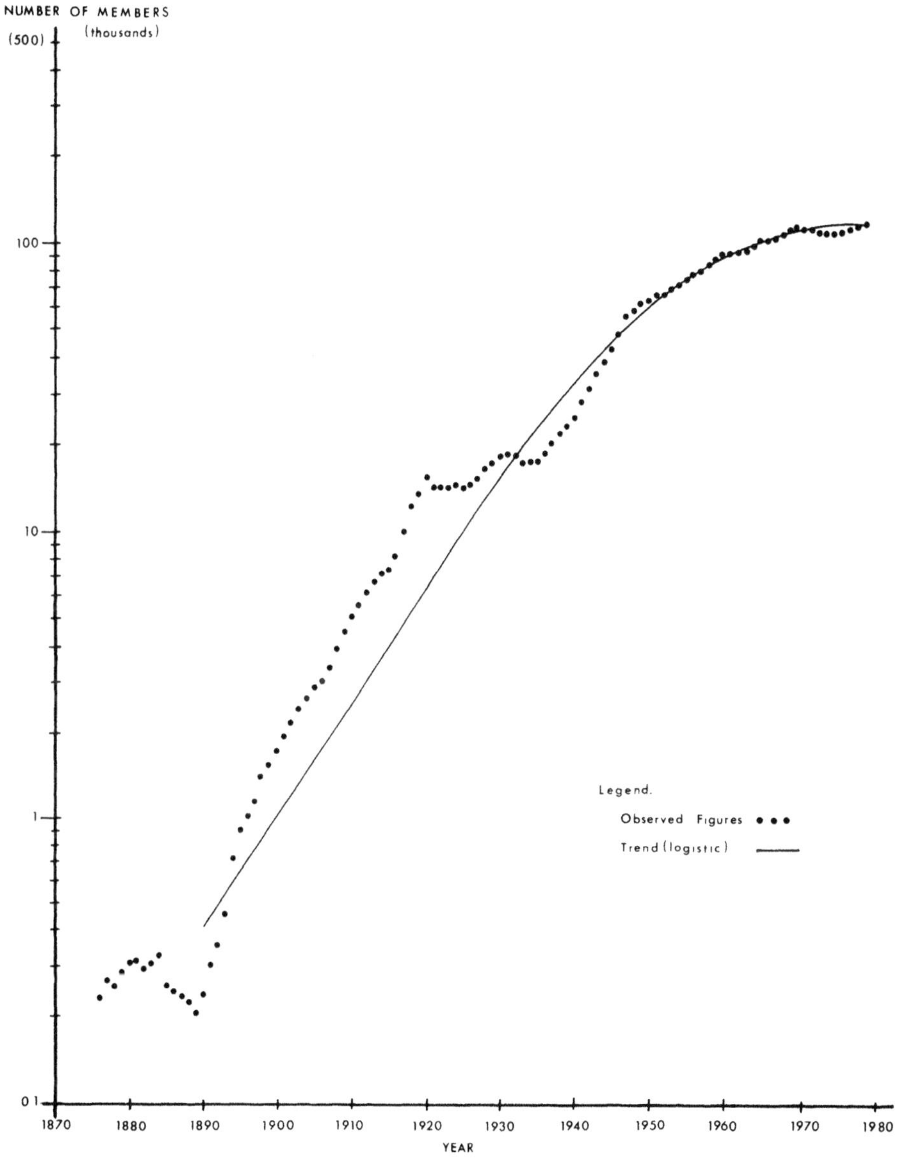

Fig. 2.2–4. Logistic trend in annual membership in the American Chemical Society, 1889–1979. (See Table 2.4, columns 1 and 3.)

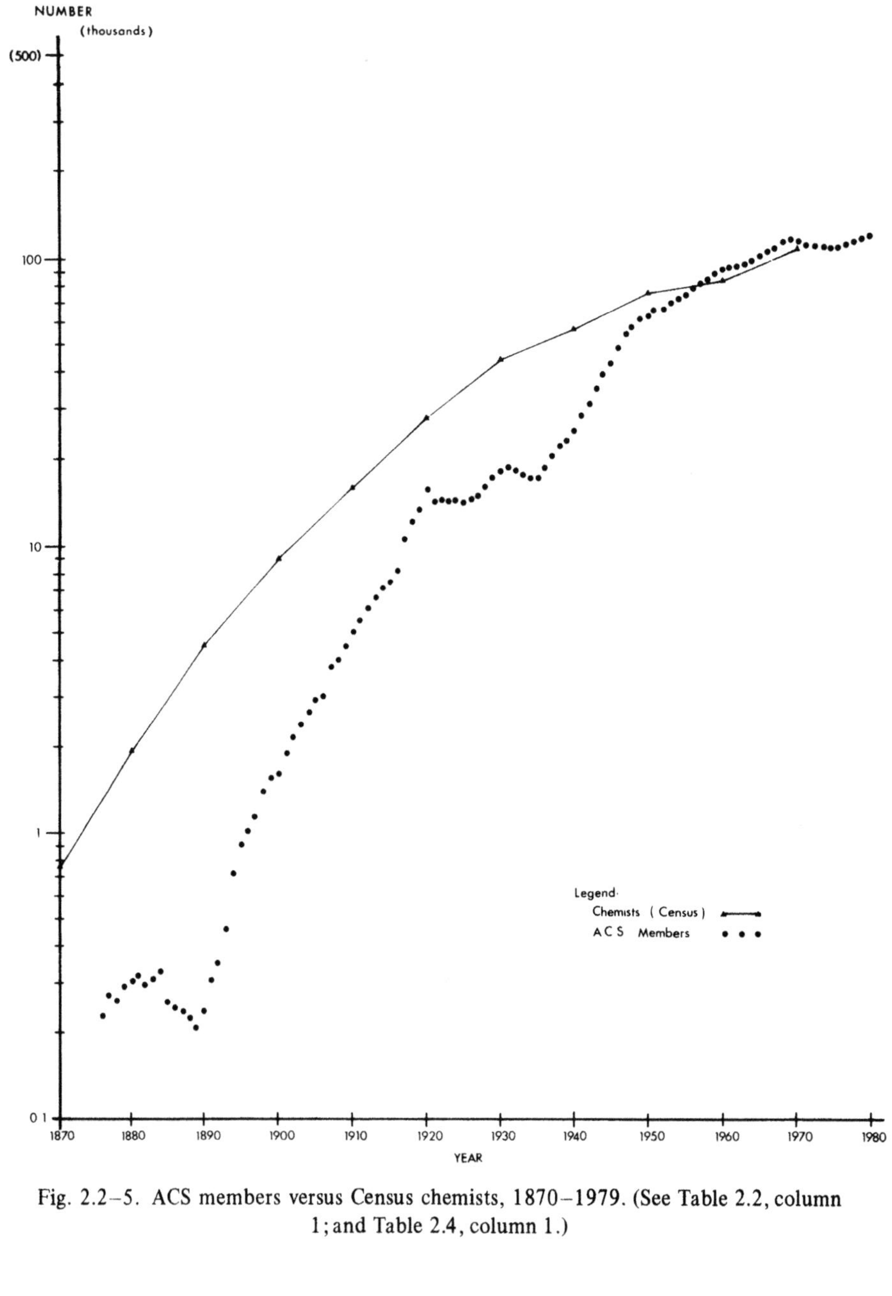

Fig. 2.2–5. ACS members versus Census chemists, 1870–1979. (See Table 2.2, column 1; and Table 2.4, column 1.)

reported in the U.S. Census and membership in the American Chemical Society are plotted together over time. Figure 2.2–6 displays the ratio of these two series. This latter indicator illustrates the transformation of chemistry from a non-professional occupation in the 1890s to an almost wholly professional pursuit in the 1960s. While one would not expect perfect congruence between ACS membership and Census chemists – some of the latter will never be among the former, and vice versa – the steady convergence of the two series is striking.[15]

Light may also be cast upon this transformation by scrutiny of some subtle shifts in the chemistry profession which have affected ACS membership. The enduring trend of growth may be decomposed into two short-run trends, displayed in Figure 2.2–7. In the period from about 1890 to the mid-1930s, ACS membership grew logistically, achieving a level at the end of this period equivalent to that which would have resulted from exponential growth at the comparatively rapid rate of 10.2% per annum, or a doubling every 7.2 yr.[16] This short-run pattern coincided with the period in which chemists associated with the Society were building the first "discipline-oriented" academic departments and making inroads into the federal bureaucracy (as discussed in Sections 5.2 and 6.1). Barriers to ACS entry were low, as those concerned with the articulation of professional and discplinary ideals sought recruits wherever they could find them among the larger body politic of chemists.[17]

After a period of stagnation in the twenties and early thirties, membership in the American Chemical Society followed a quite different trend. Growth

[15] As noted on p. 14 above, the Census statistics for "chemist" utilize a definition which excludes chemical engineers and chemistry faculty. Both of these groups, however, are eligible for ACS membership.

[16] For exponential phenomena, the doubling time in years is equal to the natural logarithm of 2 (i.e., 0.693) divided by the natural logarithm of 1 plus the annual growth rate expressed as a decimal. For a derivation of this simple relationship, see Appendix E.

[17] The ACS Constitution was revised in 1890 to include specific requirements for membership. Persons who had published chemical research, obtained a chemistry degree, taught chemistry in an established institution, or were employed as a chemist "in technical pursuits" were eligible to join the Society. (The tension between New York industrial chemists and the academic elite lent the last clause a special significance.) By 1897, membership requirements were simplified, reading only that a prospective member needed "adequate training in chemistry." In an attempt to broaden ACS membership (and with lobbying by the New York Section), the membership requirement was relaxed in 1902 to allow "any person interested in the promotion of chemistry" to be elected to the ACS. See III, Browne and Weeks, 1952, 205–206.

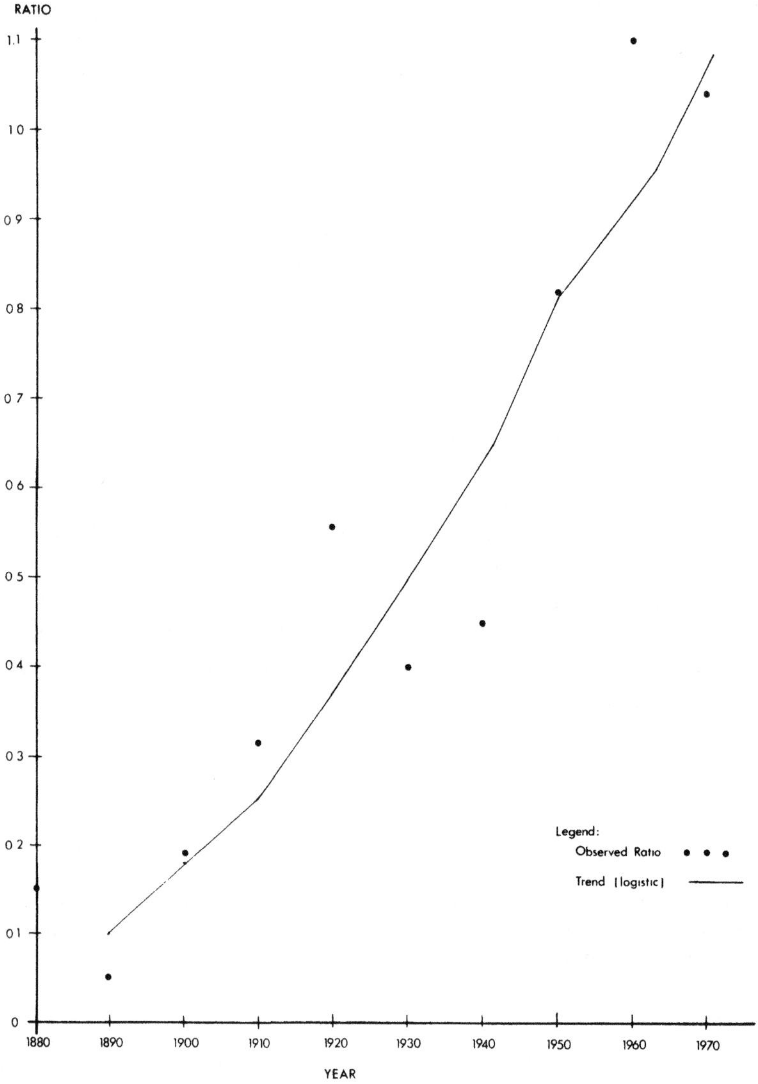

Fig. 2.2–6. Ratio of ACS members to Census chemists, 1880–1970. (See Table 2.5, columns 3 and 4.)

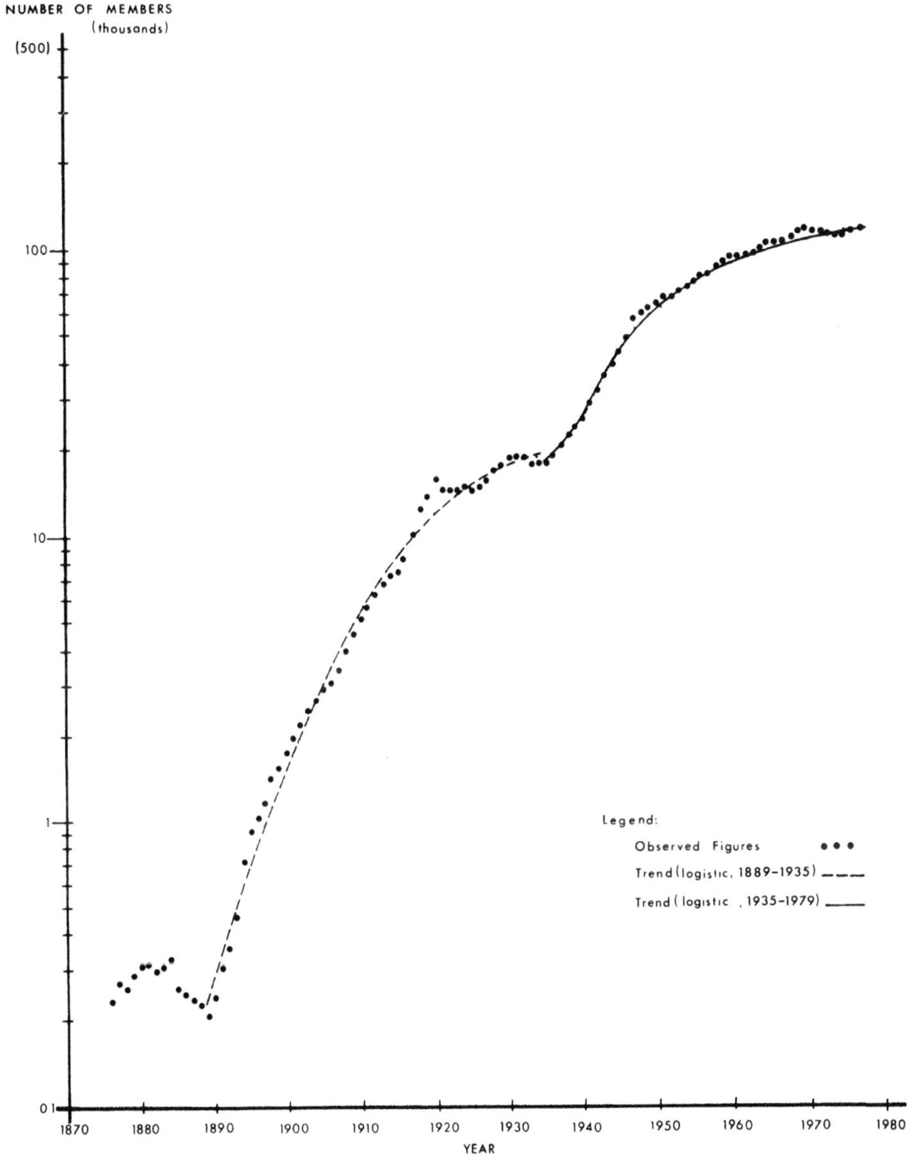

Fig. 2.2–7. Short-run nonlinear trends in annual membership in the American Chemical Society, 1889–1979. (See Table 2.4, columns 1, 4, and 5.)

proceeded more-or-less according to a new logistic curve, taking off rapidly in the late thirties, then growing at a steadily declining rate. In effect, if not in the route taken, the increase in ACS membership over this latter period was equivalent to exponential growth at the rate of 4.4% per annum, or a doubling every 16 yr. The transition to this period was related to a shift in the early 1930s in the criteria employed for electing ACS members.

From the turn of the century, any person with an "interest in the promotion of chemistry" had been eligible to join, although in practice the Society was dominated by chemists with post-baccalaureate training. By the mid-1920s, there was some dissatisfaction with this standard, as chemists began to be more directly concerned about their autonomy and professional status. One indication of this concern was the formation in 1923 of the American Institute of Chemists, an avowedly professional body requiring both certified chemical training and work experience for admission.[18] Another trend, which led to the adoption of new guidelines for ACS membership, was the gradual narrowing of modes of entry into chemical occupations: college education became the standard route into chemical careers.

These changes in the chemical community were visible to participants. In 1933 this awareness, coupled with the economic anxieties of the Depression (and widespread unemployment of chemists), prompted change in the criteria for ACS membership. Formal barriers to entry were raised: educational qualification and at least two years of work experience were substituted for mere "interest in the promotion of chemistry" as a requirement for membership. Paradoxically this restriction resulted in rapid short-term growth, since the American Chemical Society now attracted a broad, new audience of baccalaureate chemists. With the establishment in 1936 of a Committee on

[18] Much of the impetus for an alternative to the American Chemical Society stemmed from the status consciousness of many (mainly industrial) chemists after World War I. Comparing their social role to that of the doctor or lawyer (who enjoyed greater prestige), these chemists called for "professional solidarity" in order to foster public recognition of their role in society; see, for example, III, Parmelee, 1920. The debates leading to the formation of the American Institute of Chemists in 1923 may be followed in *Chemical Age*. III, Carmichael, 1974, provides an historical account of the AIC.

The American Chemical Society hardly ignored questions of professional and economic status. In 1948 there was a proposal to make the AIC a division of the ACS for professional affairs. Twelve years earlier, the ACS had formed a Committee for Improving the Professional Status of the Chemist, and the 1972 establishment of a Division of Professional Relations testifies to continuing concern among the ACS rank-and-file with the problems of professional life. For ACS activities in this area in the 1930's and 1940s, see III, Browne and Weeks, 1952, 228–248.

Accrediting Educational Institutions, the ACS sought to raise the standard of undergraduate training in chemistry. In 1941, the Society also instituted a program for the certification of graduates from approved degree programs, giving them credit toward full ACS membership.[19] The Society's reorientation notwithstanding, growth in membership slowed considerably in the post-World War II period as the number of ACS chemists approached a natural ceiling equal to the total of employed chemists, and that total itself appeared to reach a saturation level in the American labor force.

Some other indicators constructed from ACS membership data shed further light upon the changing relationship between professionalization and formal training. A plot of the ratio between Society membership and holders of terminal degrees in chemistry (top half of Figure 2.2–8) shows a steady rise through the thirties and forties, as baccalaureate-holding chemists joined the Society in significant numbers. In contrast, the ratio of holders of terminal masters and doctorates in chemistry to ACS members declined steadily in the same period (see lower curve of the figure). This latter ratio has recovered steadily in the quarter century since 1950, signaling the extent to which holders of graduate degrees in chemistry are infiltrating the profession.

2.2.3. CHEMISTRY AMONG THE PROFESSIONS

One final aspect of chemistry as occupation and profession deserves mention. In Section 2.1 it was shown that chemistry has occupied a gradually increasing fraction of the labor force. Even so, chemists have always constituted a tiny fraction of that larger group (about 15 per 10 000 in 1970). It is instructive to consider the growth of chemistry as an occupational grouping in the context of the next most general level of aggregation of Census categories, namely "professional, technical, and kindred workers". This category includes accountants, engineers, and a considerable range of occupations comparable in skills to chemists. Figure 2.2–9 shows the number of chemists per 1000 "professional, technical, and kindred workers". The steadily increasing importance of the chemist for the 80 yr before 1950 is as dramatic as the subsequent quarter-century decline.[20]

[19] This shift in membership requirements signified recognition of the role of colleges in training neophyte chemists. For an account of continuing ACS activities in this area, see III, Browne and Weeks, 1952, 207–218; and IB, Skolnik and Reese, 1976, 65–72.
[20] The professional, technical, and kindred workers category includes doctors, lawyers, clergymen, teachers (primary, secondary, college and university), and other groupings of

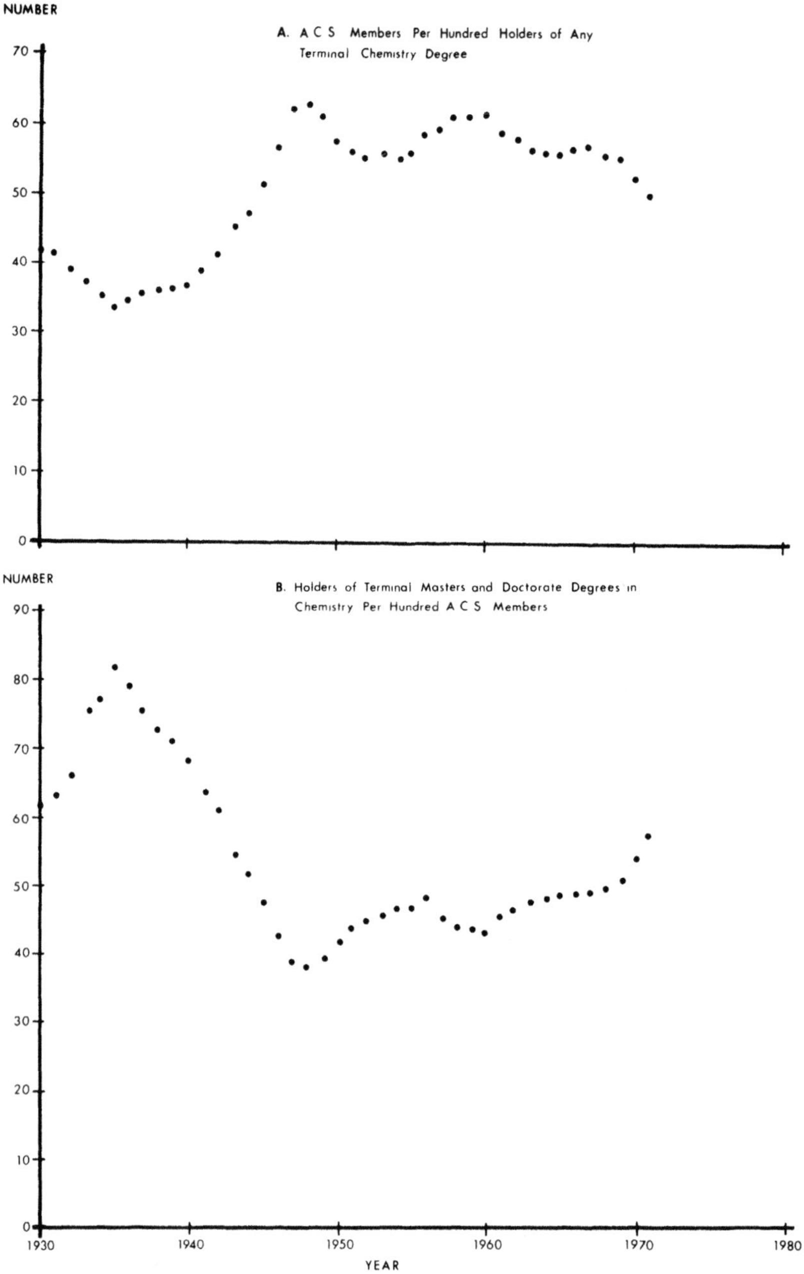

Fig. 2.2–8. Indicators of the relationship between training and professionalization in chemistry, 1930–1971. (See Table 2.6, columns 4 and 5.)

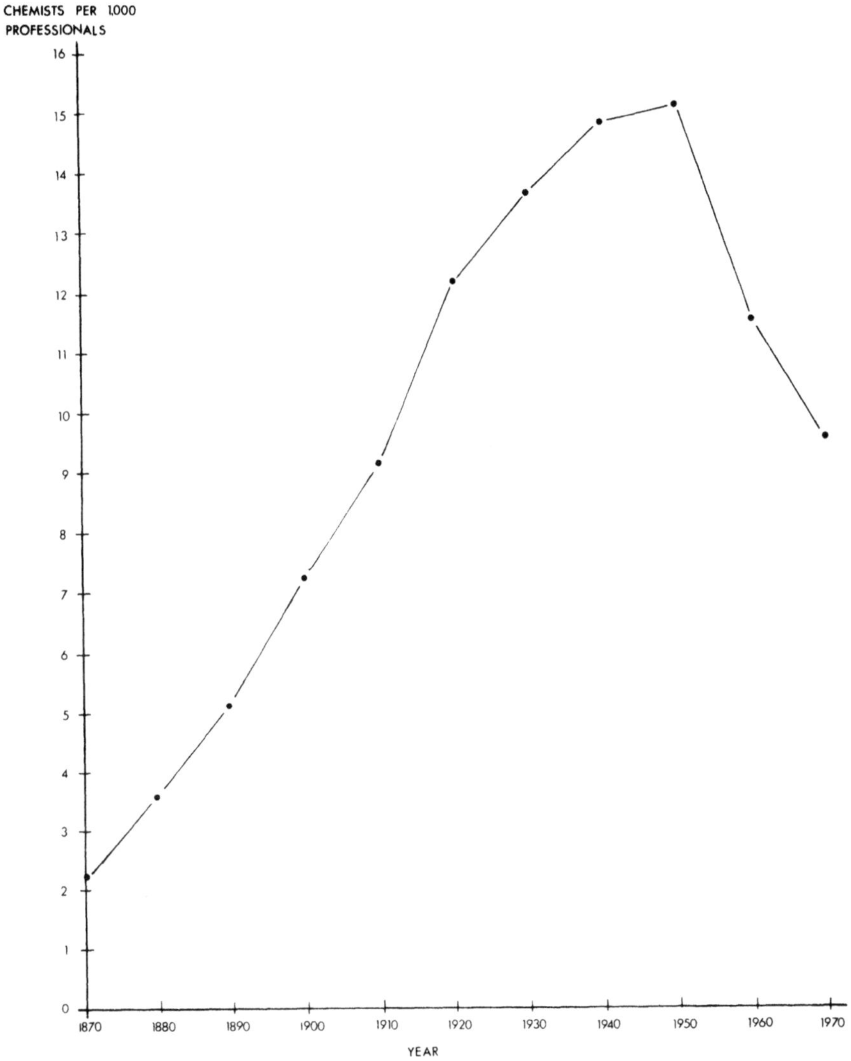

Fig. 2.2–9. Chemists per thousand professional, technical, and kindred workers, 1870–1970. (See Table 2.7, column 3.)

The relationships among the profession, the larger group of those occupied as chemists, and the still larger group of those with chemistry degrees have changed significantly during this century. To elucidate the patterns of change, Figure 2.2–10 provides an overview of the explicit-discovered indicators presented above. This graph, like Figure 2.2–5, shows that the profession (measured by ACS membership) has moved toward parity with the occupation (represented by Census estimates of the number of chemists in the labor force).

As discussed above, the time series of ACS membership exhibits a great variation in slope from one era to the next. The shift in the 1930s reflects how the reorganization of the Society linked its fortunes to the process of training certified chemists. Another uncertain transition period began about a decade ago, as ACS membership reached a ceiling which seems to correspond to a constant fraction of the entire population of practicing chemists in the labor force. This ceiling is something quite separate from the total of *credentialled* chemists (i.e., those holding a terminal chemistry degree but not necessarily practicing the science). That total has exceeded the number of chemists in the labor force for about 40 yr. Consequently, the number of *credentialled* chemists in non-chemical occupations is now considerable – a dramatic illustration of what may be termed the "decoupling" of chemical education from the occupation of chemistry. (A fuller treatment of this topic appears in Section 3.1.)

In summary, the enduring trend in chemistry as occupation has been one of dramatic growth, with the increase following a sigmoid curve. Over the past century that growth has been equivalent to an annual rate of 5% per year. Chemists have increased their importance both within the working population at large and among the burgeoning professional classes. Chemistry as occupation has continued to increase its share of the U.S. labor force as a whole, but since around 1950 its growth rate has been slackening. For chemistry as a profession the short-run trend of the past 25 yr is one of *shrinking* representation among professional occupations.

The trend toward a professional coloration for the occupation of chemistry, as well as the more recent decline in the relative importance of the

---

scientists. For occupations included in this category over time, see II, USBC, 1915, 1921, 1930, 1940, 1950, 1960b, 1971. On the professionalization of the U.S. labor force in recent years, see III, Ginzberg, 1979.

Figure 2.2–9 is based on chemists alone. However, the adjusted time series in Appendix A.2 shows that the trend after 1950 is still downward, even if one includes chemical engineers (or chemistry faculty).

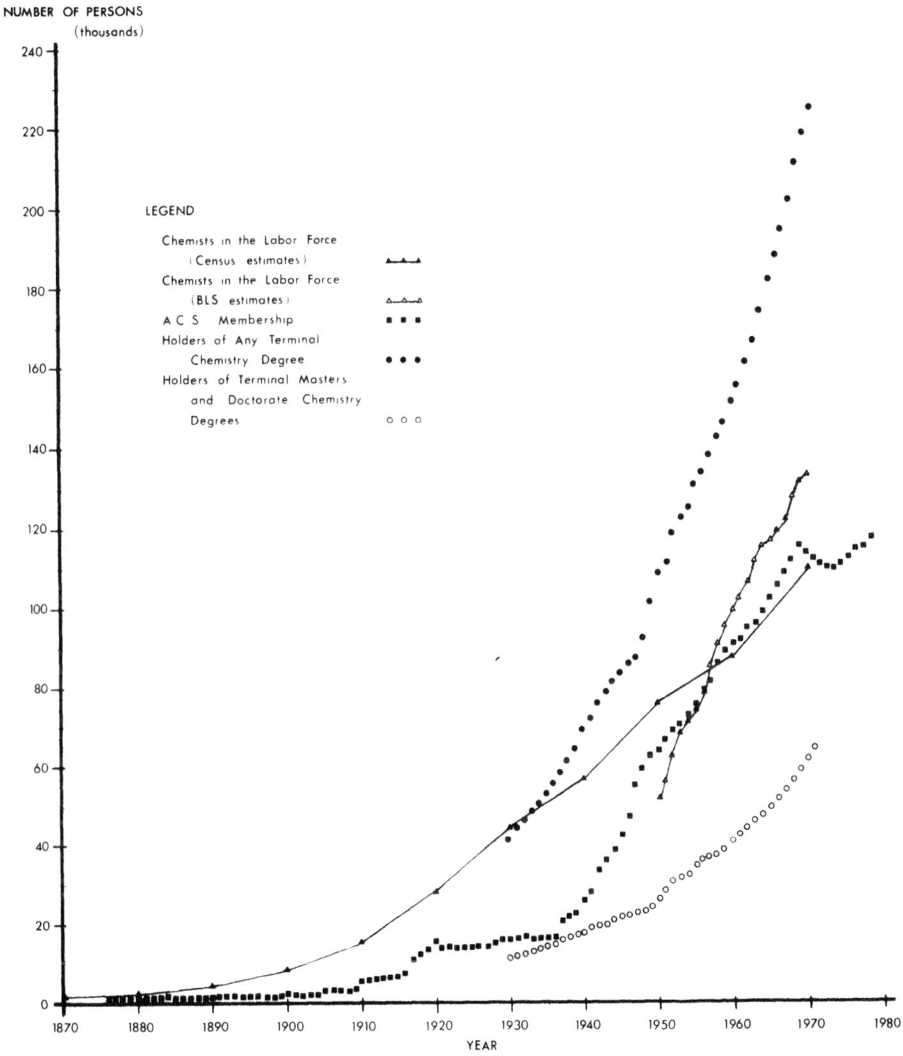

Fig. 2.2–10. The chemical community, 1870–1979. (See Table 2.2, column 1; Table 2.4, column 1; Table 2.6, columns 2 and 3; and Table 5.1, column 2.)

profession, may be linked to the changing fortunes and functions of chemical education. Those fortunes and functions require exploration, for they show some of the ways in which chemistry serves not simply as an occupation and a profession, but also as an institution and a cultural resource.

CHAPTER 3

CHEMICAL EDUCATION AS CONTEXT

A demographic approach will be used in the first section of this chapter.[1] Credentialled chemists will be treated as a subgroup of the population of persons with academic degrees. Attention will be paid to the "birth" statistics of degrees conferred and to the "stock" statistics of living degree holders. These data will be broken down to elucidate the many functions of chemical education.[2]

The progressive decoupling of chemistry as education from chemistry as profession was noted in Section 2.2. That decoupling points to how higher education in chemistry serves not only as vocational preparation but also as a cultural instrument with varied utilities.[3] While "chemical education as

[1] On the use of demography in historical research see II, Wrigley, 1969. On demographic methods generally, see II, Keyfitz, 1977.

[2] The view of the "great American degree machine" in this chapter is adopted from IB, Adkins, 1975. We have used Adkins' time series where appropriate, since they are readily accessible, amply documented, comparable over time and subject, and compatible with recent federal compilations. See also IB, ACS, 1976, 32–34; IB, ACS, 1977b, 31–34; and III, Brode, 1971.

We suspect that Adkins' figures are high for the early years of his series, since they are based on U.S. Bureau of Education surveys. For the decade 1890–1899, Adkins reports 2554 doctorates in all fields, and 464 doctorates in chemistry (IB, Adkins, 1975, Table A–2, 190–194; and Table A–5.5, 270–273). The *Comprehensive Dissertation Index*, in contrast, includes 1765 doctorates in all fields and 191 in chemistry for that period (see Table 6.1, column 4). J. McKeen Cattell, who compiled lists of doctorates from 1898 to 1915 for *Science*, called the official figures "valueless", noting that the "report of the commissioner records 343 doctorates conferred in 1901, but the table shows that the largest number of degrees conferred on examination was by Taylor University at Upland, Ind., which gave the doctorate of philosophy to no less than 45 candidates. One may well wish to learn something in regard to a university with such a remarkable record, and the information is given a few pages farther on in the report. Taylor 'University' has no productive funds; its income from tuition and other fees was $4000 and from other sources $500!" (II, Cattell, 1903, 257–258; see also II, Cattell, 1898, 198, and II, Cattell, 1911, 194).

[3] The decoupling of education from career has not passed unnoticed by members of the chemical profession. A recent report of the ACS noted that "the probability of a BS degree in chemistry leading directly to a chemical job is steadily declining" (IB, ACS, 1977b, iii). Many baccalaureate chemists, headed for medical careers, do not intend to

vocational preparation" offers immediate, direct support to the institution of chemistry, the forms of sustenance provided by "chemical education as cultural instrument" are no less important in any analysis of the institution. Both themes are also apposite in understanding the role of chemistry in the high school curriculum. And both may be used in illuminating the place of chemistry in mass culture.

3.1. Higher Education

INDICATOR HIGHLIGHTS:

About 400 000 baccalaureates and 40 000 doctorates in chemistry have been awarded on American campuses in the past century. The enduring trend in the number of degrees conferred has been one of exponential increase. This trend is in no way unique to chemistry but rather is characteristic of the American education system. The rate of growth is greater than the exponential rate of increase of the population.

While undergoing exponential growth, the conferral of chemistry degrees has also been subject to a second trend: relative decline within academe. Baccalaureates awarded annually in chemistry have dropped from roughly a tenth to a hundredth of all bachelors degrees conferred.

A third, related trend has been the steady shift away from chemical education as vocational preparation and toward chemical education as cultural instrument.

3.1.1. Exponential Growth and Relative Decline

Figures 3.1–1 to 3.1–3 plot, among other things, the number of bachelors, masters, and doctorate degrees of all kinds awarded annually; Figure 3.1–4 plots the same data on semilogarithmic axes. For the past century the trend of degree conferral in *all* subjects has been one of exponential growth on the bachelors, masters, and doctoral levels.[4] This exponential growth is a prime

become practicing chemists at all, a trend characterized as "well-established" in another ACS report (IB, ACS, 1973, 424; see also IB, ACS, 1977b, 41).

[4] Bachelors degrees are first-level degrees. The category of second-level degrees, in addition to masters degrees, includes first professional degrees such as the MD and JD.

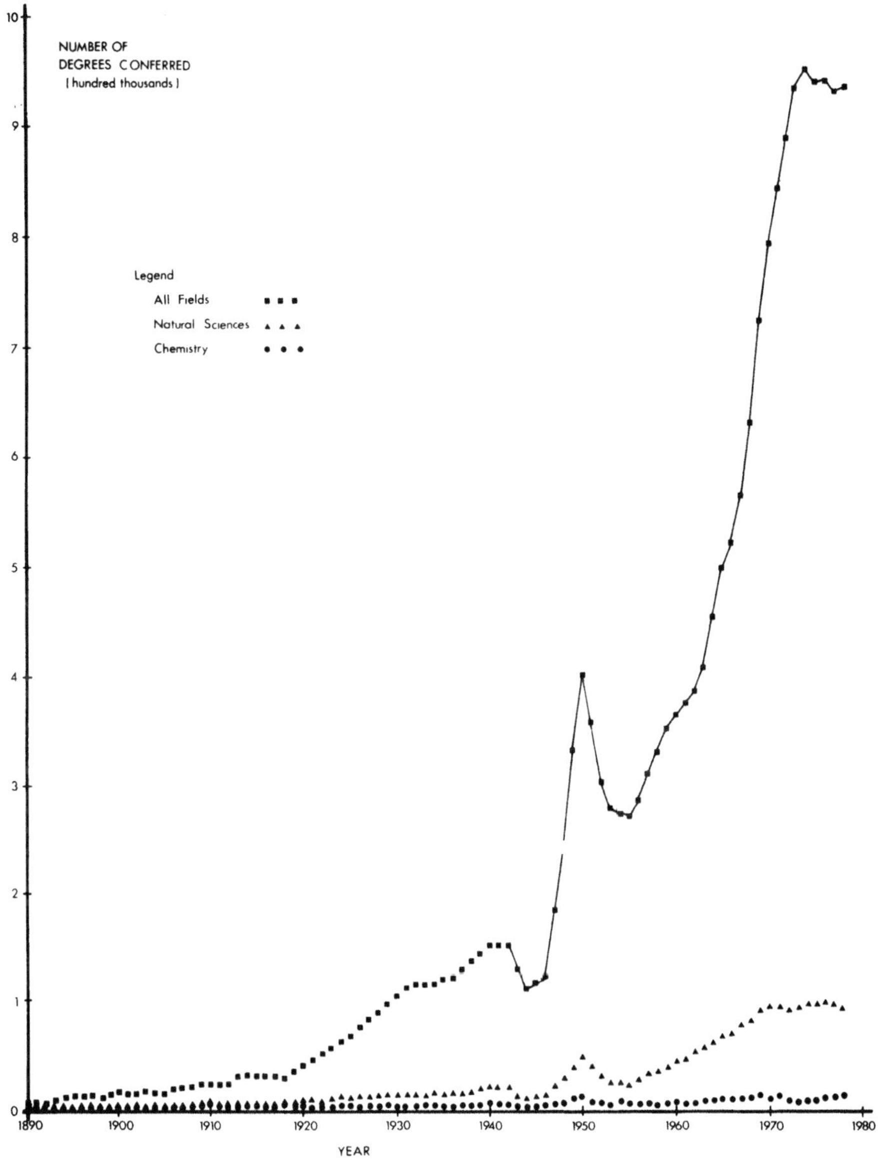

Fig. 3.1–1. Bachelors degrees conferred in selected fields, 1890–1978. (See Table 3.1, columns 1, 2, and 4.)

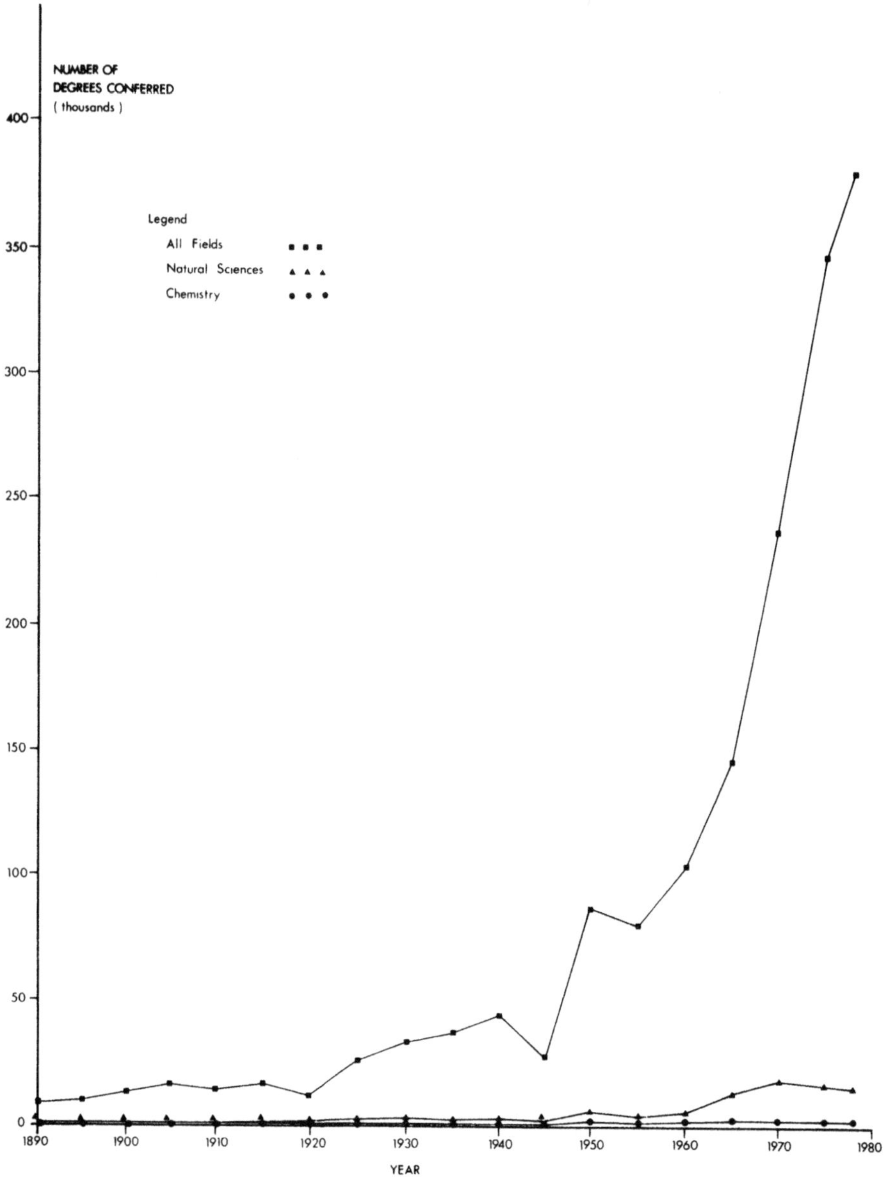

Fig. 3.1–2. Second-level degrees conferred in selected fields, 1890–1978. (See Table 3.2, columns 1, 2, and 4.)

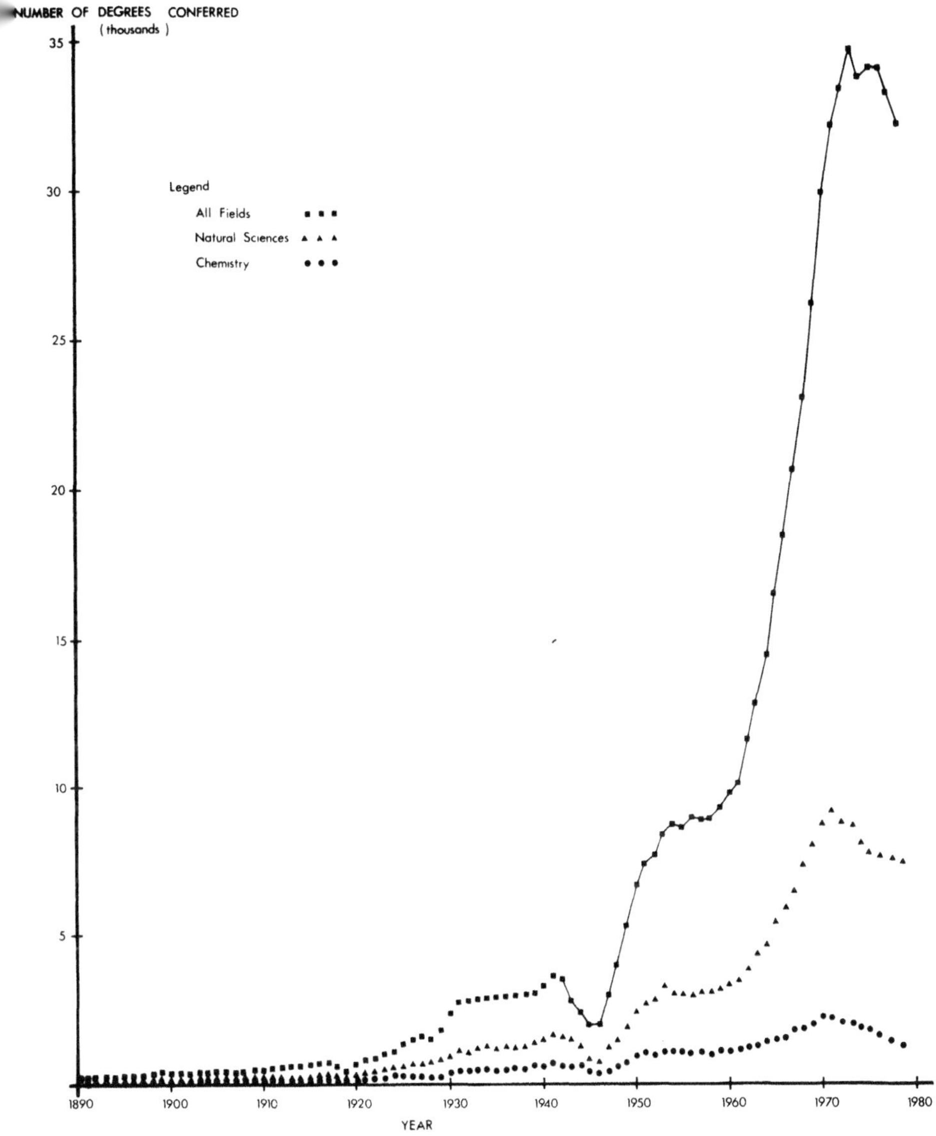

Fig. 3.1–3. Doctorate degrees conferred in selected fields, 1890–1978. (See Table 3.3, columns 1, 2, and 4.)

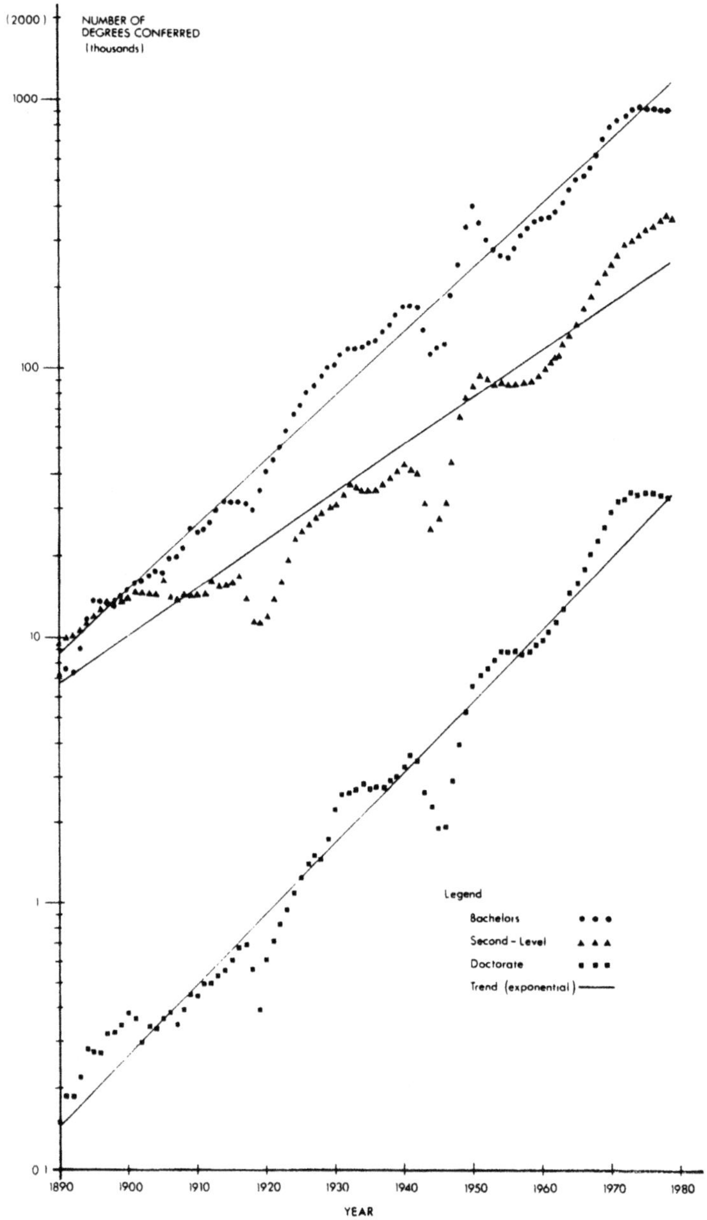

Fig. 3.1–4. Trends in degrees conferred in all fields, by level, 1890–1978. (See Table 3.1, column 1; Table 3.2, column 1; and Table 3.3, column 1.)

enduring trend, characteristic of American higher education. For the purposes of chemical indicators the growth is a "background phenomenon" obviously not explicable in terms of anything unique to the natural sciences, although its effect on those sciences has been profound.

This background phenomenon reflects more than just the increase in the American population. Calculations *per unit of population* make the point clear (see Tables 3.4 and 3.5). Between 1890 and 1970, the number of degrees conferred in all fields per million persons of college age climbed from about 2700 to over 65 000 – a twentyfold increase. Similarly, the stock of persons of *all* ages holding any college degree, per million resident U.S. population, has jumped from almost 12 000 in 1930 to almost 57 000 in 1970. Unlike the Red Queen in *Through the Looking-Glass*, who ran and ran but remained in the same place, the degree machine has outpaced population growth for nearly a century. The United States is an increasingly academic nation. It is important to remember this when discussing the place of the sciences in American society.

Identical calculations for chemistry degrees alone indicate similar growth at a slower pace. Chemistry has partaken of the education boom, but not as effectively as some other fields. Figure 3.1–3 shows doctorates conferred in chemistry as well as in all fields and the natural sciences. Growth in the conferral of chemistry degrees has been exponential (doubling time, about 13 yr), as Figure 3.1–5 demonstrates by plotting the same data on semi-logarithmic axes. Similar graphs can be drawn for bachelors and masters degrees – the doubling times being 21 and 18 yr, respectively (Figures 3.1–6 and 3.1–7).

While exponential increases characterize the American academic system, there are dangers in being transfixed by the dramatic phenomenon of absolute growth. As Figure 3.1–8 reveals, chemistry has been subject to an *enduring trend of relative decline* for over half a century. In that time undergraduate chemistry degrees decreased in relative standing *by an order of magnitude* –

The time series for doctorates are limited to PhDs (see IB, Adkins, 1975, 12–14). Since the category of first professional degree does not apply to most scientific disciplines, the discussion below uses second-level and masters as interchangeable labels when referring to degrees in these disciplines.

For Adkins' definitions of subject fields (which are compatible with the specialty codes of the National Center for Education Statistics), see *ibid.*, Table A–1, 181–190. The trends reported in this chapter employ Adkins' system of classification. See Appendix A for a discussion of possible adjustments which incorporate biochemistry and chemical-materials engineering.

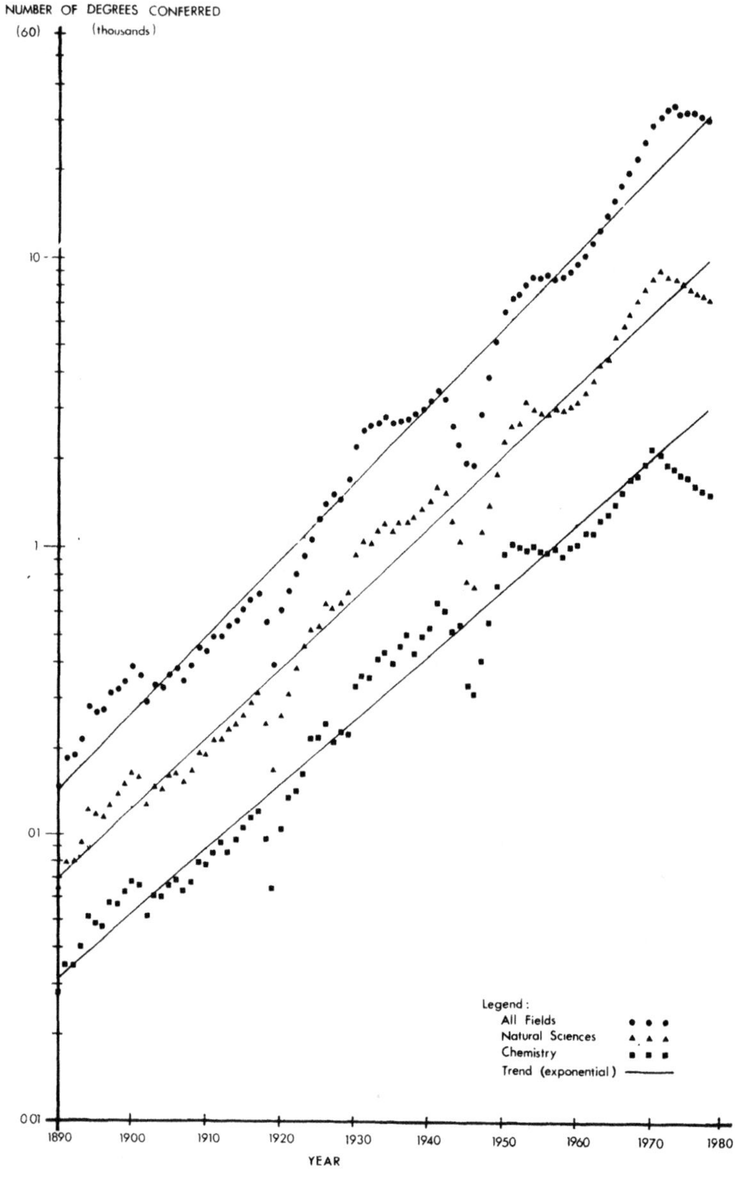

Fig. 3.1–5. Trends in doctorate degrees conferred in selected fields, 1890–1978. (See Table 3.3, columns 1, 2, and 4.)

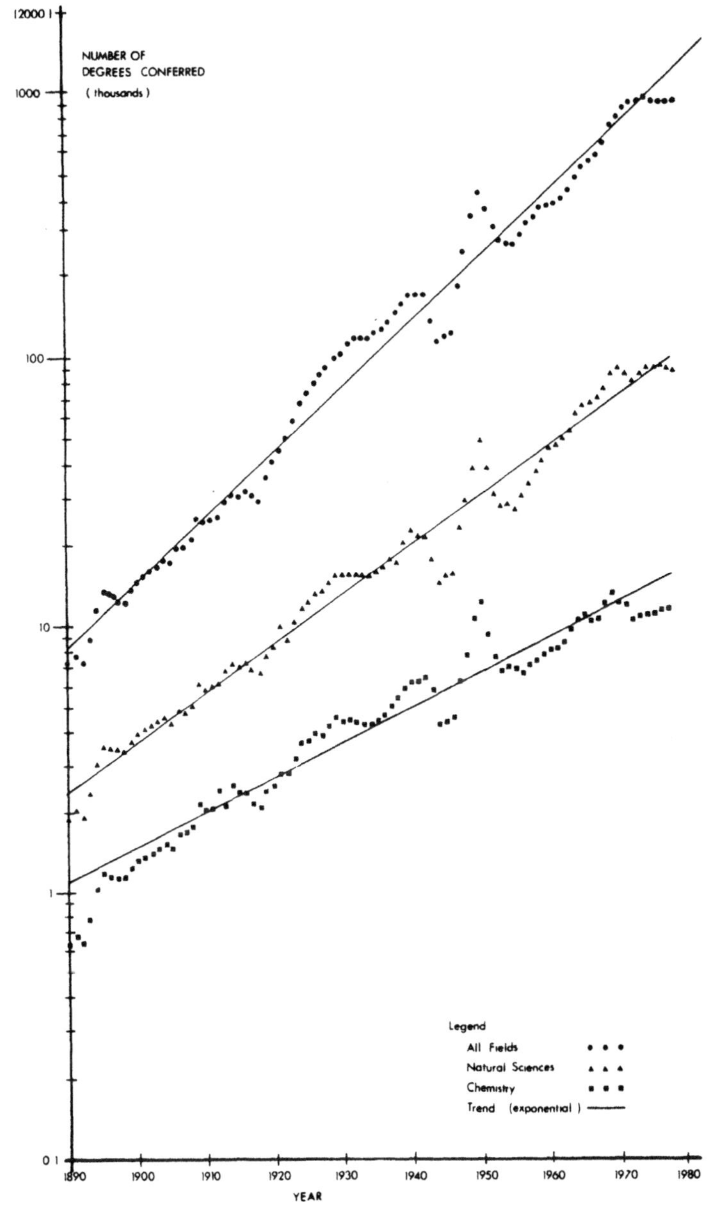

Fig. 3.1–6. Trends in bachelors degrees conferred in selected fields, 1890–1978. (See Table 3.1, columns 1, 2, and 4.)

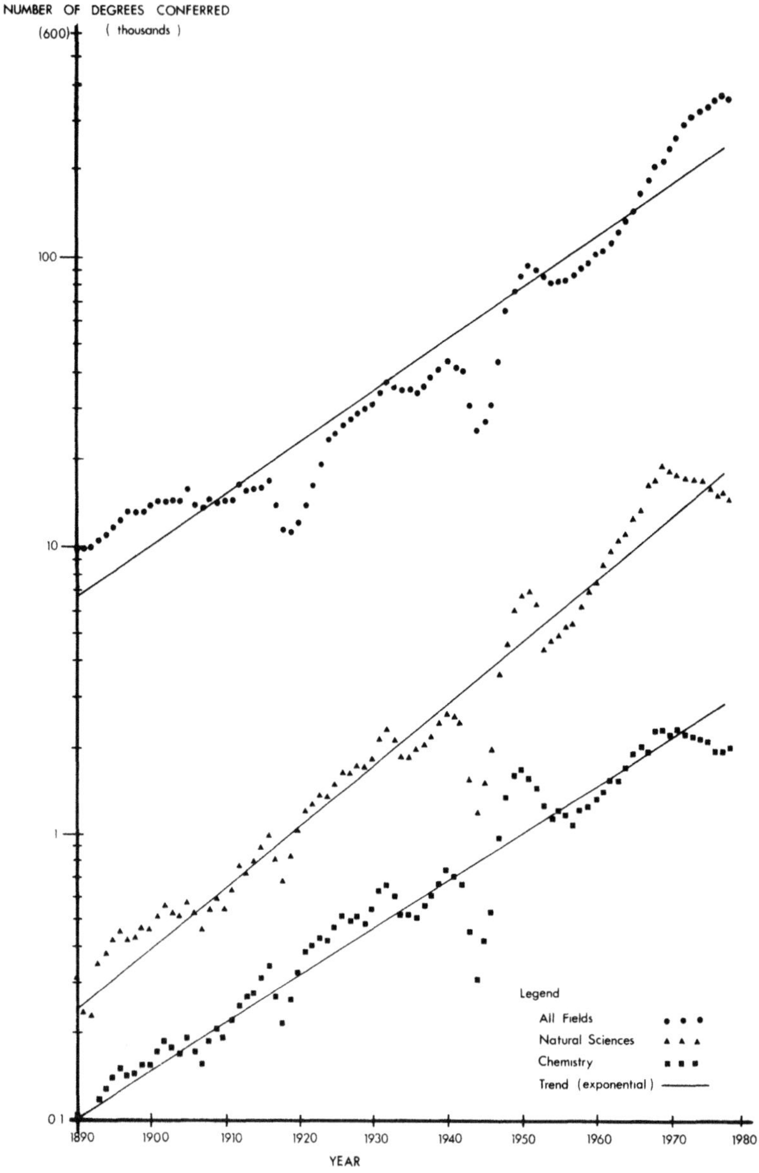

Fig. 3.1–7. Trends in second-level degrees conferred in selected fields, 1890–1978. (See Table 3.2, columns 1, 2, and 4.)

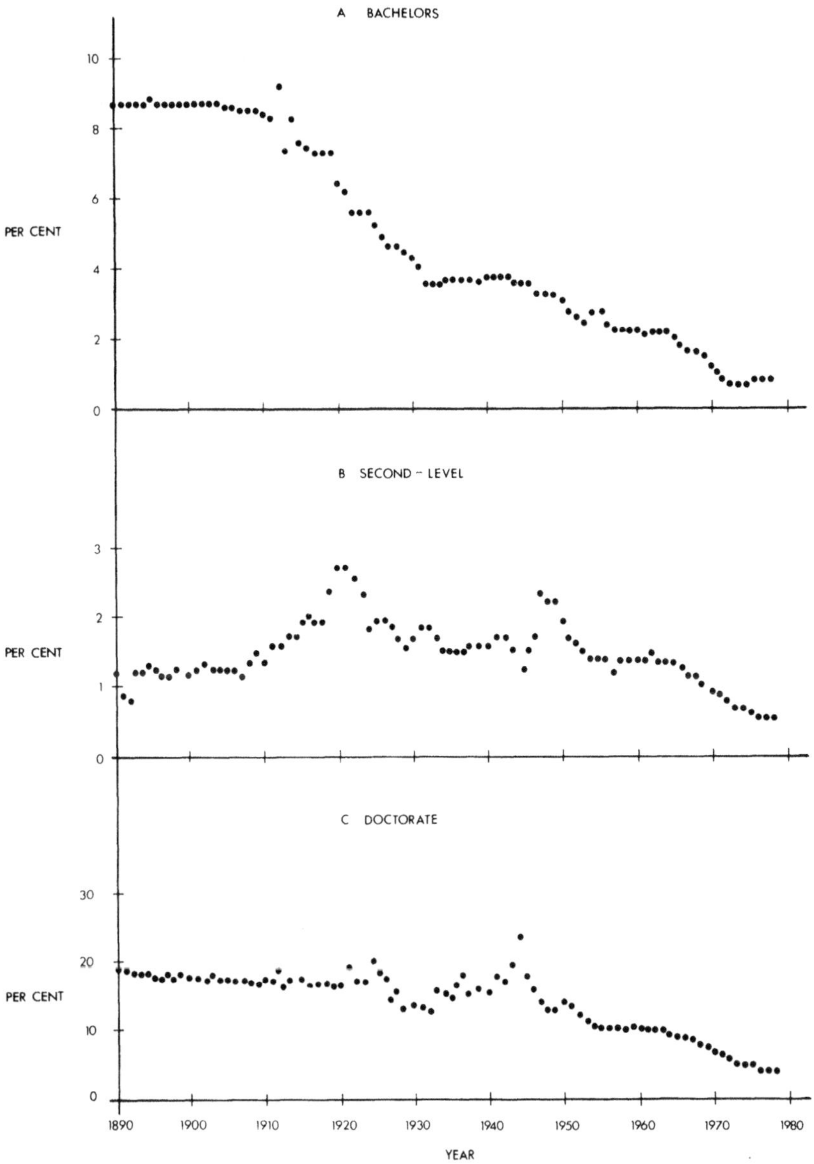

Fig. 3.1–8. Chemistry as percentage of all degree conferrals, by level, 1890–1978. (See Table 3.1, column 5; Table 3.2, column 5; and Table 3.3, column 5.)

from one in ten baccalaureates conferred to one in a hundred. A similar, more modern trend is apparent on the doctoral level. As recently as 1940, chemistry departments conferred almost one-fifth of all earned doctorates. By the early 1970s the proportion had declined to about one-fifteenth.

These relative declines could be caused by the rise of new, chemistry-related disciplines. Little further investigation is needed to establish that this is not the case. Adding the comparatively miniscule figures for biochemistry exacerbates the relative decline of chemistry on the doctoral level, and has no noticeable impact upon the trend in bachelors degrees. Including chemical-materials engineering degrees (i.e., degrees in chemical engineering, metallurgical engineering, materials engineering, and ceramic engineering) alters the fraction of degrees attributable to chemistry, but it does not change the trend. (See Appendix A.3.) The relative decline in degrees conferred is real, whether chemistry is broadly or narrowly defined.

This decline is not unique to chemistry, but characterizes the natural sciences as a whole, as Figure 3.1–9 demonstrates. The enduring trend of relative decline of the natural sciences is as important to any adequate analysis as the more familiar concept of the absolute exponential growth of those sciences. Chemistry simply provides the most extreme example of a more widespread phenomenon. In this it pays the penalty of the pioneer that comes from its early dominant role among the academic sciences.[5]

The existence of a relative decline for science implies a relative advance for one or more non-scientific fields. The arts and humanities do not hold the answer. Rather, administration, education, and the various social sciences are the fields most responsible for the slippage of the natural sciences, as Figure 3.1–10 illustrates. (Data for engineering are also included on the graphs. Comparable series for chemistry and for the natural sciences appeared in the two preceding figures. The data for the arts and humanities, while not plotted, are readily available.) On the bachelors level engineering has hovered around 10% of all degrees, and the humanities have declined modestly (29%

[5] Charles V. Kidd, in his report on *American Universities and Federal Research* (1959), noted a relative decline of PhDs in the natural sciences beginning in the mid-1920s. He attributed this trend to a swing away from the basic sciences toward applied fields such as engineering, business, education, and health (III, Kidd, 1959, 74–76). This concept of relative decline of the natural sciences in degree conferrals was adumbrated recently in II, Duncan, 1978, 31. Duncan points out that the number of degrees in the natural sciences grew at a slower rate from 1952 to 1972 than the number in social sciences, humanities, or "education and other". The trend of relative decline has also received attention within the chemical community; see III, Sanders, 1977, 23–24.

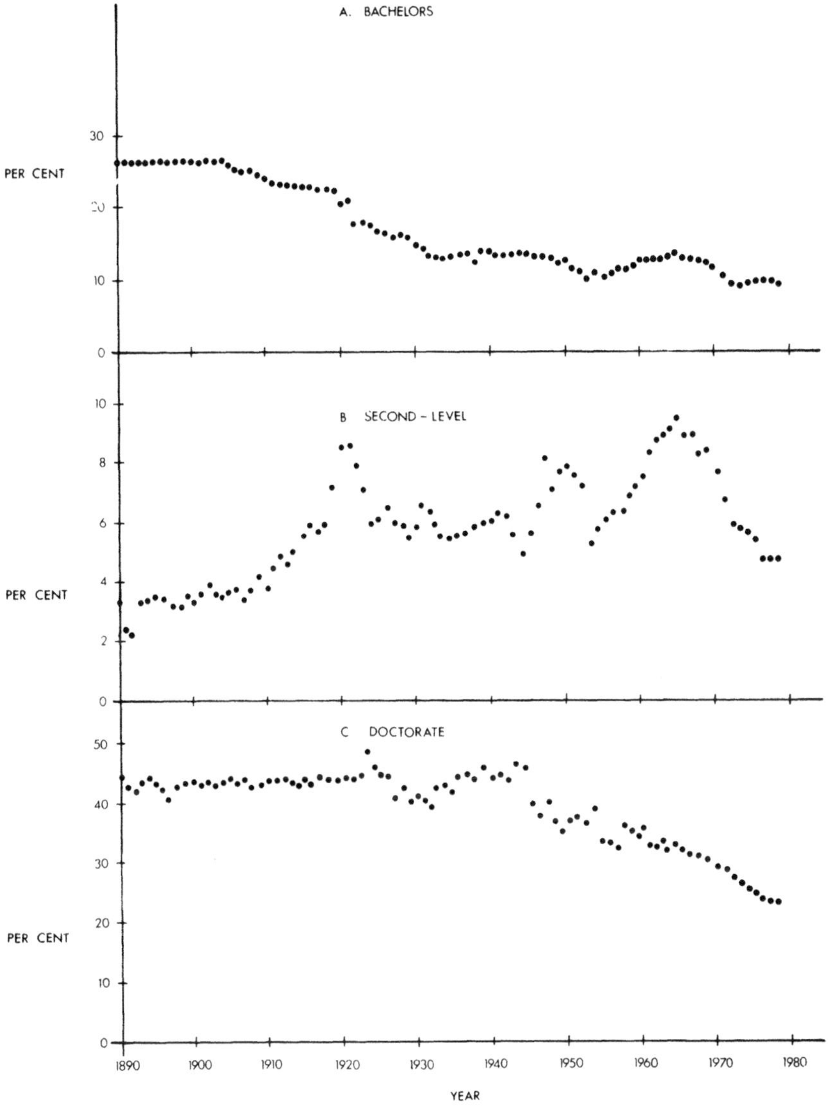

Fig. 3.1–9. Natural sciences as percentage of all degree conferrals, by level, 1890–1978. (See Table 3.1, column 3; Table 3.2, column 3; and Table 3.3, column 3.)

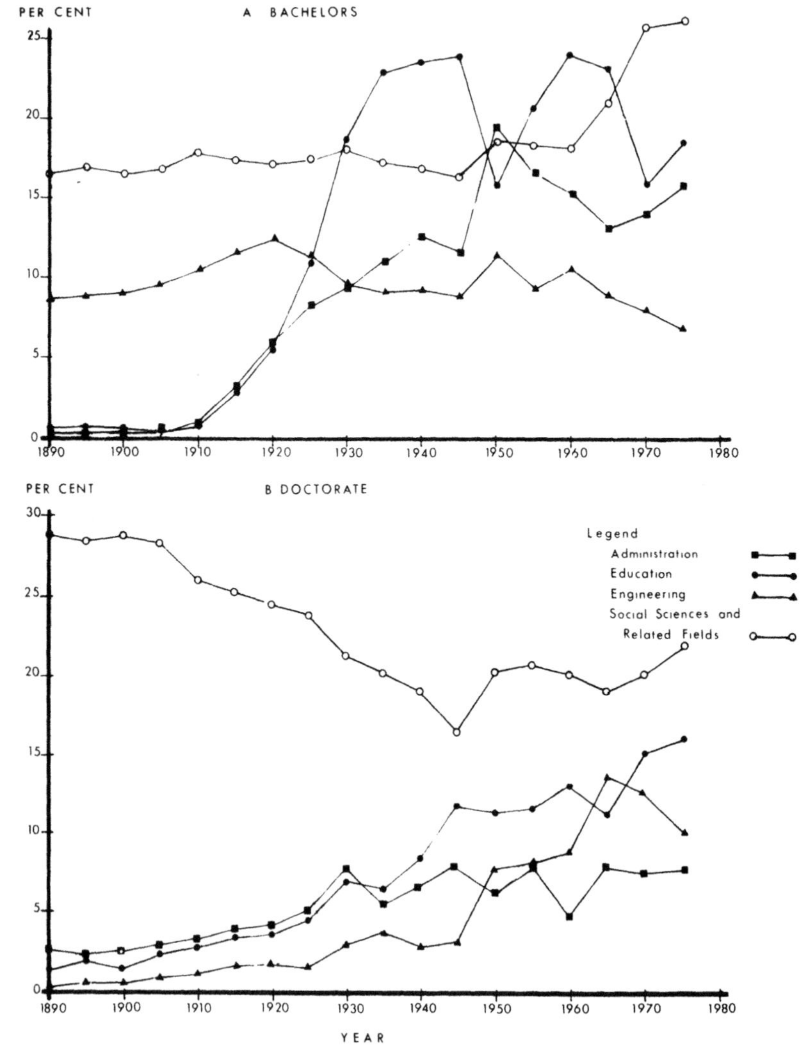

Fig. 3.1–10. Degree conferrals in selected fields as percentage of all degree conferrals, by level, 1890–1975. (See Table 3.6B and Table 3.7B.)

in 1900, 16% in 1970). In contrast, the natural sciences have dropped from a fourth to a tenth, and chemistry in particular has plummeted from a tenth to a hundredth. Education and administration have soared, especially during the interwar decades. And just as the latter two fields began to stabilize at new, higher levels following World War II, the social sciences – which had remained at about 16% for some sixty years or so – started to climb.

The situation for doctorates is similar, but not identical, as the bottom half of the figure demonstrates. While education and administration have gained proportionately, their growth has not been as spectacular with doctorates as with bachelors. Engineering doctorates have shown significant relative growth. (Far from sparking this increase, chemical engineering has suffered its own relative decline in the last fifty years. The discipline conferred about half of all engineering doctorates in the 1920s, but manages to produce only about a fifth of them today.) In contrast, the social sciences dropped in relative productivity of doctorates from 1890 through World War II and have done little more than stabilize since then. In short, education, administration, and the social sciences generally are responsible for the eclipse of the natural sciences on the bachelors level, and education, administration, and engineering have been responsible for the phenomenon at the doctorate level. The slide of the natural sciences has not unseated them from a central position in America's graduate schools, since even today they account for nearly a third of all new doctorates. But it might if the trend of decline continues unabated.[6]

The picture shows chemistry as a steadily-growing academic activity which has also suffered a sustained relative decline. On the one hand, ***growth*** has been the enduring context in which chemistry has functioned as academic discipline. That growth has had, and continues to have, important consequences with respect to scientists' expectations concerning available resources and proper procedures.[7] In its early stages growth was sustained by direct,

[6] For data on arts and humanities, and engineering, see IB, Adkins, 1975, Tables A–4.2, A–4.9, and A–5.10, 214–219, 244–247, 290–293.

Physics doctorates also suffered a slight relative decline, but this occurred prior to 1930. In 1890, PhDs in physics constituted 6.0% of all doctorates conferred; by 1930, this figure had declined to 4.7%. Since World War II, physics has regained some of this loss; in 1960 and 1970, doctorates in physics were a constant 5.1% of all doctorates conferred. See *ibid.*, Tables A–2 and A–5.7, 190–195, 278–281.

[7] The growth in federal funding for academic science during the early 1960s helps to explain the tone of the Westheimer Report on *Chemistry: Opportunities and Needs* (IB, NAS, 1965). In his letter of transmittal George B. Kistiakowsky warned that chemistry was "not as well supported as is research in other natural sciences, if qualified requests for funds and the training of scientific personnel are proper yardsticks" (*ibid.*, iii).

vocational linkages between higher education in chemistry and employment in the chemical profession. On the other hand, the same growth has concealed decline in the importance of chemistry within higher education. That decline may relate to the ways in which the linkages between chemical education and subsequent employment have steadily attenuated, without chemistry being able to avoid a "vocational" label.

3.1.2. DECOUPLING: VOCATION AND CULTURE

As early as 1926 one informed commentator, N. M. Grier, noted that the "first year of biological, chemical, and physical science is becoming increasingly popular with [college] students, but for some reason within or without their control, [college science] departments fail to hold students for advanced work." While chemistry was of considerable appeal to students, employment opportunities were lacking; Grier complained that "many professors of chemistry in the colleges and universities are inclined to discourage students from majoring in their science unless they wish to train themselves for teaching, believing that the field is too crowded for such students to arise above routine positions". Studies of announcements of "situations open and wanted" in this period suggest that there was indeed a shortage of opportunities in chemistry. This was the era in which the stock of those with educational credentials in chemistry first approached the total of chemists employed in the United States. College chemistry, perceived in vocational terms, was moving out of phase with the job market on one side and the high school situation on the other.

Far more students entering college since the 1920s *intended* to major in chemistry than actually did so.[8] It is not simply that relatively fewer college freshmen with dreams of chemistry manage to receive a chemistry degree. Even those who complete their programs are less likely to become chemists. Chemical education and chemical occupations have "decoupled" from one another. Not only is this science losing its share of the student population, and thereby experiencing a relative decline on the campus, but even those students it does capture are slipping away from chemistry after graduation.

[8] See, for example, III, Reitz, 1973. Grier's influential report on secondary science instruction (III, Grier, 1926; quotes from 932, 876) was incorporated into later discussions of the National Society for the Study of Education (III, NSSE, 1932, 305). On the oversupply of chemists, see III, Ginnings, 1925, 1929, 1930, 1934; and III, Peaselee, 1929. III, Badger, 1929, paints a more optimistic picture of the job market for chemical engineers at the time; see also III, White, 1931, 240–244.

One way to examine this decoupling phenomenon is to observe the absolute and relative trends of both the number of persons with chemical degrees (*credentialled* chemists) and the number of persons who consider chemistry to be their occupation (*practicing* chemists). Time series of the absolute size of each of these groups were displayed in Figure 2.2–10 as part of an overview of the entire chemical community. They are presented alone in Figure 3.1–11, with credentialled chemists represented by a dashed line and practicing chemists by a solid one. Since around 1930, more people have held terminal chemistry degrees than have practiced chemistry. The chasm has widened considerably since that time. The ratio of the one series to the other expresses this divergence via a single indicator (see Figure 3.1–12). There are now almost two persons with a terminal degree in chemistry for every practicing chemist. Thus not only are there now proportionately fewer chemistry majors on college campuses but those majors have proportionately fewer future chemists among them.

A number of adjustments to these time series are possible. Only two will be attempted here. In each case, the boundaries for the populations being counted will be changed to see if the trends are affected. First, one could argue that the widening of the gap between training and practice has occurred because chemistry has produced spin-off fields such as chemical engineering and biochemistry. According to this view, more and more people are receiving chemistry degrees, then taking jobs in the spin-off fields, and consequently not calling themselves chemists. This hypothesis would be confirmed if a plot similar to Figure 3.1–11 – but this time of persons credentialled versus persons practicing not just in chemistry but also in all its related fields – failed to display the divergence so obvious in that earlier figure. The second possible adjustment involves how many degrees those credentialled in chemistry receive. The two figures on decoupling already presented define credentialled chemists as only those persons who possess a *terminal* degree in chemistry. Some people receive a first degree in chemistry and then a further degree in some other field. An obvious example here is the physician who majored in chemistry as an undergraduate. If they are included, there will be an upward shift in the line representing the number of persons credentialled in chemistry.[9] The result will be an even more vivid decoupling of education and profession.

The adjusted indicators are revealing. Combining chemistry and chemical

[9] See Appendix A.4 for a technical discussion of the various modifications and their limitations.

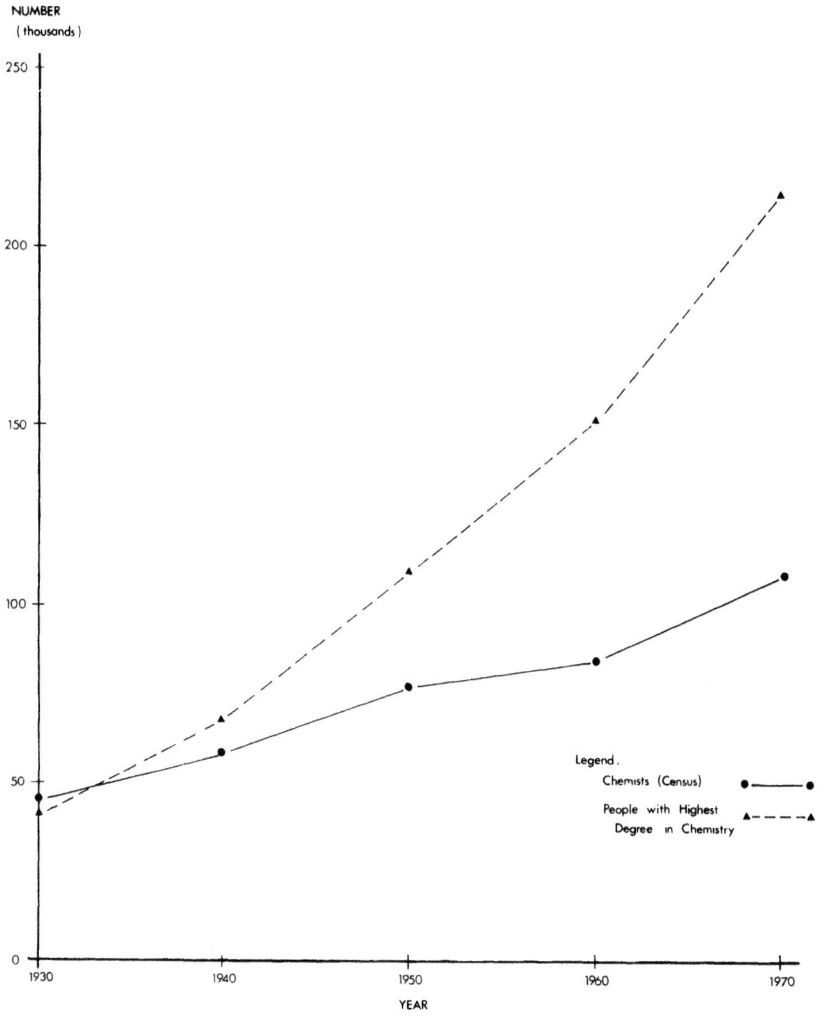

Fig. 3.1–11. Practicing versus credentialled chemists, 1930–1970. (See Table 3.8, columns 1 and 2.)

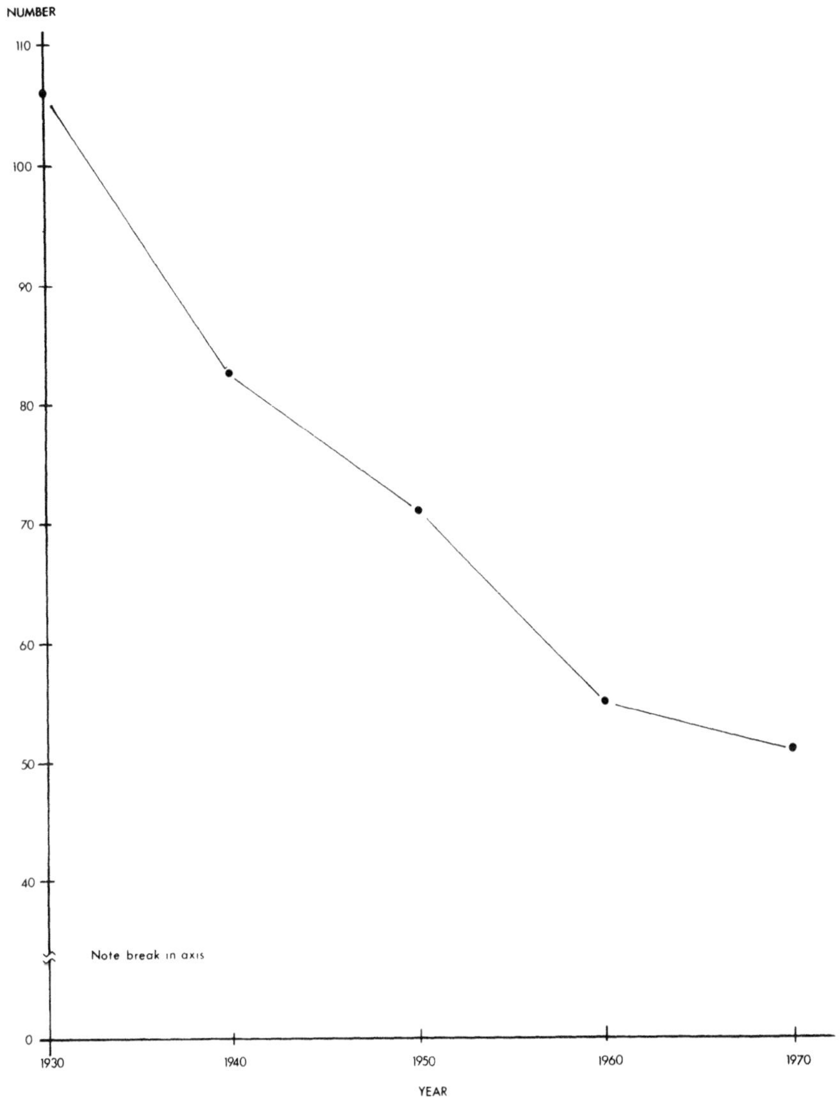

Fig. 3.1–12. Number of chemists per hundred persons with highest degree in chemistry, 1930–1970. (See Table 3.8, column 3.)

engineering, Figure 3.1–13 gives condensed, revised versions of both of the preceding figures. Additional data have been included to approximate the trends for holders of *all* degrees, whether terminal or not. In the upper graph, the bottom dashed line shows the number of practicing chemists and chemical engineers in the labor force, and the next line (solid dots) traces persons credentialled with a *terminal* degree in these fields. The divergence between them indicates that the decoupling phenomenon is still present. (While other evidence suggests much switching between chemistry and chemical engineering, such moves have already been accounted for in this graph.) The divergence of the two higher lines away from the next lower series displays the growing *additional* tendency of credentialled people to move completely away from chemistry-related degrees – terminal or not. The widening of the distances between these and the solid dots (only terminal degree holders) shows growth in the tendency for people to take a chemistry degree only as a stepping-stone to additional education and a career outside chemistry. From such a perspective, chemical education begins to look like a sieve, leaking credentialled chemists in all directions.[10]

The bottom half of the figure is no less revealing. The solid line shows the ratio of persons practicing in any chemistry-related discipline to those holding a *terminal* degree in the same group of fields. As is evident, the decline revealed in Figure 3.1–12 persists after one has accounted for migration from chemistry to the chemical spin-off disciplines. An increasing minority of persons terminally credentialled in chemistry *and related fields* are finding careers elsewhere. The dashed line shows the ratio of practicing to credentialled persons in chemistry and related fields, with an approximation to the number of holders of *non*-terminal degrees also included. This shifts the curve down from its position when only terminal degrees are counted (solid line). Decline is not as obvious in this indicator, but this is almost certainly a result of our inability to make the necessary adjustments for mortality among credentialled chemists with non-terminal degrees.

[10] While the absolute number of persons with non-terminal chemistry degrees (represented by the wedge between the solid dots and triangles in Figure 3.1–13A) doubled between 1930 and 1970, this group is a shrinking fraction of *all* persons credentialled in chemistry.

One should not conclude from this observation that decoupling is diminishing, as further inspection of Figure 3.1–13A shows. By comparing the space between the dashed line and triangles (non-chemists with chemistry degrees) with the space between the triangles and the horizontal axis (all chemistry degree-holders), one finds that the proportion of *non-chemists* among chemistry degree-holders has increased steadily since 1930.

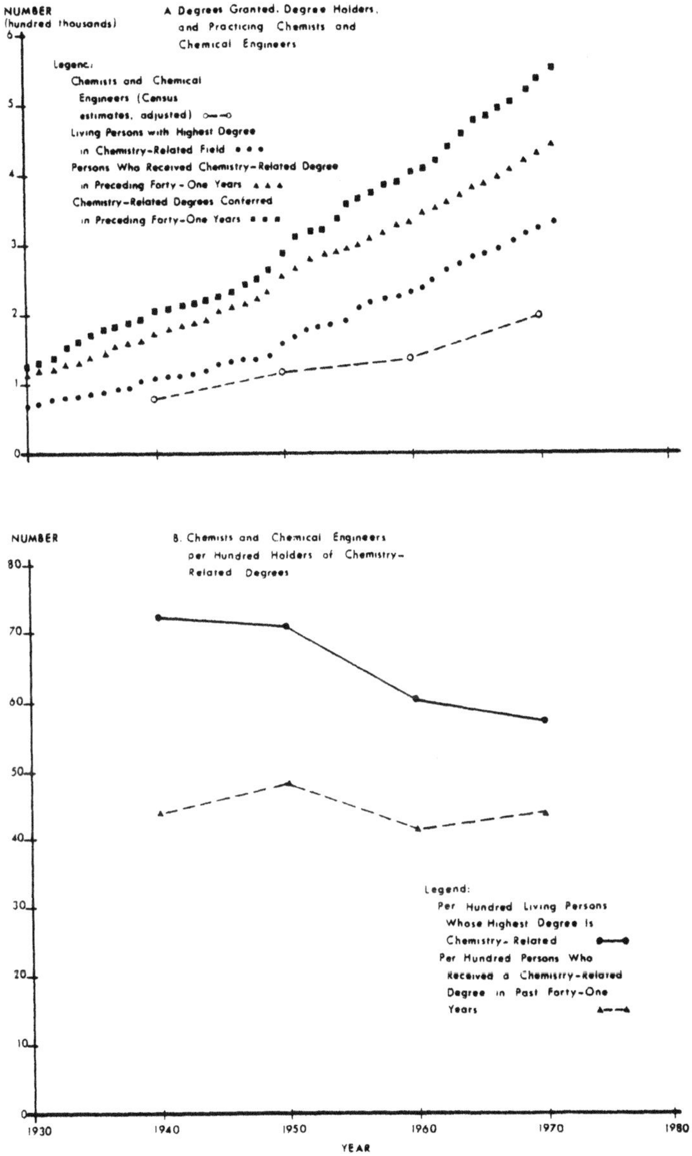

Fig. 3.1–13. Chemists and chemical engineers versus holders of chemistry-related degrees, 1930–1971. (See Table 3.10, columns 2, 3, and 4; Table 3.11, column 5; and Table 3.12, columns 2 and 3.)

In conjunction with the extended treatment in Appendix A, this evidence suggests that, at the bachelors level, there has always been a small tendency for majors in chemistry and related fields to find careers outside these disciplines. Some have done so with a terminal bachelors degree only. Others have taken the chemistry-related bachelors as an intermediate step in their education and have continued with graduate study in an extra-chemical discipline. The size of this group, relative to the chemical community defined in its broadest terms, has been growing steadily. More recently, there has been a stronger trend of graduate students in chemistry and related fields taking a chemistry-related *graduate* degree before departing for an extra-chemical career. Examples include those who take terminal graduate degrees in chemistry (such as PhD chemists who become corporate executives) and those who supplement their graduate chemistry-related degree with another post-baccalaureate degree (such as masters-level chemists who become patent attorneys). The numbers involved in this decoupling of chemical education from the occupation of chemistry have grown significantly over the years. Today, only *half* of those with a terminal degree in chemistry are practicing chemists.[11]

Such decoupling is not necessarily undesirable for the institution of chemistry. It can be interpreted as the successful expansion of chemical education, fueled by the recruitment of new students to growing chemical departments and by the concurrent need to place all graduates, however unsuited to or uninterested in chemical careers. The decoupling of college education in chemistry from careers in the field is further illustrated by information on the first employment status of new holders of the bachelors degree in recent years. Over the eighteen years from 1958 to 1976, the percentage employed in chemistry nearly halved, dropping from an initially low 38% to about 20%.[12]

It is also instructive to contrast the fate of chemistry with that of history, an admittedly idiosyncratic example of a non-natural-science field of interest to the authors. Figure 3.1–14, showing the trends of degree conferral in history and in chemistry, illustrates that the former field, like the latter, has

[11] Many of these speculations assume that the *sole* reason for the upward slope of the dashed line in Figure 3.1–13B is our failure to compensate for mortality at all times. Further analysis may undermine this presumption; see Appendix A.

[12] IB, ACS, 1977b, 43. This trend cannot be explained by an increase in those pursuing advanced work in chemistry. According to an ACS survey, about 31% of chemists in the class of 1959 planned to study chemistry in graduate school. In 1976 this proportion stood at about 32% (*ibid.*).

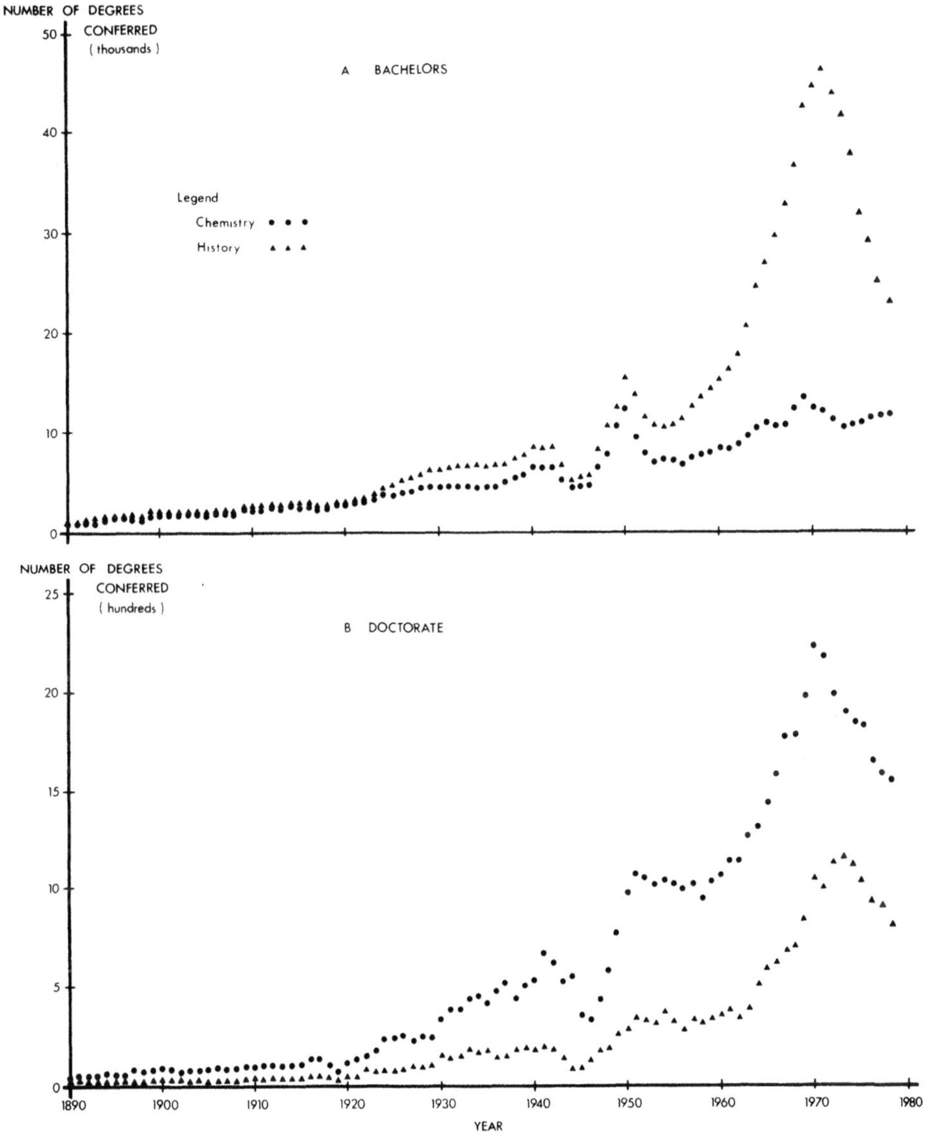

Fig. 3.1–14. Bachelors and doctorate degrees conferred in chemistry and in history, by level, 1890–1978. (See Table 3.1, column 4; Table 3.3, column 4; and Table 3.13, columns 1 and 2.)

experienced dramatic growth, from about 600 bachelors degrees conferred in 1890 to about 45 000 per annum in the early 1970s; during the same period, production of history doctorates expanded from about 10 to 1000 annually. The figure also hints at how ubiquitous the retrenchment of the 1970s has been. But there are differences between the two disciplines. The numbers of degrees conferred in chemistry and in history have diverged markedly in recent decades. While history has been the field conferring the greater number of bachelors degrees, chemistry has awarded the larger number of doctorates. (Note the transposition of dots and triangles between the upper and lower parts of the figure.) One interpretation of these indicators is that, for some time, choice of first degree subject in the mainstream liberal arts (including both chemistry and history) has been increasingly removed from career decisions. As a non-vocational field – one in which people major, but *not* in which they stake out their careers – history has been able to profit from this trend. Undergraduate chemistry has had more obvious vocational connotations; this may help explain its resilience during the 1970s. The appeal of chemistry has diverged steadily from that of history. The divergence became especially marked in the 1950s and 1960s as vocational preparation retreated fully into the graduate school. An explicit-invented indicator of this divergence between "vocational" and "abstract" appeal is the ratio of baccalaureates in chemistry to those in history (Figure 3.1–15).[13]

The situation for doctorate degrees has been quite different. For half a century following the routinization of graduate training in America, degrees in chemistry steadily outpaced those in history by a ratio of better than three to one. Chemistry PhDs were of obvious relevance to the demands and opportunities of government and industrial chemistry in the period. From a high point in the mid-1940s, the ratio of chemistry PhDs to history PhDs has eroded steadily (from 3.45 in 1950 to 1.76 in 1975). This indicator of relative decline in the past quarter century suggests a growing vocational role for history PhDs, and also agrees well with the trend for chemists among professional, technical, and kindred workers, presented earlier (see p. 33 and Figure 2.2–9).

The indicators presented reveal not only the vigorous growth of chemical education, but also its decline in relative importance within American

[13] This trend is evident even when one defines chemistry as broadly as possible, including degrees conferred in biochemistry and chemical-materials engineering. See Appendix A.3.

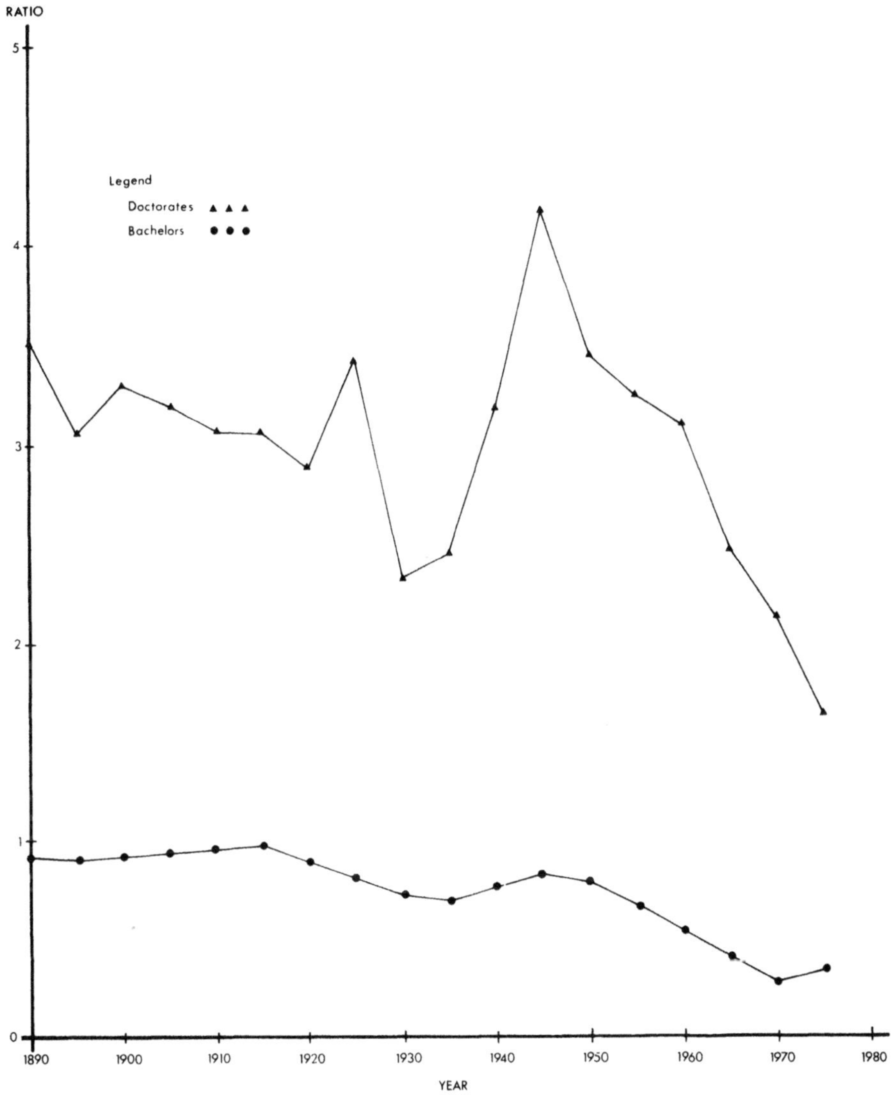

Fig. 3.1–15. Ratio of chemistry degree conferrals to history degree conferrals, by level, 1890–1975. (See Table 3.14, columns 5 and 6.)

academe. At the baccalaureate level chemical education has shifted from being a field with only vocational orientations toward serving as a significant cultural instrument of broader application, although not as broad as fields like history. Certain of these themes are repeated, and others quite different are introduced, when one turns to the place of chemistry in secondary education.

3.2. Secondary Education

INDICATOR HIGHLIGHTS:

Over the last century, the proportion of American adolescents receiving some formal chemical education has increased more than twelvefold, from 2.2 to about 28%. Enrollment in high school chemistry classes has increased from about 30 000 to over 1 000 000 students each year.

The proportion of high school chemistry students who later major in chemistry at college has declined steadily. Thc nature of the chemical curriculum in secondary schools has changed accordingly.

About 2% of instructors at the secondary level specialize in chemistry, implying about 2000 such persons in 1920 and about ten times that many in 1970. The ratio of high school chemistry teachers to persons in either the occupation or the profession of chemistry has risen steadily, though the number of such teachers with college chemistry degrees is small.

The enduring trend of the past century has been for a growing fraction of an increasing adolescent population to attend high school. Secondary school enrollment in the United States multiplied 15 times between 1910 and 1973, an attendance boom in which chemistry shared (Tables 3.15 and 3.16). During the period from 1890 to 1973, the number of students taking chemistry in public high schools rose from about 30 000 to over 1 000 000 (Figure 3.2–1). The majority of these students were juniors and seniors (i.e., 16 or 17 yr old).[14] The proportion of each cohort exposed to some chemistry teaching has increased over the years from about 2.2% to 27.9% (Figure 3.2–2). While in the nineteenth century an insignificant fraction of the

[14] A study of curricular organization in 1908, 1923, and 1930 shows that chemistry ceased to be an exclusively twelfth grade subject by 1930, when enrollments by eleventh and twelfth grade students in the science were about equal. This was also the case in 1950. III, Hunter, 1933; III, P. Johnson, 1950, 15.

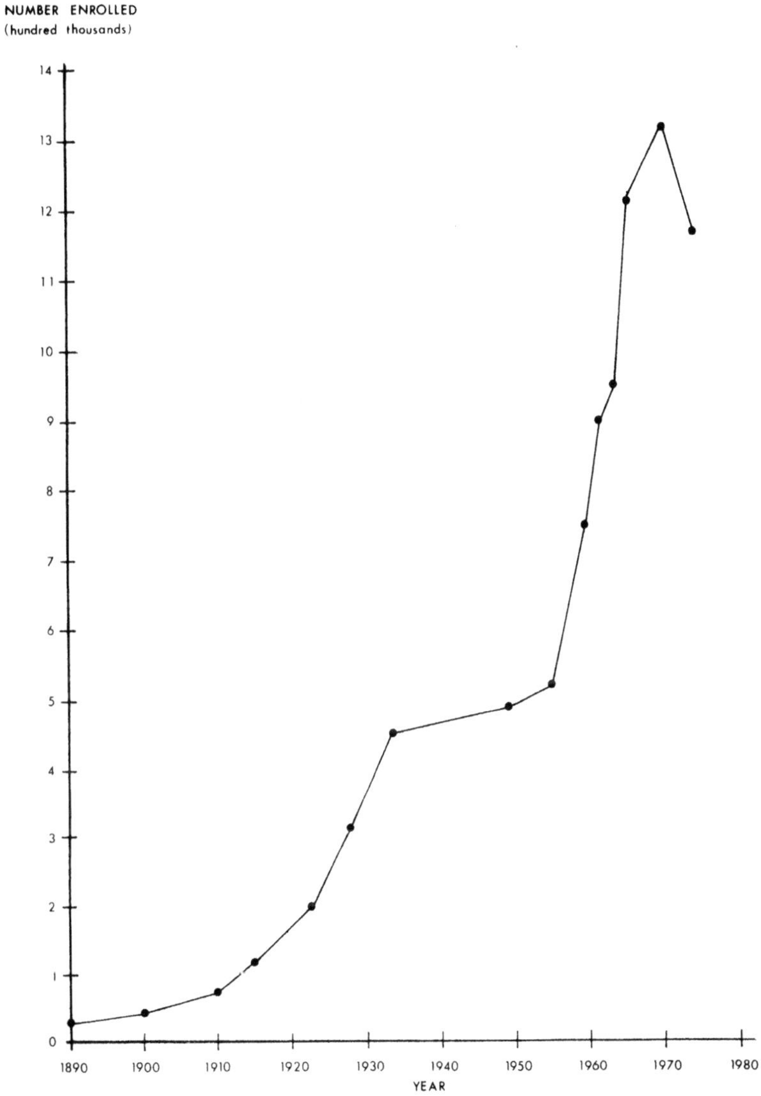

Fig. 3.2–1. High school chemistry enrollment, 1890–1973. (See Table 3.16, column 3.)

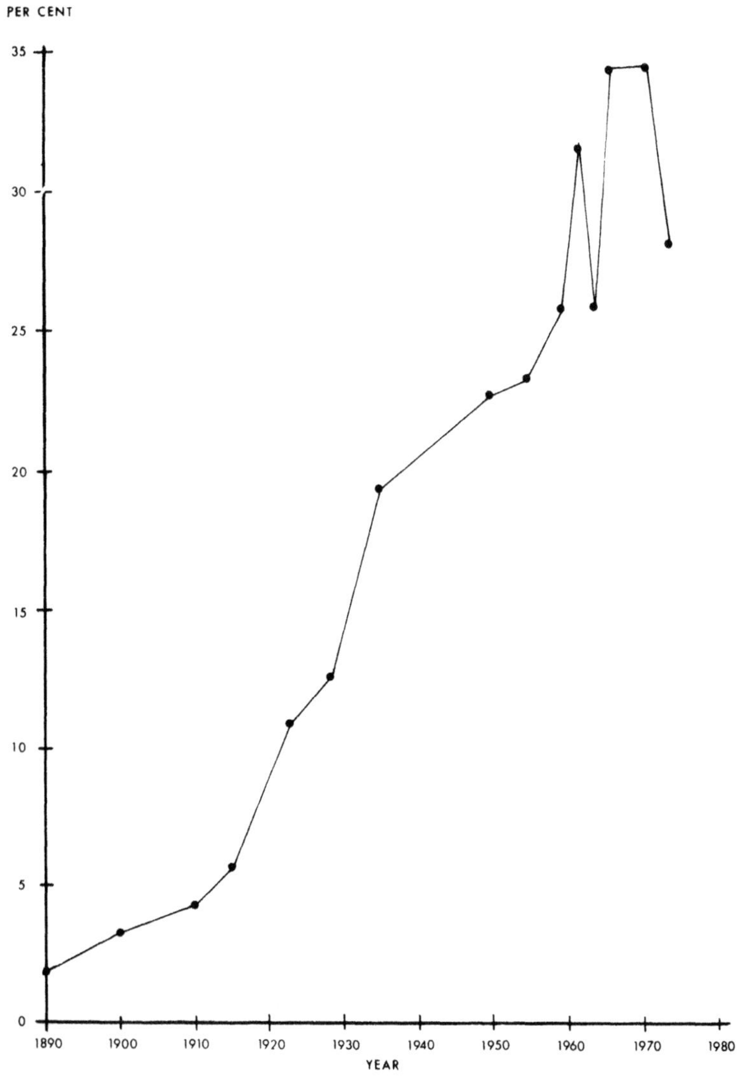

Fig. 3.2–2. High school chemistry enrollment as percentage of population aged sixteen, 1890–1973. (See Table 3.16, column 3.)

community was instructed in chemistry, today a considerable segment of the adult population has been introduced to the subject. This development carried implications with respect to public understanding of science, employment demands for high school chemistry teachers, and the relationship between high school and college chemical curricula.

The fragmentary evidence on high school chemistry teachers offers clues to the nature of this introduction to the subject. Since at least the 1920s, about 2% of all high school teachers have listed chemistry as their major instructional responsibility. This proportion suggests that there were 2000 high school chemistry teachers in 1920 and, because of the general expansion of secondary education, 10 times that number in 1970.[15] These impressive numbers might conjure visions of ubiquitous chemists proselytizing within American high schools, but several factors indicate that the genre of chemical knowledge transmitted has been rather different from that practiced at the vocational and professional levels. The majority of secondary school chemistry teachers have had concurrent obligations to teach mathematics, general science, or other specialized science courses. This fact has limited the time and attention devoted to chemical instruction. Surveys of high school science teachers reveal that few such people have doctorates and less than half have masters or other second-level certification. Furthermore, since the 1920s boom in education majors (illustrated in Figure 3.1–10), most chemistry teachers in American high schools apparently have had degrees in science education, not in chemistry *per se*. In the 1920s new high school chemistry teachers took an average of only twenty semester hours of chemical instruction: in the 1970s a significant fraction of chemistry teachers lacked even that rudimentary acquaintance with the science.[16]

[15] Estimates of the number of secondary school teachers in America are easy to find, but not of chemistry teachers. Available surveys indicate that the percentage of all high school teachers who had some responsibility for teaching chemistry ranged from 1.2 to 2.8% between the 1920s and the 1960s (IB, Kirby, 1925, 25; IB, Buckingham, 1926, 90; IB, NEA, 1950, Table III–A; IB, NEA, 1954, Table 4; IB, NEA, 1958, Table 7; IB, NEA, 1962, Table 7; IB, NEA, 1965, Table 10). Two per cent thus seems a reasonable estimate for chemistry teachers among all high school teachers.

U.S. secondary school teachers increased from about 100 000 in 1920 (IB, Ferriss, 1969, Series B6, 383) to about 1 000 000 in 1970 (IA, USOE, 1972, 6). Applying the 2% estimate for chemistry teachers yields the figures quoted above. See Table 3.18.

[16] A 1922 study showed that the median number of semester hours of chemical instruction taken ranged from 15.8 to 23.5, depending upon the size of the high school at which the chemistry teacher was employed (IB, Hutson, 1923, Table V, 431). Although teachers in this study were better prepared than other secondary science teachers (*ibid.*,

While some high school students have taken quite rigorous chemistry courses, most pupils have had little more than a pedestrian introduction to "cookbook" chemistry. One could hardly expect something different, given that the chemistry teachers themselves have not been professionally certified in the subject. High school chemistry has long been part of an educational culture geared toward producing model citizens rather than fledgling scientists. As a result, chemistry classrooms in American high schools have never been filled with budding chemists. The proportion of students who have subsequently specialized in chemistry at college has always been quite small. In 1910 college chemistry majors were only 3.4% of the number of students in the given age cohort who had enrolled in high school chemistry classes. The proportion has declined slowly since then; by 1970, it had reached 0.8% (Figure 3.2–3).[17] This decline is all the more dramatic since it occurred

---

436), their preparation was less rigorous than that of their contemporaries who majored in chemistry.

Information on the concurrent teaching obligations and qualifications of high school chemistry teachers is fragmentary. Hutson's study reports that nearly all chemistry teachers in his sample (95.5%) had obligations to teach other subjects (*ibid.*, Table III, 427). A 1923–1924 survey of new teachers in Ohio found that 85.3% of those who taught chemistry also taught other subjects (IB, Buckingham, 1926, Table 30, 90). A 1965 National Education Association survey shows much the same situation. Only 38% of new chemistry teachers were responsible for that science alone, while 62% taught other subjects as well, usually physics, general science, mathematics, or biology (IB, NEA, 1965, Table 10, 32).

Other recent NEA investigations of the population of high school teachers demonstrate that since 1960 less than 40% have held masters or doctorate degrees (IB, NEA, 1970, Table 24, 53). One study in the mid-1960s found that only one-eighth of all high school *science* teachers had received masters or doctorate degrees (IB, NEA, 1967, Table 13, 19).

[17] Although the U.S. Office of Education has published data on subject enrollments in American high schools since 1890, enrollments were not analyzed by grade level until 1907, and the figures are not reliable until about 1910; see IA, USBE, 1912, II, 1185.

Our estimates of enrollment are based upon the assumption that the percentage of private high school students enrolled in chemistry courses was the same as that in public high schools. Comparison of the chemistry enrollment in public high schools (Table 3.16, column 2) with private high school chemistry enrollment in years for which separate data are available (IA, Jessen *et al.*, 1937; IA, Gertler, 1962) shows that this assumption introduces only minor perturbations.

Other assumptions underlying the correlation of high school and college chemistry enrollments for given age cohorts (Figures 3.2–3 and 3.2–4) are discussed in the notes to Tables 3.15–3.17 and 3.19–3.25.

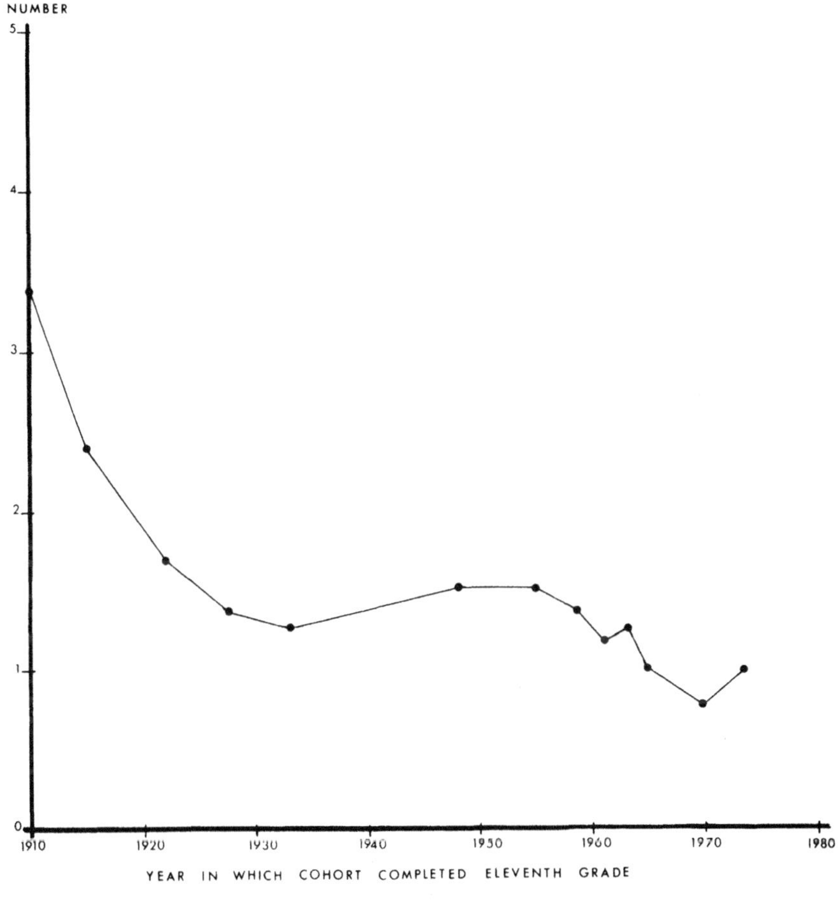

Fig. 3.2–3. Baccalaureates conferred in chemistry per hundred students of the same age cohort who had enrolled in high school chemistry, 1910–1973. (See Table 3.19, column 3.)

during a period when there was an increasing propensity for students to complete college.

Assuming that most of those who graduated from college as chemistry majors had first taken chemistry in high school, then before World War I a strong link connected secondary and later stages of chemical education. One of every six high school chemistry students in 1910 who later entered college graduated as a chemistry major. By 1970, this connection had eroded to a point where only one in 36 college-bound students in high school chemistry classes majored in chemistry at college (Figure 3.2–4). Changes in the secondary school chemistry curriculum must be understood against the background of these two enduring trends: ever more students in high school chemistry classes, and an ever weaker linkage between the enrollment in those classes and subsequent specialization.

In the fifty years preceding the 1920s, high school curricula were designed primarily to inculcate mental discipline, an orientation which was well adapted to preparing students for a contemporary college education. In the case of chemistry, a curriculum organized around a mass of descriptive material and the performance of large numbers of experiments served both mental discipline and vocational ends. It did not, however, serve a world in which chemistry was part of an emergent industrial culture and less and less significant as training for particular occupations.[18]

Characteristic of calls for reform was the 1920 report of a U.S. Bureau of Education committee formed to study the high school science curriculum. In its report on the *Reorganization of Science in Secondary Schools*, the committee, chaired by Otis W. Caldwell, placed new emphasis on the student's personal development and on the "relevance of chemistry to everyday life". The stress upon cultural aspects of the science and upon the socialization of future citizens implied less attention to "preparation of the student for college", "acquisition of factual knowledge", and "ideas of formal discipline". Such a reorientation was not unopposed. The newly-established Division of Chemical Education of the American Chemical Society took a lively interest in questions of curriculum. Its 1924 "Standard Minimum High School Course

[18] On the history of American secondary education prior to 1920, see III, Krug, 1964. For the role of mental discipline in this period, see III, Kolesnik, 1958. On the place of science in the secondary curriculum, see III, Hurd, 1949, and III, Krug, 1960, 320–339; and III, Krug, 1964, 368–375. The orientation of high school chemistry curricula before 1920 have been discussed in III, Fay, 1931; III, Sharp, 1940; III, Rosen, 1956; III, Summerlin and Craig, 1966; and III, Ogden, 1974.

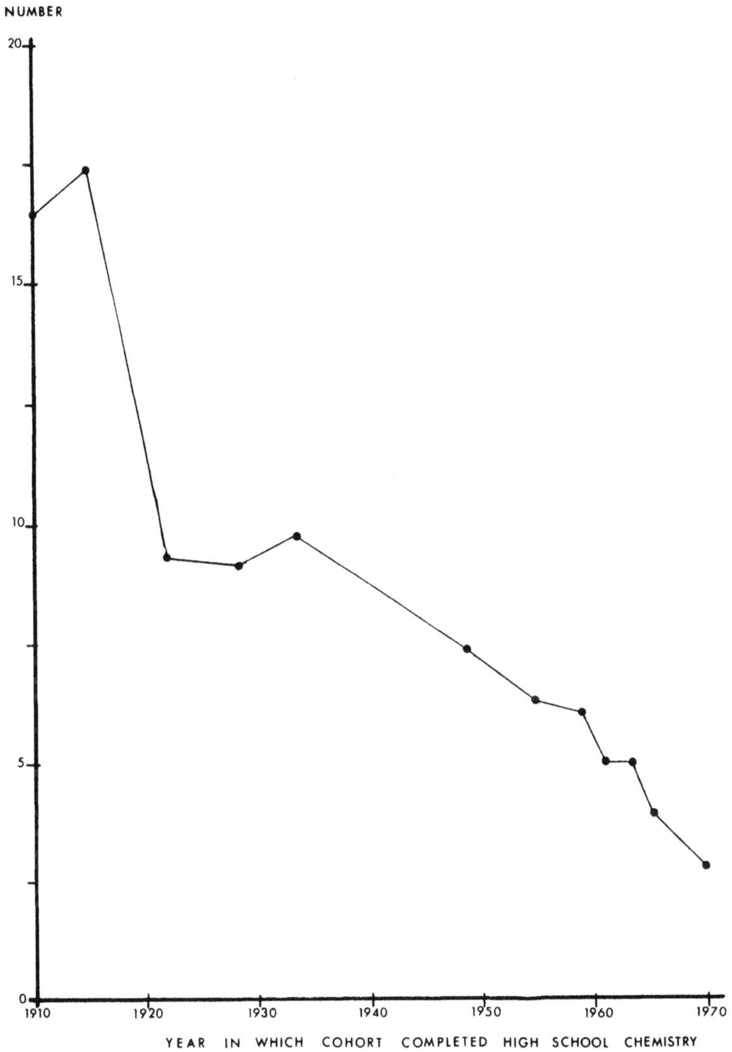

Fig. 3.3–4. Baccalaureates conferred in chemistry per hundred *college-bound* students of the same age cohort who had enrolled in high school chemistry, 1910–1970. (See Table 3.25, column 5.)

in Chemistry" presented the case for the subject as the introduction to a vocation.[19]

The debate continued sporadically for the next thirty years. On one hand, textbooks were slow to change. On the other, teaching necessarily took place in the broader cultural context of general education, rather than the specialized one of training in pure science and preparation for college. The visible event of Sputnik served to rekindle the debate, and much effort was expended in creating textbooks aimed at quickening the flow of college chemistry majors.[20] There is little evidence to suggest that this activity significantly affected either the secular trend or the contemporary situation – in which approximately one million students enroll in high school chemistry each year, but only one in a hundred chooses to major in the subject at college (see Table 3.19).

Even at the college level, instruction in chemistry has been slowly but steadily dissociated from the choice of chemistry as a career. The shift in the purpose of high school chemistry has been more dramatic and – apparently – more successful. Whereas college chemistry has declined steadily over the last half century (in terms of the proportion of all students who choose to major in it), high school chemistry has held its own. Students enrolled in chemistry courses constituted 7.7% of all high school students

[19] The Caldwell Report (III, CRSE, 1920) is discussed in III, Hall, 1939 (quotes from 117). The proposed ACS standard high school chemistry course is given in III, ACS, 1924, 87–93; see also III, Hopkins *et al.*, 1936.

Members of the ACS Division of Chemical Education were active in trying to foster cooperation among high school and college chemistry teachers. The Division's *Journal of Chemical Education* (1924) became a forum for the discussion of the chemistry curriculum; see, for example, III, ACS, 1925, 1927, and III, Mattern, 1928. On the origins and activities of the Division of Chemical Education, see III, Gordon, 1943; III, Kessel, 1973; and IB, Skolnik and Reese, 1976, 263–267.

[20] The slow transformation of high school chemistry texts before the 1960s is discussed in III, Hall, 1939; III, Sharp, 1940; and III, Summerlin and Craig, 1966. Among the reports portraying chemistry as an element of general education are III, Kaufman, 1927; III, NSSE, 1932; III, Hopkins, 1935; III, AAAS, 1941; and III, Ashford, 1942; see also III, Ogden, 1976.

On the post-Sputnik reorientation of secondary science education and its implications for chemical education, see III, Dede and Hardin, 1973; and III, Summerlin and Craig, 1966. The most widely-known of the chemistry curriculum improvement projects in this period was CHEM Study, an NSF-supported project begun in 1959 under the direction of J. Arthur Campbell (Harvey Mudd College); see III, Campbell, 1964, and III, Merrill and Ridgway, 1969. For a comprehensive review of NSF involvement in the reform of secondary science education, see III, CRS, 1975. See also III, Rhodes, 1977.

in 1900 and 9.3% in 1965. The fluctuation over the years has been small. Such comparative constancy of chemistry's position in the curriculum is all the more interesting, given the dramatic absolute increase in enrollments and the rather different experience of the other natural sciences, shown in Figure 3.2–5. Physics has steadily lost its appeal with successive generations of high school students, and biology has gained in popularity. An explicit-invented indicator of the "importance" of chemistry within the context of secondary science education may be constructed by comparing enrollment in chemistry with that in the most popular science subject at any time (Figure 3.2–6). Between the 1890s and the 1920s, physics and earth science, traditionally the preferred choices, failed to retain their appeal, and chemistry became correspondingly more prominent. Biology has attracted an increasing share of students since the early years of the century, while chemistry's appeal has been constant. As a result, chemistry has declined slowly in prominence within high school science over the last fifty years.

Figures 3.2–5 and 3.2–6 demonstrate that chemistry has never enjoyed the great appeal that has characterized either biology in recent decades or physics and earth science in the nineteenth century. On the popular level (and also on the elite level of the academic discipline, as will emerge later), chemistry may be said to have lacked in those elements that foster a sense of fashion and style. However the science has maintained its standing within the secondary school curriculum for almost a century. The contrast in the long run trend between chemistry and physics is especially striking in this regard (Figure 3.2–5). The stable standing implies a steady increase in the number of chemistry teachers and chemistry texts required by American high schools over the decades.[21] The dramatic expansion of the high school system also means that, in recent years, about one quarter of the young adult population has been exposed at least on a rudimentary level to chemical ideas and chemical theory. In this sense chemistry has become part of the cultural experience of a significant proportion of Americans.

[21] There are indications that, although chemistry has maintained its standing within the curriculum, interest in the subject has waned among bright college-bound high school seniors. A survey of National Merit Scholars shows that the proportion who intended to major in chemistry at college declined from 6% in 1966 to 2.6% in 1975. During the same decade, the preference for premedical studies and the social sciences among these students increased at the expense of physical sciences and mathematics. In 1966, 28.3% of the National Merit Scholars expressed an interest in physical sciences and mathematics, while only 17.5% intended to major in social sciences or premedical studies; ten years later, these figures had shifted to 16.6 and 24.1%, respectively (IA, NSB, 1977, Table 5–20, 288).

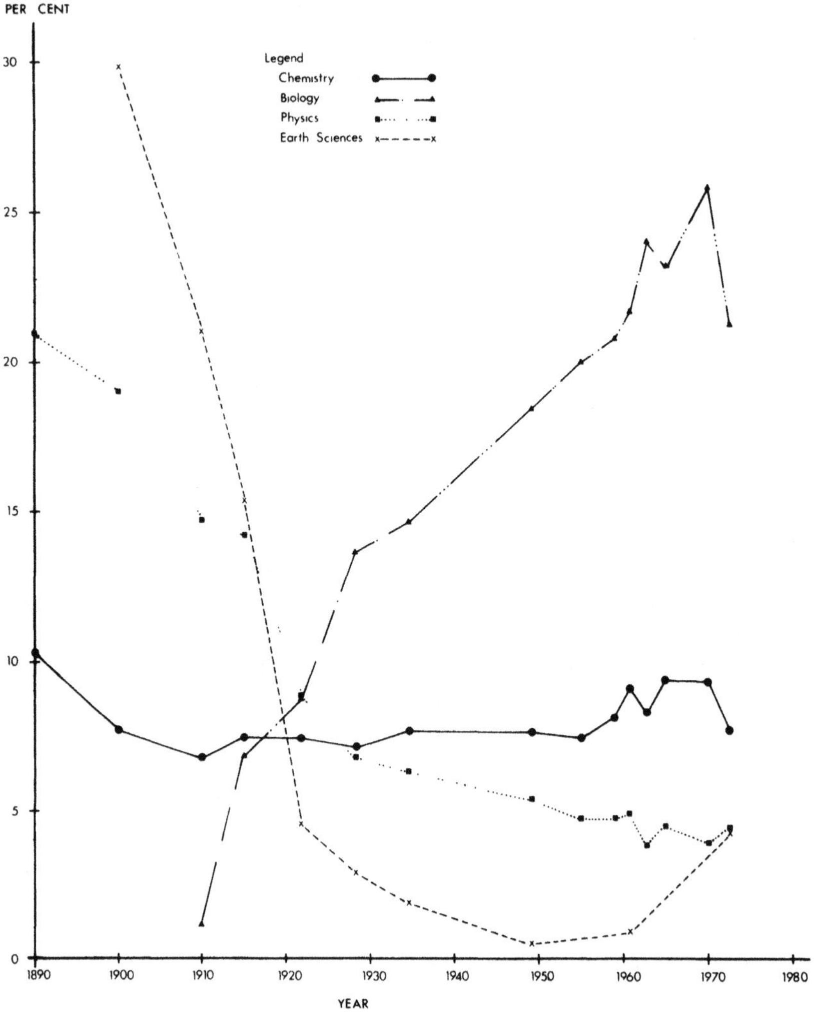

Fig. 3.2–5. Percentage of high school students enrolled in selected scientific subjects, 1890–1973. (See Table 3.26, columns 2–5.)

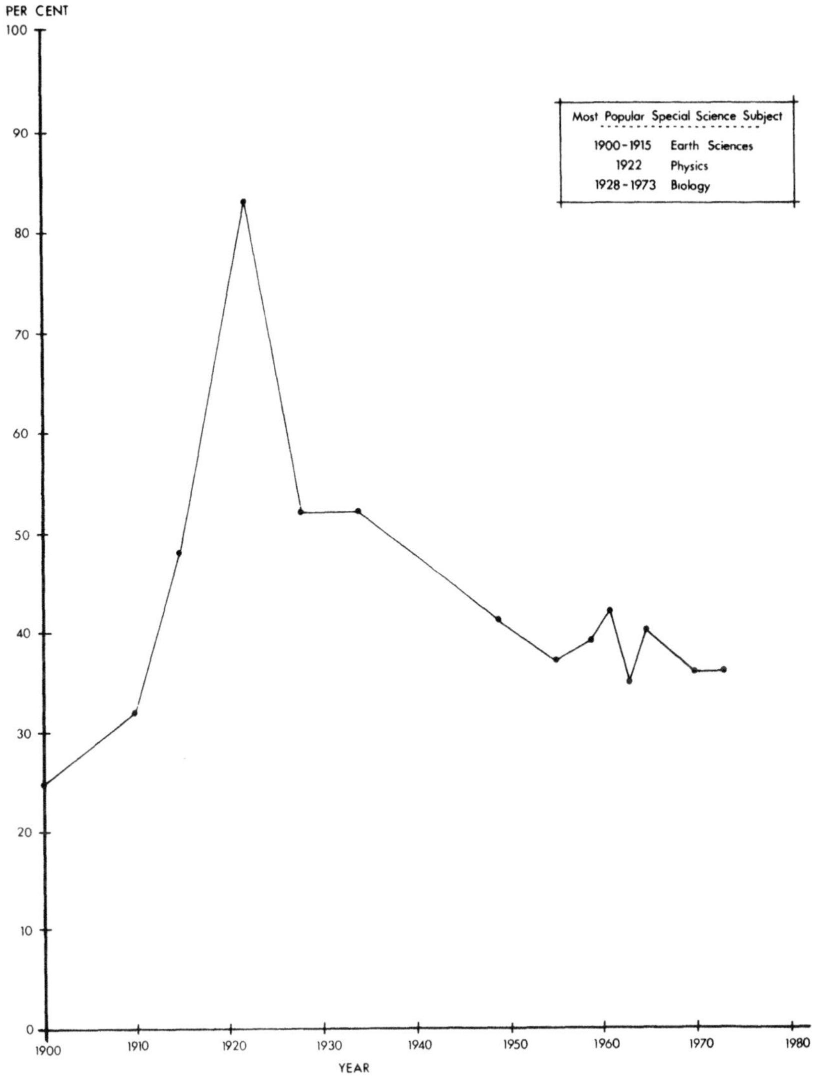

Fig. 3.2–6. High school enrollment in chemistry as percentage of high school enrollmant in most popular special science subject, 1900–1973. (See Table 3.26, column 6.)

From the standpoint of the human resources required to sustain them, the cultural aspects of chemical education have grown proportionately faster than the occupational or professional aspects of chemistry. Between 1920 and 1970, the number of teachers of high school chemistry grew from about a fourteenth to a fifth of the number of chemists in the labor force. A similar relative increase results when one compares the estimates of high school chemistry teachers and members of the ACS.[22] Such developments symbolize the extent to which "science is the mode of cognition of industrial society".

3.3. Mass Culture

INDICATOR HIGHLIGHTS:

Despite the efforts of promoters, chemistry has not held a prominent place in popular (mass) culture. Between the two world wars, only a small fraction of the content of general science courses and of popular periodicals was devoted to this particular science. Physics and the health and biological sciences were stressed instead.

The editors of two major encyclopedia yearbooks have never seen fit to devote more than 1% of yearbook pages to chemistry: since World War II the already small fraction of these publications concerned with chemistry has diminished further.

Surveys of newspaper editors show that their interest in chemistry has declined since the 1940s.

Surveys of popular opinion show that, in both 1947 and 1963, chemists were accorded less esteem than either scientists or college professors.

Over the years the high school curriculum has become an instrument of popular – or mass – culture. The rise in high school enrollment early in the twentieth century was associated with demands for teaching oriented to "the practical and the social". It was in this context that a "general science" curriculum was developed, to meet the needs of students entering an often bewildering technical society. The 1920 report on the *Reorganization of Science in Secondary Schools* (mentioned above) argued that general science should provide "a basis of discovery of interest in special sciences and of

[22] Compare the estimates for the number of high school chemistry teachers given on p. 67 with Table 2.5, columns 1–2.

vocational opportunity. It should prove to be the best training for any pupils who can take only one [science] course in high school".[23]

General science was predominantly a ninth grade subject, though often followed by electives in other sciences. By 1922, nearly one-fifth of all high school students were enrolled in general science – more than double the enrollment in physics, the next most popular science subject in the secondary curriculum. Considerable interest attaches to the relative standing of individual sciences *within* the general science course. Measures of the role of chemistry offer one implicit-discovered indicator of its perceived cultural utility compared to other sciences. These measures, though erratic and uncertain, are suggestive.

A study of twenty general science textbooks published between 1915 and 1925 ordered the various subjects according to the number of words devoted to each. Chemistry ranked last, after physics; biology; geography, astronomy, and meteorology; physiology; and geology. An analysis of nine general science texts conducted in 1937 showed similar results: physics averaged 32.0% of the available space, biology 24.5%, and chemistry only 8.9%. A 1939 investigation of 38 texts in current use showed that 23% of their space was devoted to physics, 17% to biology, and 12% to chemistry.[24]

Studies comparable to these were not conducted following World War II, in part because the structure of general science curricula changed in the 1950s from a disciplinary to a thematic orientation. In 1956, for example,

[23] III, Krug, 1964, 370; III, CRSE, 1920, 25. On the early general science movement, see III, Webb, 1959. For a contemporary discussion of American education between the wars, see III, Judd, 1933. On science education during this period, see III, Hurd, 1949, and III, Krug, 1972, 97–104.

In its 1918 report on the Cardinal Principles of Secondary Education, the Commission on the Reorganization of Secondary Education stated that the school "is the one agency that may be controlled definitely and consciously by our democracy for the purpose of unifying its people" (III, CRSE, 1918, 22). Two recent historical studies which argue that public education has served as an instrument of social control in an industrial state are III, Spring, 1972, and III, Bowles and Gintis, 1976.

[24] IB, Heineman, 1928; IB, Klopp, 1927; IB, Benner, 1940, 79–90. These studies are part of a genre. Educators interested in standardizing the contents of general science syllabuses conducted surveys of what science was being taught and how well available texts covered specific subjects. For reviews of this literature, see III, Graham, 1941; III, Hunter and Parker, 1942; and III, Henry, 1947, 155–165.

The conclusion that general science was usually taken by ninth grade students derives from III, Hunter, 1925, 1932, 1933. For the proportion of public high school students enrolled in general science, see IA, USBC, 1975b, I, 377, Series H546.

the New York City public school system recommended such general science topics as "getting acquainted with yourself", "increasing and improving our food supply", "improving our clothing and housing", "speedier transportation", and "our atomic world". While chemistry obviously figures in such topics, an exact percentage for its participation remains, understandably, an elusive measurement. There is no unambiguous finding of continued low status for chemistry in this data, but there is little to suggest that the standing of chemistry has *increased* since the 1920s and 1930s. Moreover, the one postwar general science syllabus arranged along disciplinary lines that did come to light in the course of our research (a late 1950s curriculum for freshmen in New York State high schools) devoted only one out of ten discrete subject areas to chemistry. While the niche for chemistry in the general science curriculum has been persistent, it has also been small.[25]

A similar situation prevails for chemistry in mass culture, as reflected in topics discussed in newspaper and periodical articles. A 1924 analysis of four newspapers and eight magazines most popular with Denver high school students demonstrated that biology received by far the greatest attention. Articles on chemistry and physics were almost entirely concentrated in two of the magazines (*Scientific American* and *Popular Mechanics*), in which 5.5% of the available space discussed chemistry, but more than 70% dealt with physics. A more ambitious study of material published in eight popular magazines from 1914 to 1923 showed similar results: of the words devoted to natural science in these periodicals, some three-fifths concerned biological topics, one-quarter discussed physics, and only one-fiftieth dealt with chemistry. Other studies of this genre confirm the general picture.[26]

The sparse coverage of chemical topics in mass-circulation periodicals and newspapers during the 1920s and 1930s provides an interesting counterpoint to the vigor with which members of the chemical community at the

[25] On postwar trends in general science curricula, see III, Henry, 1960, and III, Hurd and Rowe, 1964. For the New York City curriculum, see III, Beck, 1956. *Science 7–8–9*, the general science course developed by the New York State Department of Education in the mid-1950s, is discussed in III, Stollberg *et al.*, 1960, 86–87.

[26] IB, Hopkins, 1925; IB, Searle and Ruch, 1926. Hopkins' study was concerned with curriculum design. Since students spend a great deal of time reading periodicals and newspapers, Hopkins and his colleagues were interested in ascertaining the amount of scientific information one needed in order to comprehend the average newspaper or magazine (IB, Hopkins, 1925, 783). Similarly motivated studies include IB, Bobbitt, 1926, and IB, Novak, 1942. IB, Hart, 1933, part of the influential report on *Recent Social Trends*, showed that the proportion of articles devoted to chemistry in the periodial literature decreased dramatically from 1922 to 1930 (*ibid.*, 393).

time publicized advances in chemical science and industry. The American Chemical Society established a News Service in 1919 for the purpose of "effectively bringing together the public and the chemist". The ACS News Service distributed weekly bulletins on events in chemistry to over 900 daily newspapers and, under a succession of directors over the next few decades, energetically expanded its effort to spread an interest in chemistry. An ally in the struggle to bring the message of chemical progress to the public was Edwin E. Slosson, editor of Science Service. His books and voluminous other writings, notably *Creative Chemistry* (1919) and *Sermons of a Chemist* (1925), reveal the intensity of his commitment to this chemical quest. Despite the efforts of these dedicated publicists, chemistry retained at best a low profile in popular culture. However, as noted earlier, chemistry did retain its popularity as a high school subject.[27]

More recent investigations show little change in media interest. An analysis of the scientific content of the *New York Times* for 1950 revealed 126 articles on zoology and 109 on physics, compared to 71 on chemistry. Health-related sciences predominated. Biochemistry and physiology accounted for 396 articles, which implies a broad appeal for certain special aspects of chemistry. *Time* and *Newsweek* in the same year discussed chemistry in about 4% of their science articles, compared with 28% on physics. A survey of 22 local newspapers and nine major periodicals for 1962 and 1963 established that chemistry was the subject of 4% of the newspaper science articles, and 3% of the magazine science articles. Once again, this coverage was considerably less than that for biology (43% in newspapers, 59% in periodicals) or physics (43% in newspapers, 28% in periodicals). Finally, surveys of newspaper editors in 1951, 1958, and 1965 reveal a steady drop in the editors' collective interest in physics and chemistry, with dramatic increases in their taste for news of social science and of satellites and outer space.[28]

Studies of the importance of chemistry in textbooks, periodicals, and

[27] On the ACS News Service, see III, Browne and Weeks, 1952, 219–228. Public relations of this sort remain an important part of ACS activities; see, for example, III, ACS, 1979. Slosson's activities are discussed in III, Tobey, 1971, 62–95.

Slosson's *Creative Chemistry* (III, 1919) was one of the most successful books on chemistry published between World Wars I and II. After its first appearance in 1919, new editions followed in 1920, 1921, 1930, and 1938. The last two were revised by Harrison E. Howe, editor of *Industrial and Engineering Chemistry*. (See pp. 100–101 below.)

[28] IB, Bergen, 1955; IB, Bergen, 1952; IB, Koelsche and Morgan, 1964; IB, Krieghbaum, 1967, 76–78; see also IB, Sorenson and Sorenson, 1973.

newspapers reflect the forms of culture offered by its purveyors – that is, by editors and authors with particular predilections. Given the disjunction between the rhetoric of chemical journalists and the realities of editorial coverage, one may suspect a considerable gap between the cultural concerns of those producing the material and the interests of their audiences. Fortunately there have also been studies that focus directly on the interests of the consumers of culture. In 1929 Douglas Waples and Ralph Tyler investigated *What People Want to Read About*, polling 16 carefully chosen groups of men and women. The lack of interest in chemistry among the various groups was quite striking. It was even more marked in a follow-up survey done ten years later.[29]

The investigations of the place of chemistry in mass culture cited above are suggestive at best. They use limited methods to appraise a complex subject. The individual studies are largely incommensurable. They use different types of samples in static situations or over very short time periods. Even so, their agreement on the minor role of chemistry in mass culture is clear. Compared to medicine, biology, or physics, chemistry has apparently been lacking in those elements of romance, drama, and high intellectual adventure which sustain popular interest and curiosity. A rather different indicator which appears compatible with this judgment is that of the prestige of occupations, as measured in various surveys. Rankings for "chemist", "scientist", and "college professor" are available for 1947 and 1963. On each occasion, "chemist" ranked well below "scientist" and "professor". (Table 3.27 gives the detailed rankings of chemists and some other professionals selected for comparison.)[30] From all accounts, then, the public prestige of chemistry has been chronically low.

One measure suggests the recent condition of that already-low prestige. Figure 3.3–1 plots the explicit-invented indicator of the space devoted to chemistry and related subjects from the mid-1920s to the late 1970s in two

[29] IB, Waples and Tyler, 1931; IB, Moreland, 1940.

[30] Studies of occupational prestige in America may be traced to II, Counts, 1925. See II, M. Smith, 1943, and II, Siegel, 1971, for reviews of this genre. The first survey to include chemists as a separate category, conducted by the National Opinion Research Center in 1947 (II, NORC, 1947), was replicated by Hodge, Siegel, and Rossi in 1963 (IB, Hodge *et al.*, 1964).

II, Treiman, 1977, elaborates a theory in which the relative prestige of occupations derives from differences in the power and privilege accorded to occupational roles. He also discusses the problems of measurement in prestige surveys (*ibid.*, 5–29).

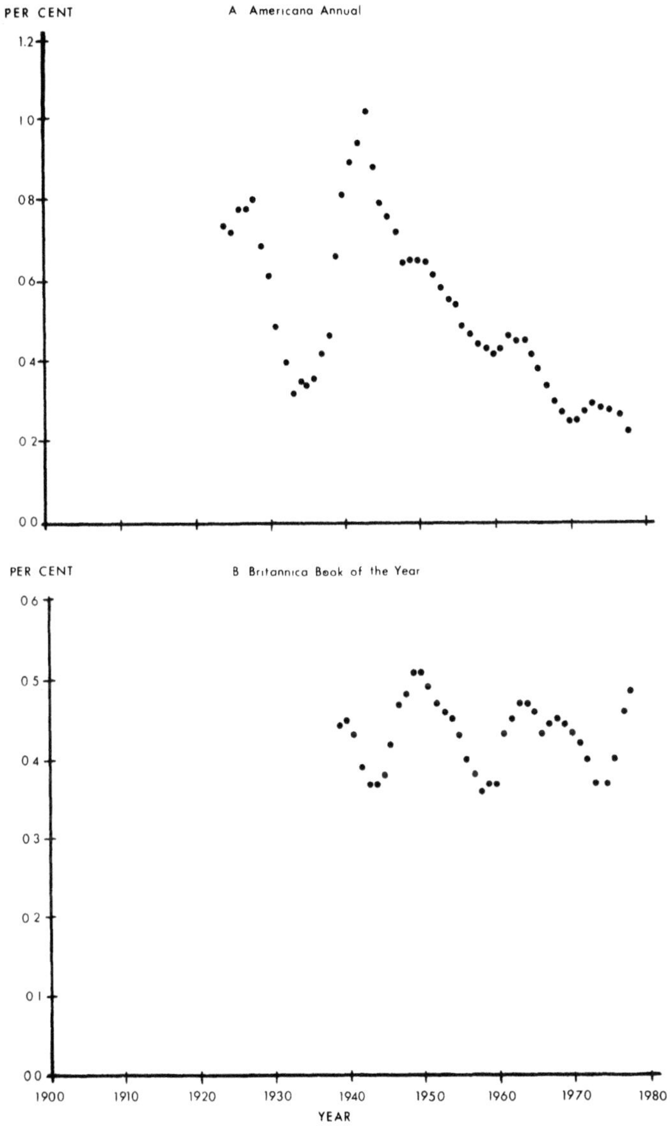

Fig. 3.3–1. Percentage of encyclopedia yearbooks devoted to chemistry, five-year moving averages, 1924–1977. (See Table 3.28, columns 4 and 8.)

prominent encyclopedia yearbooks.[31] (The figures are expressed as five-year moving averages to smooth minor perturbations.) At no time has chemistry merited much more than a single per cent of the space available. In one case, that of the *Encyclopedia Americana*, the amount that has been devoted to chemistry has dropped over two-thirds since the outbreak of World War II. In contrast, the *Britannica Book of the Year* has traditionally given chemistry little notice, despite a recent sharp increase in coverage. Such evidence combines with that of declining interest in chemistry among newspaper editors to support the conclusion that, at least in recent decades, chemistry has not been perceived as something that sells publications in the American mass market.

The cluster of indicators presented in this section suggests considerable indifference to chemistry in mass culture, providing an intriguing contrast to the long-run success of chemistry in the high school curriculum. That a fourth of the adolescent population should today study a subject which the adult world apparently finds of no great interest is a revealing situation. On the one hand, this suggests that the institution of chemistry is – and has been – more narrowly based within American culture than its adherents might wish. On the other hand, the situation may indicate a broad consensus on the utility of chemical knowledge within our civilization – and hence within the school system. That utility is partly a cultural matter involving the symbolic significance of scientific forms of cognition to modern modes of life: like the calculus, chemistry is deemed a "good thing" to know. It is also a more mundane matter of the pervasiveness of chemical products and chemical technology within industrial society. That pervasiveness finds one expression in the importance of the chemical industry as an employer of chemists, as a provider of funds for chemical research, and as a continuing source of support for the institution of chemistry.

31 The yearbooks analyzed were the *Americana Annual* (1923–1980) and the *Britannica Book of the Year* (1938–1980). These yearbooks are more suitable than the parent encyclopedias as a source for trend data on popular interest in chemistry.

A model for this study is provided by II, Bloch, 1949. Bloch compared the subject distribution of books published in 1920 with those published in 1940 in order to illuminate cultural trends.

CHAPTER 4

CHEMICAL INDUSTRY AS CONTEXT

4.1. Diversities and Definitions

The chemical industry in the United States has served as a major source of economic and social support for the chemical community. It has provided occupational niches for the majority of American chemists, contributed funds and apparatus for research, sponsored educational activity and, on an ideological level, offered to the general public tangible proof of the contribution of chemistry to national economic welfare. It is therefore of the greatest significance for the institution of chemistry that the enduring trend in chemical industry over the past hundred years has been one of dramatic growth. Capital invested in the chemical industry increased a hundredfold between the 1870s and the 1950s, and the labor force employed in the production of the panoply of chemical products increased from about 7400 persons in 1870 to over 850 000 in the early 1970s.[1]

The history of chemical industry in America is not merely a story of progress from meager beginnings to an age of global chemical conglomerates. Shifting economic conditions, changes in business organization, and the impact of new technologies are all reflected in the complex patterns of growth exhibited by the aggregate of chemical manufactures through the twentieth century. And if it were not difficult enough to unravel the evolving interactions of chemistry, business, and economics, there is the further complication that chemical processes and chemical skills are of central importance to a broad spectrum of non-chemical industries. Thus, any attempt to gauge the economic importance of chemistry in American industry must begin with a wide focus.[2]

[1] Unless otherwise specified, mention of the "chemical industry" in this report refers to the "chemicals and allied products" group (SIC 28) of the U.S. Office of Management and Budget's Standard Industrial Classification. As discussed below, fixing boundaries for the chemical industry is a difficult task.

[2] The indispensable source for the history of the U.S. chemical industry is Williams Haynes' compendious *American Chemical Industry: A History* (III, Haynes, 1945a, 1945b, 1948, 1949, 1954a, 1954b). See also III, Wigglesworth, 1928; III, Haynes, 1933; III, Haynes, 1936; III, Harkins and Wallace, 1959; III, Kahn, 1961; III, H. Johnson, 1968;

Unlike industries which sell a limited variety of products (e.g., automobiles or cigarettes) or serve common markets (like household appliances or medical supplies), the chemical industry is an amalgam of highly diverse enterprises, manufacturing an astonishing array of intermediates and final products for a multitude of markets (the chemical industry is itself a major market for its own products.) In a situation analogous to that encountered earlier with the definition of "chemist", it is a perplexing task to fix boundaries for chemical industry. We have used two major criteria discussed fully in Appendix B: the centrality of chemical processes to the manufacture of an industry's final products; and the degree to which chemists are present in an industry's work force. Application of these rough measures yields a tripartite division of American manufacturing: chemical process industries, "chemicals and allied products", and other manufacturing industries.

The chemical process industries may be defined as those in which chemistry plays a major role in the transformation of raw materials into final products. We include in this group those industrial concerns manufacturing foods, paper, petroleum, rubber, stone, clay and glass, primary metals, and their related products.[3] The group also includes within it the particular subset of manufactures of "chemicals and allied products", discussed below. Trends in the growth of the chemical process industries, which have employed two-thirds of the chemists in American industry, provide indicators of the considerable impact of chemistry in the American economy.

We shall restrict much of our discussion to that congeries of industries which manufacture chemical *products* rather than merely utilize chemical *processes* to produce consumer goods (such as paper or foods). Producers of "chemicals and allied products" have been grouped together by the U.S. Bureau of the Census since 1890. The group includes both those industries producing basic chemicals such as acids, alkalies, and organic reagents, and those which manufacture chemical products ranging from pesticides to perfumes. The diversity of the group is reflected in the malleability of the

---

and IB, Backman, 1970. III, L. F. Haber, 1958, and III, L. F. Haber, 1971, place American developments in the context of the European chemical industry.

Useful background on the business and economic history of American industry since the late nineteenth century may be obtained from III, Cochran, 1972; III, Davis *et al.*, 1972; III, N. Rosenberg, 1972; III, Porter, 1973; III, Bruchey, 1975, Chapters 3–4; and III, Chandler, 1977.

[3] See, for example, IB, ACS, 1973, 489. For a detailed exposition of the panoply of products and manufacturing methods in the chemical process industries, see III, Shreve, 1967.

Census definition of the chemical industry during the past century; from as few as seven constituent industries in 1899, the "chemicals and allied products" group has expanded to include the present 28 classes of chemical companies (see Appendix B.2). The importance of the group is suggested by its role as employer: since 1950 more than half the chemists in the chemical process industries have been concentrated in "chemicals and allied products".

Needless to say, the heterogeneity and shifting definitions of the chemicals and allied products industry require the analyst to adopt a cautious stance when interpreting the data. In what follows, we shall look at enduring trends and short-run variations in major aspects of the chemical industry defined in this way, comparing the industry's position in the domestic economy with that of the larger aggregate of chemical process industries where appropriate.

4.2. Chemicals and Allied Products

INDICATOR HIGHLIGHTS:

The American chemical industry has grown dramatically since the late nineteenth century, in absolute and relative terms. The chemical industry has become an increasingly important sector of the economy when compared to either the chemical process industries or manufacturing as a whole. This relative growth – especially during the interwar years – is evident when one considers such factors as output, capital productivity, total capital investment, and national income originating in the industry.

Chemical companies have improved their collective standing among America's top 100 industrial corporations. The chemical industry moved from seventh rank among industrial groups in the top 100 in 1909 to third rank in 1948, a position the industry still maintained in 1975. (Petroleum, a chemical process industry, has dominated the top 100 since the late 1920s.)

The secular trend in chemical production from the end of the last century to the mid-1950s was one of exponential growth with a doubling every 12 yr, as shown in Figure 4.2–1. Chemical output grew more rapidly than that of manufacturing industry as a whole, which doubled only every 19 yr. The period from 1929 to 1957 includes two phases of short-run variation. From 1929 to 1937, output of the chemicals and allied products industry grew at only 2.7% per annum, which represents one doubling every 26 yr. This figure increased dramatically between 1937 and 1957 – a period when

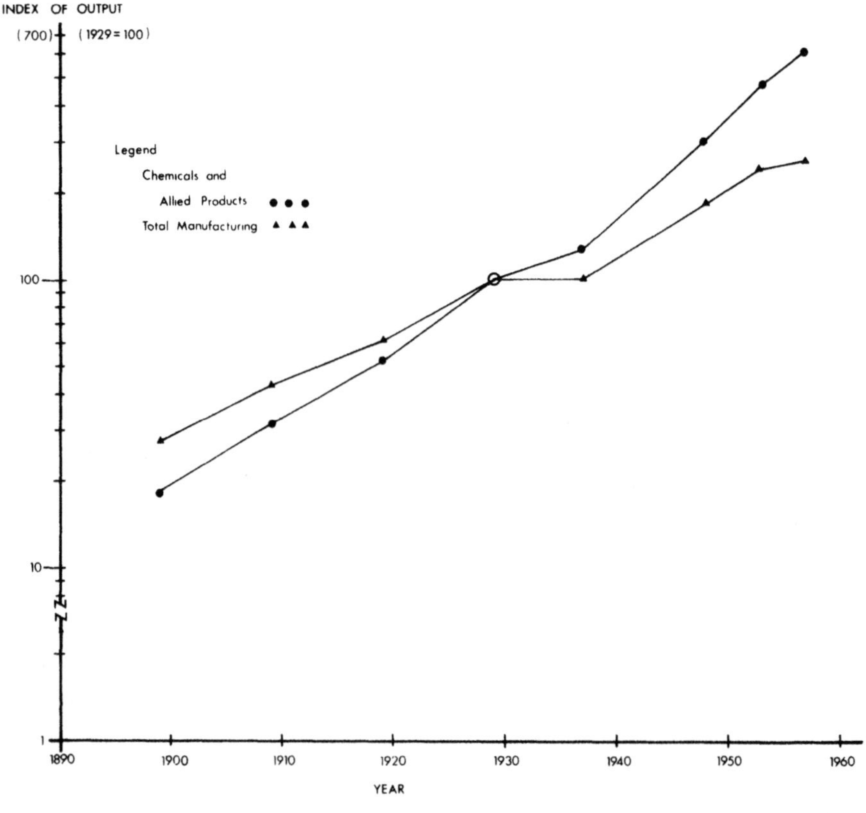

Fig. 4.2–1. Production of the chemical and of all manufacturing industries, 1899–1957. (See Table 4.1, columns 1 and 2.)

the chemical industry underwent considerable diversification – to an annual growth rate of 8.3%, or one doubling every nine years.

A similar pattern emerges from Figure 4.2–2, which displays the comparative growth of capital investment in all U.S. manufacturing, in the chemical process industries, and in the chemicals and allied products industry between 1879 and 1957. In the twentieth-century portion of this figure, there are two periods of growth punctuated by a brief stagnation between 1937 and 1948. During these periods (prior to 1937 and after World War II), the chemicals and allied products industry grew more rapidly than either the chemical process industries or manufacturing as a whole. (In general, capital investment in these industries accumulated at a quickened pace after World War II.) Capital investment in the chemicals and allied products industry grew from 869 million dollars in 1899 to some 10.6 billion dollars in 1957 (measured in constant 1929 dollars). Overall, this was equivalent to one doubling every 16.1 yr. By contrast, the chemical process industries and manufacturing as a whole had doubling times of 19.2 and 21.8 yr.

An indicator which provides a measure of the efficiency with which capital investment is used in industrial production is capital productivity (or output per unit of capital input). As illustrated in Figure 4.2–3, output per unit capital in the chemical industry increased over 80% between 1929 and 1948 alone. Economic historians have attributed the high capital productivity of chemical industry to a wave of "capital-improving" innovations initiated during the 1920s which included methods of continuous processing, automated materials handling and linkage of unit processes, increased equipment size, and the development of new products (especially such synthetic organic chemicals as phenolformaldehyde resins and nitrocellulose lacquers).[4]

Both the American chemical industry and U.S. manufacturing in general have grown tremendously since the late nineteenth century. This conclusion is hardly surprising. But if one analyzes the growth of chemical indstury *relative* to the larger system, some interesting observations emerge. Capital invested in the aggregate of chemical process industries increased from about two-fifths to three-fifths of all capital investment in U.S. manufacturing between 1879 and 1957. The distribution of capital within the chemical process industries has changed significantly since the 1870s, as Tables 4.3 and 4.4 demonstrate. Since the 1890s, capital investment in the chemical process industries has shifted from food and primary metals to the more chemically-oriented group of rubber, chemicals, and petroleum. The most

[4] III, Lorant, 1967 and 1975, 147–191, 279–288.

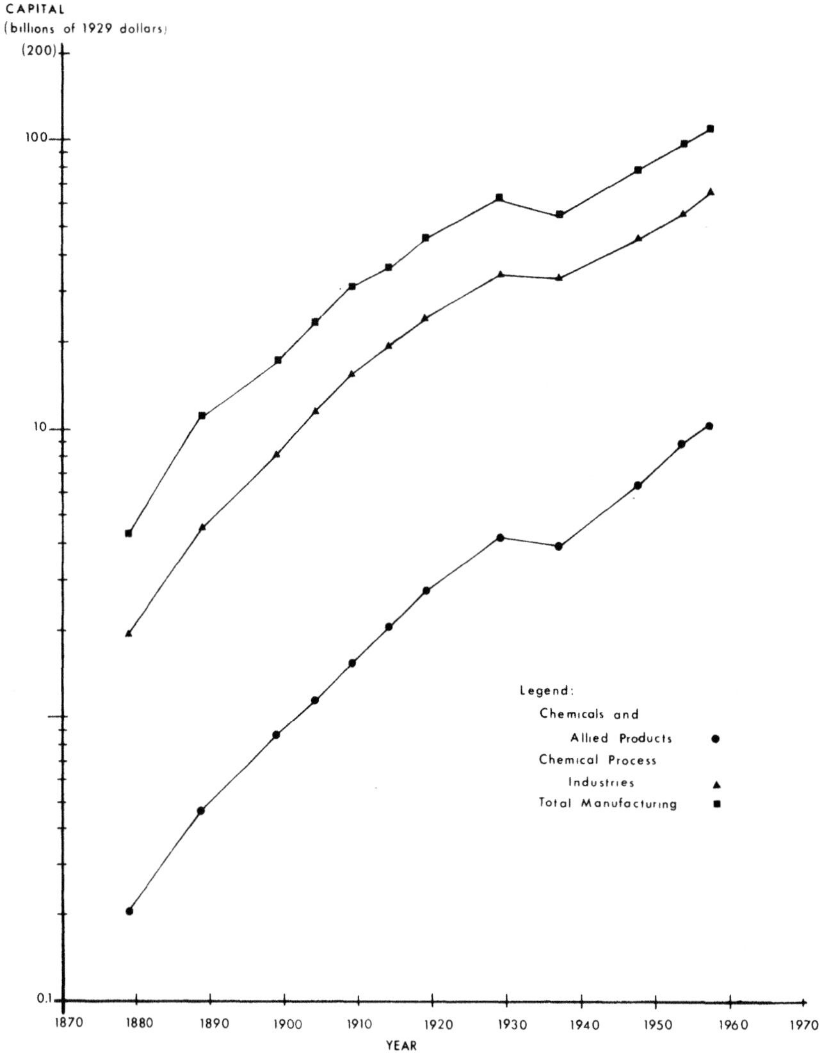

Fig. 4.2–2. Capital in the chemical and the chemical process industries, 1879–1957. (See Table 4.2, columns 1–3.)

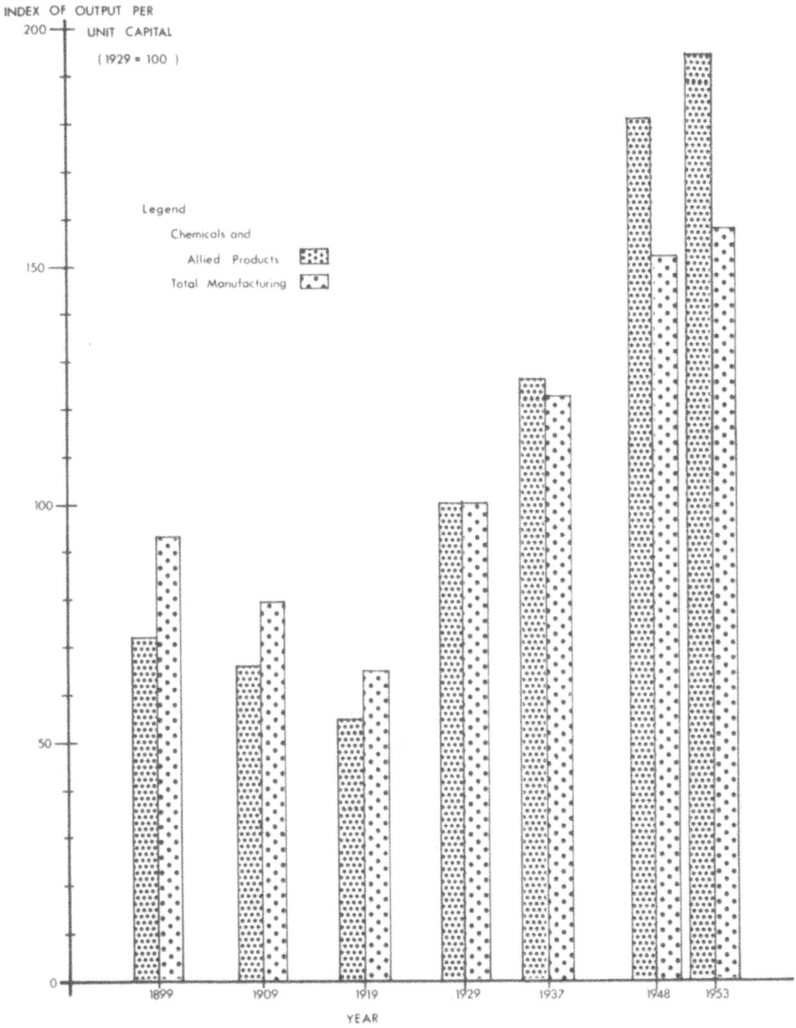

Fig. 4.2–3. Output per unit capital in the chemical and in all manufacturing industries, 1899–1953. (See Table 4.1, columns 3 and 4.)

spectacular growth occurred in petroleum, which expanded its share of capital invested in the chemical process industries from one-fortieth to one-fourth between 1899 and 1957.[5]

While not increasing quite as quickly as petroleum, capital investment in chemicals and allied products grew from one-tenth to one-sixth of the total for chemical process industries in the same period. As a proportion of all capital investment in U.S. manufacturing, capital in chemicals and allied products more than doubled, from 4.3% in 1879 to 9.6% in 1957 (as shown in Figure 4.2–4). A similar indicator of the relative growth of chemicals and allied products for more recent years is given in Table 4.5, which charts the percentage of U.S. manufacturing income originating in the chemical process industries. This percentage declined slightly from 44.3% in 1929 to 36.4% in 1978 (after dipping below 35% in the late 1960s). The proportion of U.S. manufacturing income generated by the chemicals and allied products industry, on the other hand, increased steadily from 5.0% in 1929 to 7.1% (nearly 24 billion dollars) in 1978.

Another indicator of the importance of the chemical industry to big business during the twentieth century is provided by a look at the shifting composition of the top 100 industrial corporations (see Table 4.6). In 1909 the top 100 were dominated by companies in the iron and steel, petroleum, nonferrous metals, and food products industries – all of which were members of the chemical process industries. Eight chemical corporations – chiefly in agricultural products, mineral oils, and salt – represented 4.1% (about 360 million dollars) of the total assets of the top 100. This was equivalent to one-fifth of the assets of U.S. Steel, which was the leader of the top 100 in 1909. By 1929, the petroleum industry had risen to the top position, symbolizing the "chemicalization" of industry which was said to have taken place between 1910 and 1950. Ten chemical corporations – mainly firms built on a new synthetic chemistry (such as Celanese, American Viscose, and Monsanto), in addition to the triumvirate of Du Pont, Allied Chemical, and Union Carbide – accounted for 8.7% (4.3 billion dollars) of the total assets of the top 100 industrials in 1948. Du Pont alone (ranked seventh among the top 100) had more than 1.3 billion dollars in assets, due largely to an aggressive program of diversification begun in the 1920s. The chemicals and allied products industry was truly big business, and had

[5] On the technical innovations in thermal and catalytic cracking which spurred growth in the petroleum industry, itself a response to the growing use of the automobile, see III, Enos, 1962, and III, Williamson *et al.*, 1963, 110–166, 373–422, 603–644.

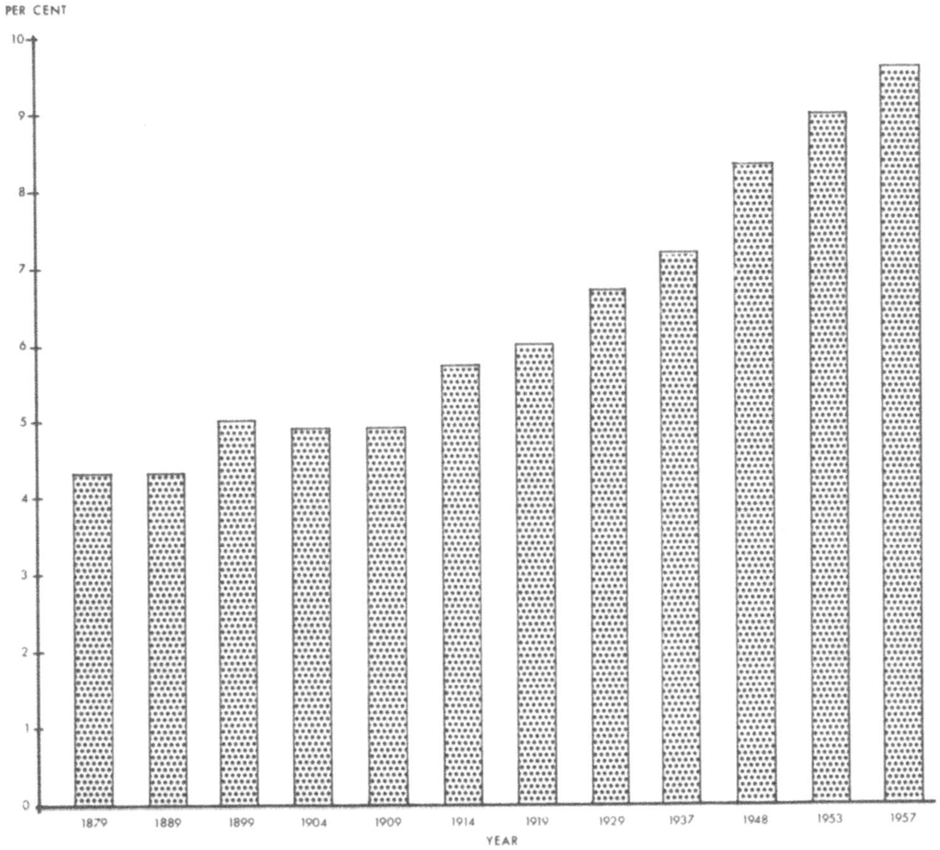

Fig. 4.2–4. Total capital in the chemical industries as a percentage of total capital in all manufacturing industries, 1879–1957. (See Table 4.2, column 4.)

shifted in rank from seventh to fourth among industrial groups in the top 100.[6]

By 1960, Dow, Monsanto, and Union Carbide had also joined the "billion dollar club", and chemical companies among the top 100 in that year included W. R. Grace and Co. and National Distillers & Chemical Corporation, two conglomerates which had only recently entered the chemical industry. In 1975, the chemical industry ranked third, accounting for 9.3% (40.8 billion dollars) of the assets of the top 100. Du Pont, Dow, and Union Carbide were the largest corporations in the chemical industry, and all placed within the top 20 industrials. But even with more than five billion dollars in assets each, they were still a fraction of the size of Exxon, General Motors, Texaco, IBM, or Mobil – the top five corporations in 1975. The petroleum industry was still the front-runner in that year, with 35.5% (more than 156 billion dollars) of the assets of the top 100.

4.3. Oligopoly and Patents

INDICATOR HIGHLIGHTS:

Oligopoly, often a concomitant of big business, has been a characteristic of the chemicals and allied products industry during this century.

The strong economic position of chemical companies enabled them to avail themselves of research and development as a means of protecting markets. One indicator of this activity is the fraction of chemical patents among total patent issues, which increased steadily from less than 5% prior to World War I to more than 20% by the late 1960s.

"Big business" implies the concentration of economic activity within an industry in the hands of a few firms. Oligopoly has been of considerable

[6] On the rise of big business in the American chemical industry, see III, Haynes, 1948, 31–47; III, Nelson, 1959; III, Kahn, 1961, 236–246; III, L. F. Haber, 1971, 310–318; and III, Chandler, 1977, 354–356, 438–450, 473–476. The "chemicalization" of industry, a term which originated with J. E. Teeple, a well-known consulting chemist of the 1920s, is discussed in III, Haynes, 1945b, 353–370. Du Pont's diversification program, which initially followed opportunities based on the relationships among explosives, raw materials, and other chemical industries, is explained in III, E. I. du Pont, 1927, and III, Stine, 1925. See also III, Dutton, 1942, 270–362; III, Chandler, 1962, 78–91; and III, Mueller, 1962.

importance in the growth of chemical industry in America. Figure 4.3–1 shows an increase of oligopoly in the industry from the 1920s through the 1950s. In 1929, 30% of the industries in the chemicals and allied products group were controlled by oligopolies whose product value was 17% of the group total. By 1947, 68% of the overall industry was oligopolistic and the firms in question controlled 44% of total product value.[7]

This trend toward concentration in large chemical corporations was complemented by another toward diversification. Corporations placed faith in the utility of chemical research (see Section 4.4 below), taking advantage of opportunities deriving from such innovations as catalytic cracking, synthetic fibers, and synthetic rubber to move into new product industries. The technical basis of change in chemical manufacturing was striking: *Fortune* magazine observed in 1937 that "the chemical industry is one in which scientists and engineers are not just tolerated and not just employed; they are in the saddle; their ideas are the controlling ideas. They may not be the actual entrepreneurs, but an entrepreneur in the industry who could not speak the language would be helpless."[8]

Companies in obligopolistic positions were able to employ research and development to protect their privileges, the use of patents being one obvious possibility. Statistics on the assignment of patents (displayed in Figure 4.3–2) provide a rough measure of the impact of research upon the character of industrial development. Private individuals obtained 82% of the patents granted in 1901, while only 17% went to domestic corporations; by 1932, the proportions were 49 and 46%, respectively. This trend in the assignment of intellectual property – which has continued to the present – reflects the battle for control of commercial territory: companies perceived

[7] For an interpretation of patterns of concentration in twentieth-century American industry based upon trends in the integration and diversification of corporate enterprise, see III, Chandler, 1969.

By the early 1950s most big corporations in American industry had adopted a strategy of diversification and decentralization. As non-chemical companies began moving into chemical markets, one result was the decrease in concentration within the chemical industry shown in Figure 4.3–1 (*ibid.*, 278).

[8] III, "Chemical Industry", 1937, 158. On diversification as a growth strategy see III, Chandler, 1962, 374–378; and III, Chandler, 1977, 473–476. Michael Gort's survey of diversification in American industry shows that the chemical industry added more new products than any other industry from 1939 to 1950 (IB, Gort, 1962, Tables 14 and 15, 42–45). III, Haynes, 1948, and III, Haynes, 1954b, provide a detailed survey of new products and markets in the chemical industry during the interwar years.

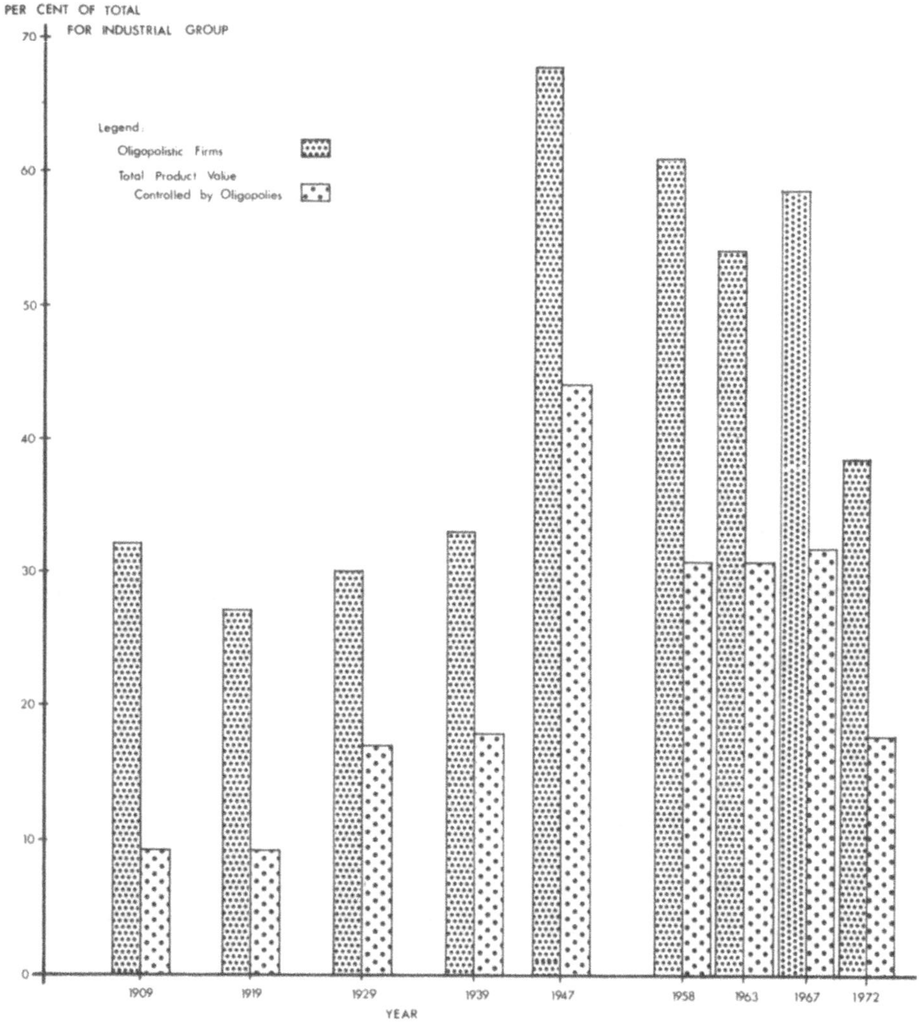

Fig. 4.3–1. Extent of oligopoly in chemicals and allied products industrial group, 1909–1972. (See Table 4.7, columns 3 and 6.)

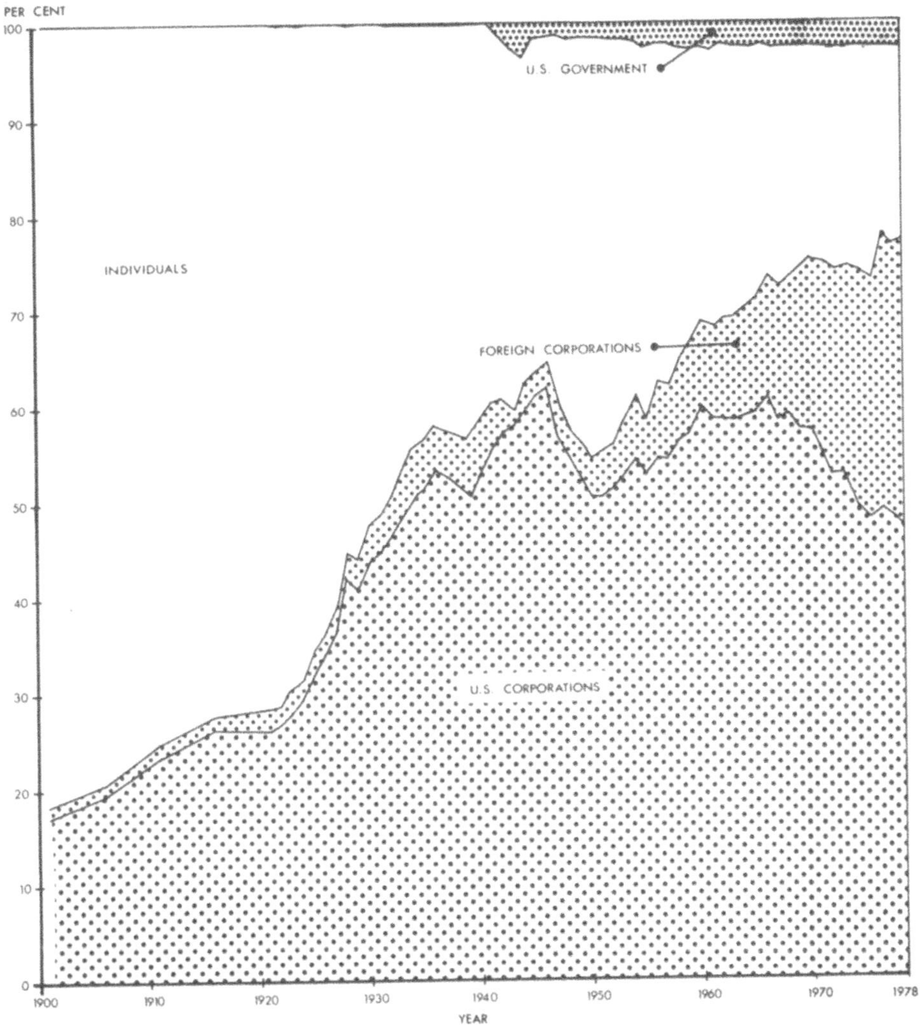

Fig. 4.3–2. Assignment of U.S. patents for inventions, by patentee, 1901–1978. (See Table 4.9.)

the research department as an increasingly important unit in the war for markets.[9]

Chemical research was not confined to the chemicals and allied products or chemical process industries. An extravagant but not atypical expression of the belief in the centrality of chemistry to all industry, which had become widespread by the 1920s, may be found in a popular book entitled *What Price Progress*?, which opened with the following counsel for investors:

> Are you the owner of the stocks, bonds or other securities of any of the more than 200 000 corporations engaged in manufacturing, mining, transportation and various other pursuits in this country? If so, you had better begin to make inquiries with respect to what the companies in which you are interested are doing in the matter of keeping in step with progress in chemistry and chemical engineering. The chemist is making a new world .... There is no business today whose welfare and interest are not bound up with chemistry .... There is no industry – not one, that is not in danger of waking up tomorrow and finding that the chemist had made a discovery that might revolutionize it.[10]

One indicator which reinforces this perception is that of chemical patent issues as a proportion of total patent issues, illustrated in Figure 4.3–3. The proportion increased steadily from about 5% in the period immediately following World War I to just over 20% in the mid-1960s.

[9] For the concept of scientific research as a weapon in the arsenal of corporate strategy, see III, "Judging", 1928. A recent historical study which analyzes corporate uses of patents in preserving control of markets is III, Reich, 1977.

The military metaphor was adopted by numerous contemporary observers of the industrial scene. Readers of *Scientific American* in 1929 were alerted to the significance of industrial research in vivid terms: "The mortality of corporations that have become victims of organized research is appalling. The war for markets today is either won or lost in the research department. Research brooks no armistice, perfection and improvement is [*sic*] its watch-word, and the corporation which does not keep on its toes in research is doomed to ultimate failure .... It is the research worker who is the Power Behind the Throne. It is he who supplies the sinews and ammunition of modern commercial warfare. The Big Berthas of business boom with his ideas." (III, Yates, 1929, 383).

Another commentator stated the theme more explicitly, asserting that "the shock troops in the army of progress are the men of science"; see III, Pound, 1936, 75, 88–89 (quoted from 89).

[10] III, Farrell, 1926, 3, 7. Another financial analyst observed that "it is a rare company that is divorced from the influence of the chemist. An examination of the stocks on the exchange quickly reveals the tremendous power which lies in the hands of the chemist" (III, E. L. Smith, 1934, 608).

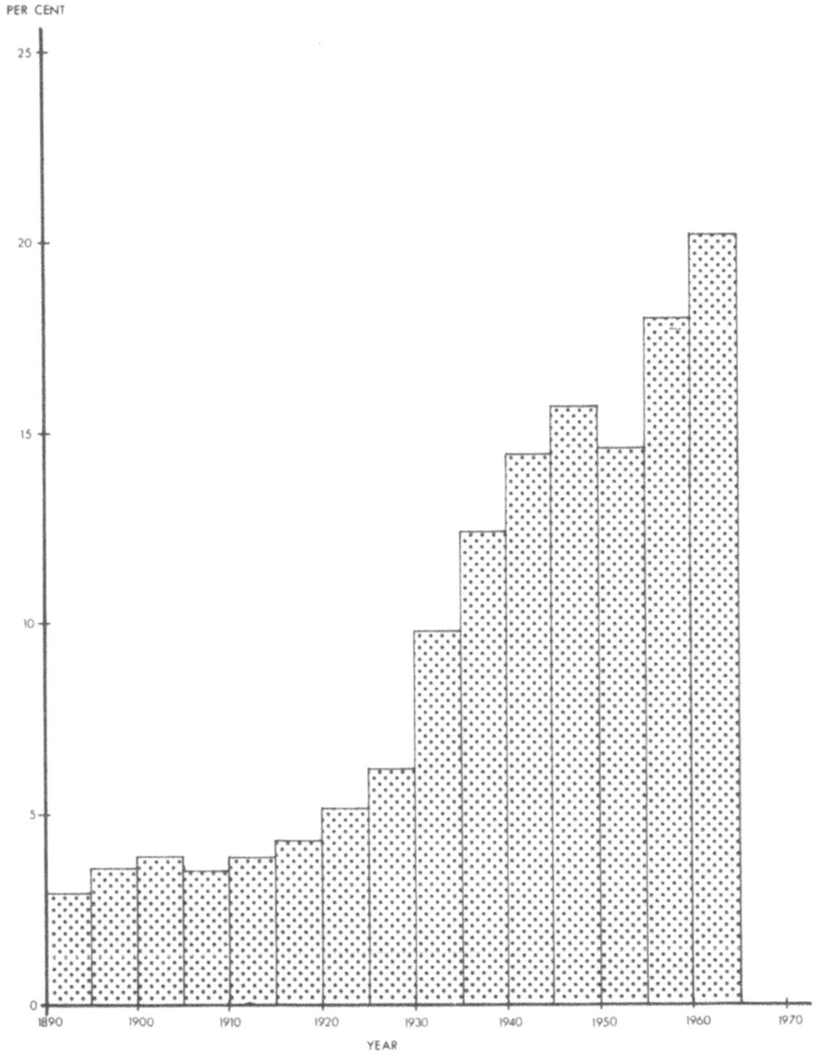

Fig. 4.3–3. Chemical patent issues as percentage of total U.S. patent issues, by pentade, 1891–1965. (See Table 4.10, column 3.)

4.4. Industry, Progress, and Boosterism

INDICATOR HIGHLIGHTS:

The rapid development of the chemical industry dominated impressions of the American chemical community between World Wars I and II. A cadre of "chemical boosters" publicized chemistry as the key to economic progress and a guarantee of national security.

This ebullient attitude cooled during the late 1950s and early 1960s as overcapacity, increasing competition, and the invasion of chemical markets by oil and rubber companies slowed growth within the more narrowly defined chemical industry. Also, chemical research and development were eclipsed by the burgeoning aerospace and electronics industries.

The chemical industry functions as both implicit and explicit support for the institution of chemistry. Explicitly, the industry provides work for the majority of American chemists (see Section 5.1 below), as well as direct financial assistance to academic chemistry in the form of fellowships, scholarships, and research grants. Implicitly, the chemical industry helps to provide legitimacy for the chemical community: its economic centrality confers authority on members of the fraternity, ratifying their claims to expertise. Promoters of chemistry have invoked the economic contribution of chemical industry as a potent symbol to aid them in making arguments on behalf of the chemical profession – a political move with obvious utilities for garnering recruits and resources. This chapter concludes with a sketch of the historical context of this process within the American chemical community.

As long as the domestic chemical industry was limited in scale, claims for pure science based on an appeal to the power of contemporary chemical technology lacked credibility. American chemists found themselves in a more advantageous position as World War I drew to a close. Already formed into an influential national community with a firm foothold in academe and substantial representation in government and industry, chemists experienced a heightened sense of their indispensability in national life as a result of wartime mobilization. Julius Stieglitz, professor of chemistry at the University of Chicago and coordinator of Chemical Warfare Service activities, expressed this sentiment in a medical metaphor: "[Chemistry's] wizardry permeates the whole life of the nation as a vitalizing, protective and constructive agent very much in the same way as our blood, coursing through our veins and arteries, carries the constructive, defensive and life-bringing materials to every organ in the body."[11]

World War I gave new impetus to the domestic chemical industry, especially in those areas related to wartime needs. Demand from home markets, coupled with the loss of crucial chemical imports, raised wartime production to unparalleled levels. Federal support helped to fuel economic expansion and the movement of chemical firms into new products, especially in nitrogen fixation and synthetic dyestuffs. When military markets dried up after the war, these new areas of production were bolstered both by the licensing of seized German dye patents under the Chemical Foundation (1919) and by the imposition of high tariffs upon chemical imports (1922). The American Dyes Institute (1918) and the Synthetic Organic Chemical Manufacturers' Association (1921) were the trade organizations most active in the crusade for legislative recognition of the importance of chemical industry to the postwar U.S. economy. The chemical provisions included in the Fordney-McCumber Tariff were partly the result of a three-year lobbying effort mounted by the newly self-conscious group of organic chemical manufacturers.[12]

These provisions supported the transformation and continued expansion of American chemical industry in the 1920s. Diversification mergers – involving such giant corporations as Allied Chemical and Dye (1920) and a reorganized Du Pont – provided opportunities for oligopolistic control over broader portions of the market. Wartime stimulation of chemical research in industry itself led to the creation of companies relying on novel methods and products. Faith in research paid off handsomely for chemical manufacturers, as developments in areas such as synthetic resins, lacquers, and petrochemical

[11] III, Stieglitz, 1919, [xvi]. On the role of chemists and chemical industry in World War I mobilization efforts, see III, Yerkes, 1920, 123–174; III, Clarkson, 1923, 387–419; III, Haynes, 1945a and 1945b; and III, Jones, 1969. III, Whittemore, 1975, discusses the postwar history of the Chemical Warfare Service in light of the Progressive ideal of public service adopted by influential members of the chemical community.

[12] The 1919 "Symposium on Maintenance and Preservation of Our Chemical Industries", sponsored by the American Institute of Chemical Engineers, illustrates the concerns of chemical manufacturers facing postwar readjustment (III, Thompson *et al.*, 1920). A survey of activity in the chemical industry after World War I is in III, Haynes, 1945b, 353–427. For a detailed statistical portrait of the transformation in American chemical industry by 1925, consult IA, Swift, 1929. On federal support for nitrogen fixation processes, see III, M. Clarke, 1977. On the coal-tar dyestuffs industry, and the postwar campaign for protective tariffs to forestall German competition, see III, Haynes, 1945b, 257–278. III, Herty, 1924, illuminates the impact of the Tariff Act of 1922 on the domestic dyestuffs industry. On the Chemical Foundation, see III, Muh, 1954.

derivatives led to the profitable exploitation of new consumer markets. As one observer commented:

[A] war-born chemical consciousness changed the public's attitude towards the industry from indifference or suspicion to a rather fulsome, patriotic appreciation. The chemical industry, long regarded as either a sort of glorified drug store or else as the producer of cheap and nasty substitutes, became suddenly the creator of "new values". Manufacturers in other fields shook off ancient prejudices against chemicals and chemical processes. In financial circles, where scientific discoveries had been viewed traditionally with distrust as the destroyer of invested capital and disturber of industrial *status quo*, [chemical] research became almost a fetish, blindly worshipped as a sort of magical charm in the competitive battle between industries.[13]

Despite some retrenchment in the early years of the Depression, the chemical industry expanded rapidly throughout the 1930s and 1940s, as the indicators discussed earlier in this chapter make clear. Wall Street financiers hailed the chemical industry as "depression-proof", an apt, if exaggerated, characterization. For instance, from 1929 to 1937, 38 of 45 leading chemical companies continued to pay regular stock dividends; of these 38 companies, 35 paid *higher* dividends in 1937 than they had in 1929. While overall industrial production dropped 10% from 1931 to 1939, chemical production increased more than 40% during the same period – from 2.6 billion to 3.7 billion dollars per year.[14]

The rapid growth and modernization of chemical industry dominated public impressions of the American chemical community during the interwar years. Spokesmen for chemistry were quick to link advances in chemical science to technical and industrial progress. In viewing technology simply as applied science, they were drawing on a widely-accepted tradition. In the 1920s and 1930s, influential segments of the American public began to be persuaded that chemistry and chemical industry were not only crucial to national defense, but also provided a key to economic progress. This realization was due in part to the vigorous promotional activities of a cadre of chemists (such as Arthur D. Little, William J. Hale, Harrison Hale, and Charles H. Herty), chemical journalists (Harrison E. Howe, Williams Haynes) and science popularizers (especially Edwin E. Slosson) who produced cascades

[13] III, Haynes, 1933, 231. Holland and Pringle's hagiographic sketches of such "industrial explorers" as Willis Whitney, Leo Baekeland, Arthur D. Little, C. E. K. Mees, and Charles L. Reese give the flavor of contemporary faith in industrial research; see III, Holland and Pringle, 1928.

[14] III, Haynes, 1954b, 31; III, H. Johnson, 1968, 120. For a popular account of developments in industrial chemistry during the 1930s and early 1940s, see III, Haynes, 1943.

of celebratory literature. Publicists for the chemical profession transformed the intense loyalty to community and sense of fierce competition with other communities which characterized an earlier tradition of civic boosterism into a similar spirit of "chemical boosterism" in which the chemist became the cynosure of the scientific community. The booster's job was to spread the word to the public at large.[15]

Francis P. Garvan's activities as head of the Chemical Foundation provided the most visible evidence of chemical boosterism during these decades.[16]

[15] On civic boosterism in nineteenth- and early twentieth-century America, see III, Boorstin, 1965, 113–168. Sinclair Lewis' *Babbitt* presents a satirical view of an enterprising booster. The shift from island communities to a national urban culture in the early twentieth century, and the associated importance of professional groups in preserving a sense of community (see III, Wiebe, 1967), suggest the concept of "chemical boosterism" discussed below.

In addition to Slosson's *Creative Chemistry* (III, Slosson, 1919), see the following widely distributed samples of the genre of chemical boosterism between the wars; III, Cushman, 1920; III, H. Hale, 1921; III, Howe, 1926; III, Little, 1928, III, W. J. Hale, 1932; and III, Haynes, 1936. On Herty, see III, Oden, 1977; III, Reed, 1983; and III, Parascandola, 1983.

[16] Garvan (1875–1937) was a New York attorney who succeeded A. Mitchell Palmer as Alien Property Custodian in 1919 and shortly thereafter established the Chemical Foundation. (See III, Hixson, 1937, for biographical details; and III, Haynes, 1945b, 260–262, 482–483, on the formation of the Chemical Foundation.) A veteran of the struggle for protective tariffs for the U.S. chemical industry just after the First World War, Garvan devoted his considerable talents and the financial resources of the Foundation to promoting the chemical industry. (On Garvan's booster activities, see III, Herty, 1929, and III, Haynes, 1945b, esp. 270–271.) Along with W. J. Hale, Haynes, and Herty, he took the case to chambers of commerce, legislative committees, social clubs, and other organizations, arguing that a strong chemical industry was in the public interest (see, for example, III, Garvan, 1919, 1921, 1922, 1929).

Garvan sponsored a series of popular essays demonstrating the utility of chemistry in industry, agriculture, and medicine (III, Chemical Foundation, 1923; III, Howe (ed.), 1924–25; III, Chamberlain and Browne, 1926; and III, Stieglitz, 1928). These books, along with Slosson's *Creative Chemistry* (1919) and other works, were distributed by the Chemical Foundation to thousands of secondary schools and colleges throughout the United States during the 1920s, to be used as reference books in connection with an essay competition privately underwritten by Garvan and his wife (see III, Gordon *et al.*, 1931). Administered by the American Chemical Society from 1923 through 1931, these essay contests were an educational project designed to promote among American youth "an intelligent appreciation of the vital relationship of the development of chemistry to our national defense, to the development of industry and our national resources, to agriculture and forestry, to health and disease, to the home, and to the enrichment of life" (III, "Report", 1925, 3). In the first year alone, an estimated 500 000 high school students participated. Garvan's interest in the training of chemists extended to his endowment of a chair of chemical education at Johns Hopkins University in 1929 (III, Frazer, 1930).

Other examples abound. In the early 1930s, "chemurgy" emerged as a utopian chemical science geared toward ushering in an age of agricultural prosperity; its small results did not daunt its proponents, the most persistent of whom was William J. Hale. A similarly melioristic vision of the future was offered in A. C. Morrison's technocratic *Man in a Chemical World: The Service of Chemical Industry* (III, 1937). More pragmatic than Morrison's paean to chemical industry was the Du Pont slogan "Better Things for Better Living Though Chemistry", coined in a decade when the company laboratories produced nylon and neoprene. This theme – deftly captured in the title of Walter S. Landis' *Your Servant the Molecule* (III, 1944) – was emphasized in countless public lectures and publications intended for general audiences.[17]

These examples suggest the sorts of activity in which chemical boosters were engaged throughout the 1920s, 1930s, and 1940s. Economic prosperity, a secure national defense, and an improved standard of living – all were possible through expanding the support of chemical industry. This ideological posture buttressed the chemist's social role in subtle and not so subtle ways. Chemical boosters were indefatigable in promoting their cause, and their identification of chemical science with its industrial manifestations was reinforced by the emerging interactions of academic chemistry departments and corporate patrons.

The cultural meanings attached to chemistry by participants and observers were also affected by the comparatively rapid growth of the American chemical industry between World Wars I and II. This striking record of growth did not escape the notice of investors. As early as the mid-1930s, the chemical industry was characterized as a progressive supporter of research, as a consequence of which it had proved particularly depression-resistant. Throughout the 1930s and 1940s, the chemical industry continued its expansion into the newer areas of plastics and synthetic fibers, even pushing forward into consumer markets. By the mid-1940s, investors and corporate executives were

[17] On the development of chemurgy, see III, Borth, 1942; III, McMillen, 1946; III, W. J. Hale, 1949, 105–115; III, Haynes, 1954b, 38–40, 226–228, 356–357, 486–490; and III, Pursell, 1969. For Hale's visionary prescriptions see III, W. J. Hale, 1934, 1949.

Du Pont adopted their slogan in 1935, informing stockholders that: "Chemical and engineering research have been major activities of your company for many years. By coordinating these activities, important new processes and improvements in existing processes have been developed and substantial savings in manufacturing costs effected, all contributing to provide "Better Things for Better Living – Through Chemistry" (III, E. I. du Pont, 1935, 15). See also III, E. I. du Pont 1936, 17; and III, E. I. du Pont, 1937, 20–22.

increasingly "bullish" toward the industry.[18] This attitude was evident in a 1944 survey of executives about postwar opportunities for young people starting on business careers: more than half of those polled pointed to chemical industry as the most promising sector of the economy. *Fortune* captured the climate of opinion in 1950, declaring (with customary hyperbole) that this was the "Chemical Century", and noting that the "chemical industry cannot be matched by any other in dynamics, growth, earnings and potential for the future". This ebullience persisted through the early 1950s. Financial analysts saw chemical research as a barometer of future possibilities for the industry, and economists cited the chemicals and allied products industry as a paradigmatic example of spectacular economic growth.[19]

Enthusiasm over the chemical industry cooled during the late 1950s and early 1960s, as growth in chemicals and allied products slowed from the pace of the immediate postwar period. Paradoxically, this situation was caused by the reputation the industry had earned for sustaining high earnings and profits. The glamorous profile of chemical producers on the stock market began to attract companies from other industries, whose directors saw an opportunity to increase profits by invading chemical markets. Oil companies such as Shell and Standard Oil of New Jersey had manufactured petrochemical products since the 1920s, but after World War II they adopted policies of expanding their chemical operations to produce increasing amounts of petrochemicals from refinery by-products. Tire and rubber companies, formerly major customers of chemical manufacturers, integrated backward to produce their own synthetic rubber and other chemicals. Other corporations, lured by the high return on chemicals, diversified by acquiring chemical subsidiaries. W. R. Grace and Co., for example, entered the chemical industry in 1950 by buying Davison Chemical Company and Dewey & Almy Chemical Company; by the early 1960s, agricultural and specialty chemicals accounted for more than 50% of Grace's net sales. The long-run significance of this trend is apparent when one considers that the top twenty American chemical

[18] In a survey of chemical research in the Depression, Maurice Holland and W. Spraragen of the National Research Council reported that "the chemical industry has been largely founded on the results of the research laboratory. It is not surprising, therefore, that even in times of adverse business conditions the leaders of this industry should turn to research as a useful tool in maintaining their profits and reducing losses to a minimum" (III, Holland and Spraragen, 1932, 956). For an indication of the favorable attitude of investors toward the chemical industry in the late 1940s, see III, Naess, 1948.

[19] See IB, "Fortune", 1944, 36. On the "chemical century", see III, "Chemical", 1950, 69. See also III, Sherman, 1951; III, Hirt, 1954; and III, Vatter, 1963, 168–174.

producers in 1976 included six oil companies, for which chemical sales were less than 25% of total sales.[20]

The invasion of chemicals and allied products by companies from neighboring chemical process and other industries created serious problems for chemical manufacturers. Increased competition and overcapacity in commodity chemicals led to a severe cost-price squeeze, which decreased the high profit margins that had attracted outsiders in the first place. In addition, the chemical industry began to face stiff competition for its claim to being the most progressive supporter of research and development activity. Aided by an influx of federal funds, the aircraft and missiles, electronics, and other industries soon surpassed chemicals and allied products in this regard.[21] By the late 1950s Wall Street investors were no longer enamored of chemical

[20] The six oil firms were Exxon, Shell, Occidental, Standard of Indiana, Gulf, and Mobil, whose combined chemical sales totaled more than 9.38 billion dollars. Du Pont, the leading chemical company, reported 6.4 billion dollars in chemical sales (IB, Kiefer, 1977, 41). The slowdown of growth in the chemical industry during the 1950s is discussed in III, Semple, 1963, and III, Soule, 1963.

The petrochemical industry reveals how rapid postwar growth attracted entrants from other industries; see III, "Petrochemicals", 1955, and III, Chandler, 1962, 350–362. On the Shell and Standard Oil petrochemical ventures, see III, Beaton, 1957, 502–547, 615–617, 676–684; and III, Larson *et al.*, 1971, 765–770. Another company which diversified into petrochemicals in the early 1950s was National Distillers & Chemical Corporation; see III, Berenson, 1963, 134–141. On the history of Goodyear's chemical division, see III, Dietz, 1955. W. R. Grace's postwar transformation into one of the largest chemical corporations in America is discussed in III, Bernstein, 1978.

[21] See e.g. total funds for research and development (IA, NSF, 1972, Table 3, 34–35; IA, NSF, 1976c, Table B–3, 26), the number of research scientists and engineers engaged by the industry (IA, NSF, 1972, Table 23, 58–59; IA, NSF, 1976c, Table B–24, 44), research and development expenditures as a percentage of sales (IA, NSF, 1972, Table 40, 76–77; IA, NSF, 1976c, Table B–36, 55), and research and development personnel per 1000 employees (IA, NSF, 1972, Table 32, 66; IA, NSF, 1976c, Table B–29, 49). Chemicals and allied products have accounted for a relatively constant 10% of industrial research and development funds since the late 1950s. "Aircraft and missiles" and "electrical equipment and communications" have dominated, accounting for nearly half of all expenditures during the same period.

Company funds for research and development reveal a different picture. By this measure, the chemical industry was still the most research-intensive industry in American manufacturing in 1957, and has since been second only to electrical equipment (IA, NSF, 1972, Table 9, 41; IA, NSF, 1976c, Table B–9, 31). Company funds as a percentage of sales also show the chemical industry faring much better. Through the early 1970s, only "electrical equipment and communications" and "professional and scientific instruments" led the chemical industry in this respect (IA, NSF, 1972, Table 42, 79–80; IA, NSF, 1976c, Table B–37, 56; cf. IA, NSB, 1977, 106).

companies. One observer announced in 1961 that for chemicals "the ball is over".[22] Interestingly enough, while perceptions of the health of the chemical industry shifted dramatically during the late 1950s and 1960s, there is evidence to suggest that chemicals were increasingly important to the American economy after World War II. Input-output models for the U.S. economy indicate that chemical products contributed proportionately more to the output of end products in all consuming industries in 1958 than in 1947, with a 31% increase in overall requirements for chemicals during the period.[23]

This chapter has provided a sketch of the considerable wealth and prestige which have accrued to the chemical industry over the past century. The economic significance of chemistry and the role of chemists and chemical engineers in applying the science successfully to industrial uses have been potent symbols for the chemical profession in America. The chemical industry has provided implicit social support for American chemists. The explicit patronage of chemists through employment in industry is considered in the next chapter, along with other areas of chemistry as occupation.

[22] III, Stryker, 1961. Another observer noted that "it is no news that the chemical industry's investment appeal has declined notably in the last decade. Leading chemical stocks still command a loyal following, but they have lost the halo that once adorned them" (III, Soule, 1963, 13). *Business Week*, as early as 1957, had pronounced that chemicals had lost their "old-time fizz" (III, "Chemicals", 1957, 89).

[23] On chemical requirements in the U.S. economy, see IB, Carter, 1966, 31; and IB, Carter, 1970, 37–39, 78–82. For further details on the 1958 interindustry structure of the economy, see IA, Goldman *et al.*, 1964.

CHAPTER 5

A SECOND LOOK AT EMPLOYMENT

The aggregate trends in chemistry as occupation and as profession were considered in Chapter 2. The present discussion will focus upon disaggregated trends for those three sectors of the American economy that have employed significant numbers of chemists: industry, government, and academe. Some mention will also be made of the employment of chemists in other sectors. In every case, the necessary statistics are available only for widely scattered dates or for short runs of time. The varying definitions employed make the compilation of chronological series a more than usually hazardous enterprise. Nonetheless, certain indicators can be assembled.

5.1. Industry

INDICATOR HIGHLIGHTS:

Seventy per cent of American chemists work in industry. This proportion has been roughly constant since World War I, during which time the number of industrial chemists increased from about 13 000 to more than 93 000.

Two-thirds of all chemists in industry have been employed in the chemical process industries over the past quarter century. Between 1950 and 1970, the proportion of industrial chemists employed by the chemicals and allied products industry alone increased from one-third to about one-half; at the same time, chemists constituted a growing fraction of the total labor force in the chemical industry.

For more than half a century, significant numbers of chemists have worked in industrial research. Between 1920 and 1950, there were dramatic increases in the number of research laboratories within the chemical process industries. During the same period, the number of chemists employed in industrial research rose from about 3800 to more than 23 000 persons. A growing proportion of all industrial chemists was engaged in research and development activity during the interwar years.

Chemists have been a shrinking fraction of all scientists and engineers in

American industry since 1950. Nonetheless, chemists still represent the largest disciplinary group among American industrial scientists.

The most direct way in which industrial activity has served as a support for chemistry as an institution is through the employment of chemists and chemical engineers. Considerable difficulties are involved in any attempt to construct indicators of that employment. It is not just in the chemicals and allied products industries or the chemical process industries that chemists have found employment. Other manufacturing and even nonmanufacturing industries have called upon the services of trained chemists. In 1950, for instance, two hundred chemists were employed by American railroads, one hundred by medical and dental laboratories, and two hundred in engineering and architectural services.[1] Chemical industry employed only a third of all chemists in private industry in 1950. While tracing variations in the employment of chemists in the multiplicity of employing industries is an obvious desideratum, it is not possible here.

Considerable ambiguities attach to any attempt to specify the work in which chemists have been engaged. There are no clear points of demarcation between research, development, and routine analysis, nor between research and administration; nor is it possible to differentiate sharply between industrial chemistry and chemical engineering. These problems come into focus when the available indicators are presented. There are irritating gaps within, and serious discrepancies between, alternative series of measures.

5.1.1. CHEMISTS IN INDUSTRY

Figure 5.1–1 displays the employment of chemists in American industry, showing an increase from roughly 12 700 persons in 1917 to over 93 000 in 1970. (Census and Bureau of Labor Statistics estimates of *all* chemists in the labor force – not just chemists in industry – are also provided for comparative purposes).[2] The paucity of information prior to 1950 is apparent.

[1] Not to mention the 96 members of the Association of Official Racing Chemists (1947), who analyze the urine and saliva of thoroughbreds at American racetracks to guard against the unauthorized use of drugs (IB, Pair *et al.*, 1978, I, 408).

For estimates of chemists employed in selected nonmanufacturing industries in 1950, see IA, BLS, 1973, 50–51.

[2] The discrepancies between Census and BLS estimates of the number of chemists arise from differences in the extent of survey coverage. The BLS figures probably

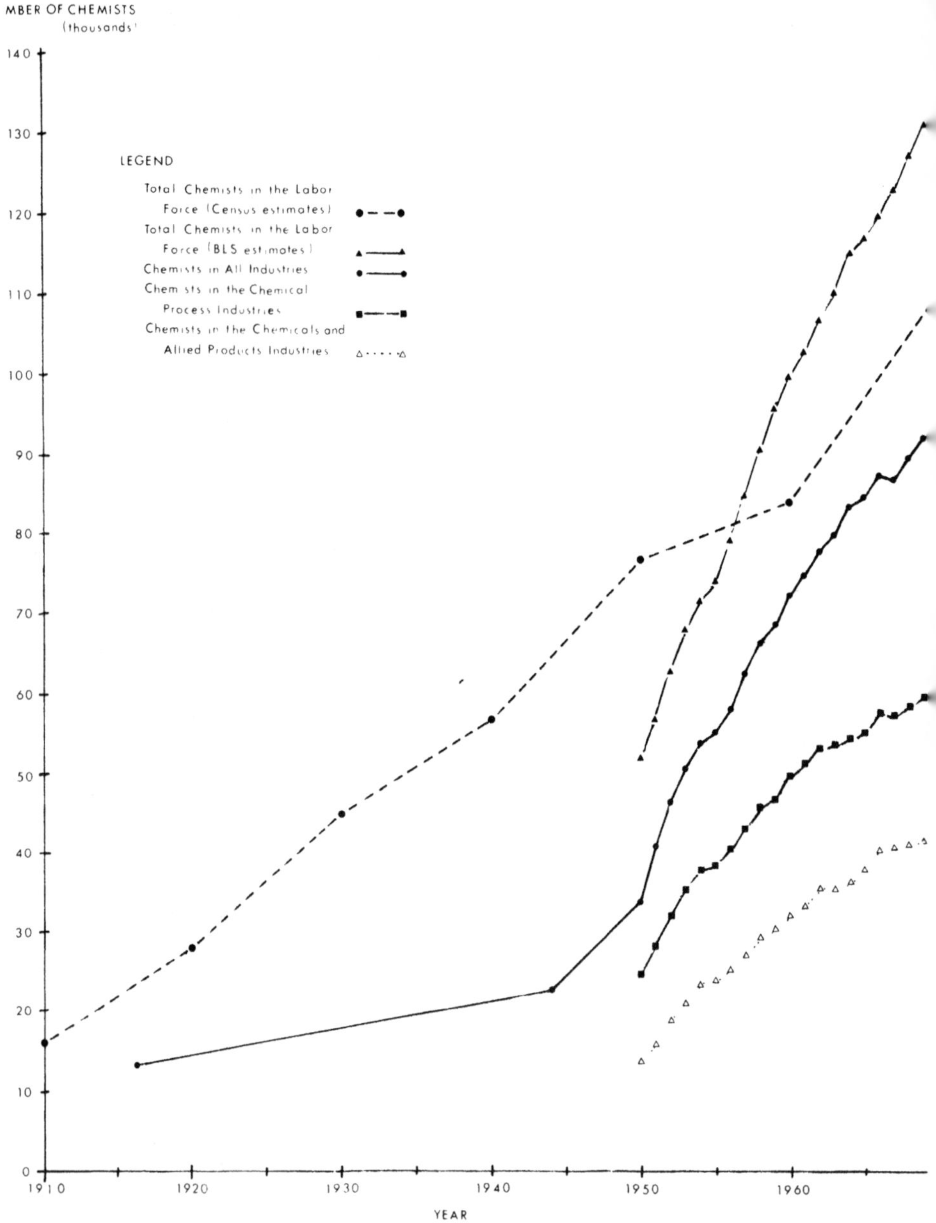

Fig. 5.1–1. Rough estimates of the number of chemists in industry, 1917–1970. (See Table 5.1.)

So too is the long-term importance of industrial employment to the American chemical community. According to Bureau of Labor Statistics estimates, industry was the primary sector of employment for American chemists from at least 1950 to 1970, providing jobs for more than 70% of the chemical community over that period. Evidence from earlier surveys suggests that this distribution of chemists has prevailed since World War I, except for a 10% dip in favor of academic employment lasting from the mid-1920s to the mid-1930s.[3]

The chemical process industries have played a major role in the employment of industrial chemists in recent years, providing jobs for two-thirds of those chemists. Within the chemical process industries, the chemicals and allied products group has become increasingly important: the proportion of *all* industrial chemists employed in the latter group increased from one-third to one-half in the years from 1950 to 1970. Chemicals and allied products accounted for 52% of the growth in industrial employment of chemists over the twenty-year period, and for 81% of the increase in chemists employed in the chemical process industries (see Tables 5.1 and 5.2).

The growing importance of chemists within the chemicals and allied products group is confirmed by an examination of total employment in the industry. Figure 5.1–2 compares overall levels of employment in chemicals and allied products to the number of chemists employed in this industry. Chemists constituted a growing fraction of a growing labor force, increasing from about 208 per ten thousand employees in 1950 to over 480 in the mid-1960s.[4] This one indicator displays in microcosm the increasing importance of chemical skills and knowledge within the industrial economy. A different indicator of the role of such skills may be found by considering the category of "research and development" in industry.

---

underestimate the size of the chemical community during the 1950s, but overestimate the number of chemists by 1970. See IB, ACS, 1975, 2–3, 48, for a comparison of the BLS and Census trend data.

The BLS estimates of the distribution of chemists by industry have been adopted as the basis of the discussion below for two reasons. First, they provide *comparable* estimates over a twenty-year period. Second, although the BLS figures may yield inflated absolute growth rates, there is no reason to suppose that the picture of *relative* distribution of chemists and other scientists in industry derived from these data is overly distorted.

[3] See IA, Fay, 1917, and IB, Fraser, 1942c, 1567, Table 17.

[4] Compare columns 1 and 4 in Table 5.3.

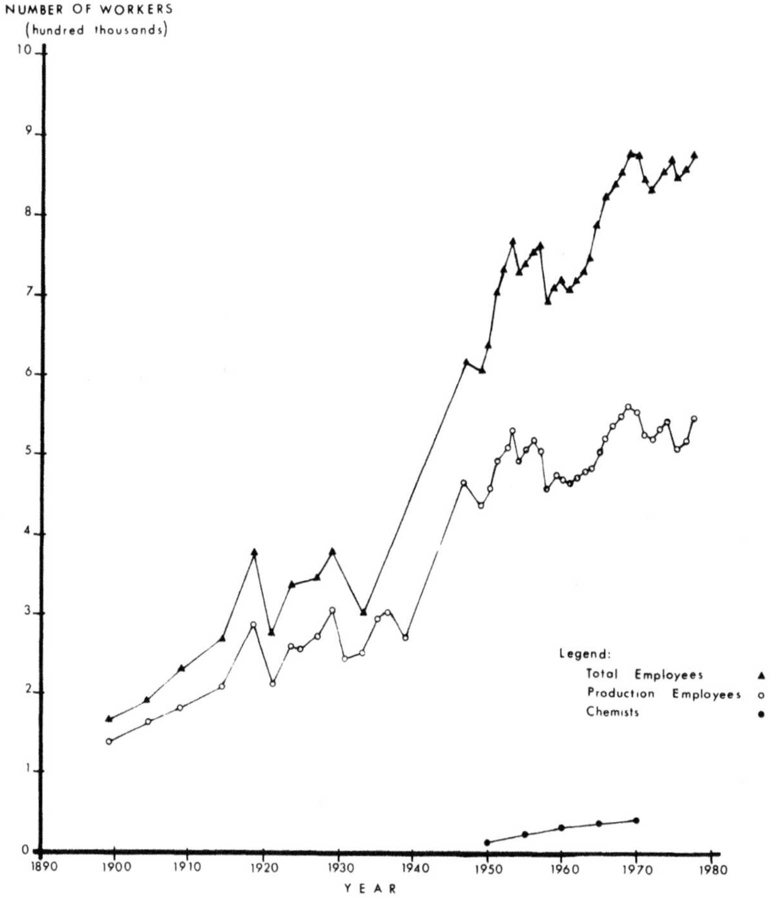

Fig. 5.1–2. Employment levels in the chemicals and allied products industries, 1899 1977. (See Table 5.3, columns 1, 2, and 4.)

5.1.2. Research Laboratories and Research Workers

The institutionalization of research in industry has been one of the most striking features of the social history of twentieth century American science and technology. The establishment of corporate research laboratories began around the turn of the century. Adopting a utilitarian rhetoric which resonated strongly with a Progressive era faith in social progress through science, scientists appealed successfully to the captains of an expanding industrial community, who perceived continuous innovation as a new weapon in corporate strategy. First used in the electrical and chemical industries, this new weapon spread rapidly prior to World War I; more than 300 research laboratories had been organized by 1915. The industrial research movement gained impetus from the wartime experience of cooperative effort (as indicated by the rate of formation of new laboratories, shown in Figure 5.1–3), and science became inextricably linked with industry during the boom years of Coolidge prosperity.[5]

The cumulative trend in the formation of laboratories from 1890 to 1940 is illustrated in Figure 5.1–4, which represents a 59% sample of companies reporting research activity to the National Research Council in 1940. Growth in the cumulative number of *laboratories* established in American industry began to slow by the 1920s. Rapid growth in the number of *companies* supporting research activity continued. During the twenties both the number of companies maintaining research laboratories and the number of people employed in industrial research tripled (see Table 5.5). The almost 200% increase in industrial researchers is dramatic when considered against the 28% increase in all persons employed in U.S. manufacturing between 1921 and the onset of the Depression. The rate of growth of industrial research laboratories slowed still further during the early 1930s, but the number of research workers still doubled during each of the next two decades. Between 1921 and 1950, the increase in industrial researchers (9.6% per annum)

[5] On the strategic role for industrial research which emerged in the 1890s, see III, Chandler, 1977, 374–375. III, Mees, 1920, is an early practical guide by the research director of a successful corporate laboratory; see also III, Jenkins, 1975, 300–318.

Among recent discussions on the history of American industrial research, see III, Birr, 1979; III, Galambos, 1979; III, Noble, 1977, 110–166; III, Rae, 1979; and III, Reich, 1980. A classic article which remains useful is III, Bartlett, 1941; see III, Birr, 1966; III, Lewis, 1967; and III, Pursell, 1972. III, Vagtborg, 1976, focuses on independent research organizations such as the Mellon Institute. See also III, Reich, 1983; III, Wise, 1980 and 1983; and III, Mowery, 1983a and 1983b.

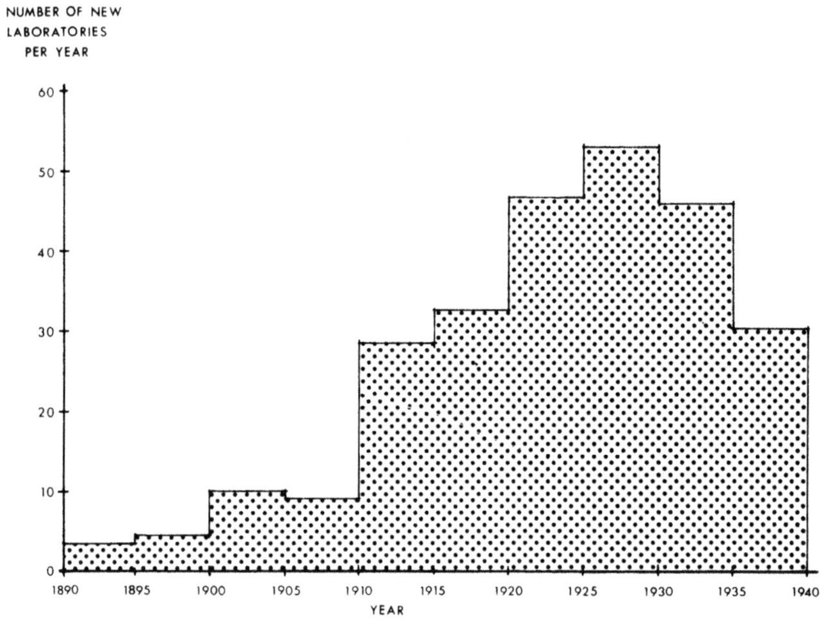

Fig. 5.1–3. Average annual rate of formation of industrial research laboratories, by pentade, 1890–1940. (See Table 5.4, column 2.)

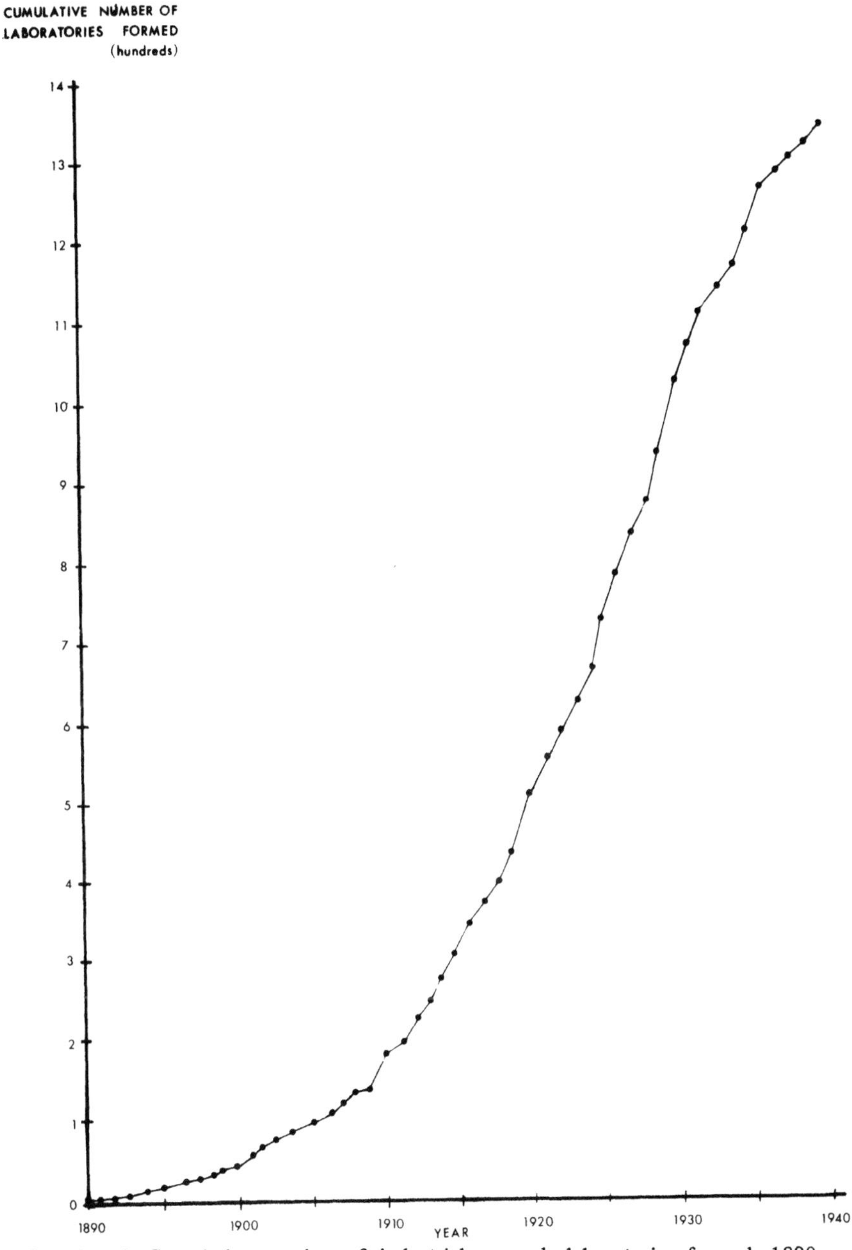

Fig. 5.1–4. Cumulative number of industrial research laboratories formed, 1890–1940. (See Table 5.4, column 3.)

was more than four times that of all employees in U.S. manufacturing (2.3% per annum).[6]

The chemical process industries were strong supporters of research and development throughout this period. Companies such as General Chemical (1899), Dow (1901), Du Pont (1902), Standard Oil of Indiana (1906), Goodyear (1909), Eastman Kodak (1912), and American Cyanamid (1912) were among the first to establish central research and development laboratories. By the late 1920s over half of the industrial laboratories reporting to the National Research Council were located in the chemical process industries. Between 1927 and 1938 the proportion of all research personnel employed in this group increased from 41.0 to 52.2%. Within the chemical process industries, companies in chemicals and allied products, petroleum, primary metals, and rubber employed the largest research staffs, accounting for nearly 85% of the research workers in the group by the late 1930s. The chemicals and allied products industry had more researchers than any other industry in the late 1920s and 1930s, employing one in five research workers (see Table 5.6). Figures 5.1–5 and 5.1–6 present detailed employment information for three components of the chemical process industries – industrial chemicals (the largest subgroup of chemicals and allied products), petroleum, and rubber. These figures reveal the expanding opportunities for chemists and other scientists.

Much of the growth after the late 1920s can be attributed to the expansion of existing laboratories, especially in industrial chemicals. For example, between 1927 and 1938 the number of research workers employed at Dow Chemical increased from about 100 to more than 500; at Du Pont, from about 850 to over 2500 persons. In industrial chemicals and rubber, research was directed mainly toward the improvement of manufacturing processes and the exploration of new applications for products. In petroleum, breakthroughs in catalytic cracking and polymerization technology, along with early movement into the field of synthetic organic chemicals, help to explain the growth in the 1920s and 1930s of both the number of laboratories and the number of researchers in the industry. Research workers in the petroleum industry increased from a few hundred in the 1920s to more than 5000 in 1938, accounting for one in nine industrial research personnel and placing

[6] Excluding partners and proprietors, 7.557 million people were employed in U.S. manufacturing in 1921, 9.66 million in 1929, and 14.467 million in 1950 (IA, USBC, 1975b, II, 666, Series P4, P5). We thank Richard O'Connor for Figure 5.1–4.

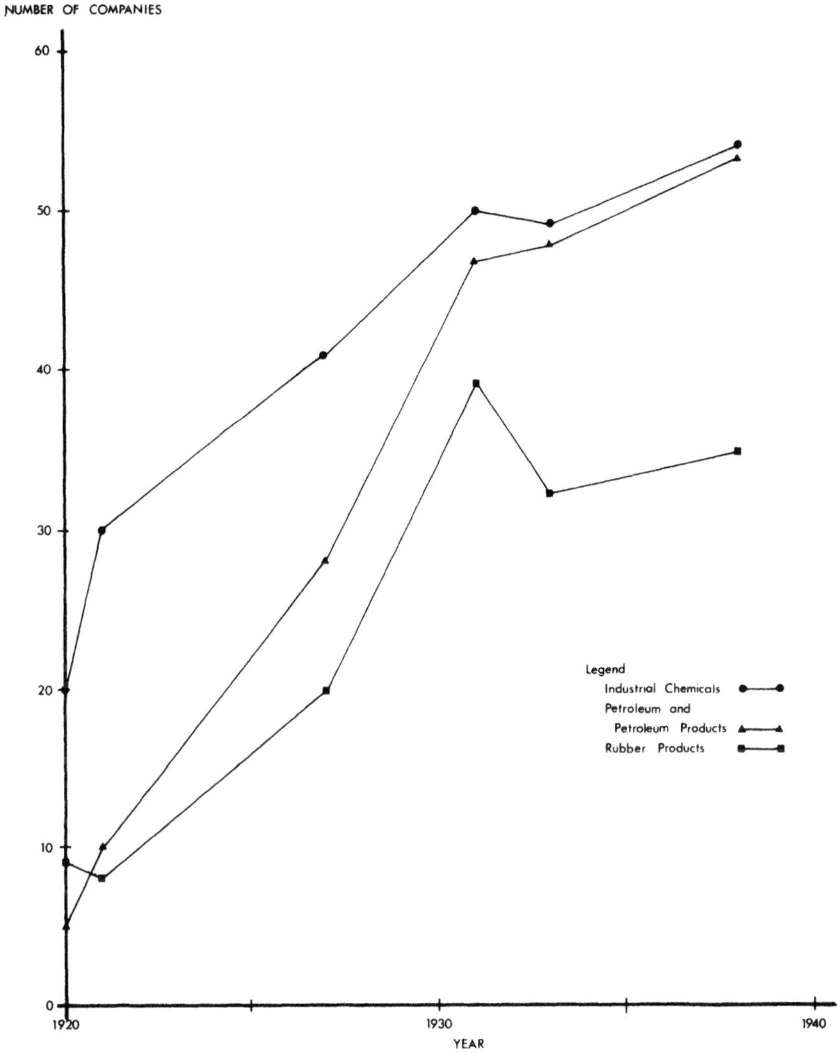

Fig. 5.1–5. Number of companies maintaining research laboratories in selected industrial groups, 1920–1938. (See Table 5.7.)

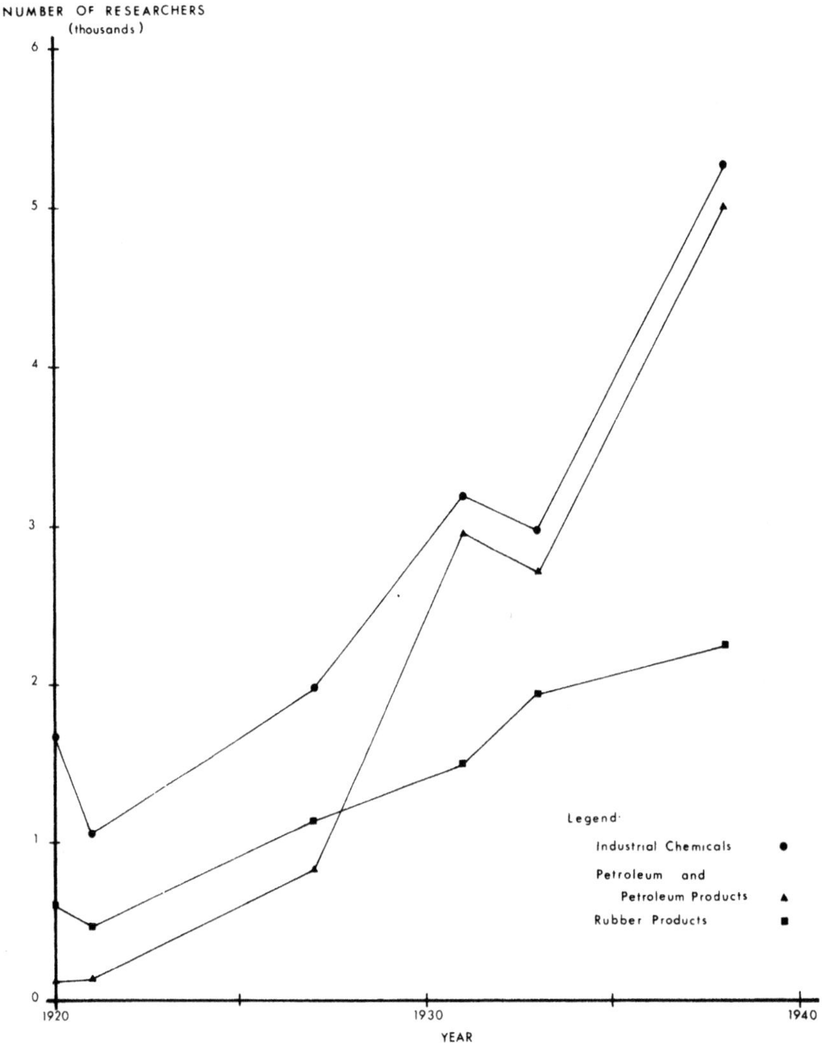

Fig. 5.1–6. Research personnel in selected industrial groups, 1920–1938. (See Table 5.8.)

the industry second only to chemicals and allied products in the extent of its research activity.[7]

Since industry-by-industry comparisons of the proportion of chemists among company research workers are not readily available, we must return to aggregate data to obtain information on the disciplinary breakdown of scientists engaged in industrial research. Figure 5.1–7 presents information on the employment of chemists in American industrial research from 1921 to 1960 (with physicists and engineers included for comparative purposes).[8] During this period, the number of chemists in industrial research laboratories increased elevenfold, from about 3800 to 42 800. Making no adjustment for the 11% decline at the beginning of the Depression, this represents a doubling every 11 yr. By comparison, the total number of chemists reported by the Census grew more slowly, from about 28 000 in 1920 to 84 000 in 1960. This is equivalent to a doubling every 25 yr (Figure 5.1–8). The relatively rapid shift in the deployment of chemists transformed the contours of the chemical community, a situation reflected in the occupational composition of the American Chemical Society. A 1941 survey of ACS membership revealed that 25% of the respondents were engaged in industrial research, while 15% taught in secondary schools and colleges. By contrast, of those respondents working in 1926, 24% had been teaching at the secondary and

[7] Figures on research workers at Dow and Du Pont are from IB, West and Risher, 1927, 37–39; and IB, Hull, 1938, 69–72.

For research activity in chemicals, petroleum, and rubber during this period, see IB, Perazich and Field, 1940, 29–36. III, Dietz, 1943, describes research in a lab central to the rubber industry. On cracking processes in petroleum refining, see III, Enos, 1962. Petrochemicals research in the 1920s and 1930s is discussed in III, Haynes, 1954b, 208–225; and III, Williamson *et al.*, 1963, 423–430.

The contributions of chemical research (especially in the dehydration and desulfurization of crudes and the depolymerization of heavy oils) to the progress of the petroleum industry were noted as early as 1916: " . . . [S]ome of the largest refiners now concede that chemistry is the intelligence department of the petroleum industry and gratefully acknowledge that the efficiency of their plants has resulted largely from research" (III, Bacon and Hamor, 1916, II, 799).

[8] The data are derived from periodic surveys of industrial research laboratories conducted by the National Research Council since 1920. Although these surveys have continued – the most recent appearing in 1982 – the last one for which comparable tabulations on particular professional groups are available is the 1950 survey. A number of ambiguities in the NRC surveys should be kept in mind, e.g., the scope of activities denoted by the protean term "research", the criteria by which individuals were designated as research scientists or engineers, and the difficulty of determining the extent of coverage. (See Appendix B. 3). III, Mowery, 1983a, also uses NRC data.

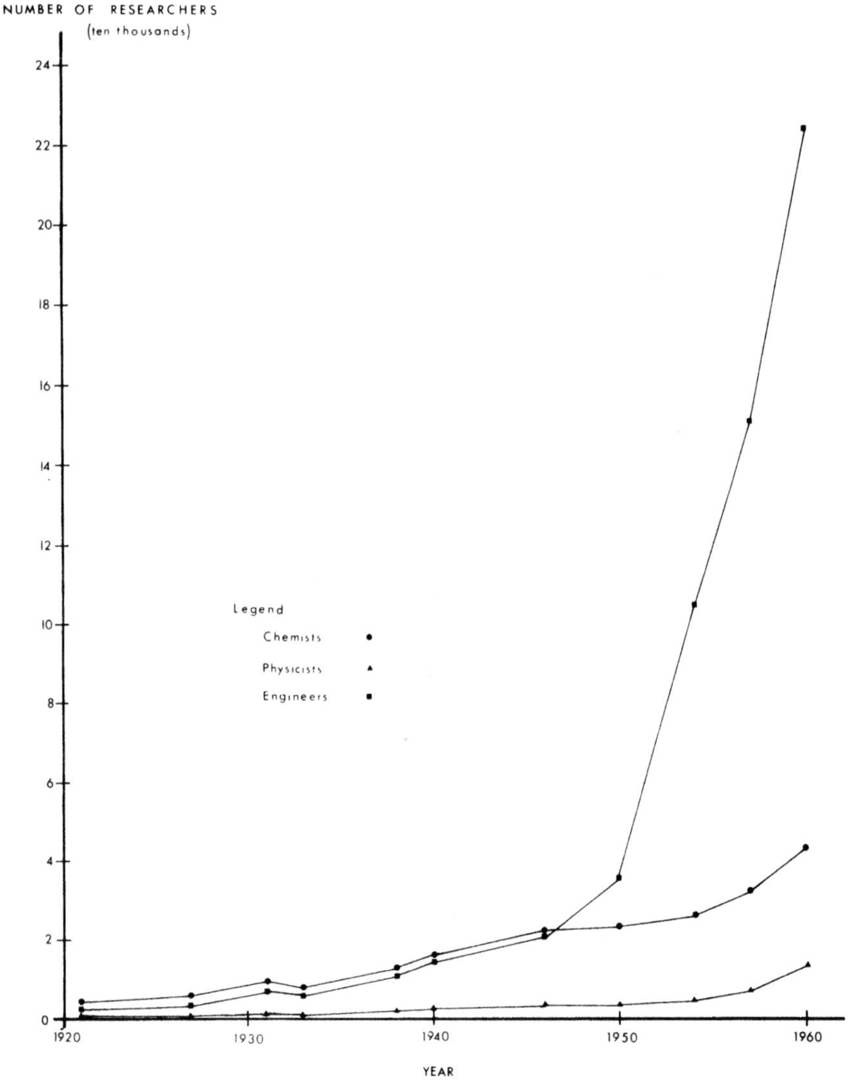

Fig. 5.1–7. Chemists, physicists, and engineers in industrial research, 1921–1960. (See Table 5.5, columns 3–5.)

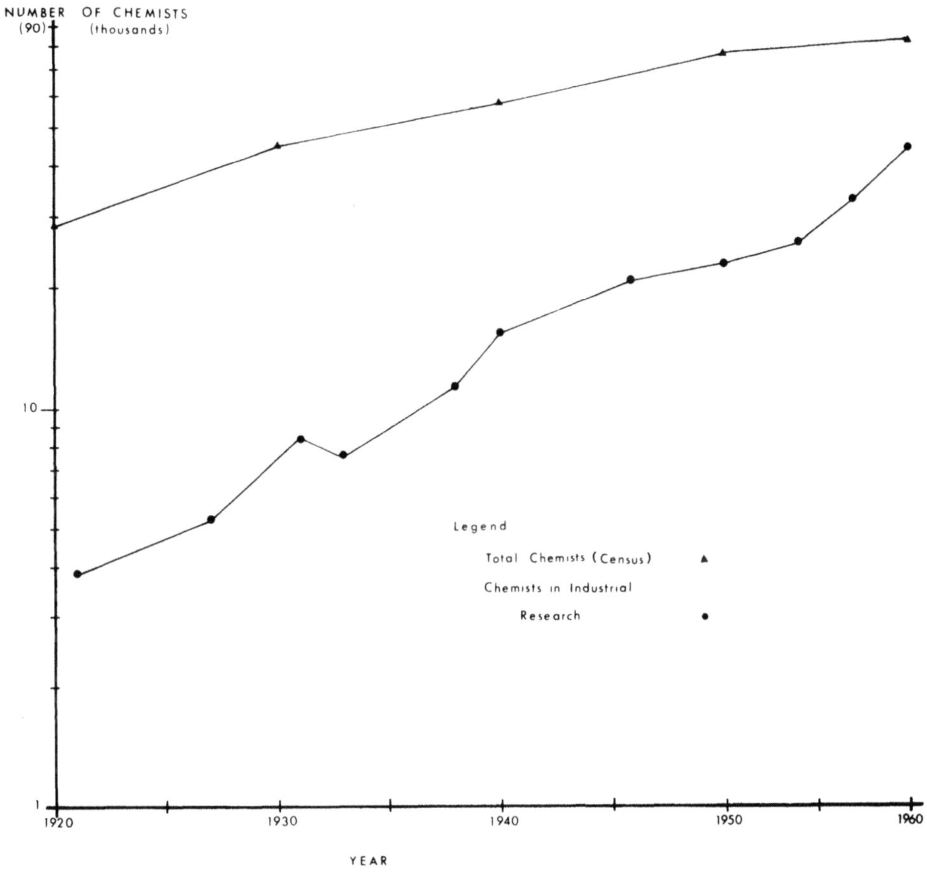

Fig. 5.1–8. Industrial research chemists versus Census chemists, 1920–1960. (See Table 2.2, column 1; and Table 5.5, column 5.)

college levels, while only 16% had been employed as industrial research personnel (Figure 5.1–9).

Among those engaged in industrial research, chemists were the predominant group during the interwar years. As Table 5.5 shows, one of every three persons in industrial research in 1921 was a chemist.[9] By the late 1920s, as a variety of industries jumped on the research bandwagon, the proportion of chemists dropped slightly, to one in four. But chemists maintained their central position throughout the 1930s and 1940s. Only after World War II did engineers as a group eclipse the chemists among professionals in industrial research (see Figure 5.1–7). By 1957 the proportion of chemists had dropped to one in nineteen, while that for engineers had reached one in four industrial research workers. Chemical engineers probably constituted only a modest proportion of these people, since the federally-funded expansion of aerospace and other defense-related research in the 1950s – which accounts for the trend – demanded engineering expertise mainly in non-chemical specialties. Engineers had thus displaced chemists as the central figures in industrial research by the late 1950s.[10]

Data comparable to those presented in Table 5.5 are not available on chemists in industrial *research* after 1960, but some time series data do exist for chemists in industry generally. Compilations by the Bureau of Labor Statistics show that, from 1950 to 1970, chemists were the largest group of scientists employed in American industry (see Figure 5.1–10). The growth rate of 4.8% per year was slower than that for all scientists in industry (5.3% per year): the proportion of chemists among industrial scientists thus fell slightly from 48.2% in 1950 to 43.9% in 1970. Concomitantly, there was a rapid influx of mathematicians into industry (Table 5.11).

Table 5.2 indicates the distribution of chemists in the chemical process industries, where roughly two-thirds of all chemists in industry were located in this period. The fraction of chemists employed in the chemicals and allied products industry increased steadily from about one-third in 1950 to about one-half of all chemists in industry in 1970. Since the mid-1950s, aside from the chemical process industries, the greatest number of industrial chemists

[9] The occupational breakdown in Table 5.5 is based on data from 45 to 75% of the companies in the NRC surveys of industrial research labs between 1921 and 1938 (IB, Perazich and Field, 1940, 11–12, 62–63, 78) and 62.5% of the companies in 1940 (IB, Cooper, 1941, 176).

[10] For contemporary comment on the relative decline of the chemists in industiral research, see III, Perazich, 1951, 21; III, Blank and Stigler, 1957, 47–74; and IB, Terleckyj, 1963, 72–73.

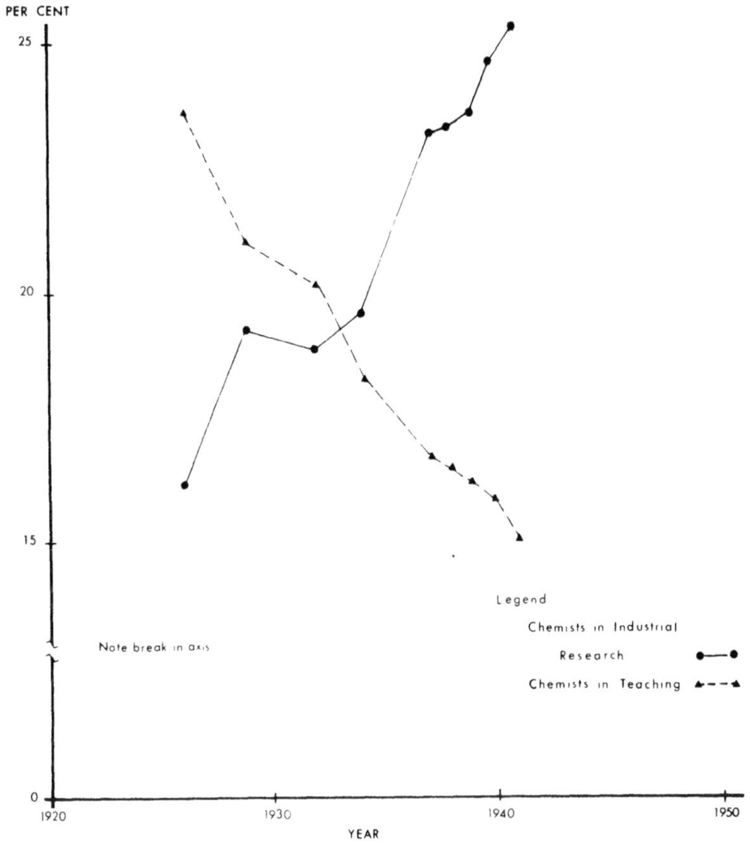

Fig. 5.1–9. Chemists in industrial research versus chemists in teaching as percentage of all respondents to 1941 ACS survey, 1926–1941. (See Table 5.9, columns 3 and 7.)

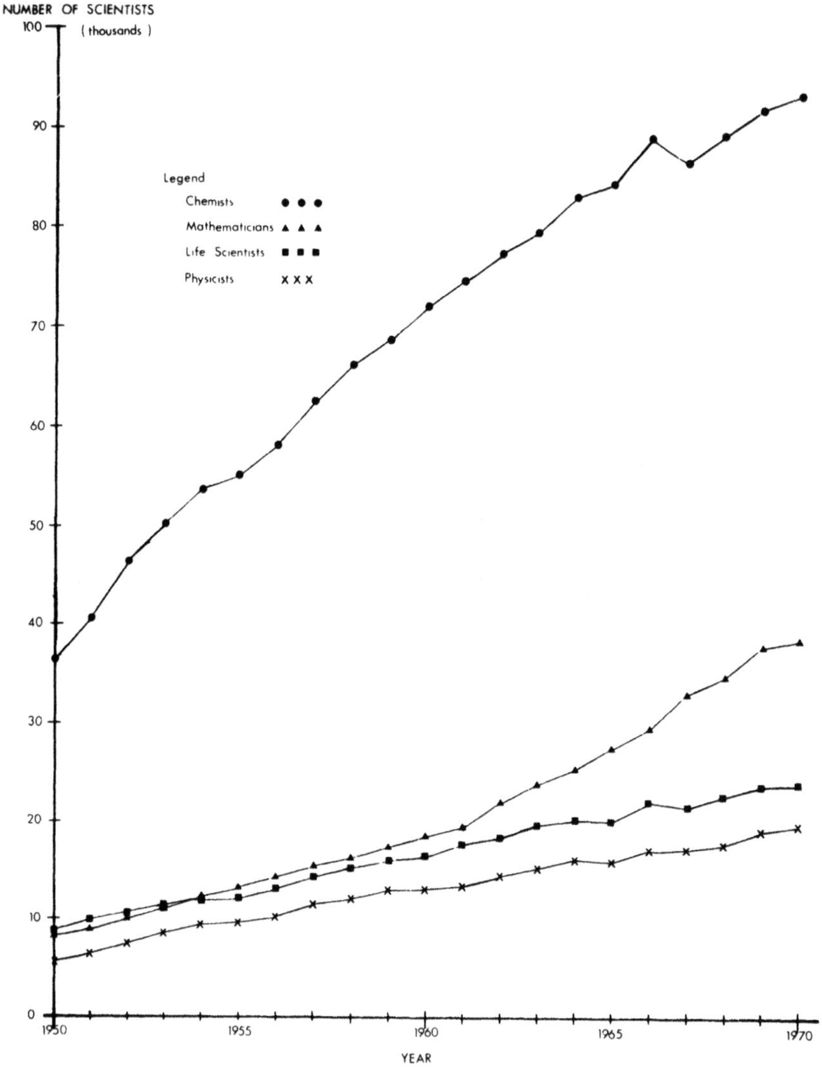

Fig. 5.1–10. Scientists in industry, selected fields, 1950–1970. (See Table 5.10, columns 1, 2, 4, and 5.)

has been employed in "miscellaneous business services", a category which includes commercial research and development laboratories and similar consulting services.[11] The proportion of all industrial chemists in this category increased from one in every 15 in 1955 to one in every 11 chemists in 1970 (see Table 5.12).

Why did chemists have such an important place among industrial scientific personnel? To publicists for the chemical profession between the wars, the reasons were self-evident: since all manufacturing involved some form of material transformation, *all* industries were seen as *chemical* industries. Though this position was exaggerated, chemists did fill a variety of roles in a broad spectrum of industries, ranging from petroleum and rubber to textiles and cellulose and wood products. Chemists provided analytical skills which promised direct economic benefit to companies through product control, and chemical research offered the lure of future profits through the utilization of wastes, substitution of cheaper raw materials in existing processes, and development of new products.[12]

[11] The growth of the population of mathematicians and biomedical scientists from 1950 to 1970 meant that the proportion of chemists among the *total* U.S. scientific population fell from 34.9% in 1950 to 26.8% in 1970 (IA, BLS, 1973, 13–16).

Miscellaneous business services (SIC 739) also includes commercial testing laboratories; see II, OMB, 1972, 305, 307.

[12] Statements that chemistry is basic to all manufacturing processes – a recurrent theme of chemical boosterism (see Section 4.4) – are easily gleaned from the popular scientific and trade literature of the period. As one enthusiastic commentator wrote just after World War I:

> [W]ith chemical industries cropping out all over the land, with chemical control coming into vogue for nearly all the industries, with a change in the ways of trade under both statutes and custom so that *caveat emptor* is becoming obsolete and the rule is developing that the seller must beware that his goods are exactly as they are represented to be – the day of the chemist is at hand. He will be needed everywhere; in making things, in keeping them, in buying and selling them, and in the disposal of that which is not used .... [Chemistry] is the coming profession. It must determine for us in the future what we shall eat and drink and wherewithal we shall be clothed. Whether we grow as an industrial nation or sink into decrepitude is in large part dependent upon our chemists (III, Hendrick, 1919, 220–221, 226–227).

Through the 1920s, manufacturers were urged to view chemistry as a sound investment (III, Little, 1928, 33–60; III, Farrell, 1926) and a "source of new wealth" (III, Griffith, 1924). In the mid-1930s, synthetic organic chemistry was said to be opening new vistas by replacing natural products; as a result, all industries promised to be affected by chemistry (III, Howe, 1937, 289–90, 312).

The status of chemists in the eyes of competition-conscious executives was undoubtedly boosted by the successes of chemists in such fields as plastics, petrochemicals, and synthetic textiles. As a result of the industry's growing demand for trained chemists, a new relationship emerged between industry and the curriculum and research interests of many academic chemistry departments. Universities supplied industry with scientists and with fundamental research to supplement work done in industrial laboratories. In turn industry aided chemistry departments. Many of the increasing number of chemistry students in American universities were supported by pre- and postdoctoral fellowships from chemical corporations. Du Pont alone offered fellowships worth eighteen thousand dollars at 12 institutions in 1936–1937; between 1918 and 1932, the company awarded 326 fellowships and 34 scholarships to young academic chemists.[13]

The claim that pure science – the province of the university chemists – provided the basis of industrial innovation was perhaps best stated in Roger Adams' peroration to a 1935 address on "The Relation of the University Scientist to the Chemical Industries":

> The chemical companies in the United States, progressive, efficient, well-managed, have now reached that level of success where they must consider not merely the building of powerful research and producing units, but must give equal consideration to the conditions of those organizations in which the technical leaders of the future are to be trained .... The American chemical industries will recognize the importance of the situation, and will do their part to maintain the prestige of pure science, as well as of applied science.[14]

Earlier tributes to the chemist's role are catalogued in II, West, 1920; see also III, Howe (ed.), 1924–1925. For an extensive survey of occupational niches for chemists in post-World War I industry (emphasizing opportunities for women), see III, BVI, 1922, 44–154. On the economic returns of chemical research through by-product utilization, quality control, and other means, see III, Howe, 1925a, 1925b, 1925c; and III, Howe, 1941, esp. 227–234.

Chemical engineering achievements also contributed to the perceived status of chemists in industry. Chemical engineering was seen as the "engineering of the future" (III, Hammond, 1929, 114). C. M. A. Stine, director of chemical research at Du Pont, stressed the importance of the chemical engineer's expertise in dealing with the "pocketbook reaction" (III, Stine, 1928, 46); see also III, Kirkpatrick, 1941, 307. For a synoptic account of chemical engineering's contributions to various industries, see III, Kirkpatrick (ed.), 1933.

13 IB, "Du Pont Fellowships", 1936.

14 III, Adams, 1935, 367. Another prominent spokesman for the advantages of closer relations between business and academe was James Flack Norris, flamboyant professor of chemistry at the Massachusetts Institute of Technology; see, for example, his talks

How seriously one should take rhetoric about the "chemicalization of industry" in constructing explanations of the chemist's role in commerce remains uncertain. In any case, as Figures 5.1–7 and 5.1–10 illustrate, chemists did play a significant part in the development of American industrial research, and they remain the largest disciplinary group among scientists in American industry.

5.2. Government

INDICATOR HIGHLIGHTS:

Civilian chemists employed by the federal government increased from a handful in the 1870s to about 8000 in the early 1970s, following a logistic growth pattern from about 1906.

Three qualitatively different phases of growth in the number of federal chemists can be distinguished in this pattern: from the 1870s to just after the turn of the century; from roughly 1906 to World War II; and from World War II to the present.

Before the Second World War, the Department of Agriculture was the largest employer of chemists in the federal government. Since the late 1940s, the Departments of Defense and of Health, Education, and Welfare have become the dominant loci for chemists.

About 1000 chemists worked for state and local governments as the United States entered World War I. This number held steady during the interwar years, but increased after 1950. Expressed as a proportion of all chemists, those employed by state and local governments dropped from about 8% in 1910 to 3% in 1970.

5.2.1. THE FEDERAL GOVERNMENT

To discuss the role of chemists within the federal government, data must be assembled on a great variety of chemical niches within a complex

on "Academic Research and Industry" (III, Norris, 1925) and "Research and Industrial Organic Chemistry" (III, Norris, 1932). Suggestive parallels with the chemists' experience emerge from a recent account of the encounter of academic physicists with American industry prior to World War II; see III, Weart, 1979. See also III, Carroll, 1982.

organization. One indication of the magnitude of the undertaking is provided by a 1916 survey of federal agencies, which listed 71 separate job titles for chemical employees. The fact that each federal department had an idiosyncratic employee classification scheme prior to 1923 (when the number of civilians working for the federal government totaled more than 500 000) compounds the problem. The most practical solution is to rely upon the government's own definitions of "chemist" as a distinct bureaucratic category. Accordingly, the data reported in this section derive from a variety of official sources: the *Official Register* of federal employees (prior to 1914); employment brochures designed to recruit chemists for the Civil Service; and (since 1931) Civil Service Commission figures generated for government use.[15]

This pragmatic criterion does not guarantee that our data are free from inconsistencies. Caution is also advised in comparing the number of government chemists (according to the definition adopted here) with Census estimates of chemists. Nonetheless, it is clear that the federal government has employed a growing proportion of American chemists. In 1911, roughly three hundred men and women (3% of a Census total of 9000 chemists) were employed in chemical positions at various federal agencies. By 1970, the number of chemists in the federal government had increased to over 8100 – about 7% of all American chemists in that year.[16]

Figure 5.2–1 exhibits the growth in the number of chemists employed by the U.S. government since the late nineteenth century. This growth conforms to a logistic pattern since about 1906. The data of Figure 5.2–1 are plotted on a semi-logarithmic scale in Figure 5.2–2. The trend lines illustrate

[15] On the occupational standing of chemists in the federal civil service hierarchy following passage of the Classification Act of 1949, see II, USCSC, 1951–; II, USCSC, 1950–; and II, USCSC, 1966, 25–31.

Even adopting internally-generated figures compiled by the Civil Service Commission does not obviate problems caused by inconsistent definitions of occupational categories (see IA, NSF, 1957b, 212–213). Margins of error are difficult to estimate; accordingly, caution is advised in interpreting trends in the employment of chemists by the federal government.

The figures reported below exclude chemists within the Armed Forces. The only available source states that 417 officers and enlisted personnel conducted chemical research for the Department of Defense in 1954 (IA, NSF, 1957b, Table VI–9, 175), which is less than 10% of the number of civilian chemists in the federal government in that year (see Table 5.13).

[16] The Bureau of Labor Statistics estimates that 6.5% of all chemists worked for the federal government between 1950 and 1970 (IA, BLS, 1973, 52–53). The discrepancy between this static picture and that presented here is the consequence of BLS underestimation of the number of chemists in the 1950s (see IB, ACS, 1975, 2–3, 48).

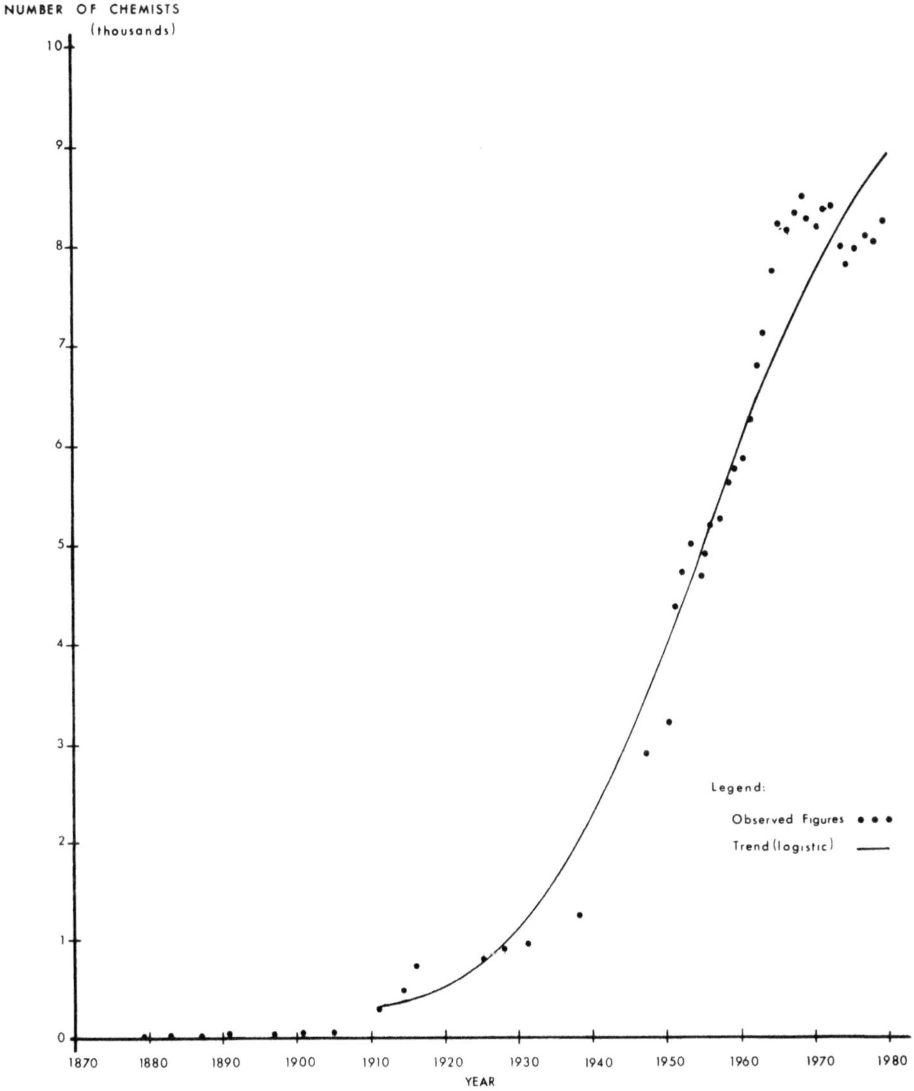

Fig. 5.2–1. The number of civilian chemists in the federal government, 1879–1978. (See Table 5.13, columns 1 and 2.)

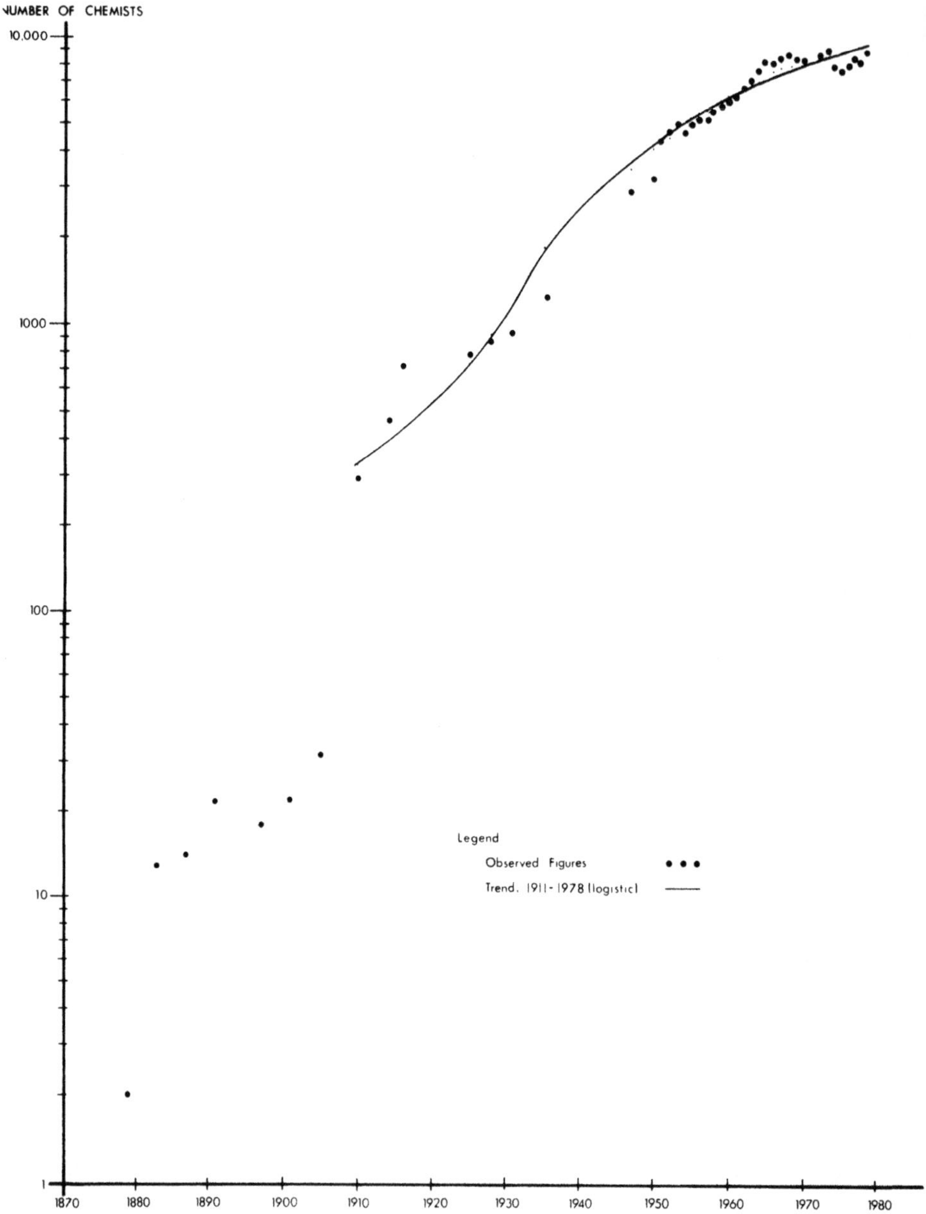

Fig. 5.2–2. Nonlinear trend in the number of civilian chemists in the federal government, 1911–1978. (See Table 5.13.)

a long-run "saturation" effect (the saturation level being roughly 9 900 people). The rate at which chemists have increased within the federal government in recent years has declined steadily, despite the increase in absolute numbers of chemists employed. In 1920 the ranks of federal chemists were expanding at a rate of about 9% per year; 40 yr later, this figure had dropped to just over 3%.[17]

The trends in Figure 5.2–2 may be further decomposed in light of historical knowledge of the context of chemical work in the federal government. The past century saw three qualitatively different periods of growth. The first phase, covering the last third of the nineteenth century, marked the entry of chemists into federal service. In the second phase (from roughly 1905 until World War II), chemical positions in the federal government increased rapidly in connection with the elaboration of government regulatory functions. Passage of the Pure Food and Drug Act in 1906 was especially important. During these two phases chemists were centered in the Department of Agriculture. The situation changed after World War II as government became more concerned with national security interests and the promotion of social welfare. This third phase, lasting from just after World War II until the present, saw the Departments of Defense and of Health, Education, and Welfare (now Education and Health and Human Services) replace Agriculture as the leading employer of civilian chemists, while the total of chemists in all federal agencies grew more slowly than during the interwar years.

5.2.2. CONTEXTS OF FEDERAL EMPLOYMENT

The category of "chemist" was established in the federal bureaucracy within the newly created Department of Agriculture in 1862, when Charles M. Wetherill was appointed to conduct chemical analyses of soils, manures, and a wide variety of plants and vegetables as an aid to farmers eager to raise agricultural productivity. As the most conspicuous laboratory in Washington, the Division of Chemistry in the Department of Agriculture quickly became the locus of government chemical activity. The Division inherited numerous routine testing assignments from other departments (such as investigating

[17] These growth rates were obtained by application of the compound growth formula to predicted values of the logistic curve shown in Figures 5.2–1 and 5.2–2. The saturation level of 9 900 chemists is the asymptote of the logistic curve (see Appendix E).

printing inks for the Post Office).[18] As the scale and complexity of administrative activity gradually increased, other federal agencies hired their own chemical personnel. In addition to agricultural research, the corps of civilian chemists in the federal establishment became occupied with such diverse tasks as the analysis of water samples and mineral specimens for the Geological Survey, the determination of the purity of alcohol, margarine, and other foodstuffs for the Bureau of Internal Revenue, and the inspection of supplies provided to the Department of Commerce, to ensure that contract specifications had been met. A total of twenty-two chemists, mainly concentrated in the Geological Survey and the Division of Chemistry of the Department of Agriculture, were employed by federal agencies at the end of the nineteenth century (see Table 5.14).[19]

Early in the twentieth century, the federal government adopted many novel regulatory and service functions in order to grapple more effectively with the complex industrial economy of the United States. New functions for government were associated with fresh institutional mechanisms. Several of the nascent agencies required chemical workers in order to meet their statutory obligations. Research chemists were employed by the Bureau of Standards (1901) to provide reference samples of grades of steel and iron ores, to test varieties of structural materials, to establish calorimetric and electrochemical standards, and to promote the use of accurate methods of analysis. Passage of the Pure Food and Drug Act in 1906 assured a greatly

[18] III, Beardsley, 1964, 55–60, provides a convenient sketch of the opportunities for chemists in the federal government during the late nineteenth century. The problems faced by the first generation of government chemists as they tried to strengthen a precarious foothold in the federal bureaucracy are suggested by the careers of Charles Wetherill, Henri Erni, and Thomas Antisell (see III, E. F. Smith, 1929; III, Miles and Kuslan, 1969; and III, Browne, 1938a). Antisell, who succeeded Wetherill and Erni as the chemist at the Department of Agriculture, had been employed by the Patent Office in the 1850s as a chemical patent examiner. In a letter to Wetherill, he described the job as "a place not very suitable to me but which has kept the wolf from the door until in Micawber [*sic*] phrase something better turns up" (Antisell to Wetherill, 15 November 1857, E. F. Smith Memorial Collection, University of Pennsylvania; quoted in III, Browne, 1938a, 215).

[19] The expansion of the federal chemical establishment during the 1880s is described in III, F. W. Clarke, 1885. On Wiley's Division of Chemistry in the Department of Agriculture see III, Wiley, 1899; III, Weber, 1928; III, Ball, 1938, I, 267–336; III, Dupree, 1957, 176–181; and III, Anderson, 1958, 32–119. The increasing amount of chemical work at the Geological Survey (discussed in III, F. W. Clarke, 1909) coincided with this agency's rise to hegemony in government scientific affairs in the late-nineteenth century (see III, Dupree, 1963, and III, Manning, 1967).

expanded role for chemical analysis and for Harvey Wiley's Bureau of Chemistry in enforcing the new legislation. Chemists were also active in the Bureau of Mines (1910), which embarked on an ambitious program of eliminating wasteful practices in the mining and petroleum industries.[20] On the eve of American entry into World War I, the number of civilian chemists in federal employ (displayed in Figure 5.2–1) had expanded to about one thousand.

Joining other professional groups in volunteering their services for the cause of national defense in World War I, some 4000 chemists were drafted during 1917 and 1918. Prior to the war, the U.S. military establishment had employed only a few chemists in ordnance investigations. Most of the chemists drafted and assigned to work on toxic gases and other war-related areas were demobilized soon after the armistice, leaving behind a considerably smaller Chemical Warfare Service (which attained permanent status in the postwar years). After this brief wartime episode, the Department of Agriculture maintained its central place in the organization of government chemistry throughout the 1920s and 1930s, employing from one-half to one-third of all civilian chemists in the federal government in this period.[21]

Faith in the progressive power of scientific research spurred the growth of the federal government's chemical establishment in the latter years of the Depression. Following the passage of the Bankhead-Jones Act of 1935 and the Agricultural Adjustment Act of 1938, the Department of Agriculture expanded its chemical (and other scientific) activities by establishing a national research center at Beltsville, Maryland, and a network of regional laboratories. But, as Figure 5.2–3 demonstrates, Agriculture soon lost the position of hegemony it had long enjoyed. The shift of the Food and Drug Administration from the Department of Agriculture to the Federal Security Agency in 1940 signalled a loss of some two hundred chemists. A more

[20] III, Bigelow, 1908, surveys the variety of chemical positions in the federal government at the turn of the century. On chemistry at the Bureau of Standards during this period, see III, Hillebrand, 1910; III, Weber, 1925, 43, 45; and III, Cochrane, 1966, 81–82. Harvey Wiley's role in the fight for pure food, and the regulatory activities of the Bureau of Chemistry after 1906, are treated in III, Weber, 1928, 59–84; and III, Anderson, 1958, Chapters 7–12. On chemical work at the Bureau of Mines, see III, Clement, 1911; and III, Powell, 1922, 16–29, 45–50.

[21] Figures on chemists mobilized during World War I are in III, Bogert *et al.*, 1919, 415. On the Chemical Warfare Service, see III, "Chemical Warfare Service", 1918; III, Jones, 1969; and III, Whittemore, 1975. Opportunities for chemists in the federal government during the 1920s are surveyed in IA, USCSC, 1926, and IA, USCSC, 1929. The concentration of chemists in the Department of Agriculture is noted in III, O'Rourke, 1940, 183. See also the informative III, Marshall, 1938–1941.

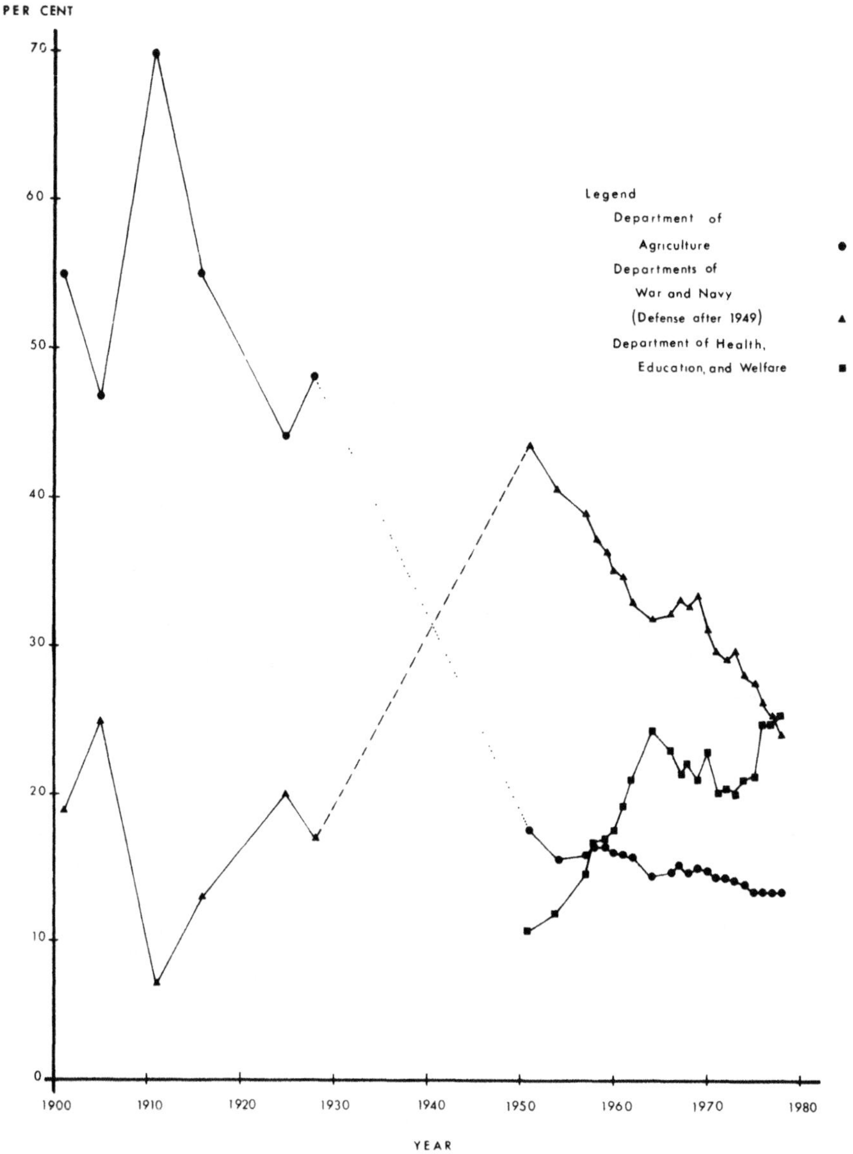

Fig. 5.2–3. Percentage of civilian chemists in the federal government employed by selected departments, 1901–1978. (See Table 5.14, columns 3, 5, and 7.)

fundamental transformation in institutional structure occurred as a result of the military's changed commitment to research during and after World War II. By the early 1950s, the new Department of Defense had become by far the largest employer of civilian chemists in the federal government, with more than twice as many as the Department of Agriculture.[22]

The Department of Defense has maintained its leading position as an employer of chemists throughout the post-World War II period, although its share of the total has declined from 43% to 28%. The Department of Health, Education, and Welfare emerged as a third major patron of government science during the 1950s and 1960s. The laboratories of the new department (which incorporated both the Food and Drug Administration and the National Institutes of Health) exemplified the postwar style of government-sponsored "big science". The number of chemists employed by HEW was correspondingly large. By 1958, the department had eclipsed Agriculture in the utilization of trained chemists. Other federal agencies (including the Veterans Administration; the Department of Energy and its predecessors, the Energy Research and Development Administration and the Atomic Energy Commission; and the Commerce and Interior Departments) have employed significant numbers of chemists over the past few decades, but since the early 1950s, the Departments of Defense, of Health, Education, and Welfare (now the two Departments of Education and of Health and Human Services), and of Agriculture have accounted for 60% of federal chemists.[23]

One other aspect of the growth in numbers of civilian chemists in federal

[22] On the development of regional agricultural research centers in the 1930s and on the reorientation of Department of Agriculture activities, see III, Harding, 1947, 37–39, 53–57; III, Jackson, 1970; and III, Pursell, 1968. See III, O'Rourke, 1940, 184, for the number of chemists involved in the transfer of the Food and Drug Administration.

[23] For an overview of postwar federal scientific activity, see III, Mansfield, 1968; III, NSB, 1976, 9–19; and III, NSB, 1978, esp. 323–366. Alvin Weinberg has remarked that the HEW research laboratories are an exemplar of "big science" (III, Weinberg, 1967, 123). On research support in the National Institutes of Health and the Department of Agriculture, see III, H. Rosenberg, 1960, and III, Shaw, 1960.

There are no long-run time series on the changing work activities of chemists in federal scientific agencies. Available evidence shows that roughly seven of ten chemists in the federal government were engaged in research and development between 1954 and 1962 (IA, NSF, 1957b, Table III–5, 75; IA, NSF, 1965, Table 4, 8). According to a later classification of work activities, only 58% of the chemists employed by the federal government were involved in research and development in 1967; testing and evaluation occupied 19.6% of these chemists, and data collection and analysis was the major occupation for another 6.9% (IA, NSF, 1969, Table B–2, 36).

agencies deserves comment. Not only have there been shifts in the importance of chemists to the mission of different agencies since World War II, but there have also been changes in the means used to meet research objectives. Instead of attempting to generate innovations entirely through their own efforts, government agencies began to rely on contract research in the mid-1940s. A 1946 survey of American Chemical Society members revealed that nearly half of those surveyed who had participated in wartime government chemical activities had conducted contract research, compared with about one-third who worked on in-house projects. Continuation of the practice of contracting for a significant proportion of federal research and development needs was a factor leading to the postwar decline of the rate of increase in the number of government chemists illustrated in Figure 5.2–2. This trend was accentuated by the slowdown in support for "in-house" science during the late 1960s and early 1970s. As a result, the number of civilian chemists employed by the federal government in 1975 was still 500 less than the peak of about 8500 persons reached in 1968.[24]

5.2.3. State and Local Government

The network of regulatory and administrative agencies that has evolved in the United States involves some duplication of federal functions at the state level. Likewise, municipal departments, which provide for the daily maintenance of public services, often have responsibilities similar to those of state agencies.[25] Such myriad state and local agencies provide even more potential opportunities for chemical employment than the federal government. But investigation of the number of chemists in these sectors of government is hampered by the

[24] On the turn to contract research and the organization of the government's wartime scientific research in general, see III, U.S. Congress, 1945; III, Stewart, 1945; III, Kevles, 1978, 287–301; and III, Pursell, 1979. The ACS survey is reported in IB, "Utilization", 1947. By 1973, some 118 000 scientists and engineers were involved in industrial research and development using federal funds, compared to 48 000 engaged in federal intramural research and development (IA, NSB, 1975, 126, 212, Table 4–4).

Changes in classification schemes for federal workers exacerbate the problem of comparing the number of government chemists with other occupational groups in federal service over time.

[25] On the scientific activities sponsored by state governments, see III, Cleaveland, 1959. On the coordination of federal, state, and local work in agricultural research, see III, Ball, 1938.

paucity of long-run quantitative data and the lack of any systematic information prior to 1950.

Occupational niches for chemists in state government were established earliest in the geological surveys begun in the 1830s to conduct censuses of the mineral resources of individual states. Other opportunities for chemists in public service at the state level included analysis of ores for assay offices, testing of illuminating gas from coal for gas commissions, and analysis of fertilizers, soils, and other products for state boards of agriculture. The range of positions available to chemists expanded considerably in the 1860s and 1870s, with the establishment of state boards of health, and agricultural experiment stations. Boards of health were a response to the social ills of urbanization, and public health chemists used their analytic skills to help ensure the provision of pure food, potable water, and adequate sewage disposal for a rapidly growing population. Agricultural experiment stations were to bring scientific expertise to bear in augmenting farm productivity, thus providing a food supply essential to the urban populace. Chemists in the experiment stations performed analyses of fertilizers, soils, mineral waters, and a broad spectrum of products for farmers, as well as conducting research into problems of crop nutrition and similar questions. In 1889, two years after passage of the Hatch Act, more than 100 chemists worked in agricultural experiment stations; sixteen years later, this number had increased to 166.[26]

Agriculture and public health were the primary concerns of state chemists in the late nineteenth and early twentieth centuries. A similar situation prevailed at the local level. As one participant aptly commented, municipal chemical work was divided into the two categories of "protecting the city's purse, and . . . protecting the city's health". The municipal chemist inspected foodstuffs and milk for adulteration, maintained the quality of public water supplies, and supervised sewage disposal processes – functions essential to

[26] III, Beardsley, 1964, 49–54, surveys opportunities for chemists in state government work during the nineteenth century. On state geological surveys, see III, Hayes, 1911; III, Hendrickson, 1961; and III, Nash, 1963. Charles Chandler's activities as a consultant and member of the New York State Board of Health suggest the possibilities for chemists in this context (III, Larson, 1950, 143–190); see also III, Mason, 1911. On the scientist's role in the agricultural experiment station, see III, C. Rosenberg, 1972, 1977; and III, Rossiter, 1979. III, True, 1937, and III, Knoblauch *et al.*, 1962, are comprehensive guides to the experiment station movement, while III, Rossiter, 1975, 160–176, gives a detailed portrait of S. W. Johnson's work at the Connecticut Agricultural Experiment Station. The figures on chemists in agricultural experiment stations come from III, True, 1937, 137.

preserving the urban order. The municipal chemist also tested the materials supplied to the city for use in the construction and maintenance of public works, buildings, and thoroughfares, in order to insure against the purchase of inferior commodities. Fuels, soaps, disinfectants, paper, iron and steel, cements, asphalt, and paints were only a few of the panoply of products which required analysis.[27]

The increasing importance of the chemist's role in public service did not pass unnoticed. It prompted commentary on both the number of chemists involved and their salary levels relative to other state and municipal civil servants. Two 1916 estimates of the number of state and local government chemists survive. Harvey Wiley estimated 650 chemists in state employ, nationwide. Adding an equal number for local government chemists (which he acknowledged was probably an underestimate) yielded a total of 1300 chemists in "non-federal" government positions. The second estimate was given in a comprehensive survey of the "status and compensation of the chemist in public service", in which Frederick Breithut calculated that there were about 2100 chemists at the state and local levels.[28]

Time series prepared by the U.S. Bureau of Labor Statistics on the number of state and local government chemists are available beginning in 1950 (see Figure 5.2–4). The BLS figure of twelve hundred chemists in such positions in 1950 may mean that the chemical services on the state and local levels did not grow between World Wars I and II. During the 1950s and 1960s, however, the number of chemists at the state level grew at the exponential rate of 4.6% per year, while the number at the local level grew at the linear rate of about 60 new jobs per year. Comparison with the data presented earlier shows that during this period, the number of chemists employed by state and local governments combined was one-third to one-half the number employed by the federal government.

In the 1960s employment still followed the pattern established in the nineteenth century: 70% of chemists in state government agencies had responsibilities for agriculture or public health areas.[29] But the relative

[27] III, Klein, 1917, 79. For surveys of the activities of municipal chemists, see III, Mahr, 1914; IB, Breithut, 1917, 74–79 (on chemists employed by New York City); and IB, Wiley, 1917, 81–83. III, Baskerville (ed.), 1911, is indispensable.

[28] IB, Wiley, 1917, 83; IB, Breithut, 1917, 78. Breithut estimated a total of 2800 chemists employed at the federal, state, and local levels. Subtracting the 716 chemists employed by the federal government (*ibid.*, 66) gives 2100 in state and local government.

[29] See III, Cleaveland, 1959, 49–94; IA, NSF, 1961a; IA, BLS, 1964; and IA, Andrews and Moylan, 1969.

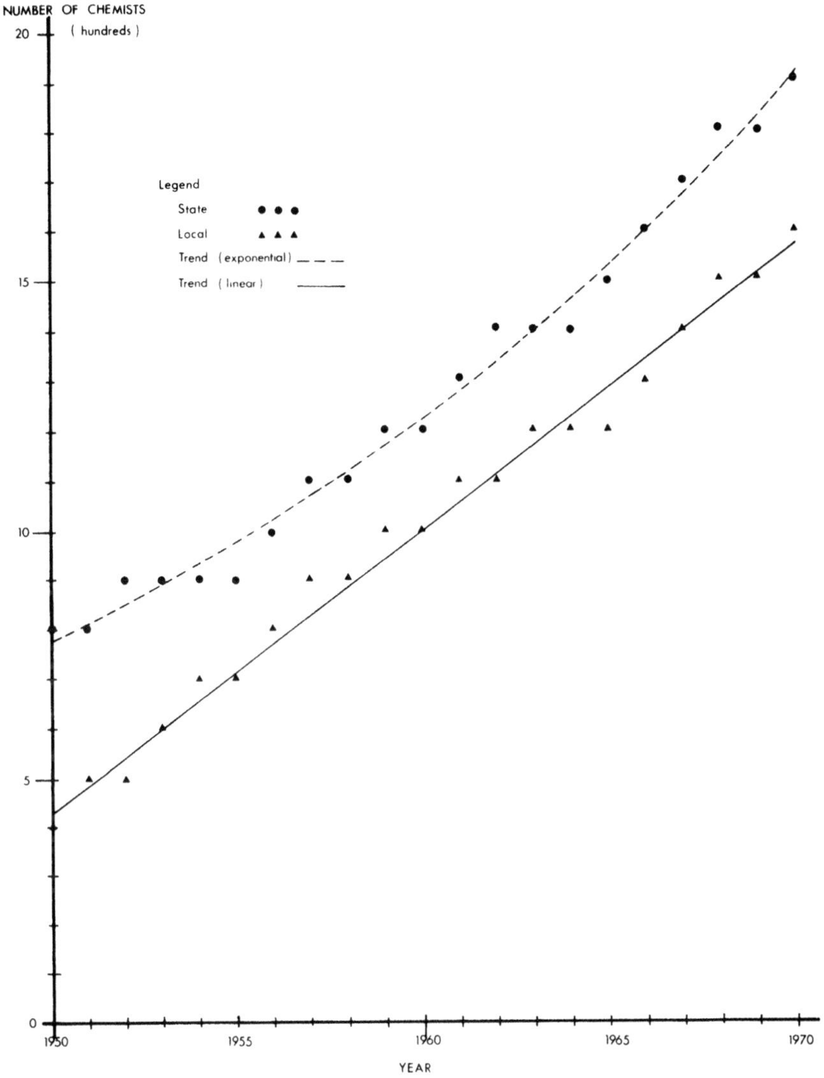

Fig. 5.2–4. The number of chemists working for state and local governments, 1950–1970. (See Table 5.15.)

importance of state and local governments in the employment of American chemists has changed over the past two generations. In 1910, state and municipal chemists were about 8% of the total number of Census chemists. By contrast, the 3500 chemists employed by state and local governments in 1970 represented only 3% of all American chemists.

5.3. Academe

INDICATOR HIGHLIGHTS:

In 1870, American colleges and universities employed about one hundred chemistry faculty. By 1970, that total was in the neighborhood of 10 000. Exponential growth at a rate of 4 to 6% per year has been the enduring trend.

Since at least 1870, chemists have constituted 1 to 2% of all college-level faculty in America. However the percentage appears subject to a slow erosion.

Since at least World War I, chemistry faculty have comprised between 5 and 10% of all professional chemists – an indication of the strong link between profession and formal training.

It is hard to obtain reliable historical information on the number of chemists in academe. A massive endeavor to count academic chemists one institution at a time would be needed to improve our tentative conclusions.[30]

Those concerned with counting *academics* have not seen fit to collect statistics upon academic *chemists*. For example, the Bureau of the Census persisted until 1970 in lumping academic chemists in the heterogeneous group known as "college presidents, professors, and instructors". Only recently has it disaggregated this category by discipline for the 1970 returns and retroactively developed comparative data for the 1960 census. The

30 The National Center for Education Statistics and its predecessor organizations have rarely published disciplinary affiliations for college and university faculty; a classic example is IA, Dunham *et al.*, 1966. Their published series on faculty are normally disaggregated only by major disciplinary groups (e.g., physical sciences), by institutional variables (e.g., public or private support), or by geographic region; for a representative list of such tabulations, see the sources for Series E–5 in IB, Ferriss, 1969, 334–335. The traditional practices of the Census Bureau in this area are explained in its series of *Alphabetical and Classified Indexes of Occupations* (see above, p. 10, note 2). Their recent change of policy is detailed in II, USBC, 1972, esp. Table 1, 21.

situation is complicated by differences in terminology among enumerators. Labels like "faculty" and "teachers" can mean tenured professors only; tenured and non-tenured professors; all professors plus instructors; professors, instructors, postdoctoral fellows, and teaching assistants; or professors, instructors, assistants, laboratory curators, and even chemical librarians. Treatment of adjunct and emeritus professors is not standardized. Sometimes only research faculty are counted, and other times only teaching faculty. And since the techniques of the early surveys are often unknown, attempts to rectify any discrepancies via retroactive readjustments are usually precluded.

Especially in the earlier years studied, a further problem involves the credentials of chemistry faculty at small, obscure institutions. The best-known nineteenth-century survey of academic chemists in America – Frank W. Clarke's *Report on the Teaching of Chemistry and Physics in the United States* (1881) – justifiably counts such notables as Ira Remsen, Josiah Parsons Cooke, and Charles F. Chandler among the nation's chemistry faculty. It includes in the same totals such dubious figures as the two unnamed teachers of chemistry at Simpson Centenary College in Indianola, Iowa. (The chemistry curriculum at Simpson is described – *in toto* – as one-third of a year of general chemistry for juniors, with experiments by the teacher, but no regular laboratory work for pupils. In addition, the survey reports that the teachers were required to teach physics and other subjects besides chemistry.)[31]

Despite problems of this kind, some usable estimates do exist of the number of academic chemists before World War II. The best come from American Chemical Society surveys in 1876 and 1901, and from the censuses of graduate research in chemistry which were conducted annually between 1924 and 1935 by the National Research Council. Like Clarke's study, but unlike most other early surveys, these enumerations are accompanied by *some* intimations of survey techniques and breadth of coverage. More important, these studies seem to count faculty at universities and at four-year colleges only, eliminating most troublesome small institutions like Simpson Centenary and making the data comparable with recent tabulations. When tied to the more rigorous data collected since the Second World War, the totals allow identification of the approximate trend in employment of academic chemists at the professorial level at four-year institutions. That trend is plotted in Figure 5.3–1; a comparable trend for faculty in

[31] IB, Clarke, 1881; for Simpson, see *ibid.*, Table II, 570–571.

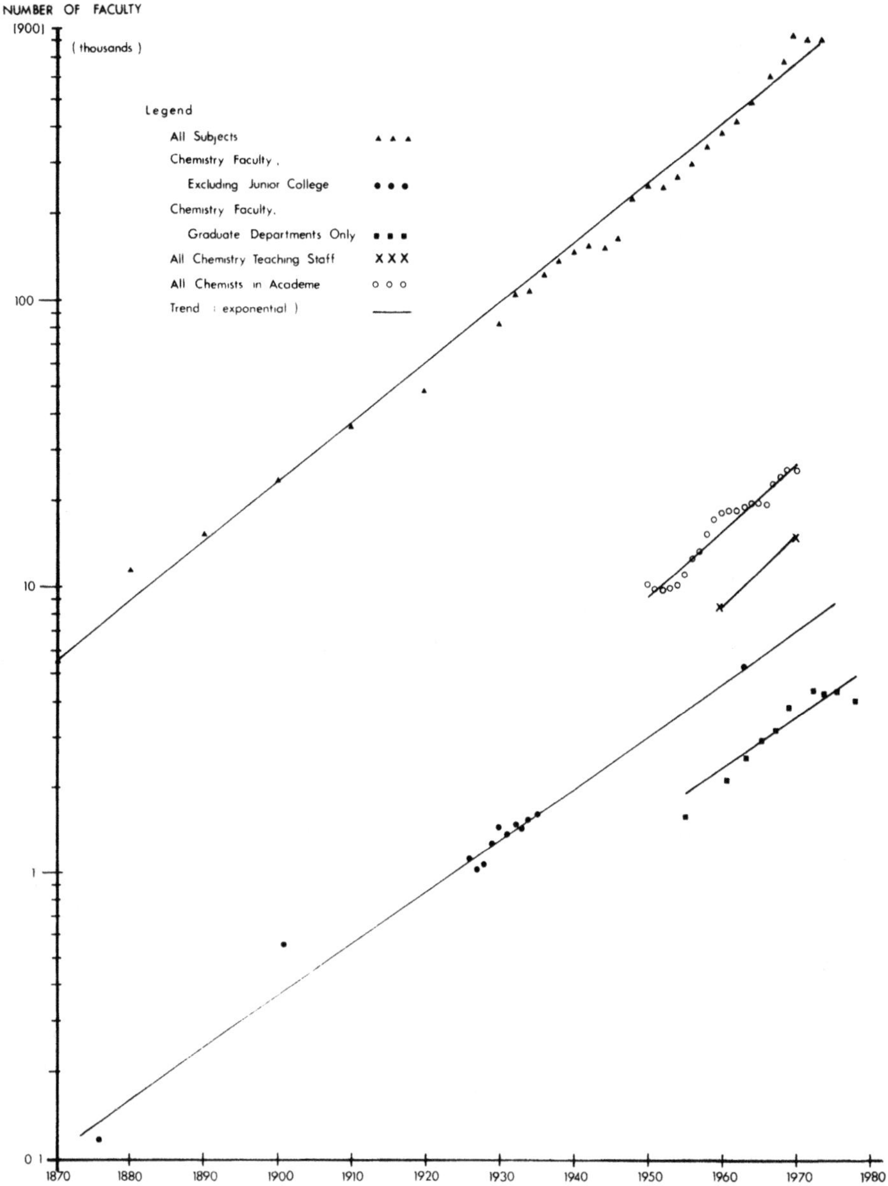

Fig. 5.3–1. Faculty in higher education: all subjects and chemistry, 1870–1978. (See Table 5.16.)

all subjects and some additional data on chemists since World War II are also included.[32]

The disparities among the additional estimates highlight the softness of the data and the continuing ambiguities of terms like "faculty" and "college teacher". The disparities also provide an envelope for the employment of chemists in academe. From an estimated hundred or so in the 1870s, academic chemists have grown steadily until they now number in the neighborhood of 10 000. Placement of the upper endpoint for the series is difficult because of the different types of academic posts for chemists in existence today. Selection of a compromise estimate yields a century-long growth rate of 4.3% per year. This rate of increase exceeds the 3.3% per year rate of growth in the number of chemistry bachelors degrees conferred annually, but it falls short of the 5.3% rate of growth for chemistry doctorate conferrals. In all probability, there are today more chemistry faculty members per undergraduate student majoring in chemistry – and fewer per graduate student – than a hundred years ago.[33]

The exponential increase in the absolute number of academic chemists is not so impressive as it first appears. As shown in Figure 5.3–2, chemistry faculty have constituted between 0.9 and 2% of all professorial personnel at American colleges and universities during the past century. The data show a decline over the century, but given ambiguities in the estimates, it is not clear what to make of this trend. The employment of chemists in American higher education certainly shows no evidence of the precipitous relative decline that surfaced when degree conferrals were examined.[34]

32 For the ACS surveys and the NRC censuses, see the sources for Table 5.16. The first and second NRC censuses did not generate totals for faculty; see IB, Zanetti, 1924, and IB, Norris, 1925.

33 The growth rate for faculty derives from the assumption of exponential growth over the century, and the use of the 1876 and calculated 1970 figures for chemistry faculty in Table 5.16. The growth rates for degree conferrals, based on the 1890 and 1972 entries in Table 3.1 and 3.3, also assume exponential growth. Using one of the other estimates in Table 5.16 for the recent number of chemistry faculty might necessitate a modification of this claim. By comparing the calculated number of chemistry faculty given in Table 5.16 with the number of chemistry bachelors and doctorate degrees in Tables 3.1 and 3.3, one finds that the number of teachers per 100 baccalaureates conferred grew from about 40 in 1890 to 60 in 1970. The faculty-to-doctorate ratio declined during this period from about nine to about three.

34 Adding chemical engineering faculty hardly eliminates the decline. The Bureau of the Census estimates that there were 10 527 college and university teachers of engineering in the United States in 1960, and 16 537 in 1970 (IA, USBC–COP, 1973, Table 221, 718).

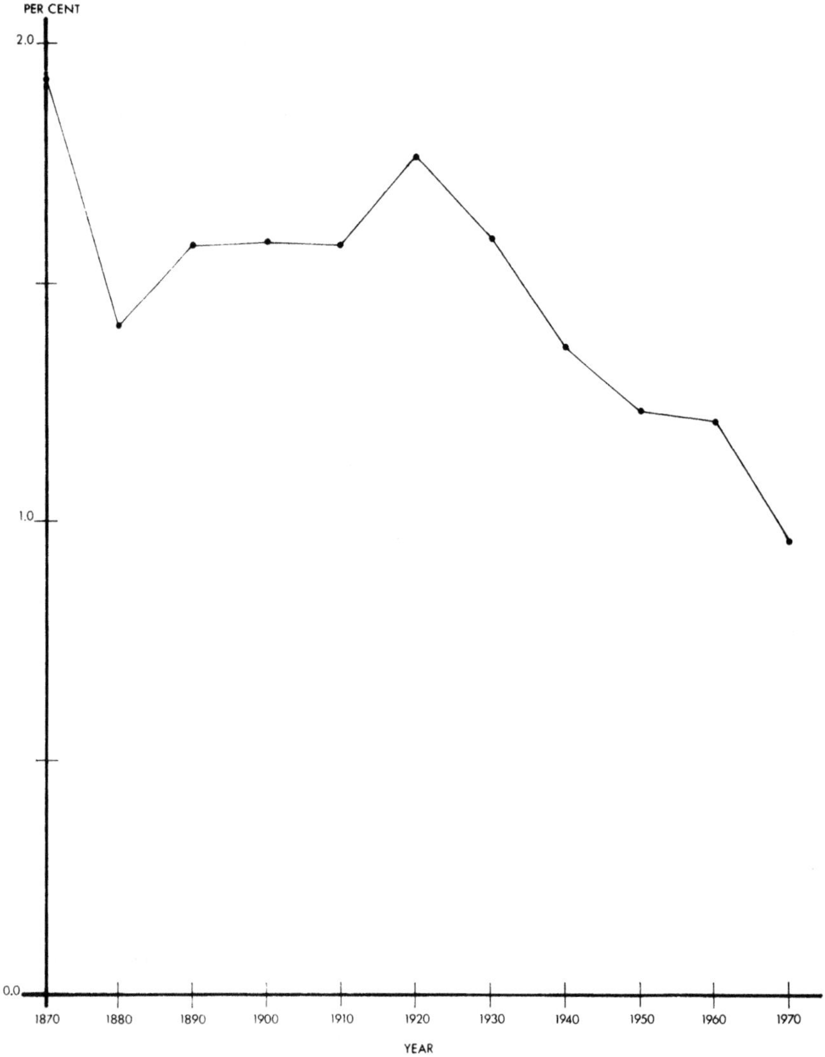

Fig. 5.3–2. Chemistry faculty as percentage of total faculty, 1870–1970. (See Table 5.17, column 3.)

Comparisons on the level of chemistry as profession attest to the staying power of the academic side of chemical employment. Following the uncertain early period of the American Chemical Society, the number of chemistry faculty in America, expressed as a fraction of the number of members of the ACS, has remained roughly constant (see Figure 5.3–3). Since World War I faculty have equalled between 5 and 10% of ACS membership. Not all chemistry faculty have been ACS members, but the relationship between the size of the profession and the number of those charged with the training of new entrants to the profession has been stable.

5.4. Other Contexts

It is conceivable that chemists have been employed in large numbers in private service sectors and in other places outside the obvious contexts of industry, government, and academe, but a look at the data for 1970 suggests otherwise. The Bureau of the Census counted about 110 000 chemists in 1970, a total which covers industry and government as well as any residual locales but does *not* include high school teachers of chemistry or chemists employed in academe; nor does it include those in *chemistry-related* fields such as biochemistry or chemical engineering. Consequently, if the 1970 Bureau of Labor Statistics estimates for industrial and government employment of chemists (presented in the preceding sections) are subtracted from the Census total, the remainder should approximate the number of chemists employed *outside* the three contexts of industry, government, and academe. That remainder equals about 4100 chemists, or roughly 4% of the Census total (see Table 5.19). Over 96% of non-academic chemists employed in 1970 (as estimated by the Bureau of the Census) can be found in private industry and in the public sector alone. In addition, about seventeen hundred of the "missing" 4% were employed in nonprofit institutions and around

---

At least in 1963, and presumably also in more recent years, some 7.1% of these individuals were chemical engineers (see IA, Dunham *et al.*, 1966, 15, Table 8, which lists 669 chemical engineers among 9455 engineering faculty surveyed). Multiplying the Census totals by this percentage yields 726 chemical engineering faculty in 1960 and 1170 in 1970. Combining these persons with the chemistry faculty figures for 1960 and 1970 in Table 5.17, column 2, then converting these totals to percentages of all faculty, yields 1.41% for 1960 and 1.12% for 1970. This is a decline of 20.6% during the sixties, compared with the 20.7% decline represented by the last two points in Figure 5.3–2.

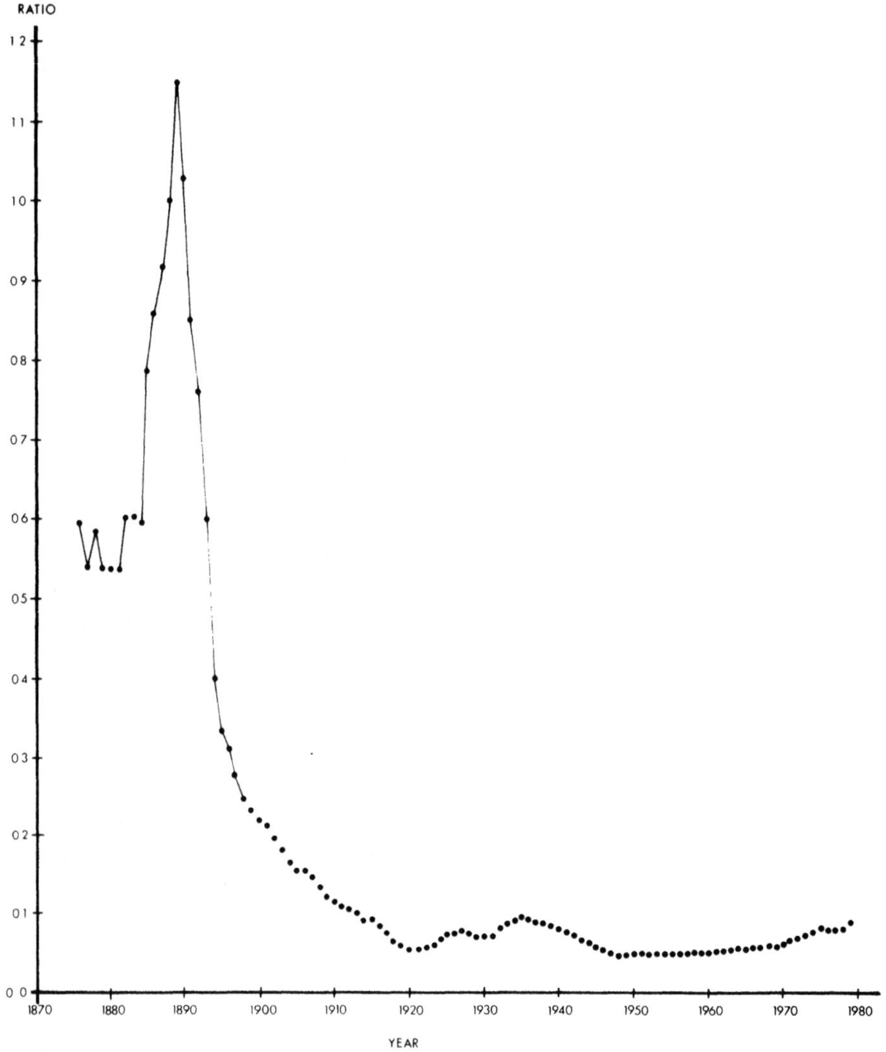

Fig. 5.3–3. Ratio of chemistry faculty to ACS membership, 1876–1979. (See Table 5.18, column 3.)

four hundred worked for the military, leaving about two thousand chemists unaccounted for. Given the errors in the data, it would be folly to speculate further on these "missing" souls.[35]

The farther back in time one looks, however, the more difficult it becomes to reconcile available statistics. The best available tabulations of the industrial and government employment of chemists in 1920 account for only about 60% of the Census estimate of 28 000 chemists employed in that year. A likely explanation is that the data for industrial and government chemists in 1920 are defective. The industry estimate was derived from a military preparedness survey conducted by the U.S. Bureau of Mines to locate Americans with metallurgical skills; it counted a total of only 15 000 chemists *in all sectors* in 1917, an estimate that compares poorly with the Census figure of 28 000 chemists for 1920. The 1920 tabulations of chemists in the federal government and in state and local governments are also approximate and may need revision upward. Wherever the fault lies, it seems sensible to conclude that industry, government, and academe have accounted for nearly all chemical employment during the last one hundred years.[36]

It does not follow that almost all persons with formal exposure to chemistry may be found in industry, government, and academe. As the discussion of "decoupling" in Section 3.1 made clear, many more people have been *trained* in chemistry than have been employed in it. A great many of these chemically literate non-chemists (for example, physicians in private practice) work outside the industrial, academic, and public sectors.[37] There is great diversity in the occupational niches filled by chemists. There is even greater diversity in the people exposed to chemistry in a non-vocational way, and in the degrees and types of exposure they have had.

[35] It is *not* necessarily that there are two thousand chemists in exotic jobs. Rather, the available surveys have slightly different totals because of systematic and random errors. On chemists in the military, see note 15 on page 126, above. The BLS estimate for chemists in nonprofit institutions is form IA, BLS, 1973, Table A–16, 51. The other sources of data are listed in Table 5.19.

[36] For the sources of the early estimates of chemists in industry, federal government, and state and local governments, see Tables 5.1, 5.13, and 5.15. The tabulations of the Bureau of Mines survey appear in IA, Fay, 1917. The 1920 Census figure is from Table 2.2, column 1.

[37] The path from chemical education to medical practice has become so well-trodden that students are now declaring undergraduate majors in chemistry with the explicit intention of becoming physicians. This tendency has disrupted traditional chemical education in subtle ways, and educators have voiced concern over it; see III, B. Smith and Karlesky, 1977, I, 94–96.

The common denominator of chemical employment continues to be the subject matter of chemistry – the "theoretic parts" of the science. Because those "theoretic parts" serve as a core for the enterprise, and also because many observers identify chemistry solely in terms of that subject matter, it is necessary to consider more carefully the articulation and advancement of those theoretic parts – that is, to take a look at chemistry as discipline – while keeping in mind the cultural context that has been adumbrated above.

CHAPTER 6

CHEMISTRY AS DISCIPLINE

The members of the chemical discipline are not only trained and skilled in the theoretic parts of chemistry, but also committed to the articulation and advancement of those theoretic parts considered as an autonomous intellectual system. Depending on whether stress is placed on "articulation" or "advancement", the boundaries of membership will be more or less tightly drawn. Either way, those boundaries coincide roughly with the boundaries of academe. The academic community of chemists possesses no monopoly on the articulation and advancement of the concepts of chemistry. It does, however, represent the central core – and probably the overwhelming mass – of those engaged in the activities of the chemical discipline.

6.1. The Chemical Discipline and the Research University

INDICATOR HIGHLIGHTS:

The chemical discipline was of central importance to the emergence of the American research university. During the nineteenth century, American universities awarded more doctorates in chemistry than in any other single field.

This early importance and visibility of chemists was translated into the promotion of chemists to university leadership at the turn of the century, a movement which reached a peak soon after 1910.

In subsequent decades, a trend of absolute growth and relative decline has marked chemistry's place in the internal economy of universities.

The emergence of the chemical discipline in America was coincident with the development of an academic community of chemists. As at earlier periods in both Germany and Britain, academic chemists were central actors in the movement to redefine the purposes of the university in terms of research. That redefinition implied the creation of clearer boundaries between disciplines, and a firmer sense of identity within them. The strategic role of the chemists may be seen in the fact that, in the closing decades of the nineteenth century, American universities awarded more doctorate degrees in chemistry than in any other discipline. The 251 PhDs awarded in chemistry between 1863 and 1900 accounted for 11% of all doctorate degrees. Only the broad, diffuse category of "languages and literature, general" took more (17.2%). The chemistry degrees provided over a quarter of the total for all natural sciences. As Figure 6.1–1 displays, chemistry was a dominant motif within the new definition of the American university as a research university. In its first decades, this new definition of the American university was narrowly based. Five institutions were responsible for over four-fifths of the chemistry PhDs between 1863 and 1900 (Figure 6.1–2). Only a dozen schools conferred more than one chemistry doctorate before the turn of the century, and only 19 were in the business at all. In subsequent years the base has widened steadily: by 1976, 62 schools had awarded at least 10 chemistry PhDs in the previous year.[1]

An additional indicator of the important place of chemistry among the academic disciplines in the late nineteenth century may be obtained by focusing on the Johns Hopkins University alone. Johns Hopkins was the acknowledged pioneer of the forms of the research university. Chemistry was correspondingly important to its internal economy. The subject accounted for 19.5% of all doctorate degrees awarded at Johns Hopkins over the years

[1] III, Veysey, 1965, is the standard account of the emergence of the American university; see also III, Rudolph, 1962, 240–286; III, Thackray, 1977; III, Guralnick, 1979; and III, Shils, 1979. On the research ideal in German universities and its adaptation to the American context, see III, Metzger, 1961, 93–109, and III, Ben-David, 1971, 108–168; for the case of physical chemistry, see III, Dolby, 1977; and III, Servos, 1979. III, H. Hale, 1932, deals with American chemical education from 1870 until World War I. For a comparative institutional analysis of chemistry, mathematics, and physics in the United States during this period, see III, Kevles, 1979. On chemical engineering education, see III, White, 1928; III, Newman, 1938; and III, Hougen, 1977.

There were European precedents for the role that chemists played in the reorientation of the American university. Liebig's laboratory at Giessen was the first in a tradition of institutes which transformed German scientific education (see III, Morrell, 1972; III, Borscheid, 1976, esp. 33–71; and III, Riese, 1977, 134–145, 216–224). Henry Roscoe's chemistry department at Owens College (later the University of Manchester) was modeled on those of Liebig and Bunsen. On Roscoe's role in the reorganization of scientific and technical education at Owens, see III, Kargon, 1977, 167–220. Roscoe's own account of his tenure is III, Roscoe, 1887, and his influence is discussed by a former student in III, Thorpe, 1916, 34–52, 77–96.

By 1979, Wisconsin and California had changed places, and Michigan State had replaced M.I.T., completing the displacement of the Eastern schools. See IB, "Number of Grads", 1980, 58.

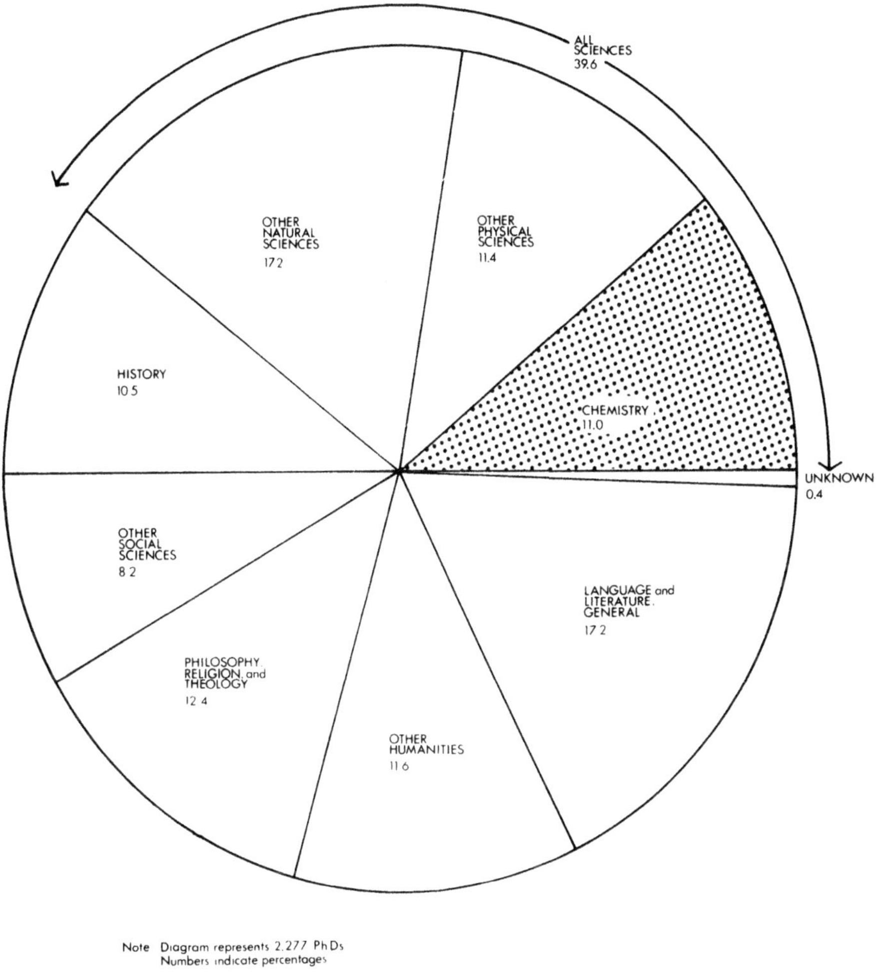

Fig. 6.1–1. Subjects of American doctoral dissertations before 1900. (See Table 6.1, column 6.)

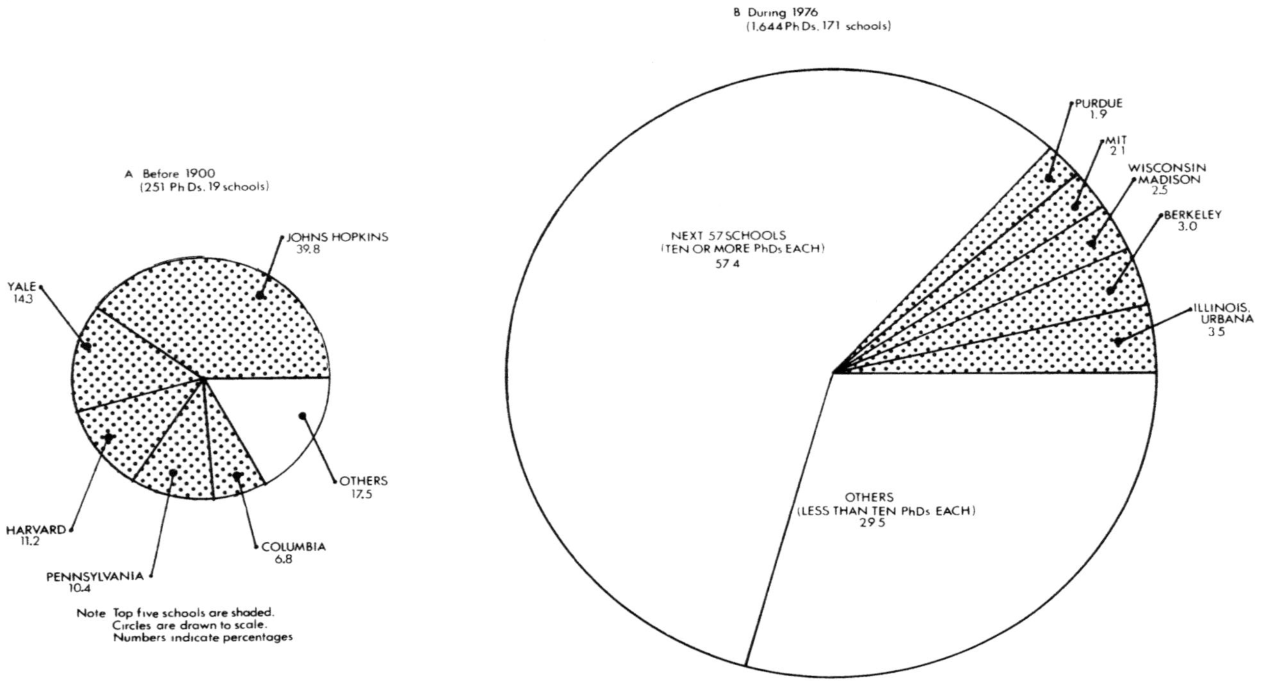

Fig. 6.1 – 2. Institutions granting doctorates in chemistry before 1900 and during 1976. (See Table 6.2, column 6; and Table 6.3.)

1878–1899. Physics, its nearest scientific rival, took a mere 7.4% (Figure 6.1–3).[2]

The early visibility and importance of the chemical discipline were reflected in the promotion of chemists to university leadership at the turn of the century. Examples include Thomas M. Drown (President, Lehigh, 1895–1904); James Mason Crafts (President, MIT, 1897–1900); Francis P. Venable (President, North Carolina, 1900–1914); Ira Remsen (President, Johns Hopkins, 1901–1913); and Edgar Fahs Smith (Provost, Pennsylvania, 1911–1920). The new currents in administration and their intellectual rationale were also studied closely by the period's outstanding president, the chemist Charles W. Eliot of Harvard.[3] The subsequent decay in the visibility of chemistry is reflected in the decline of such appointments. Despite the enormous increase in the number of colleges and universities, the record decade for the appointment of chemists to permanent presidencies was 1910–1919 (Figure 6.1–4).

The enduring trend of absolute growth and relative decline in degrees conferred in chemistry at all three levels – bachelors, masters, and doctorates – was examined in Section 3.1. Between the 1890s and 1970s baccalaureates decreased from one in ten to one in a hundred, doctorates from one in five to one in fifteen. Despite this decline, the representation of chemists among American academics has remained fairly constant over the years (see Figure 5.3–2). Stability is the term that best characterizes the emplyment of chemists in academe. Though small in comparison with industrial employment of chemists, the niche carved on American campuses is a secure one, with solid links to the profession with which it is associated. Its stability represents

[2] On the establishment and early years of Johns Hopkins, see III, French, 1946, and III, Hawkins, 1960. III, Hannaway, 1976, offers a novel interpretation of Ira Remsen's role in the development of graduate research training in chemistry at Johns Hopkins; see also III, Tarbell, Tarbell, and Joyce, 1980. For a participant's account which captures the atmosphere of the Hopkins chemistry department at the turn of the century, see III, Reid, 1972, 58–77, 99–104, 111–132.

[3] Drown's call to Lehigh is discussed in III, Bowen, 1924, 99–103. Crafts' brief tenure as MIT's president is treated in III, Prescott, 1954, 151–166. Venable presided at North Carolina during a crucial transition: see III, Battle, 1912, Vol. II, 590–763; III, Wagstaff, 1950, 77–104; and III, L. R. Wilson, 1957, 43–175. Remsen's administration at Johns Hopkins is covered in III, French, 1946, 142–152; and III, Getman, 1940, 77–86. Smith's term as provost at Pennsylvania is discussed in III, Cheyney, 1940, 360–388. (During Smith's tenure, his position was equivalent to that of university presidents elsewhere.) Charles Eliot's stellar career as president of Harvard is examined in III, Hawkins, 1972.

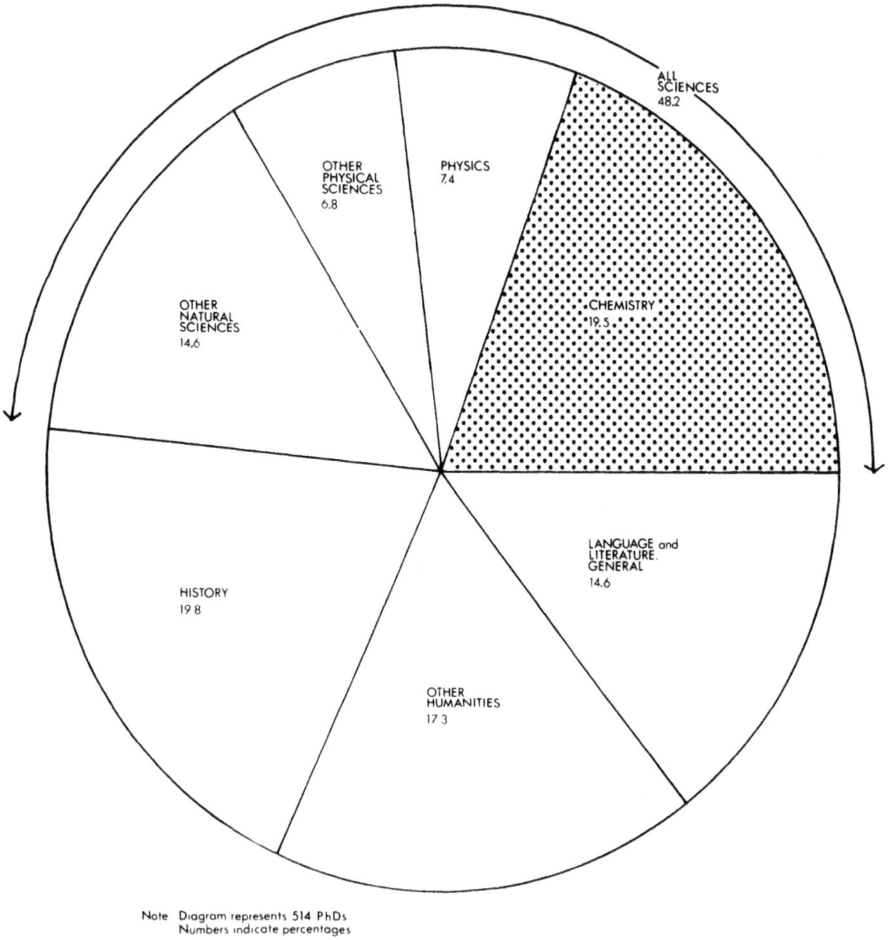

Fig. 6.1–3. Subjects of doctoral dissertations at Johns Hopkins University before 1900. (See Table 6.4, column 5.)

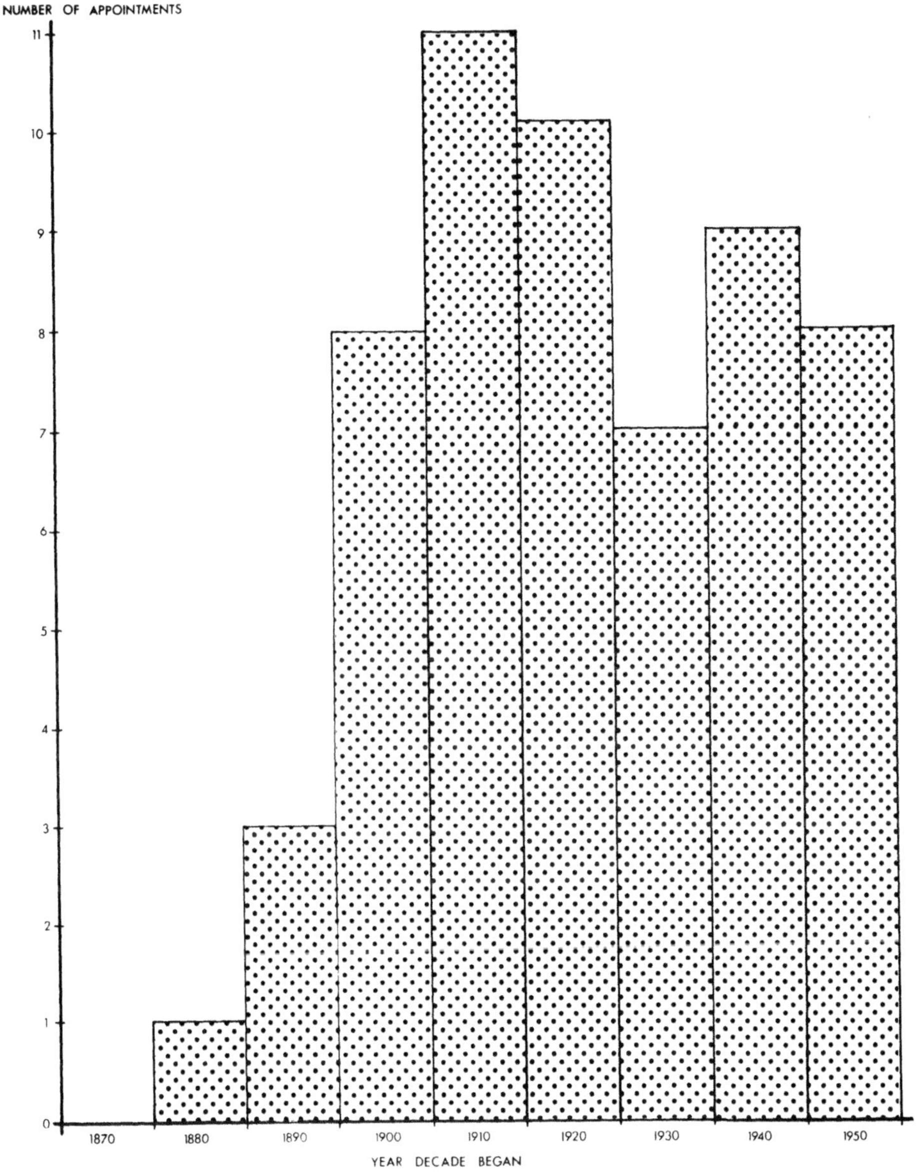

Fig. 6.1–4. Chemists appointed to permanent college presidencies, by decade of appointment, 1870–1959. (See Table 6.6.)

one way in which the chemical discipline has become "entrenched" in American academic life – a matter to which we shall return.

6.2. Papers, Prizes, and International Prestige

INDICATOR HIGHLIGHTS:

American contributions have become steadily more visible in the world output of chemical publications, attaining hegemony since World War I. Before World War II, increasing citation of American work reflected the growing volume of American chemical publication. Thereafter, continued growth in the degree of citation of American research was based instead on increased relative quality of the papers published.

As measured by Nobel Prize awards, little American chemical research before the 1920s carried universal acclaim. Since World War II, however, Americans have received a plurality of the chemistry prizes.

While the visibility of chemistry within American academe has been steadily eroded, its visibility on the world scene has equally steadily increased. The admittedly imperfect indicators of literature cited and Nobel Prizes won both point to this conclusion. By so doing they highlight how the absolute growth in size of American academic activity is more salient when it comes to international comparisons than any relative decline in the position of chemistry within the American university.

6.2.1. CITATIONS OF AMERICAN RESEARCH

The question of whether and how knowledge may be subject to measurement is one of notorious difficulty. We shall therefore content ourselves with the simpler, but not simple, question of measuring the influence of published research as reflected by its mention in the footnotes (citations) of other published work. In recent years applications of citation analysis to the study of science have achieved a high level of technical sophistication. Proponents of the technique contend that it promises to yield quantitative indicators of precise cognitive and social changes in scientific specialities.[4]

[4] A recent analysis of collagen research by Henry Small is a case in point (II, Small, 1977). The use of co-citation analysis as a tool for elucidating the dynamics of scientific

Here we adopt the more limited objective of using a rudimentary form of citation analysis to assess the shifting distribution of attention to American chemical research by German and British chemists. (Sampling procedures are discussed in Appendix C.)

Germany exercised a leading role in chemical research in the late nineteenth century. Accordingly, the importance attributed to American chemical research in *Chemische Berichte*, the central German chemical journal, is of interest. Since German chemists tend to cite German literature disproportionately, we have calculated the proportion of those citations in *Chemische Berichte* to non-German-language literature that were received by American books, journals, and patents for selected years over the past century (Figure 6.2–1).[5] As this figure shows, the relative attention paid to American work by authors in *Chemische Berichte* has increased greatly since the beginning of the twentieth century. This growing attention is a function of two quite different things: awareness of the ever-increasing quantity of American chemical papers, and appreciation of their quality. An attempt has been made to distinguish between these two effects in Figure 6.2–1, which includes a line (solid triangles) indicating the volume of American chemical publications as a proportion of the cumulative total of non-German publications over time.

While a research article draws upon the literature as an accumulated resource, the review journal offers a rather different perspective. Here the citations are only to the best of recent work; citation patterns in review journals therefore reflect informed judgments of the comparative importance

---

specialties is surveyed in II, Garfield, Malin, and Small, 1978; but cf. II, Sullivan, White, and Barboni, 1977. (See also Chapter 1, note 9.)

Another genre of quantitative study of scientific communication derives from the model adopted in Fritz Machlup's economic study of the "knowledge industry" in the United States (II, Machlup, 1962). In this genre, measurements of the volume, variety and patterns of use of scientific literature are related to the generation and dissemination of scientific and technical information, as in II, King *et al.*, 1976.

[5] The preference of German chemists for citing chemical research published in their own language was noted as early as 1919 by an ACS Committee on the Publication of Compendia of Chemical Literature. The Committee complained that "in their compendia and monographs, German chemists have been wont to ignore to a certain large extent, or to underrate, the work done in other countries, especially perhaps that done in the United States" (III, Stieglitz *et al.*, 1919, 415). This tendency has continued: in 1975, German chemists cited German work 4.1 times as often as one would expect from its relative proportion among world chemical publications. By contrast, Germans cited American work 1.2 times as often as warranted by the relative volume of American publication (IB, Narin and Carpenter, 1975, Tables 8 and 9, 90–91).

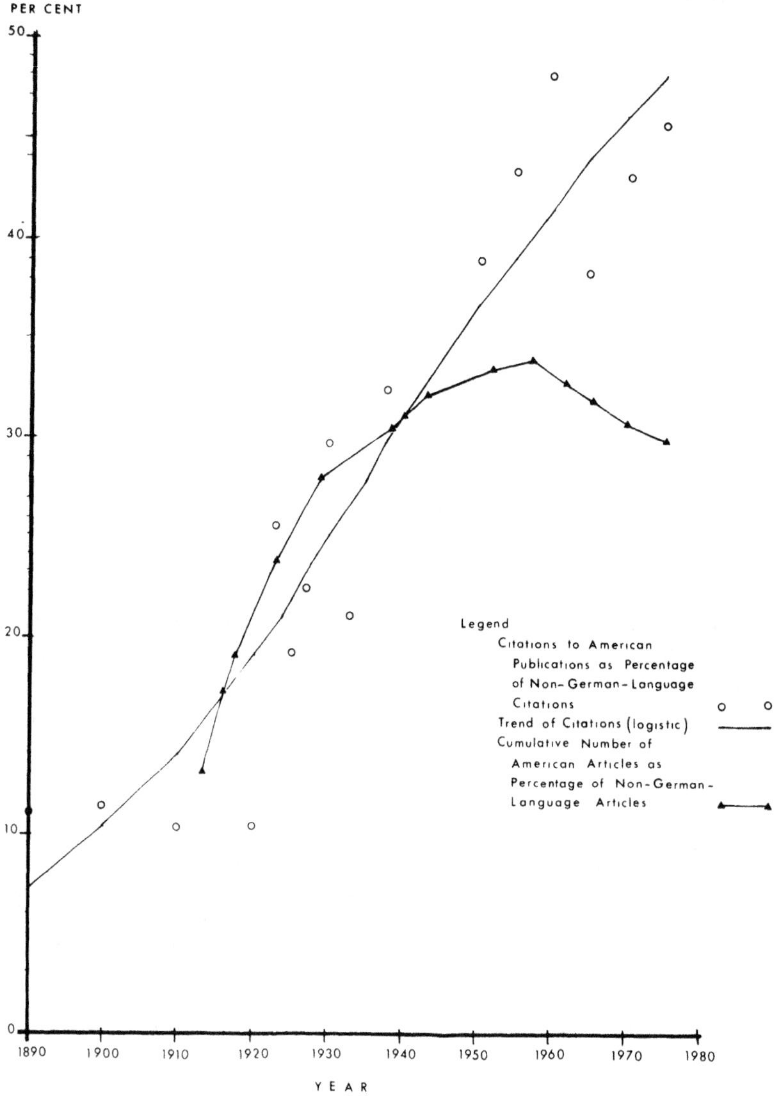

Fig. 6.2–1. Citations to American publications in *Chemische Berichte* and cumulative number of American articles as percentage of all non-German-language work, 1890–1975. (See Table 6.7, columns 5 and 6; and Table 6.8, column 7.)

of current contributions.[6] Figure 6.2–2 shows the changing importance of American chemical research as indicated by the references in *Annual Reports on the Progress of Chemistry*, a central British review journal. As with *Chemische Berichte*, the attention paid to American work by authors in *Annual Reports* has increased greatly since the early twentieth century. And once again, a line (solid triangles) in Figure 6.2–2 compares the American contribution to current literature with the proportion of references to American journals in the *Annual Reports*.

The two figures point to similar conclusions. First, from comparative obscurity before World War I, American chemistry rose steadily in esteem to a position of international dominance. Almost half the citations in the *Annual Reports* in 1975 were to American publications. Similarly, almost half the citations to non-German-language literature in *Chemische Berichte* in 1975 went to American work. It is striking that this hegemony is the culmination of a fifty-year trend of increasing presence, and not merely the result of post-World War II developments. Second, it is clear that the increasing attention received in the two decades before World War II reflected the growing *volume* of American chemistry, rather than a changed assessment of its worth. Since World War II, however, in both *Chemische Berichte* and the *Annual Reports*, American chemistry has been cited proportionately more than is warranted by increasing quantity alone. The prominence of American work within the international literature has been sustained by quality.[7]

The enduring trend of steady increase in the visibility of American chemistry is further highlighted by a comparison of citation patterns in *Annual Reports, Chemische Berichte*, and certain leading American chemical journals. Figure 6.2–3 shows linear regression (solid) lines for the proportion of total citations going to American work in the two journals we have studied (open dots and squares) and a set of leading American journals (solid dots).

[6] Even if authors cite at random, it is more likely that a particular reference will cite recent publications because scientific literature has grown exponentially. This "immediacy factor" is reinforced by the obsolescence of older literature with respect to the research front. These characteristics of scientific literature are discussed in II, Price, 1963, 1965, 1970, 1976.

[7] Narin and Carpenter found that the citation-to-publication ratio for American chemical research increased from 1.35 to 1.52 between 1965 and 1972 (IB, Narin and Carpenter, 1975, Table 6, 89). The increasing volume of American publications in organic and analytical chemistry in particular is discussed in III, Boig and Howerton, 1952a, 1952b; and III, Brooks and Smythe, 1975.

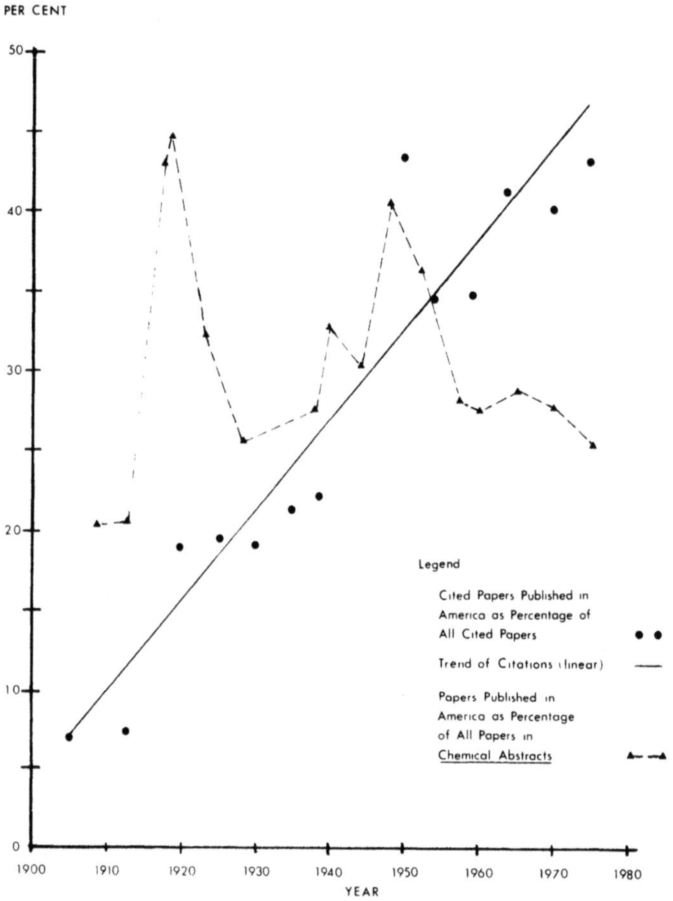

Fig. 6.2–2. Citations to American papers in *Annual Reports on the Progress of Chemistry*, 1905–1975. (See Table 6.9, columns 3 and 4; and Table 6.10, column 3.)

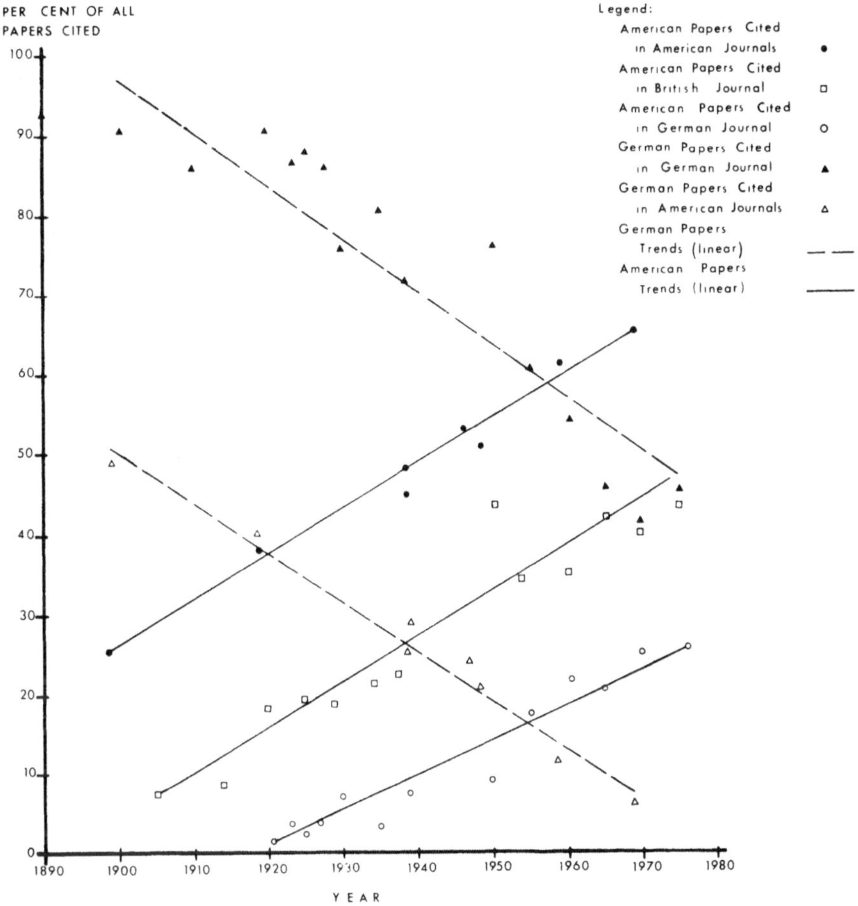

Fig. 6.2–3. Trends in citations to American and to German papers in selected American, British, and German journals, 1890–1975. (See Table 6.9, columns 3 and 4; Table 6.11; and Table 6.12.)

The proportion of citations going to German work in *Chemische Berichte* and in the American journals is also graphed (triangles, solid and open, plus dashed linear trend lines) to indicate the contrasting decline in the importance of German chemistry since 1890. Although the baseline level of interest in American work varies between countries, the parallelism of trends is striking. The share of American references seems to have increased by about 0.5% of all papers cited per year since World War I. The commanding position of American chemistry in the world chemical literature seen in this small sample of journals is corroborated by 1975 citation data from 2400 journals in the *Science Citation Index.*[8]

These indicators are compatible with logistic growth in the American contribution to the world chemical literature; with a leveling off of that growth in recent years; with the increased visibility of American chemical literature being largely a stock – and thus a "GNP and life-style" – phenomenon; and with the inference that the quality of American chemical research has increased, relative to that of other nations, in recent decades.

6.2.2. Nobel Prizes

A second, very different, indicator – but one that also points to the rising importance of American chemistry within the world scientific community – may be obtained by analyzing shifts in the nationality of recipients of the Nobel Prize in chemistry. The existence of the Nobel Prizes is itself testimony to the prestige and rewards of industrial chemistry, since the prizes were endowed from the fortune that Alfred Nobel had accumulated during a lucrative career in the explosives industry. From the inception of the prizes in 1901, selection of prizewinners has been subject to changing political, social, and contextual influences and constraints, but the symbolic significance of the awards is undeniable. Universities, industrial corporations, and even nations use the Nobel Prize "box-score" in competing for scientific prestige.

[8] Although the American share of world chemical publications in 1975 was only 22%, fully 65% of all citations in foreign chemical publications were to American work (IA, NSB, 1977, Tables 1–4 and 1–6, 11, 13).

The American chemical journals examined for selected years from 1899 to 1969 for the data graphed in Figure 6.2–3 included the following: *American Chemical Journal, Analytical Chemistry, Chemical Reviews, Journal of the American Chemical Society, Journal of Chemical Physics, Journal of Organic Chemistry, Journal of Physical Chemistry*, and *Organic Syntheses*. For further details, see Appendix C.

In one recent instance, the 1976 sweep of all Nobel Prizes by the United States was hailed by President Ford as convincing confirmation of the health of American science. Nobel Prize-winning is thus a widely-accepted – if impressionistic – indicator of the relative standing of nations with regard to recognized contributions to the advanced, esoteric knowledge of the chemical discipline.[9]

The 89 individuals who were awarded the Nobel Prize in chemistry between 1901 and 1976 were citizens of only 15 nations. Among those countries, Germany (with 24 prizes), Great Britain (20 prizes), and the United States (21 prizes) have been preeminent (Figure 6.2–4A).[10] The pattern of awards reveals several interesting features. Prior to 1930, Germany and Great Britain accounted for nearly three-fourths of the Nobel Prizes awarded in chemistry. Only *one* American was awarded the prize (Theodore William Richards in 1914), and this despite the fact that in two years (1919 and 1924) the Nobel committee declined to make any award for lack of candidates of suitable stature. Assuming that at least a few years must elapse between completion of the work and award of the Nobel Prize, it is apparent that little American chemical research in the late nineteenth and early twentieth centuries carried universal acclaim. Germany and, to a lesser extent, Great Britain were the centers of the world chemical community.

In contrast, two Americans were awarded the Nobel Prize in chemistry in the 1930s (Irving Langmuir in 1932; Harold C. Urey in 1934). Since World War II work done in the United States has been responsible for a plurality of the chemistry prizes. These results are congruent with the literature indicators discussed above, and show the increasing prominence of American chemistry since the late 1920s and 1930s. While American research has been amply rewarded by Nobel Prizes, its showing is less impressive when corrected for population differences among those countries which received awards (Figure 6.2–4B). Postwar perceptions of American hegemony in chemical

[9] For background on the establishment and administration of the Nobel Prizes, see IB, Nobel Foundation, 1972a. III, Zuckerman, 1977, is a thorough evaluation of the sociological significance of the Nobel Prize in the American scientific community. Recent work includes III, Bernhard *et al.* (eds.), 1982; and III, Küppers *et al.*, 1982.

[10] Sixteen Americans received the Nobel Prize for chemistry between 1901 and 1970, as listed in IB, Nobel Foundation, 1972a, 638–645. Adding the five prizes awarded to C. B. Anfinsen, Stanford Moore, and W. H. Stein in 1972, to P. J. Flory in 1974, and to W. N. Lipscomb in 1976, brings the total to twenty-one in 1976; see annual volumes of IB, Nobel Foundation, 1972b – . For further information on the Nobel Prize in chemistry, see Arne Westgren, "The Chemistry Prize", in IB, Nobel Foundation, 1972a, 281–385; and III, Nobel Foundation, 1964, 1966a, 1966b, and 1972.

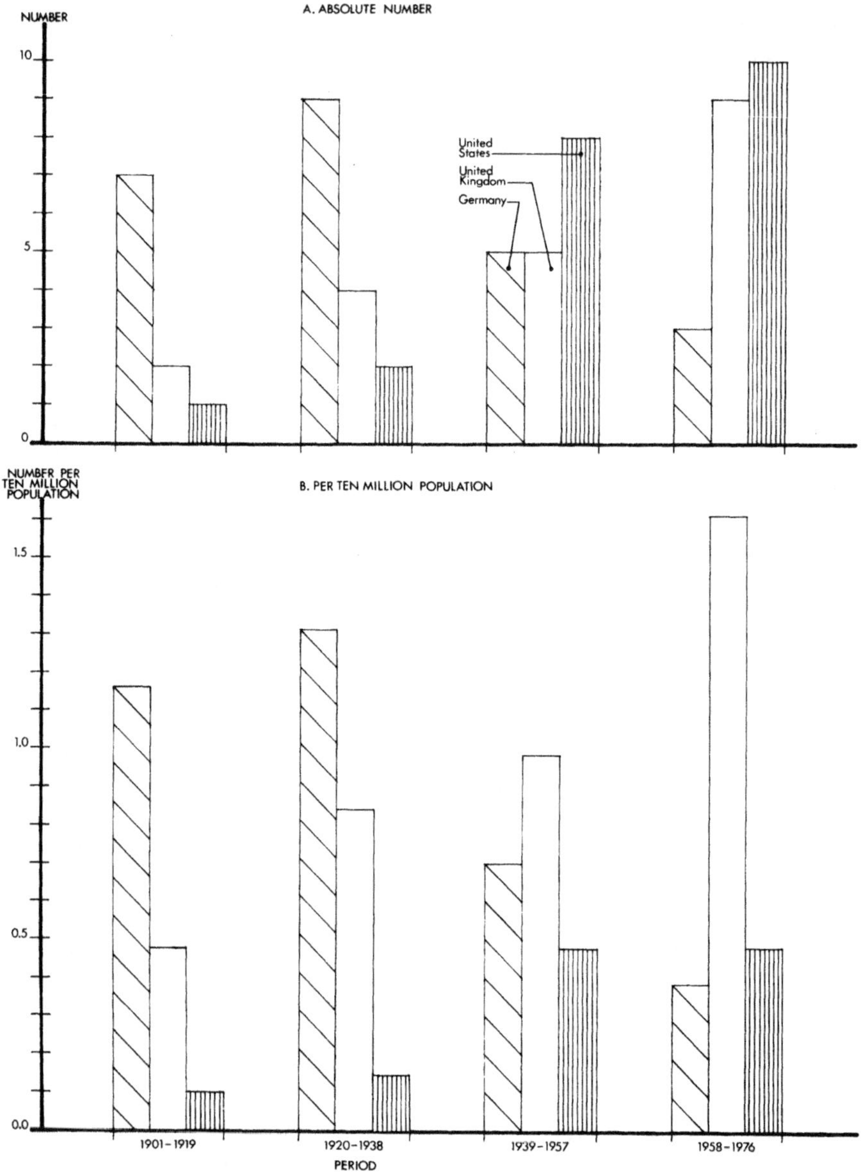

Fig. 6.2–4. Nobel Prizes awarded in chemistry, selected countries, 1901–1976. (See Table 6.13A and 6.13B, columns 2, 5, and 6.)

Table 6.13A and 6.13B, columns 2, 5, and 6.)

science have undoubtedly been affected by the large proportion of Americans among the "ultra-elite" of chemistry Nobel laureates.

6.3. The Entrenchment of Chemistry

INDICATOR HIGHLIGHTS:

Chemists have been appointed to deanships of American graduate schools more often than members of any other discipline. As the percentage of doctoral degrees in chemistry has fallen, the percentage of deanships going to chemists has risen. The number of chemist-deans appointed per thousand chemistry doctorates conferred has been growing for 50 yr and is now well above the average ratio for all fields.

Chemists have fared disproportionately well in elections to the presidency of the American Association for the Advancement of Science. The growth in size, wealth, and influence of the American Chemical Society over many decades also testifies to the entrenchment of the discipline.

Chemists have been strongly represented in awards of National Research Council fellowships and Guggenheim fellowships, and in election to the National Academy of Sciences, though in none of these cases does their showing adequately represent the size of their constituency.

Academic chemists receive substantial funding for research, but such funds represent a small and declining fraction of total obligations for academic research and development.

Certain of the indicators presented so far suggest a relative decline in the significance of the chemical discipline within America. Others point toward the burgeoning importance of American chemistry within a world context. What is unequivocal, but caught by neither, is the entrenchment of the chemical discipline within the social, economic, and power structures of American academe. It is hard to imagine a college or university of any stature without a chemistry department. Indicators of this entrenchment are by their nature unobtrusive, but not unobtainable. One to which we might point, not wholly tongue-in-cheek, is the fact that Harvard University had a chemist as its president for better than half the period from 1876 to 1976.[11]

[11] Charles W. Eliot, who taught chemistry at both Harvard and MIT between 1858 and 1869, was president of Harvard University from 1869 until 1909. James Bryant Conant, an organic chemist, presided over Harvard from 1933 to 1953.

Between 1870 and 1970, chemists have been appointed to deanships of American graduate schools more often than members of any other discipline. The early importance of chemistry in the establishment of American graduate education and the steady development in size of the chemistry faculty have already been noted. That these realities should translate into an entrenchment of chemists in graduate deanships is not surprising. Nonetheless, it is interesting that since the early 1920s the chemists' share of graduate deanships has risen steadily toward a plateau of about 10% (Figure 6.3–1).

This is entrenchment indeed, especially when viewed in relative terms. As Figure 3.1–8 indicated, chemistry has accounted for a declining fraction of doctoral degrees in recent years. And as Figure 5.3–2 showed, chemistry has accounted for a slowly declining 1 to 2% of all faculty. The trend in deanships is impressive by comparison.[12] An explicit-invented indicator of the discipline's success in capturing such positions of administrative power is displayed in Figure 6.3–2, which shows the ratio of chemist-deans appointed per thousand doctorates conferred in chemistry, divided by a similar ratio for all fields. (This measure assumes that the number of deanship appointments in a discipline should be proportional to its rate of doctorate production; above-average performance will be reflected in an index score greater than one.) Chemistry has obviously improved with age, and now occupies a more influential position in the administrative hierarchies of American universities than the average discipline.

The move to appoint chemists to graduate deanships originated in Southern and Midwestern institutions. Only in 1926, after seven chemists had already been appointed to nine terms as graduate deans, did Northeastern schools follow suit (see Tables 6.14 and 6.17). Even then, the first appointment in the region was at the Philadelphia College of Pharmacy, rather than at an elite Eastern university. Fourteen years later, Harvard became the first of what would later be known as the Ivy League schools to appoint a chemist (Arthur B. Lamb) as dean of the graduate school. This trend suggests that the entrenchment of chemistry did not depend solely on the role of chemists in the establishment of graduate education at prominent East Coast universities. The regional pattern of the initiative in appointing chemists to graduate deanships provides a counterpoint to the earlier trend

[12] From 1925 to 1969, chemists secured an average of between 9.2 and 12.6% of the deanship appointments in any given pentade. Only four other fields – education, history, English literature, and all the biological sciences taken together – exceeded the upper figure for chemistry, and only the biological sciences performed as steadily in obtaining deanships during the entire period.

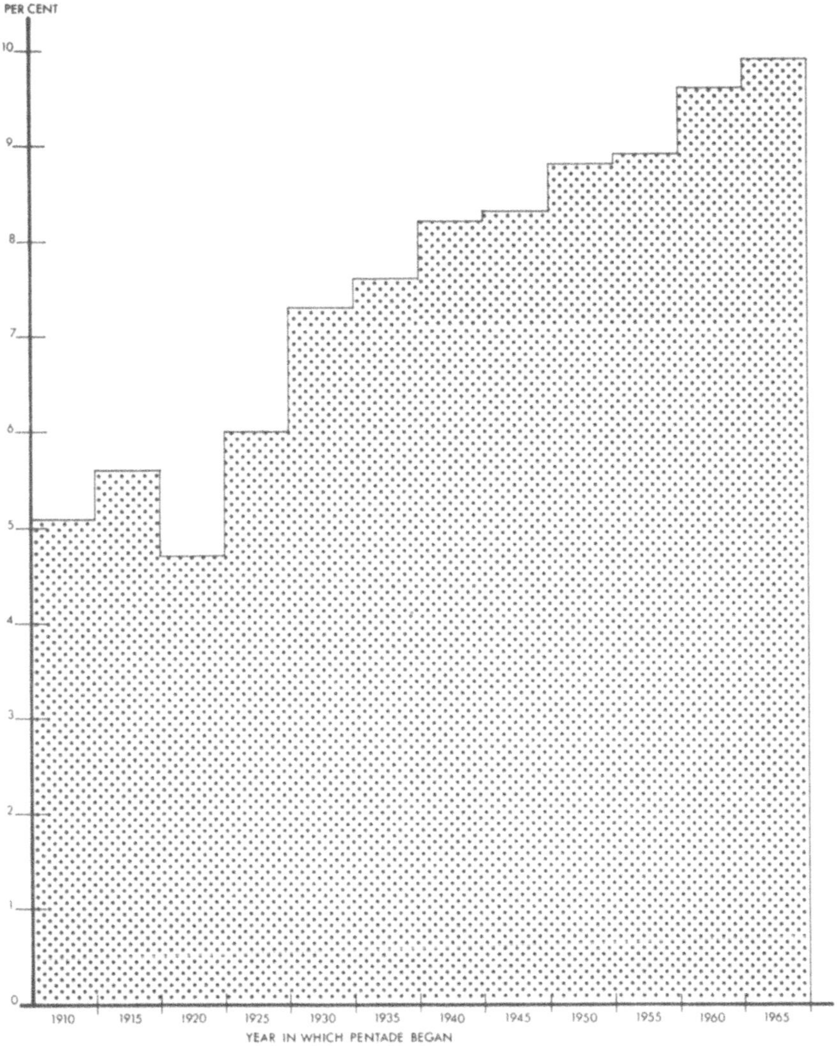

Fig. 6.3–1. Chemists as cumulative percentage of appointments to graduate deanships, by pentade of appointment, 1910–1969. (See Table 6.15, column 5.)

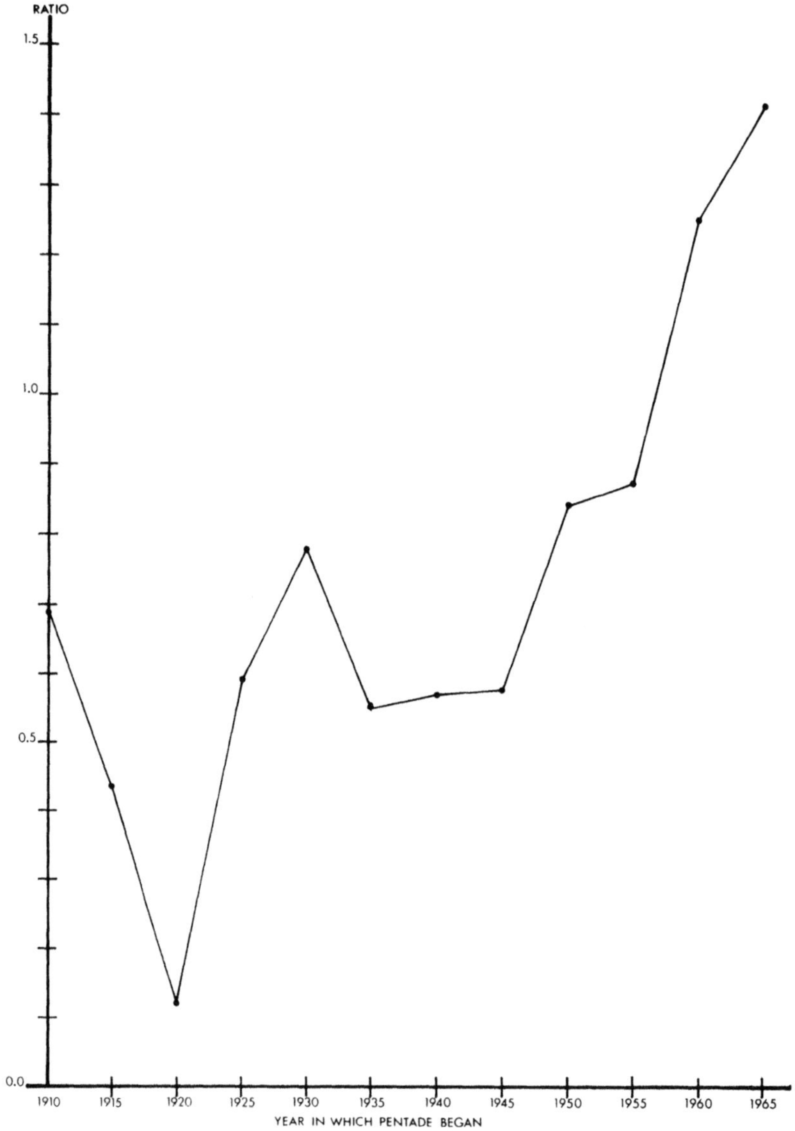

Fig. 6.3–2. A "graduate deans" indicator of the entrenchment of academic chemistry, 1910–1969. (See Table 6.16, column 5.)

in which chemistry doctorate production was first routinized in Eastern institutions.[13]

Another indicator of the entrenchment of chemistry within American science may be constructed by sorting the presidents of the American Association for the Advancement of Science according to their disciplinary affiliations (Tables A, 6.18, and 6.19). Since 1848, every decade but two has

TABLE A

Disciplinary affiliations of presidents of the American Association for the Advancement of Science, 1848–1981

Field	1848–1917	1918–1981	Total
Geology	15	2	17
Chemistry	8	8	16
Physics	7	7	14
Astronomy	9	2	11
Mathematics	4	5	9
Botany	5	3	8
Zoology	5	3	8
Anthropology	4	2	6
Geography	3	1	4
Medicine	3	1	4
Biology	1	3	4
Engineering	0	4	4
Economics	1	2	3
Genetics	0	3	3
Physiology	0	3	3
Others	2	15	17
Total	67	64	131

See Table 6.19.

[13] As the number of doctorates awarded in chemistry has grown, the center of gravity of the "top producer" departments has shifted away from the Eastern seaboard. Johns Hopkins and the Ivy League have been replaced by schools like Illinois, Wisconsin, California, and Texas, in line with demographic trends in the overall American population (see Figure 6.1–2). Other evidence suggests that this westward movement was more than just a demographic rearrangement. The trend in deanship appointments implies a new role for chemists in the Midwest, which followed closely the emergence of large state universities in the area. Knapp and Goodrich (III, Knapp and Goodrich, 1952, 278–282) and Lindsey R. Harmon (IB, NRC, 1978, Chapter 4, esp. 99–100) discuss the Midwestern context for graduate education in science and other fields. B. R. Siebring focuses upon chemists in two articles which, *inter alia*, contrast the East and the Midwest (III, Siebring, 1954, 1961). See also III, Carroll, 1982.

provided at least one chemist among AAAS presidents. Overall, chemistry is second only to geology, and for the sixty years since World War I it outranks every other science. However, it should be noted that physics – a subject with a far smaller membership base in the AAAS – runs chemistry close throughout the time series.

A quite different indicator of the entrenchment of the chemical discipline in America may be found in the growth patterns of the American Chemical Society itself. Among the oldest, largest, and most prosperous of American learned societies, the ACS was plagued by financial difficulties and flagging enthusiasm during its first decades. Since the 1890s, the Society has expanded continuously in membership and influence (see Figure 2.2–2). The scope of the American Chemical Society's activities broadened beyond the chemical community, especially after the first World War, and now includes public service and governmental advisory work conducted through a complex array of committees. This growth is reflected in figures on ACS expenditures, which grew from \$20 000 in 1906 to \$107 000 in 1915, then increased to nearly \$670 000 on the eve of World War II. During the postwar period, ACS finances entered a new era, reaching some 20 *million* dollars by the mid-1960s. In 1976, the ACS required a budget of \$37 105 000 to finance a profusion of activities, including the circulation of some 350 000 copies of Society journals and the administration of 28 specialty divisions and 175 local sections. The current membership of over 120 000 chemists places the American Chemical Society in a category comparable to the American Association for the Advancement of Science (133 000), the American Bar Association (250 000), and the American Medical Association (213 900).[14] Table B provides comparative information on these and other American professional societies. One striking aspect of this listing is the bureaucratic complexity of the ACS, indicated by its relatively high ratio of staff to members (about 1 staff person for every 79 members).

These indicators of the entrenchment of chemistry must be interpreted with caution: one should not equate presence with power. The discussion of chemistry in mass culture presented in Section 3.3 suggested that chemistry, while pervasive, did not command adulation: there are indications that the chemical discipline, though entrenched, is not as powerful as it first appears.

[14] Figures on ACS expenditures are from IB, Parsons, 1933, 122; IB, "Finances", 1940; and IB, ACS, 1977a, 56. The ACS administrative structure is anatomized in IB, Skolnik and Reese, 1976, 179–235. Comparative membership and staff figures (see Table B) are from IB, Yakes and Akey (eds.), 1980.

TABLE B
Selected American professional societies

Organization	Year established	Membership in 1979 (thousands)	Staff
American Medical Association	1847	213.9	900
American Association for the Advancement of Science	1848	133.0	171
American Society of Civil Engineers	1852	77.7	116
National Education Association	1857	1600.8	600
American Institute of Mining, Metallurgical, and Petroleum Engineers	1871	69.0	–
American Chemical Society	1876	118.2	1500
American Bar Association	1878	250.0	480
American Society of Mechanical Engineers	1880	80.0	250
Modern Language Association of America	1883	30.0	60
American Historical Association	1884	14.5	20
Institute of Electrical and Electronics Engineers	1884	180.0	250
National Science Teachers Association	1895	42.0	40
American Institute of Chemical Engineers	1908	45.0	75
American Association of University Professors	1915	68.1	55

See Table 6.20.

A good example is found in the representation of chemists in the National Academy of Sciences. Studies of members now living and of representative samples of members elected in each decade since the 1870s show that chemists constitute about 13% of the regular membership – more than for any other disciplinary group represented by a section of the Academy (Tables C and D). Standing alone, this finding is impressive. When compared with the relative presence of chemists within the American scientific community, it takes on another aspect. From one-fourth to one-third of all practicing scientists in America have been chemists since 1950, and since at least 1930 some 30 to 40% of those holding terminal doctorates in the natural sciences obtained their graduate training in chemistry (Table 6.23). From this standpoint, chemists are underrepresented with only 13% of NAS membership. Furthermore, of the 18 presidents of the National Academy, only two (Wolcott Gibbs and Ira Remsen) have been chemists, or just over 11% (Table 6.24). Physicists, in contrast, have achieved at least proportional representation within the ranks of the National Academy of Sciences. Between 11 and 17%

TABLE C
Section affiliations of living members of the National Academy of Sciences, as of 1 July 1972

Section		Members in section	
Number	Subject	Number	As percentage of total
5	chemistry	127	13.3
3	physics	120	12.6
14	biochemistry	99	10.4
4	engineering	60	6.3
1	mathematics	58	6.1
8	zoology	57	6.0
17	medical sciences	49	5.1
6	geology	42	4.4
2	astronomy	41	4.3
13	geophysics	41	4.3
7	botany	40	4.2
18	genetics	38	4.0
12	psychology	37	3.9
9	physiology	35	3.7
16	applied physical and mathematical sciences	33	3.5
11	anthropology	26	2.7
10	microbiology	23	2.4
15	applied biology	20	2.1
19	social, economic, and political sciences	8	0.8
Total		954	100.1

See Table 6.21.

of terminal doctorate holders in the natural sciences have physics PhDs (Table 6.23), and physicists constitute about 10% of all practicing scientists. Almost 13% of living members of the National Academy are physicists, while over 20% of its 18 presidents have been physicists. In entering the high elite of American science, chemists have fared poorly in comparison with physicists.[15]

[15] According to BLS estimates, between 1950 and 1970 the proportion of chemists among American scientists ranged from 35.3% to 26.8%, while the proportion of physicists during that period fluctuated only between 9.4 and 10.2% (IA, BLS, 1973, 16, Table 8). Census data yield even higher percentages for chemists, ranging from 53.6

TABLE D
Chemists and other members elected to the National Academy of Sciences, selected periods, 1870–1980

Period	Members Elected		
	Total	Chemists	
		Number	As percentage of total
1870–1872	23	4	17.4
1880–1883	10	3	30.0
1890–1895	11	1	9.1
1900–1902	14	1	7.1
1910–1911	17	3	17.6
1920	14	2	14.3
1930	15	3	20.0
1940	16	2	12.5
1950	30	3	10.0
1960	35	3	8.6
1970	50	6	12.0
1980	59	6	10.2
Total	294	37	12.6

See Table 6.22.

Similar indications of entrenchment without glamour appear in the relative proportion of chemists among natural scientists who received National Research Fellowships awarded between 1919 and 1950. Because their discipline has been allotted more fellowships – applicants from different sciences do not compete directly for each fellowship – chemists hold the largest percentage of these honors (almost 27%) among the natural scientists (Table E). Only the medical sciences taken together boast a larger number of Fellows. But the NRC can hardly be accused of favoritism since chemistry accounted for over 38% of the natural sciences doctorates conferred during those years (Table F). In contrast, those employed as physicists can lay claim to almost 23% of the fellowships, compared with that discipline's

to 63.9% between 1950 and 1970 (see Table 6.25). The wide variation in these estimates probably arises from the use of different occupational classification schemes by the two organizations. The more modest range suffices to make the point.

For sketches of the scientific careers and official activities of NAS presidents to the early 1960s, see III, Cochrane, 1978.

TABLE E

National Research Fellows in the natural sciences, 1919–1950, by field of employment in 1950

Field of employment in 1950	National Research Fellows, 1919–1950	
	Number	As percentage of total
Chemistry	229	26.7
Physics	196	22.8
Zoology	164	19.1
Mathematics	126	14.7
Botany	112	13.1
Astronomy	16	1.9
Geology and Geography	15	1.7
Total	858	100.0

See Table 6.26.

Note: Excludes the 332 Fellows with careers in the Medical Sciences, 93 in Psychology, 41 in Agriculture, 27 in Anthropology, and 8 in Forestry.

TABLE F

Doctorates conferred in the natural sciences, selected years, 1920–1950

Year	Natural sciences doctorates conferred				
		Chemistry		Physics	
	Total	Number	As percentage of total	Number	As percentage of total
1920	268	104	38.8	32	11.9
1930	945	332	35.1	109	11.5
1940	1 459	532	36.5	132	9.0
1950	2 351	967	41.1	379	16.1
Total, 1919–1950	32 026	12 443	38.9	3653	11.4

See Table 6.27.

representation of only about 11% among natural sciences doctorates conferred during the period.[16]

Chemists have consistently won fellowships from the John Simon Guggenheim Memorial Foundation, another well-known honorific award for intellectual performance. If one makes the limiting assumption that Guggenheim Fellowships should be awarded to a particular field roughly in proportion to the number of persons holding faculty appointments in that field, then it is possible to construct an explicit-invented entrenchment indicator similar to that discussed for deanship appointments. On the far more generous – but less likely – assumption that Guggenheim Fellowships should stand in proportion to terminal doctorates in the field, an alternative series of values is obtained for the entrenchment indicator. The latter series shows chemistry to be somewhat under-represented,[17] while the former – more realistic – series shows the discipline to be well entrenched, but losing ground over the past quarter century (Figure 6.3–3).

Available figures on the funding of academic research also suggest that chemistry receives a lower share of resources than might be warranted by the size of the discipline. Funds available for academic chemical research increased from 14.7 million dollars in 1954 to 117.5 million dollars in 1974, but this was a small and declining fraction of total funds available for academic science during the same period, as shown in Figure 6.3–4. This relative decline in the patronage of chemistry provides a context for understanding the tone of *Chemistry: Opportunities and Needs*, the 1965 report of the National Academy of Sciences Committee for the Survey of Chemistry (chaired by Frank Westheimer). This document presented a case for increasing the level of federal support of the discipline, stressing both the proven payoffs of basic chemical research and the lack of federal support for chemistry – a "little science" when compared with high-energy physics or biomedical research.[18]

16 On the establishment of the National Research Fellowships, see III, Coben, 1976, and III, Reingold, 1977. IB, Rand, 1951a, analyzes the patterns of fellowship support between 1919 and 1950. For the topics of National Research Fellowships in chemistry and post-fellowship employment of the recipients to 1938, see IB, NRC, 1938, 28–39. This latter source demonstrates the strong correlation between field of fellowship and field of subsequent employment.

17 This is indicated by the failure of the ratio based upon terminal doctorates to reach or exceed unity. Whether this performance has resulted from explicit Foundation policy, or from chemists not pursuing these fellowships, is unclear.

18 IB, NAS, 1965, [iv], 194. For contemporary reaction to the Westheimer Report, see III, Abelson, 1965, and III, Greenberg, 1965. The origins and impact of the COSPUP

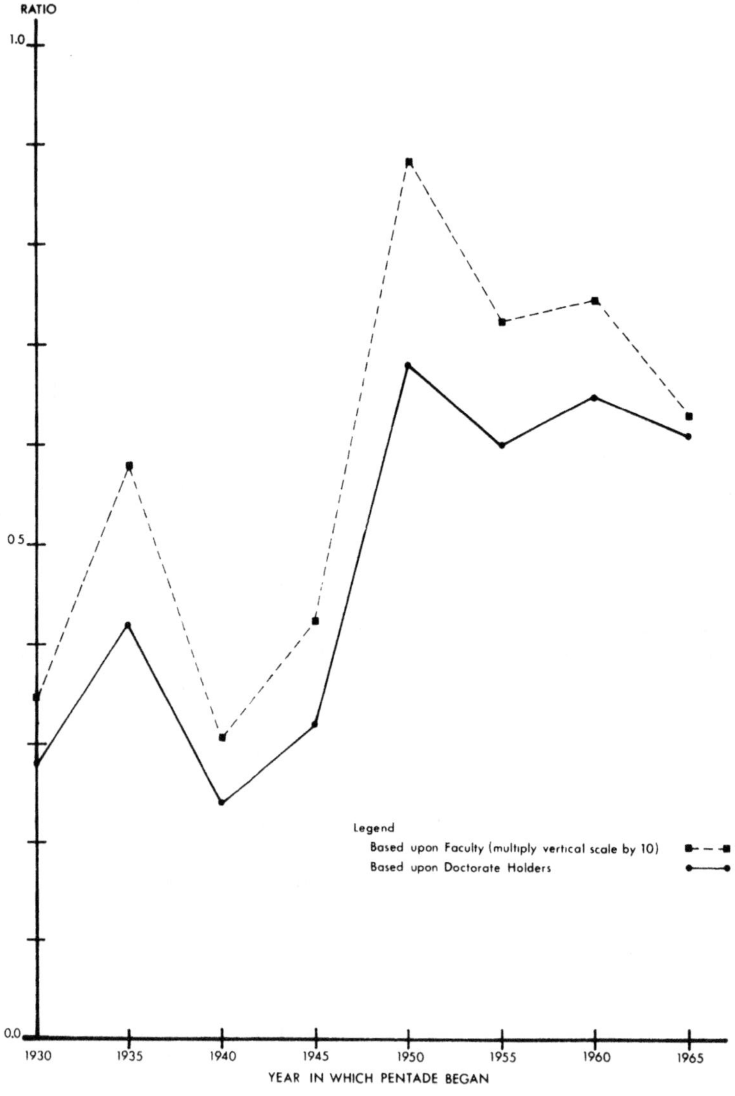

Fig. 6.3–3. Two "Guggenheim Fellows" indicators of the entrenchment of academic chemistry, 1930–1969. (See Table 6.29, columns 7 and 8.)

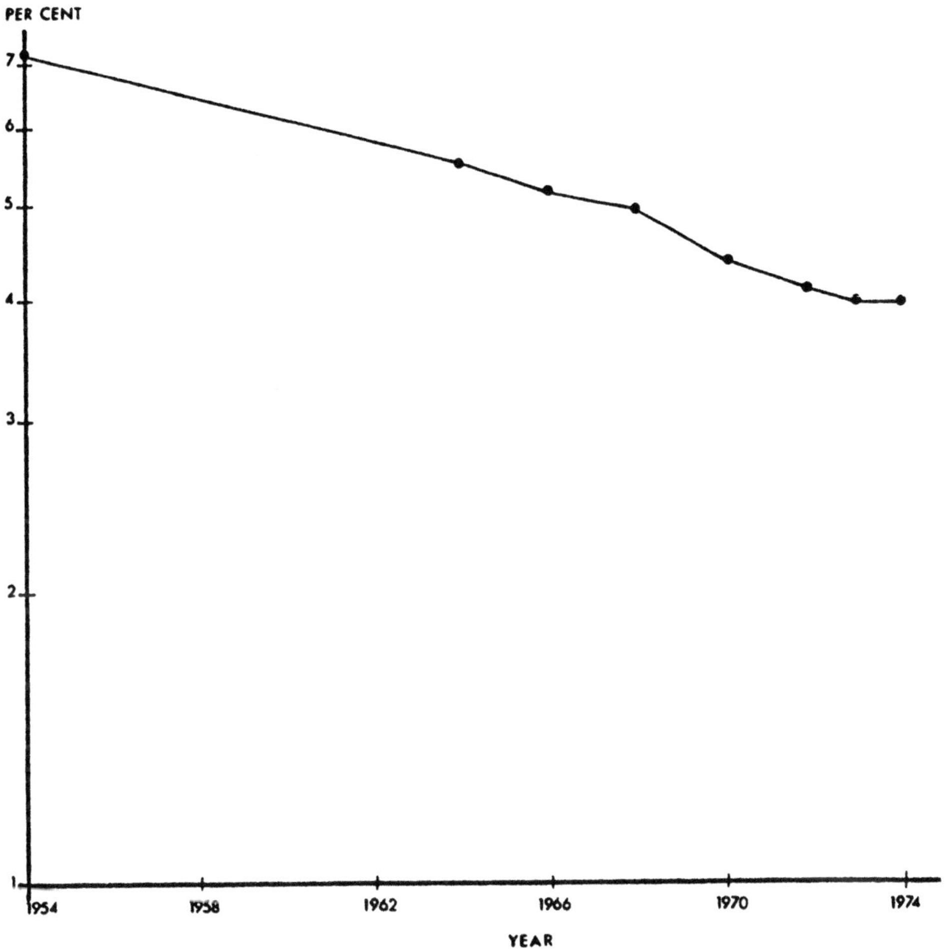

Fig. 6.3–4. Percentage of total R & D obligations to universities and colleges spent on chemistry, 1954–1974. (See Table 6.30, column 3.)

6.4. The Differentiation of Chemistry

INDICATOR HIGHLIGHTS:

The rapid differentiation of the chemical discipline is evident in the appearance of a wide range of specialty, service, and abstracting journals. This phenomenon is reflected in the variety of official American Chemical Society journals.

American chemistry is served by relatively few specialist societies. Instead, the American Chemical Society has developed an elaborate structure of specialist divisions.

The basic organization of, and relative attention to, the various academic specialties within the discipline have been stable for the last half century.

In the 1880s and 1890s academic chemists in America began to number in the hundreds (see Figure 5.3–1) and strong centers of research and training emerged in Boston, New York, Philadelphia, and Baltimore. It was in this era that a clear need was first felt for American chemical journals and societies. The enduring trend since that time has been one of differentiation and specialization.[19]

6.4.1. CHEMICAL JOURNALS

The first successful American scholarly journal devoted to chemistry was the *American Chemist*, which evolved from a supplement to an American reprint edition of the British *Chemical News* that was begun in 1868. Charles F. and William H. Chandler acquired the rights to the reprint edition and its supplement, out of which they launched the *American Chemist* in 1870. Prior to the Chandlers' journalistic venture, the major indigenous outlet for chemical research was Benjamin Silliman's *American Journal of Science*. The most prestigious route for publication had been to send papers abroad to Liebig's *Annalen der Chemie*, Erdmann's *Journal für praktische Chemie*, or,

---

surveys of fundamental research are reviewed in III, Lowrance, 1977. III, M. Wilson, 1975, discusses trends in NSF support for chemical research.

[19] The discussion of American chemical journals in this section draws on III, Beardsley, 1964, 34–42; III, Browne and Weeks, 1952, 296–426; and IB, Skolnik and Reese, 1976, 94–143. On the *Journal of Physical Chemistry*, see III, Servos, 1982.

beginning in the 1860s, to the *Zeitschrift für Chemie* (1865) or *Chemische Berichte* (1868). A significant amount of the column space in the *American Chemist* was devoted to reports of the European literature and other chemical intelligence from abroad.[20]

When the American Chemical Society was established in 1876, Charles Chandler – a central figure in the new organization – arranged for publication of the Society's proceedings in the *American Chemist*. The ACS began independent publication of its *Proceedings* in 1877, and subsumed them into the new *Journal of the American Chemical Society* in 1879. Despite the grand vision of its proponents, the *JACS* soon languished for lack of quality contributions and a sound financial basis. By the time the ACS reached its nadir in the late 1880s, publication of *JACS* had been suspended on occasion for a few months at a time. Partly as a result of this situation, *JACS* faced severe competition from journals owned and vigorously promoted by two former members, Ira Remsen and Edward Hart. Remsen's *American Chemical Journal*, founded in 1879, was a highly regarded journal devoted to pure chemistry which also served as a vehicle for the research performed by his graduate students at Johns Hopkins University. Hart, who ran a chemical manufacturing business in addition to teaching at Lafayette, first published his *Journal of Analytical Chemistry* in 1887. Reform of the American Chemical Society in the early 1890s included changes in its publications program. Hart merged his thriving journal with the *JACS* and, for the first time, American academic chemistry had a strong central voice.[21]

[20] A popular science journal titled the *Boston Journal of Chemistry* appeared in 1866, along with the *Journal of Applied Chemistry* (1866–1875), another Boston-based periodical which seems not to have been distributed very widely.

For a detailed study of the publication habits of American chemists before 1880, see IB, Siegfried, 1952, 119–134. Siegfried (*ibid.*, 120–122) reports that Liebig's *Annalen* and Erdmann's *Journal* carried 18.8% and 14.8%, respectively, of all American chemical publications in the 1850s. During the 1860s, Fresenius' *Zeitschrift für Chemie* replaced Liebig's *Annalen* as the most important German journal for American chemists, publishing 13.7% of all American contributions during the decade. *Chemische Berichte* became the preferred foreign outlet for American chemists soon after its establishment in 1868: in the 1870s, roughly one in six (17.6%) American chemical research papers were published in *Chemische Berichte*.

Kevles and Harding note that for the period 1879 to 1891, *Chemische Berichte* still accounted for 54% of all foreign publications by American chemists, followed by the *Journal für praktische Chemie* with 18% (IB, Kevles and Harding, 1977, Table 9, 17).

[21] On Remsen and the founding of the *American Chemical Journal*, see III, Getman, 1940, 48–50, and III, Hawkins, 1960, 75–76, 107. Edward Hart always supposed that

By the first decade of the twentieth century, chemical publications had begun to differentiate along specialty lines. The first set of journals to emerge as strong independent entities were those catering to industrial chemists and chemical engineers. Some of these publications were geared toward chemists in particular industries, e.g., *Electrochemical Industry* (1902), *Transactions of the American Electrochemical Society* (1902), or the *Journal of the American Leather Chemists Association* (1906). Others were designed to meet the needs of applied chemists in general, supplementing the academic slant of the *Journal of the American Chemical Society* or the *American Chemical Journal*. Examples include the *Chemical Engineer* (1904) and the *Chemist-Analyst* (1911). The desire for a separate society of industrial chemists was voiced as early as 1905 in the columns of the *Chemical Engineer*, and this movement culminated in the establishment of the American Institute of Chemical Engineers three years later. The new organization began publishing its *Transactions* that year (1908). Industrial chemists within the American Chemical Society responded to this trend with the *Journal of Industrial and Engineering Chemistry* (1909).[22]

Differentiation of chemical journals continued, paralleling specialization of interests among the many groups of chemists within the United States. The pattern of differentiation, reflected in journals published by the ACS (Table G), may be characterized by the two broad categories of *specialty*

---

the chief reason he was asked in 1893 to edit *JACS* and to merge his journal with it was to increase ACS membership through access to his larger subscription list (III, Hart, 1922, 446).

22 The journal *Electrochemical Industry*, published in Philadelphia, proposed to fill the need for "an authoritative exponent of electrochemical science and its various applications" (III, [Roeber], 1902). Later that year, J. W. Richards opened the *Transactions of the American Electrochemical Society* with the following declaration: "Differentiation and specialization are the watchword, now, of all progress, – industrial, scientific, philosophical. The day is past, we all acknowledge, when one man . . . can know all that is to be known; the day is also past when one scientific society can cover satisfactorily the whole field of scientific research . . . . [T]he analogue of the specialist in science is the *society which specializes*" (III, Richards, 1902). The American Electrochemical Society, Richards continued, was formed in response to the rapid development of the American electrochemical industry and the increasing number of electrochemists. Charles F. McKenna, addressing the inaugural meeting of the American Institute of Chemical Engineers, noted that "the followers of almost every conceivable study have some forum, some printed herald to the world" (III, McKenna, 1908, 8). In the remainder of his address, printed in the first volume of the AIChE's *Transactions*, McKenna articulated the sense of group identity which fostered the fledgling organization. See also III, Parker, 1909, and III, Noyes, 1908.

TABLE G
American Chemical Society journals

Journal	Year established
Journal of the American Chemical Society	1879
Journal of Physical Chemistry	1896
Chemical Abstracts	1907
Journal of Industrial and Engineering Chemistry	1909
Chemical and Engineering News	1923
Chemical Reviews	1924
Journal of Chemical Education	1924
SciQuest (originally Chemistry)	1927
Rubber Chemistry and Technology	1928
Analytical Chemistry	1929
Journal of Organic Chemistry	1936
Journal of Agricultural and Food Chemistry	1953
Journal of Chemical and Engineering Data	1959
Journal of Medicinal Chemistry	1959
Journal of Chemical Information and Computer Sciences	1961
Biochemistry	1962
Industrial and Engineering Chemistry –	
Process Design and Development	1962
Fundamentals	1962
Product Research and Development	1962
Inorganic Chemistry	1962
Accounts of Chemical Research	1967
Environmental Science and Technology	1967
Macromolecules	1968
Chemical Technology (CHEMTECH)	1971
Journal of Physical and Chemical Reference Data	1972

See Table 6.31.

and *service* journals. The *Journal of Physical Chemistry* (1896), owned and edited by Wilder Bancroft (Cornell), was the first American chemical journal devoted to a particular specialty, followed by the *Journal of Biological Chemistry* (1906). The next specialty journals did not appear until the late 1920s and mid-1930s, when the ACS initiated *Rubber Chemistry and Technology* (1928) and *Analytical Chemistry* (1929) and Morris Kharasch (Chicago) founded the *Journal of Organic Chemistry* (1936).

Aside from the *Journal of Agricultural and Food Chemistry*, begun in

1953, there was a hiatus in the appearance of new specialty journals until the post-Sputnik era, when the number of ACS journals doubled in just 13 yr (1959–1972). This wave of new publications included the *Journal of Medicinal Chemistry* (1959), *Biochemistry* (1962), *Inorganic Chemistry* (1962), and *Macromolecules* (1968). The appearance of *Macromolecules* is one indication of the growth in importance of polymer chemistry for industry in the 1950s and 1960s. Other signs of the differentiation of interests among industrial chemists by the 1960s are the splitting of *Industrial and Engineering Chemistry* in 1962 into three separate journals (dealing with processes, products, and theoretical studies in chemical engineering) and the establishment of *Environmental Science and Technology* in 1967.

The increasing number of specialty journals mirrors the progressive fragmentation of the research front in American chemistry. The passing of the day when every chemist could expect to be conversant with all areas of chemical knowledge was recognized early by ACS members; *Chemical Abstracts*, established in 1907, was their first corporate response toward coping with the rapidly increasing store of chemical information. *Chemical Abstracts* served this purpose admirably, and its coverage increased from about 12 000 documents in 1907 to over 475 000 documents in 1980.[23] The proliferation of chemical information at a quickening pace led to the establishment of other service journals, especially in recent years.

Shifting activity on the research front has been reported through the ACS journal *Chemical Reviews* (1924) and the *Survey of American Chemistry* (1925), an annual publication of the National Research Council's Division of Chemistry and Chemical Technology during the 1920s and 1930s. *Chemical Reviews* was supplemented by *Accounts of Chemical Research* in 1967. The need for current data on the chemical properties and physical constants of myriads of new substances was met with the *Journal of Chemical and Engineering Data* (1959) and the *Journal of Physical and Chemical Reference Data* (1972). Finally, the founding of the *Journal of Chemical Documentation* (later the *Journal of Chemical Information and Computer Sciences*) in 1961 signaled the difficulty of keeping track of chemical literature, as well as the increasing sophistication of specialists in chemical information. In the following year *Chemical Abstracts* expanded its number of specialty categories, which had been roughly constant since 1907, from 31 to 80.

The last group of service journals are those dealing with education and news of the chemical community. The *Journal of Chemical Education* (1924)

[23] IB, ACS, 1974, 28; and IB, Baker, 1981, 29.

is published by the ACS Division of Chemical Education. *SciQuest* (founded as *Chemistry* is 1927) is a periodical designed to interest secondary school students in chemistry. *Chemical & Engineering News* (1923), which began as the *News Edition* of *Industrial & Engineering Chemistry*, is a weekly magazine sent to all ACS members.

6.4.2. SPECIALIZATION AND ACS STRATEGY

As the chemical community has doubled and redoubled in size, there has been a steady growth in the range of specialist groups into which disciplinary activity has been channeled. Part of the power and authority of American chemistry as an institution derives from the way in which the American Chemical Society has been able to harness and profit from that growth.[24]

Technical chemists at the turn of the century were the first to establish their own associations, e.g., the American Section of the Society of Chemical Industry (1894) and the American Electrochemical Society (1902). These developments created anxiety among some ACS officers, who were concerned that the power of the chemical profession would be diminished by disunity among chemists and a narrow specialization of interests. Meanwhile, other chemical organizations continued to appear, and industrial chemists in the ACS began to wonder whether they benefited from membership in an academically-oriented society. The formation of the American Society of Biological Chemists at the 26 December 1906 ACS meeting – involving academic rather than industrial chemists – finally galvanized the ACS hierarchy into action.

A committee was charged with considering the need for new chemical publications and the relationship of the ACS to other chemical societies. The following year a subcommittee was formed to look into the publication

24 On the proliferation of specialist scientific societies in late-nineteenth century America, see III, Bates, 1965, 85–136.

Specialist chemical societies have yet to receive adequate historical treatment. On individual societies see, however, III, AACC, 1965 (American Association of Cereal Chemists); III, Burns and Enck, 1977 (American Electrochemical Society); III, Reynolds, 1983 (American Institute of Chemical Engineers); III, Turley, 1953, and III, Rogers, 1961 (American Leather Chemists Association); III, AOCS, 1947 (American Oil Chemists' Society); III, Chittenden, 1945 (American Society of Biological Chemists); III, AOAC, 1934 (Association of Official Agricultural Chemists); and III, SCI, 1931, 24–25 (Society of Chemical Industry, American Section).

of an industrial journal. By early 1908 the ACS had approved a plan which provided for this publication (which became the *Journal of Industrial & Engineering Chemistry*) and revived a proposal for specialist sections within the Society. The Division of Industrial Chemists and Chemical Engineers was the first to be organized. Four other Divisions (Agricultural and Food Chemistry, Fertilizer Chemistry, Organic Chemistry, and Physical and Inorganic Chemistry) quickly followed. By allowing members with similar interests to present papers to and exchange information with like-minded colleagues within the ACS, the Divisions managed to obviate the problems of specialization.[25]

The success of this strategy may be seen in the relatively small number of specialist chemical societies established in the United States (Table H). Using a broad definition, only 19 such societies have been formed since 1908, a period in which ACS membership increased from 4000 to about 110 000. The differentiation of members' interests along both industrial and research specialty lines has been accommodated by increasing the number of Divisions within the ACS, of which there are now 34 (Table I).[26] The effectiveness of this policy is illustrated by comparing the growth rates of the ACS and some early competitors (Figure 6.4–1). Growth in the membership of competitors stagnated between World Wars I and II and, for all but the AIChE, barely doubled between 1920 and 1980. ACS membership mushroomed nearly ten-fold during the same period. At once both cause and consequence of this growth, the ACS has been increasingly able to function as a professional organization for chemists throughout industry, government, *and* academe. Its umbrella role is similar to the American Bar Association or the American Medical Association. Only chemical engineers have seen fit to go it alone, as the meteoric postwar rise of the AIChE indicates.

6.4.3. Specialty Structure

The stability of the ditribution of research interest among the various

[25] The proposal for specialist sections within the American Chemical Society originated with Arthur A. Noyes of MIT, but the initiative of ACS President Marston T. Bogert (Columbia University) was instrumental in carrying out the reform in 1908; see III, Bogert, 1908. A convenient guide to ACS organizational reform in this period is III, Browne and Weeks, 1952, 68–89.

[26] The number of Divisions expanded rapidly from the original five to eighteen by the mid-1920s, reaching a plateau until after World War II. The history and activities of the ACS Divisions are discussed in IB, Skolnik and Reese, 1976, 238–381.

TABLE H
Specialist chemical societies in the United States

Society	Year established
American Water Works Association	1881
Association of Official Analytical Chemists	1884
Society of Chemical Industry, American Section	1894
American Society for Testing Materials	1898
American Ceramic Society	1899
Electrochemical Society	1902
American Leather Chemists Society	1903
American Society of Biological Chemists	1906
American Society for Pharmacology and Experimental Therapeutics	1908
American Institute of Chemical Engineers	1908
American Oil Chemists Society	1909
American Association of Cereal Chemists	1915
Societé de Chimie Industrielle, American Section	1918
American Association of Textile Chemists and Colorists	1921
Association of Consulting Chemists and Chemical Engineers	1928
American Society of Brewing Chemists	1934
American Microchemical Society	1935
Association of Analytical Chemists	1941
Association of Vitamin Chemists	1943
Society of Cosmetic Chemists	1945
American Association for Clinical Chemistry	1948
Histochemical Society	1950
Coblentz Society [molecular spectroscopy]	1954
Geochemical Society	1955
Society of Flavor Chemists	1959
Phytochemical Society of North America	1960
Catalysis Society of North America	1966
American Society for Neurochemistry	1969

See Table 6.32.

academic specialties within chemistry may be seen from Figure 6.4–2, which records PhD research. Agriculture and Food shows a decline from 6.3% to 1.4% over the last half century, but this decline may simply reflect the absorption of much of this work by the category of biochemistry. Pharmaceutical chemistry shows an increase from 1.3% to 2.2%, but the

TABLE I
Specialty divisions in the American Chemical Society

Division	Year established
Industrial and Engineering Chemistry	1908
Agricultural and Food Chemistry	1908
Fertilizer and Soil Chemistry	1908
Organic Chemistry	1908
Physical Chemistry [a]	1908
Medical Chemistry	1909
Rubber	1909
Biological Chemistry	1913
Environmental Chemistry	1913
Carbohydrate Chemistry	1919
Cellulose, Paper, and Textile	1919
Dye Chemistry [b]	1919
Leather and Gelatin Chemistry [c]	1919
Chemical Education	1921
History of Chemistry	1921
Fuel Chemistry	1922
Petroleum Chemistry	1922
Organic Coatings and Plastics Chemistry	1923
Colloid and Surface Chemistry	1926
Analytical Chemistry	1936
Chemical Information	1948
Polymer Chemistry	1950
Chemical Marketing and Economics	1952
Inorganic Chemistry	1956
Microbial and Biochemical Technology	1961
Fluorine Chemistry	1963
Nuclear Chemistry and Technology	1963
Pesticide Chemistry	1969
Professional Relations	1972
Computers in Chemistry	1974
Chemical Health and Safety	1977
Geochemistry	1978
Small Chemical Businesses	1978
Chemistry and the Law	1982

See Table 6.33.

[a] Originally Physical and Inorganic Chemistry.
[b] Merged with Organic Chemistry in 1935.
[c] Discontinued in 1938.

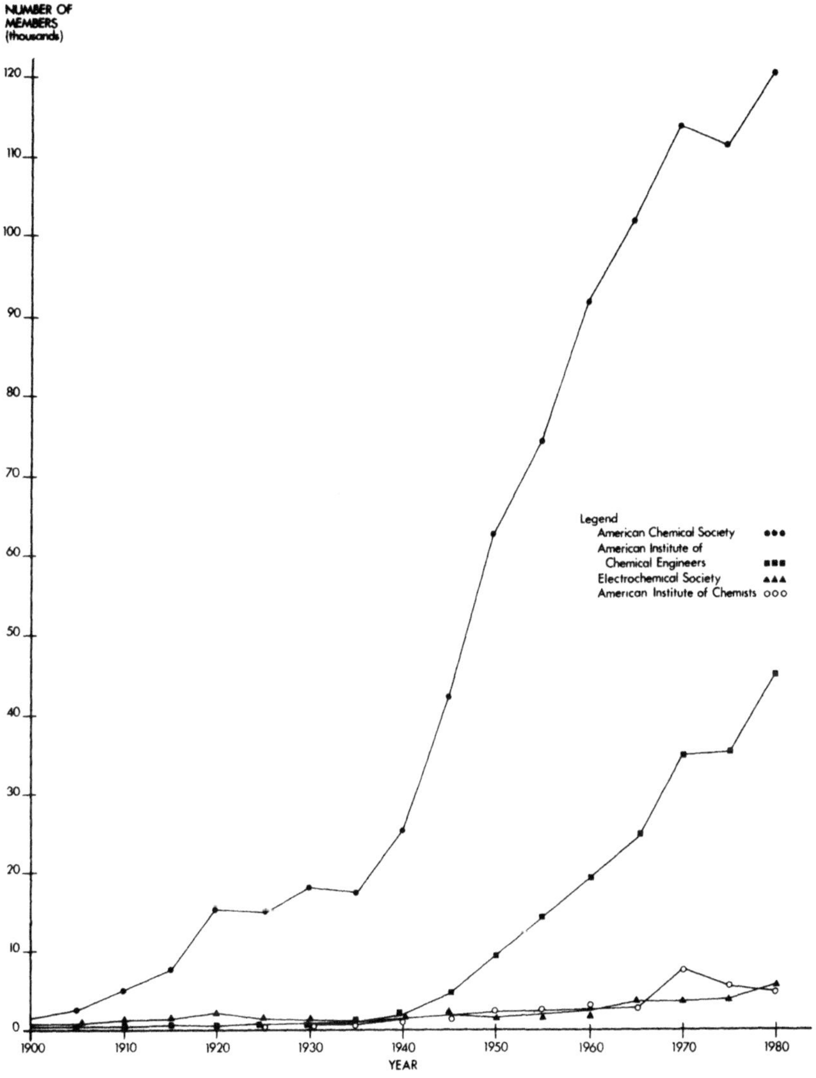

Fig. 6.4–1. American Chemical Society membership versus membership in selected specialist chemical societies, 1900–1980. (See Table 6.34.)

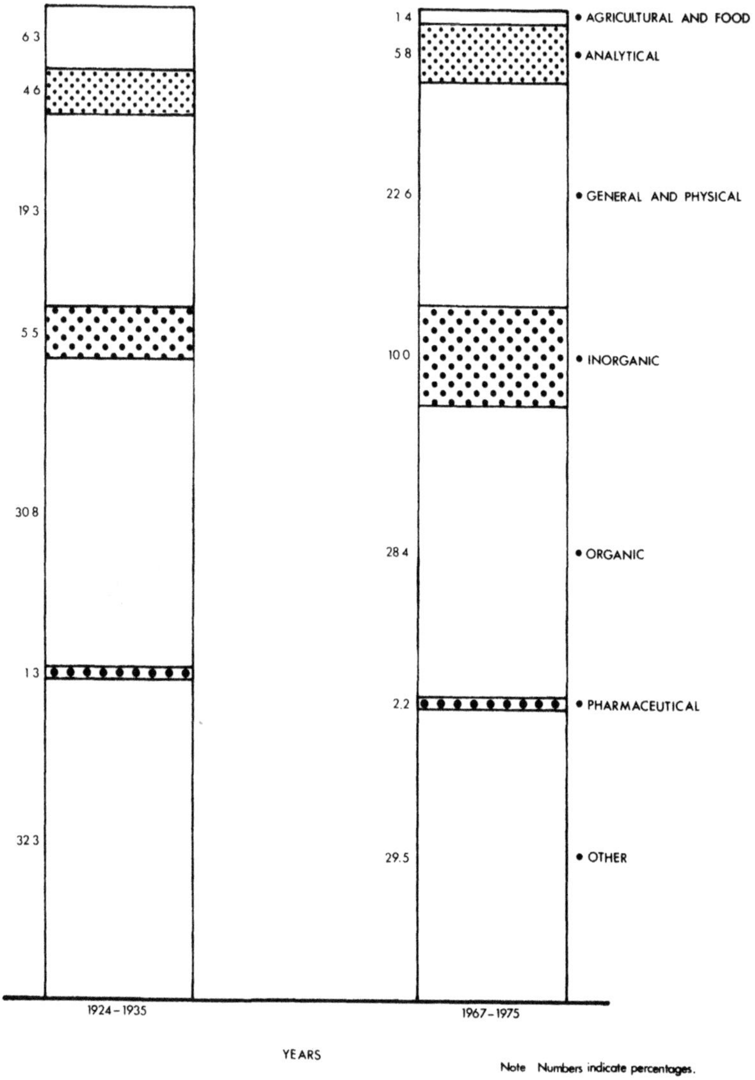

Fig. 6.4–2. Graduate research in chemistry, by selected field, 1924–1935 and 1967 1975. (See Table 6.35.)

latter fraction is still a very modest one. Inorganic chemistry shows a strong rise from 5.5% to 10%, a shift which reflects new interest in rare earths and transuranium elements and the renaissance of coordination chemistry. General and Physical, Organic, and Analytical chemistry show great stability over time. Together these categories accounted for slightly over half of graduate research both around 1930 (54.7%) and some four decades later (56.7%). This relatively static distribution suggests that the major divisions of chemistry are well entrenched. Certain aspects of the intellectual content of the discipline have been built into its very structure. The division of the discipline between specialties has changed little, either with the very great growth of the subject or with the creation of "new knowledge".[27]

A conspectus of both stability and change may be seen by comparing the original 30 sections of *Chemical Abstracts* in 1907 with the 80 sections of 1970s volumes (Table 6.36). Along the way the 1907 section on *Biological Chemistry* was divided into subsections (1911). By 1927, the *Foods* and the *Nutrition* sections had been combined; the section on *Patents* was dropped (patent information was incorporated into other sections by subject); and two new sections (one on *General Industrial Chemistry* and another on *Rubber and Allied Substances*) had been added. This system of 30 sections was still very similar to that in the first volume. By Volume 41 (1947), a new section on *Synthetic Resins and Plastics* had been added, bringing the total of the still more-or-less intact original system to 31. Each section increased in bulk as growing numbers of researchers, moving from specialist problem to specialist problem, articulated the possibilities inherent in particular areas.

The new system of 80 sections introduced in 1962 was designed to alleviate the problem of bulk. Once again, the character of the sections shows considerable stability. By Volume 66 (first half of 1967), when the new system was five years old, 72 of the 80 sections already had the exact titles they had in 1977. Six of the eight that then changed by 1977 differ in

[27] Biochemistry is included in the category "other"; see Table 6.35.

Ihde cites data which appear to contradict our claim of stability; for example, one may infer from his table that only 16.1% of the American papers in *Chemical Abstracts* during 1957–58 were in organic chemistry. In addition, the study upon which this percentage is based only counted abstracts in five fields; presumably, the percentage for organic chemistry would have been smaller still if all fields had been considered. But many specialized subunits of organic chemistry are counted separately from organic chemistry proper in *Chemical Abstracts*, the source of these aberrant data. See Table 6.36; III, Ihde, 1964, 726, Table 27.1; and III, Brooks, 1958.

minor details; for example, *Pesticides* in 1967 had become *Agrochemicals* by 1977 (a change as amusing as it is revealing), *Hormones* had become *Hormone Pharmacology*, and *Plant-Growth Regulators* had become *Biochemical Interactions*. The only important shifts were that *Petroleum, Petroleum Derivatives, and Related Products* and *Coal and Coal Derivatives* had been combined into *Fossil Fuels, Derivatives, and Related Products*; and a new section on *Electrochemical, Radiational, and Thermal Energy Technology* had appeared. More recent changes are similar.

On the basis of this evidence it would seem that the broad fields within American chemistry are more-or-less fixed by their institutional and other contextual settings. Within each of those fields there has been a tendency for increasing elaboration, as growing numbers of academic researchers have sought out possible problems and as people have jumped from one promising area to another. The broad specialties of chemistry have thus come to possess an increasingly developed fine-structure of ideas, facts, and techniques. The names of sections in the 1907 *Chemical Abstracts* well describe the areas of current graduate research, as shown in Figure 6.4–2. However, to find those names in a recent issue of *Chemical Abstracts* one must look not to individual sections but to groups of sections.

6.5. ACS Presidents: Some Micro-Indicators

INDICATOR HIGHLIGHTS:

Prosopographical analysis of the careers of American Chemical Society presidents elucidates the changing nexus among occupation, profession, and discipline in American chemistry.

Since the first World War, an elite gerontocracy has presided over American chemistry. Academic chemists have exerted disproportionate control over the governance of the ACS. Routes of entry to ACS leadership have been extremely limited. Intricate networks of social ties have linked members of the ACS presidential elite.

The concepts and constructs of a scientific discipline may be thought of as the language of the science. Academics have special skills in the articulation and use of that language. These skills give them natural advantages and major roles within the organized life of the science, a phenomenon reflected in the governance of American chemistry during the past century.

The American Chemical Society has long occupied a central position within the institution of chemistry. The choice of a president for the Society symbolizes the institution's values. The careers of presidents display something of the changing nexus between occupation, profession and discipline. Those values and that nexus may be partially exposed by a highly disaggregated analysis using what might be termed "micro-indicators". In this last section we shall turn away from macro-indicators of the institution of chemistry and instead use micro-indicators to elucidate certain matters of interest. The use of micro-indicators also provides a specimen of the possibilities inherent in the analysis of fine structure, an effort which may be mounted within the frame secured by macro-indicators of the kind presented earlier in this work.[28]

6.5.1. AGE STRUCTURE

The mean age of ACS presidents has remained remarkably constant over the past hundred years at about 60, with the exception of the period from 1886 to 1915 (Figure 6.5–1). An elite gerontocracy has presided over the chemical community since the first World War. In this respect the ACS is characteristic of American scientific and learned societies.[29]

The group of ACS presidents elected from 1886 to 1915 were ten years younger on average. At that time the ACS hierarchy was captured by a generation of German-trained PhDs which included a core group of ten presidents born between 1850 and 1870. Unlike their earlier counterparts, whose careers followed diverse lines, these men were administrators of the newly established and rapidly expanding "PhD machine" at the turn of the century. Their work experience reflects in microcosm that of a large number

[28] Information for the prosopographical study in this section has been drawn from IB, Miles (ed.), 1976, and successive editions of *American Men of Science* and *American Men and Women of Science*. We have also used the *Biographical Memoirs of the National Academy of Sciences* and IB, Haynes (ed.), 1951. (For a similar study conducted in the early 1930s, see IB, Howe and Nelson, 1933.) Table 6.37 gives a list of American Chemical Society presidents.

[29] A 1955 survey of sixty-seven American scientific and learned societies showed that presidents of these organizations were most frequently between the ages of 52 and 62 at the time of first election (III, Lehman, 1955); see also III, Zuckerman, 1970, 239. For a general discussion of the implications of age stratification in science, see III, Zuckerman and Merton, 1972.

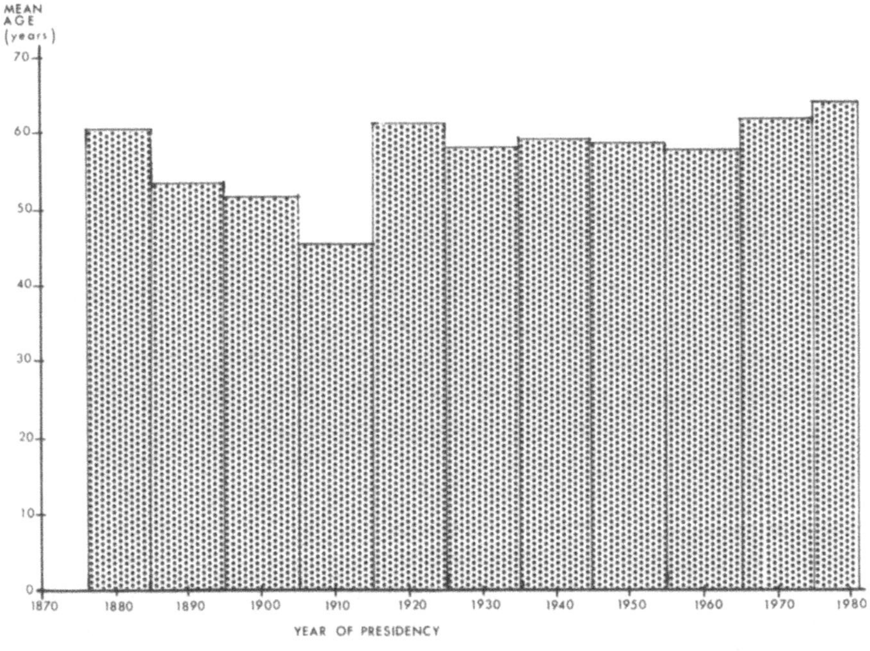

Fig. 6.5–1. Average ages of American Chemical Society presidents during tenure of office, by decade, 1876–1981. (See Table 6.38, column 1.)

of contemporary chemists during a transitional period when disciplinary training was being introduced within the institutional context of American research universities.[30]

6.5.2. EDUCATIONAL BACKGROUND

Shifts in the educational background of ACS presidents over time display vividly the decline of German hegemony in the advanced training of American chemists. Before 1896, six out of ten ACS presidents had been trained in German universities, while another three out of ten came to chemistry via medical training. By the turn of the century (1896–1905), ACS presidents were already as likely to have been trained in the United States as in Germany, and the proportion of American chemistry PhDs among ACS presidents increased steadily thereafter (as shown in Figure 6.5–2).

The institutional pattern for training of ACS presidents exhibits a high degree of concentration, centered on a small circle of ten universities which accounted for two-thirds of all ACS presidential terms from 1876 to 1977. Table J lists these top ten institutions. A mere ranking of institutions obscures long-run trends in the education of the ACS elite, since the appeal of different chemistry departments for the most ambitious students varied over time. This is exemplified in the temporal distribution of degrees conferred on future ACS presidents by the top three institutions in the table: Göttingen, Harvard, and Illinois. Although men with Göttingen backgrounds were elected to the ACS presidency as late as 1936 (Edward Bartow), this group of 11 chemists (with the exception of Irving Langmuir) was trained at Göttingen between 1832 and 1895. Harvard maintained a strong position throughout the century; the ten Harvard-trained ACS presidents held office as recently

30 The ten members of this group are: W. F. Hillebrand (1853/1906), E. F. Smith (1854/1895), J. H. Long (1856/1903), F. P. Venable (1856/1905), A. A. Noyes (1866/1904), M. T. Bogert (1868/1907–1908), W. R. Whitney (1868/1909), W. D. Bancroft (1867/1910), A. Smith (1865/1911), and T. W. Richards (1868/1914). (The numbers in parentheses indicate year of birth/year of ACS presidency.)

Of these individuals, Willis Whitney was the only one to have moved out of academe by the time he was elected president of the American Chemical Society. Although he spent most of his active career as director of the General Electric Research Laboratory, his first position upon returning from Leipzig in 1896 was in the chemistry department of the Massachusetts Institute of Technology.

T. W. Richards spent two important postdoctoral years in Germany, although his degree was actually from Harvard University (1888).

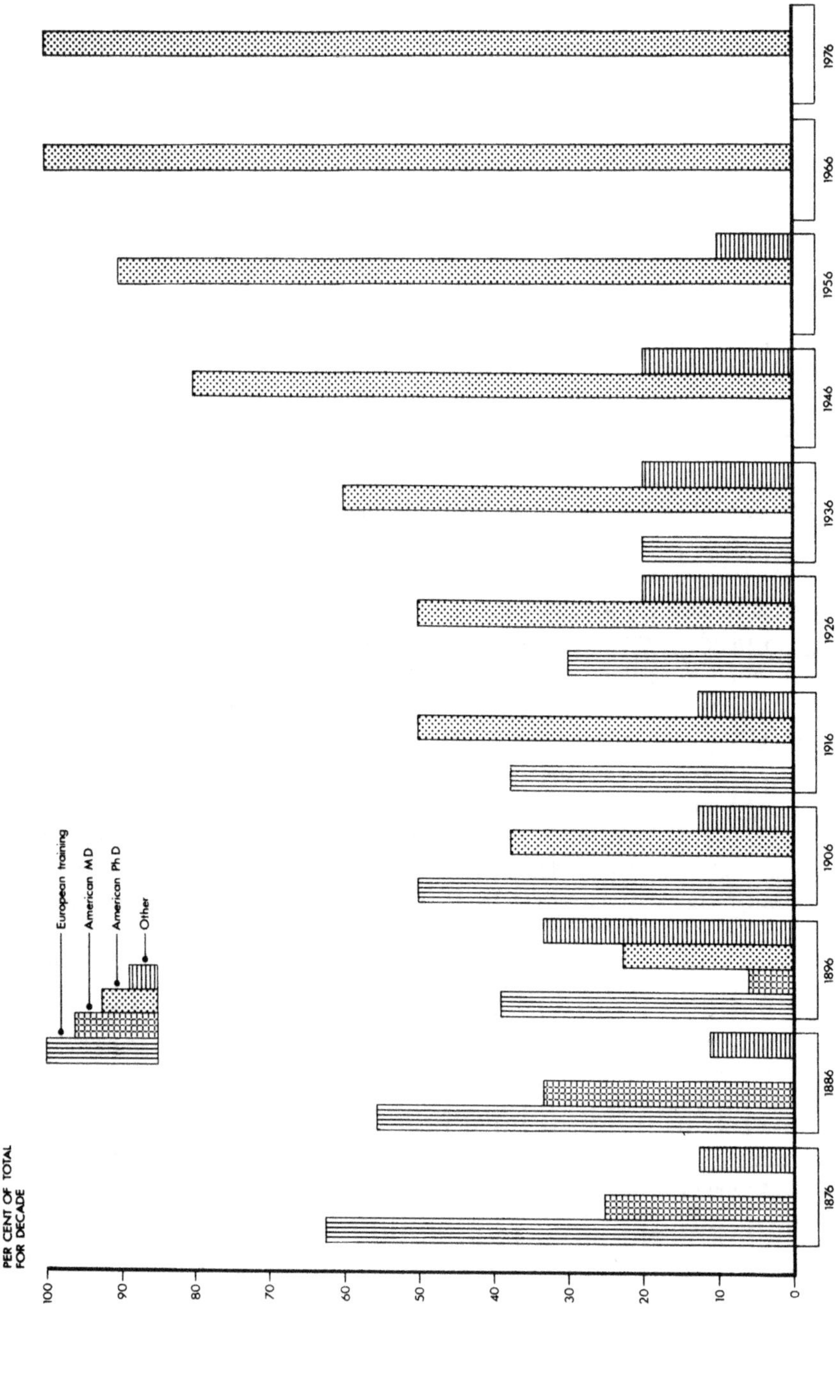

Fig. 6.5–2. Educational backgrounds of American Chemical Society presidents, by decade, 1876–1981. (See Table 6.39, columns 3, 5, 7, and 9.)

TABLE J
Leading institutions at which American Chemical Society presidents were educated, 1876–1981

Institution	Number of ACS presidents
Universität Göttingen	11
University of Illinois	11
Harvard University	10
Massachusetts Institute of Technology	7
Universität Leipzig	6
Johns Hopkins University	5
University of Michigan	4
University of California	3
University of Chicago	3
University of Wisconsin	3
Yale University	3

See Table 6.40.

as 1972, and they were trained at Harvard between the mid-1850s and late-1930s. In contrast to both Göttingen and Harvard, the University of Illinois only recently attained its position as a major producer of ACS presidents. Presidents with Illinois backgrounds received their training during the inter-war years, and they have been predominant since 1955. No fewer than nine of the twenty-six most recent leaders of the ACS were Illinois-trained.[31]

31 The Göttingen group consists of the following eleven chemists: C. A. Goessmann (1852/1887), J. W. Mallet (1852–1853/1882), C. F. Chandler (1854/1881, 1889), G. C. Caldwell (1857/1892), H. B. Nason (1857/1890), Ira Remsen (1870/1902), E. F. Smith (1876/1895, 1921–1922), F. P. Venable (1879–1881/1905), E. Bartow (1895/1936), T. W. Richards (1895/1914), I. Langmuir (1904–1906/1929). (The numbers in parentheses indicate year(s) at Göttingen/year(s) of ACS presidency.)

The Harvard-trained ACS presidents were: C. E. Munroe (1867–1871/1898), T. W. Richards (1886–1888/1914), H. W. Wiley (1872–1873/1893–1894), F. W. Clarke (1865–1868/1901), A. B. Lamb (1902–1904/1933), R. Adams (1906–1912/1935), F. C. Whitmore (1907–1914/1938), F. Daniels (1910–1914/1935), C. C. Price (1934–1936/1965), and M. Tishler (1934/1972).

ACS presidents who received their highest level of training in the chemistry department at the University of Illinois are: E. H. Volwiler (1914–1918/1950), C. S. Marvel (1915–1920/1945), A. L. Elder (1919–1928/1960), C. F. Rassweiler (1920–1924/1958), W. R. Brode (1922–1925/1969), B. Riegel (1930–1934/1970), B. S. Friedman (1930–1936/1974), W. J. Sparks (1936/1966), C. G. Overberger (1941–1944/1967), W. J. Bailey (1943–1946/1975), and G. W. Stacy (1943–1946/1979).

There is also a clear trend in the geographical distribution of centers at which ACS presidents have been trained. Prior to World War I, the chemist with high ambitions packed off to Germany, headed especially for Göttingen or Leipzig. By the turn of the century, the proportion of ACS presidents who had traveled abroad for advanced chemical training was declining steadily, reflecting the increased availability of such training at American research universities. From about 1900 to the early 1950s, most ACS presidents were recruited from prestigious private universities in the northeastern United States. Since the mid-1950s, as Table 6.41 shows, nearly three-quarters of the ACS presidents elected have been graduates of state universities (especially Illinois) in the North Central region of the United States, a shift related to an underlying demographic trend in U.S. doctoral degree production.

It is rewarding to rank the institutions with which ACS presidents were *affiliated* as students, faculty members, postdoctoral fellows, or research workers (Table K). The ranking may be compared with that given above for educational institutions *producing* future ACS presidents (Table J). There is considerable overlap between the two lists, with Illinois, Harvard, Göttingen, MIT, Johns Hopkins, Leipzig, Chicago, California, Michigan, and Wisconsin appearing on both. Illinois, Harvard, MIT, and Chicago especially emerge as even more central than indicated by their graduates' mobility into the ACS presidential elite. One may also note the presence of other Ivy League institutions (Penn, Columbia, and Cornell) and state universities (Michigan, Wisconsin, and Ohio State) in prominent positions, or the relative importance of federal government agencies (the Bureaus of Chemistry in the Department of Agriculture and U.S. Geological Survey; U.S. Bureau of Mines; National Bureau of Standards) and the pharmaceutical industry (especially Merck) in the careers of those cosmopolitan influentials who comprised the ACS presidential elite.

6.5.3. Institutional Loci and Employment

Analysis of the institutional locations of ACS presidents during their tenure of office yields several interesting observations (Figure 6.5–3). First, until the mid-1960s, academic chemists were the dominant group among ACS presidents, exerting an influence disproportionate to their size as a sector of the American chemical community. (Some evidence suggests that the over-representation of academics extends down to the level of local section office-holding.) Second, although four of the first 25 individuals to become

TABLE K
Leading institutions with which ACS presidents were affiliated, 1876–1981

Institution	Number of ACS presidents
University of Illinois	18
Harvard University	15
Massachusetts Institute of Technology	12
Universität Göttingen	11
Johns Hopkins University	7
University of Pennsylvania	7
University of Chicago	6
Columbia University	6
Cornell University	6
Federal government laboratories[a]	6
Universität Leipzig	6
University of California	5
University of Michigan	5
University of Wisconsin	5
Merck, Inc.	4
Ohio State University	4

See Table 6.44.

Note: An institution is included in this list if, at any time in his post-baccalaureate career, an ACS president was a student, faculty member, postdoctoral fellow, or research worker at that institution.

[a] Includes the Bureaus of Chemistry of the U.S. Department of Agriculture, U.S. Geological Survey, and U.S. Forest Service; the U.S. Bureau of Mines; and the National Bureau of Standards.

ACS president were employed by the federal or state governments when elected, no government chemist has been elected to the presidency of the American Chemical Society since 1906. Finally, chemists from the industrial sector were a major group among ACS presidents during the Society's first decade, but (except for a group of four industrially-connected presidents in the decade from 1926 to 1935) did not regain their position until after World War II. The picture of academic dominance given by the institutional locations of ACS presidents upon election retains its essential features when one considers instead the employment experience of ACS presidents over the course of their careers prior to taking office. As Table 6.43 shows, fully

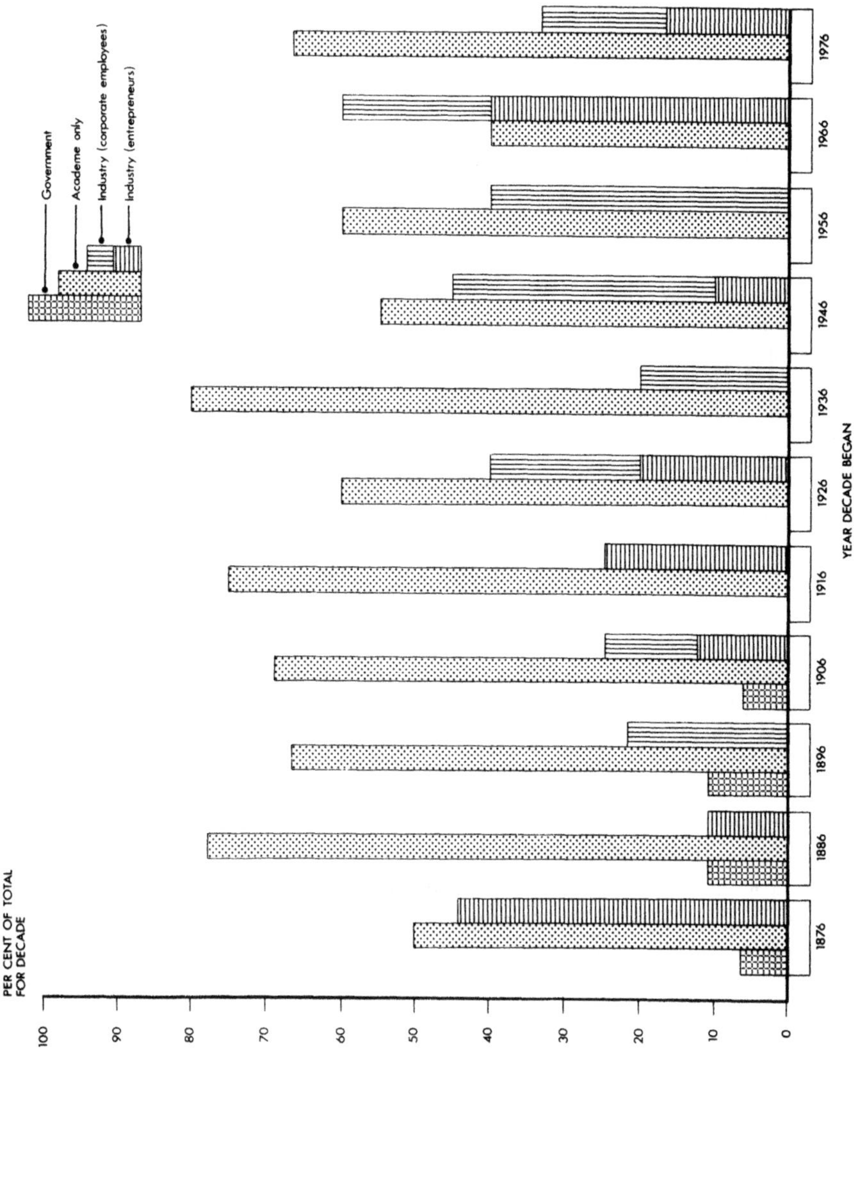

Fig. 6.5–3. Institutional locations of American Chemical Society presidents, 1876–1981. (See Table 6.42, columns 3, 5, 7, and 9.)

half of all ACS presidents from the mid-1880s to the mid-1960s held only academic positions during their pre-presidential careers.[32]

The multiplicity of occupational roles of ACS presidents in the first few decades of the Society's existence reflects the institutional bases for American chemistry in the latter part of the nineteenth century. Variegated academic niches were only one of many sources of social support for chemistry during the early years of the ACS, when individual careers characteristically encompassed a range of jobs as varied as public health work for municipalities, chemical testing and assaying for state geological surveys and agricultural experiment stations, or private consulting for industrial firms. Perhaps the best example of the diversity of professional work available to the resourceful chemist in the Gilded Age is afforded by the career of Charles F. Chandler, ACS president in 1881 and 1889. Concurrently in the 1870s, Chandler held three academic appointments in chemistry (at the Columbia University School of Mines, New York College of Pharmacy, and the New York College of Physicians and Surgeons), served as President

[32] This point is highlighted when one recalls that, at least since the late 1920s, 70% of American chemists have been employed in private industry (see Section 5.1). Strauss and Rainwater note that 70.4% of ACS members in 1960 were either self-employed or in private industry (III, Strauss and Rainwater, 1962, 37). The same survey indicated that academic chemists (and research administrators) are at least twice as active as other groups within the chemical community in local section governance (*ibid.*, 176, 187).

The resurgence of industrial chemists in the ACS presidency since World War II is connected to at least two developments in ACS policies and politics. First, in 1943 the ACS instituted a policy of alternating academics and industrialists in the presidency. Second, since the mid-1960s increasing concern among rank-and-file ACS members over such issues as unemployment and member assistance has led to a number of petition candidacies from the industrial chemical community (see IB, Skolnik and Reese, 1976, 14, 183–185).

The four government chemists (and their positions during their tenure as ACS president) were: S. W. Johnson (1878), Director, Connecticut Agricultural Experiment Station; H. W. Wiley (1893–1894), Chief, Bureau of Chemistry, U.S. Department of Agriculture; F. W. Clarke (1901), Chief, Bureau of Chemistry, U.S. Geological Survey; and W. F. Hillebrand (1906), Chemist, U.S. Geological Survey, appointed Chief Chemist, National Bureau of Standards, 1908). Although these four individuals were the only ACS presidents working for the federal government during their term of office, others have had experience as government scientists at some time during their careers. Among recent examples are Per Frolich (1943), Chief Scientist, U.S. Army Chemical Corps, 1954–1960; R. W. Cairns (1968), Deputy Assistant Secretary for Science and Technology, U.S. Department of Commerce, 1971–1972; W. R. Brode (1969), Associate Director, National Bureau of Standards, 1947–1958; and G. T. Seaborg (1976), Chairman, Atomic Energy Commission, 1961–1971.

of the New York Board of Health, performed chemical analyses on a commercial basis, provided expert testimony in numerous patent litigations, co-edited the ***American Chemist*** with his brother William, wrote extensively for *Johnson's New Universal Encyclopedia* (edited by his Columbia colleague, F. A. P. Barnard) – and still found time to play a central role in the establishment of the American Chemical Society in 1876![33]

After the turn of the century, occupational backgrounds of ACS presidents became less diversified, a change related to the routinization of careers within the chemical community at large. Edgar Fahs Smith (1895, 1921, 1922), Ira Remsen (1902), and Marston T. Bogert (1907, 1908) – Chandler's successor at Columbia – are three archetypal academic chemists of this period, just as Willis R. Whitney (1909) and William H. Nichols (1918, 1919) exemplify newly-available careers in industrial research and corporate chemical enterprise. This shift was accompanied by a general decline in the importance of state or federal government positions as a route to the ACS elite, along with the long-term displacement of "chemist-entrepreneurs" and consulting chemists among ACS presidents, in favor of executives of chemical and other industrial corporations. Recent years provide some indication that members of the elite are returning to consulting work as a livelihood (as shown in Table 6.43 by the breakdown of employment experience for ACS presidents from 1966 to 1981).

6.5.4. Social Ties

In a situation familiar to students of social stratification, high elites in science are maintained by selective processes of recruitment, socialization and allocation of resources. Thus it is not surprising to find ACS presidents linked by social ties similar to those found among other groups in the aristocracy of American science, such as members of the National Academy of Sciences or Nobel laureates.[34] This section will explore the importance

[33] III, Beardsley, 1964, 43–69, surveys occupational opportunities for chemists in late-nineteenth century America. On Chandler, see III, Bogert, 1931; III, Billinger, 1939; III, Larson, 1950; and III, Rossiter, 1977. Chandler was the most adept practitioner of *cumul* in his generation of chemists, but similar examples are provided by the careers of other ACS presidents, such as J. L. Smith (1877), T. S. Hunt (1879, 1888), F. A. Genth (1880), or J. C. Booth (1883–1885).

[34] Recent studies on social stratification and elites in science include III, Cole and Cole, 1973; III, Mulkay, 1976; III, Zuckerman, 1970, esp. 243–245; and III, Zuckerman, 1977.

of social ties through kinship, apprenticeships, colleagueship, and shared experience.

Only a few illustrations can be given. The one direct kinship tie among ACS presidents is found in the father-son team of William A. Noyes (1920) and W. Albert Noyes, Jr. (1947). Examples of apprenticeship abound. For instance, J. W. Draper (1876) taught William H. Nichols (1918, 1919) at New York University in the early 1870s; J. W. Mallet's (1882) students at the University of Virginia in the 1870s and 1880s included Francis P. Venable (1905) and Charles L. Reese (1935); and Albert Prescott's (1886) student, Moses Gomberg (1931), was himself mentor to Edgar C. Britton (1952) and John C. Bailar, Jr. (1959). The most striking case of such ties among a group of ACS presidents involves those chemists who obtained their training during the department chairmanships of T. W. Richards (1914) and Arthur B. Lamb (1933) at Harvard, and Roger Adams (1935) at Illinois (Figure 6.5–4). This "Harvard-Illinois axis" has accounted for approximately one in four ACS presidents elected since Theodore W. Richards' term of office in 1914. Illinois is even more important than the figure shows, for Folkers obtained his baccalaureate there and Price was on the faculty in the era of Adams's chairmanship.[35]

As the example of Price suggests, many ACS presidents were colleagues of other members of the ACS high elite in particular chemistry departments or industrial laboratories. For instance, when Charles Chandler (1881, 1889) was succeeded as ACS president in 1890 by Henry Nason of Rensselaer Polytechnic, he was probably reasonably sure that the former colleague of Charles A. Goessmann (1887) – one of Chandler's fellow students at Göttingen in the 1850s – would fill the post admirably well. Penn's chemistry faculty in the late 1870s and 1880s included F. A. Genth (1880), E. F. Smith (1895, 1921, 1922) and George Barker (1891), while Willis R. Whitney (1909), A. A. Noyes (1904) and James F. Norris (1925, 1926) were colleagues at MIT in the 1890s. After Whitney's move to the General Electric Research Laboratory in 1900, he recruited Irving Langmuir (1929) to the research staff. Moving to a later period, seven members of the ACS presidential group were colleagues at Illinois in the 1920s and 1930s – S. W.

[35] Richards' department at Harvard is discussed in III, Kopperl, 1976. On Roger Adams and chemistry at Illinois between World Wars I and II, see III, Beal, 1927; III, Bartow, 1941; III, Carroll, 1982; III, Tarbell and Tarbell, 1979, 1981; and III, Tarbell, Tarbell, and Joyce, 1980. For chemical genealogies of the Illinois and Wisconsin departments, see III, Bartow, 1939, and III, Rocke and Ihde, 1979. The listing in Figure 6.5–4 is arranged by year of PhD.

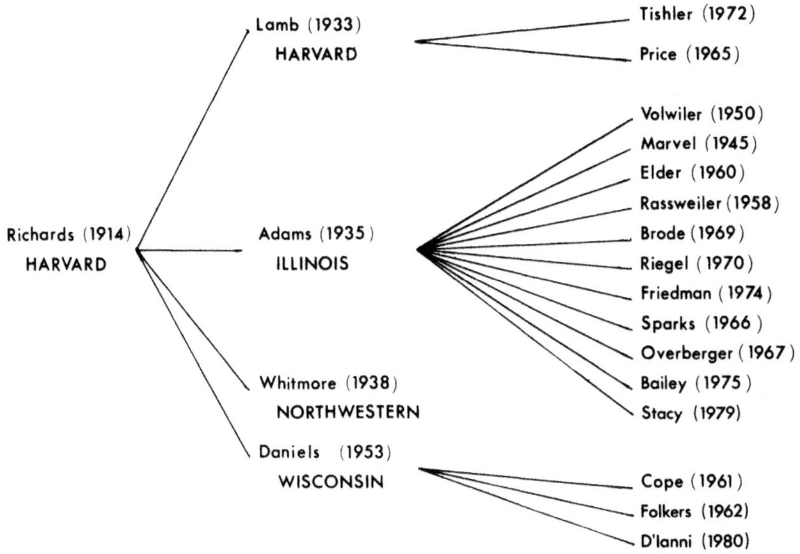

Fig. 6.5–4. The "Harvard-Illinois axis" among American Chemical Society presidents.

Parr (1928), W. A. Noyes (1920), Roger Adams (1935), Edward Bartow (1936), Carl S. Marvel (1945), B. S. Friedman (1974), and Charles C. Price (1965). And we may note that Joel Hildebrand (1955), Melvin Calvin (1971) and Glenn T. Seaborg (1976) had been colleagues at the Berkeley campus of the University of California since the late 1930s.

L. V. Redman (1932), Edward Weidlein (1937) and Leo Baekeland (1924) provide an example of similar colleagueship ties in industrial chemistry. Redman and Weidlein were two of the earliest recipients of industrial fellowships under Robert K. Duncan at the University of Kansas. Both men accompanied Duncan to Pittsburgh with the establishment of the Mellon Institute for Industrial Research in 1913. Weidlein remained at the Institute, eventually assuming the directorship, but Redman left in 1914 to set up Redmanol Chemical Products. This entrepreneurial venture in marketing phenolic resins soon brought Redman into competition with Baekeland's General Bakelite Company, which manufactured similar plastics. After considerable litigation, the two companies were merged into the Bakelite Corporation in 1922.[36]

The final class of linkages among members of the ACS presidential elite includes the strong informal ties formed by shared social experience. For instance, there is the solidarity which derived from being fellow students of great masters, such as Wöhler (Booth, Chandler, Goessmann, Caldwell, Remsen, Smith, and Nason) or Ostwald (A. A. Noyes, Bancroft, Whitney, Richards, and Lind). Also important in this regard are the local chemical communities of the 1880s and 1890s – in Boston, New York, Philadelphia, and Washington, D.C. – which served to foster a sense of common purpose among chemists of the period. The complex character of these social relationships is suggested by one final example concerning C. E. Munroe (1898), Frank W. Clarke (1901), and Harvey W. Wiley (1893, 1894).[37]

Wiley and Clarke were the leaders of the movement for a national chemical society to replace the parochial New York-centered chemical club the ACS had become by the late 1880s. They led that movement, using the Chemical Society of Washington and Section C of the American Association for the

[36] On Baekeland and the early plastics industry, see III, Haynes, 1949, 437; and III, Friedel, 1983. For descriptions of Duncan's industrial fellowships plan and its operation at both the University of Kansas and the Mellon Institute, see III, Duncan, 1907, 241–256; III, Duncan, 1913; III, Bacon, 1915 and 1916; and III, Mellon Institute, 1924.

[37] The role played by Munroe, Clarke and Wiley in the reorganization of the American Chemical Society have been well-documented by C. A. Browne; see III, Browne, 1938b, 1939; and III, Browne and Weeks, 1952, 30–39.

Advancement of Science as their institutional bases. Munroe acted as a go-between, advocating similar measures from within the New York group, and he was instrumental in organizing the first section of the ACS in Rhode Island in 1891. This much of the story is well-known, but several intriguing facts emerge from the backgrounds of Munroe, Wiley, and Clarke which help one understand why they were in particularly advantageous positions relative to ACS reform.

Their connections may be traced to the Lawrence Scientific School at Harvard, where Munroe and Clarke were fellow students in the late 1860s. Munroe stayed at Harvard as an instructor from 1871 to 1874, where one of his students was the young chemist, Harvey Wiley. During the agitation for reform of the ACS between 1888 and 1892, Clarke and Wiley were both stationed in Washington (as Chiefs of the Bureaus of Chemistry of the U.S. Geological Survey and U.S. Department of Agriculture, respectively) and Munroe was at the U.S. Torpedo Station in Newport, Rhode Island (hence his role in the establishment of the Rhode Island section). Not only could Munroe assist his old friends by attempting to win converts in New York to the new goals they advocated for the ACS, but as son-in-law to George Barker (1891) he could also find a sympathetic ear among the Philadelphia chemists. After the reorganization of the ACS in 1892–1893, Munroe moved to George Washington University, thus joining the strong Washington chemical community organized by Clarke and Wiley.

These examples suggest some of the ways in which it is possible to move between macro- and micro-indicators, and from the latter to familiar levels of historical analysis. Our report has necessarily focused on macro-indicators in an endeavor to assay the changing structure of American chemistry over a century-long period. The assembling of those indicators is one move toward the creation of a context in which more fruitful historical and policy questions may be posed. Many of those questions hinge on a detailed knowledge of the fine structure of events, of a kind that depends on the availability of many micro-indicators.

6.6. Concluding Remarks

American chemistry has grown tremendously over the past century as occupation, profession, and discipline. From a fledgling field in the 1870s, lacking cohesive institutional forms or a strong social identity, chemistry has evolved into one of the most influential scientific professions. To interpret this growth as the continuous expansion of a homogeneous system would be to misread the historical evidence. Relative rates of growth for different sectors of the American chemical community have shifted radically over the past hundred years. The chemical indicators presented above have suggested insights into otherwise hidden trends, short-run variations, and conjunctures that shaped the perceptions of participants in this historical development.

Long-run trends in aspects of American chemistry – the aggregate series on which much of this report is based – provide a background against which short-run variations in different indicators may be evaluated. Careful scrutiny of contemporary pronouncements by chemists and other observers of the enterprise may also suggest implicit indicators of the state of American chemistry at a given time. It is unwise to interpret a given indicator in isolation: one should look instead to the conjuncture of several indicators over a limited period for the context in which chemists and others reacted to events.

The judicious selection of indicator clusters, and astute shifts among different levels of trend analysis, should enable historians to enrich significantly their accounts of American chemistry. With few exceptions, the scant literature on chemistry in America is either filiopietistic biography or dry, narrow institutional history. The approach adopted in this study has potential for improving this historiographic situation. Earlier chapters simply presented a wide range of indicators, a necessary first step toward the ambitious goal of providing an adequate historical account of the changing place of chemistry in American culture. When applied further in studying micro-indicators of, for example, the changing "ecology" of the chemical discipline, the shared experiences of certified chemists from strategically-placed university departments, or the interactions of academic chemists with industrial patrons, chemical indicators can shed new light on traditional problems of both history and policy.

There is no special magic in numbers, no "royal road" to an imperial science of history. We agree with François Furet that "the division of historical reality into series leaves the historian confronted with his material broken down into different levels and subsystems, among which he is at liberty to suggest internal relationships if he chooses." This essay has concentrated on a first exploration of those levels and subsystems, while hinting at some relationships among them. We leave for another time any action on the agenda thus suggested.

A. CHEMISTRY AND CHEMISTS: ALTERNATIVE DEFINITIONS

In the text above, care was taken to indicate the criteria by which group such as "all chemists in the labor force", "chemists employed in college teaching", or "chemistry degree recipients" were defined. We elucidated the differences between ideal criteria and those used to construct the best *available* time series. For those series our usual course was to rely upon explicit-discovered indicators, which at least provide homogeneous data. Where alternative data are available, we have considered the effects on our conclusions of varying the definitions of what constituted a chemist. This appendix discusses the results of checking our procedures in this manner, emphasizing those adjustments which seemed most likely to have an adverse effect on our interpretations.

A.1. Chemists, Assayers and Metallurgists

As noted on p. 14, the Census definition of chemists in the experienced civilian labor force excludes chemistry teachers and chemical engineers, classifying members of these groups in other occupational categories. Another long-standing characteristic of Census publications of occupational statistics is the grouping of assayers and metallurgists with chemists. The data used for Table 2.2 represent Census recalulations for chemists and assayers only from 1900 to 1960, leaving us with an inconsistency for the years, 1870, 1880, 1890, and 1970. In these years, the reported number of chemists includes assayers *and* metallurgists. This presents little problem for the nineteenth-century data, since assayers and metallurgists then were likely to share similar training and work activities with the majority of chemists. The 1970 figure is more questionable, since the training and roles of metallurgists have become separated from chemistry during the past few decades. For this reason, an attempt to obtain an estimate of the number of chemists (and assayers) *alone* in 1970 is of some interest.

Metallurgists were automatically combined with chemists in the 1970 Census coding.[1] This means that even in the most disaggregated, unpublished

[1] See Table 2.1.

series of detailed occupational statistics, no number exists for metallurgists alone. Only by returning to original 1970 Census returns and counting metallurgists alone – a daunting task -- could a precise figure be determined. Thus we must be content with an approximation, for which Census publications provide a clue.

The Census Bureau changed its system of occupational codes between 1960 and 1970. Because of this, Census statisticians recalculated 1960 figures on a 1970 basis, so that the two classification schemes could be compared. This was accomplished by recoding a sample of 100 000 returns from the 1960 Census according to the 1970 occupational classification scheme.[2] Detailed analysis indicated the percentage of the 1970-style figure for chemists in 1960 which derived from each of the 1960 subcategories, resulting in the following formulas:

> Male chemists in 1960 (according to the 1970 system) X 0.87266 = Male chemists in 1960 (according to the 1960 system).
>
> Female chemists in 1960 (according to the 1970 system) X 0.97633 = Female chemists in 1960 (according to the 1960 system).

Assuming that these ratios were constant from 1960 to 1970, this provides conversion factors with which to obtain 1970 figures for the number of chemists of both sexes according to the 1960 system. This is precisely the approximation desired, since the 1960 classification scheme allocated chemists and metallurgists to different categories, counting the latter with metallurgical engineers. Since metallurgical engineers and metallurgists are the only subgroup factored into the recalculation of the 1960 data according to the 1970 system, one can assume that the difference between the 1970-style figure for chemists and the 1960-style figure produced by this method consists largely of metallurgists.

Applying the conversion factors and rounding to the nearest thousand yields a total of 98 000 chemists in the labor force in 1970, *not counting*

[2] II, USBC, 1972. The conversion percentages given below are from 19, Table 1. (See *ibid.*, 4–5, for a discussion of the standard error of these estimates.)

metallurgists.[3] This is about 13 000 below the reported figure of 110 000 chemists (Table 6.25, column 2). There is corroborative evidence that this adjustment is correct. The difference of roughly 13 000 (actually 12,641) between the observed and adjusted figures for chemists in the 1970 civilian labor force should equal the number of metallurgists in America in 1970. One means of testing this number is to compare it with the 1970 membership of the major professional society for American metallurgists, the Metallurgical Society of the American Institute of Mining, Metallurgical, and Petroleum Engineers. This society reported membership of around 13 000 in 1970 – remarkably close to the above estimate.[4] Having derived an alternative estimate of 98 000 chemists in 1970, the question becomes why did we analyze trends based on the uncorrected figure? The answer is a pragmatic one. Had we used a recalculated figure for the number of chemists in 1970, the nineteenth-century values would still have been inconsistent. The corrected value would only have accentuated the trend of relative decline shown in Figure 2.2–9. Hence the decision to utilize the data as reported.

This exercise in modification of Census estimates demonstrates the prodigious effort required to bring available data into accord with new categories and questions. The moral seems to be that the most prudent course of action is to use reported data *as discovered*, with due regard to the problems of differing definitions.

[3] Calculated as follows:

1970 Entry	x conversion factor	= same entry by 1960 system
Male chemists, 14 yr old and over, in the experienced civilian labor force	96,783 × 0.87266	84 459
Female chemists, 14 yr old and over, in the experienced civilian labor force	13,384 × 0.97633	13 067
	Total	97 526

The 1970 entries are from IA, USBC-COP, 1973, 718, Table 221.

[4] Membership figures for the Metallurgical Society are printed monthly in the *Journal of Metals.*

A.2. Chemists in the Labor Force

Another modification to the Census definition of chemist involves the addition of chemistry faculty and chemical engineers, which accords with a "common-sense" notion of what a chemist is. Figure A.2–1 displays the results of using different definitions of chemists as an occupational group in constructing time series. It presents the same time series given in Figure 2.1–2, as well as four new lines representing various adjustments to correct for the exclusion of chemistry faculty and chemical engineers from the Census data. Most of the discrepancy between Census estimates and the Bureau of Labor Statistics estimates since the late 1950s can be explained by the addition of chemistry faculty – using the broadest possible definition – to the Census data. This is plausible, since the BLS estimates include all academic chemists, even graduate students.[5]

The inclusion of chemical engineers adds significantly to the number of chemists for recent years – so much so that one might be led by the appearance of this graph to conclude that the rapid growth in the number of chemical engineers accounts for the apparent slowing in the growth of the number of chemists reported by the Census. This alternative hypothesis is undermined by Figure A.2–2, which presents the data from Figure A.2–1 on semilogarithmic axes. The straight line on this graph indicates the calculated exponential trend (about 5.5% per annum) in the Census estimates *alone* from 1900 to 1930. It is apparent that all of the data – even those adjusted to include chemistry faculty and chemical engineers – deviate from this exponential trend at each end of the period investigated. For the nineteenth-century era, growth was faster than 5.5% per annum, while the rate of increase in recent years has been slower. This pattern accords with one of Gompertz growth, as we concluded in the analysis on pp. 15–20. The only modification of our claims necessitated by these adjustments is in the range of the upper limit on the number of chemists in the labor force. Using the Census figures yields an upper limit of about 256 000 (actually, a 95% confidence interval of 232 000 to 280 000). This limit would be about 50% higher if chemical faculty and chemical engineers were included.

[5] See Tables 5.16 and 5.17 for the different measures of chemistry faculty used in these adjustments. See Table 3.11 for the estimates of chemical engineers and Table 5.1 for the BLS estimates. Figures A.2–1 and A.2–2 include graduate students only for the BLS estimates. According to the NSF, the number of chemistry graduate students in 1970 was about 9 000 – nearly exactly the difference between the highest Census figure for chemists and chemistry faculty and the BLS figure.

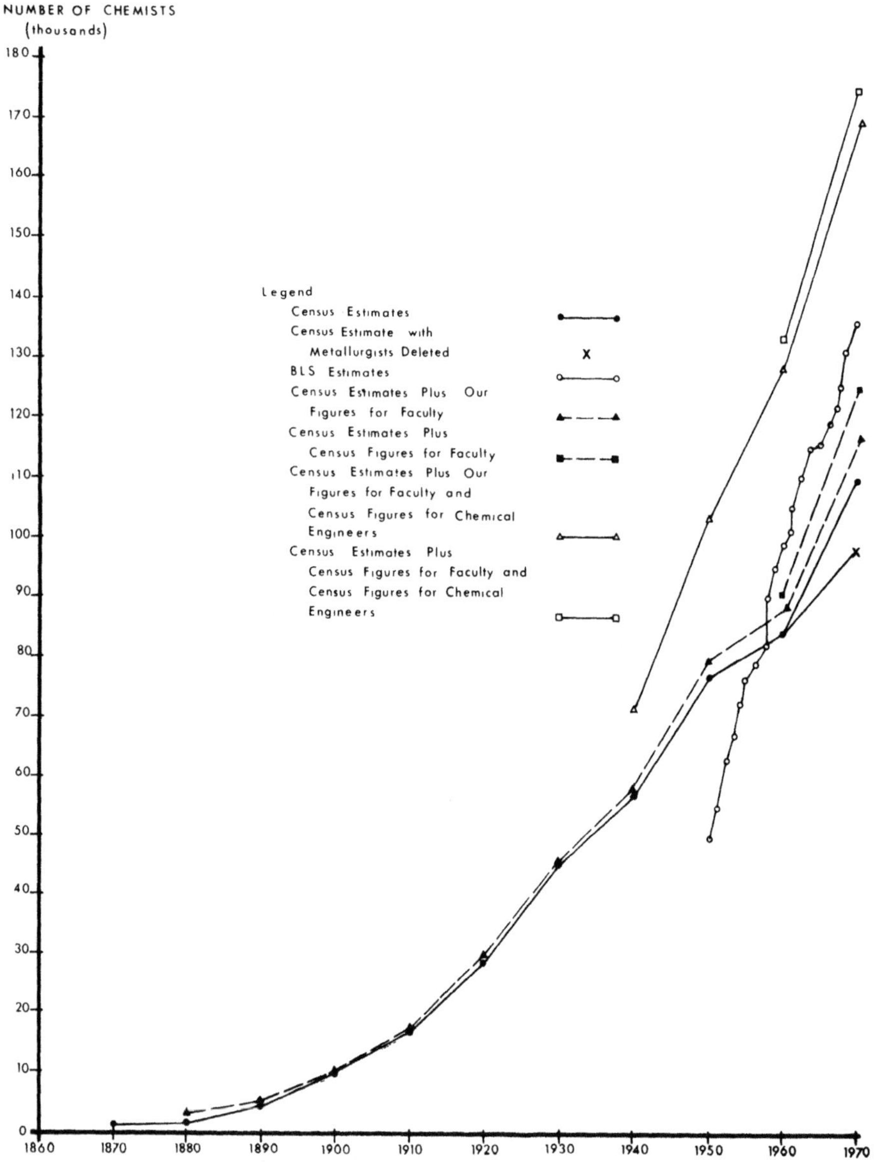

Fig. A.2–1. The number of chemists, with possible adjustments, 1870–1970. (See Table 2.2, column 1; Table 3.11, column 2; Table 5.1, column 2; Table 5.16, column 4; and Table 5.17, column 2.)

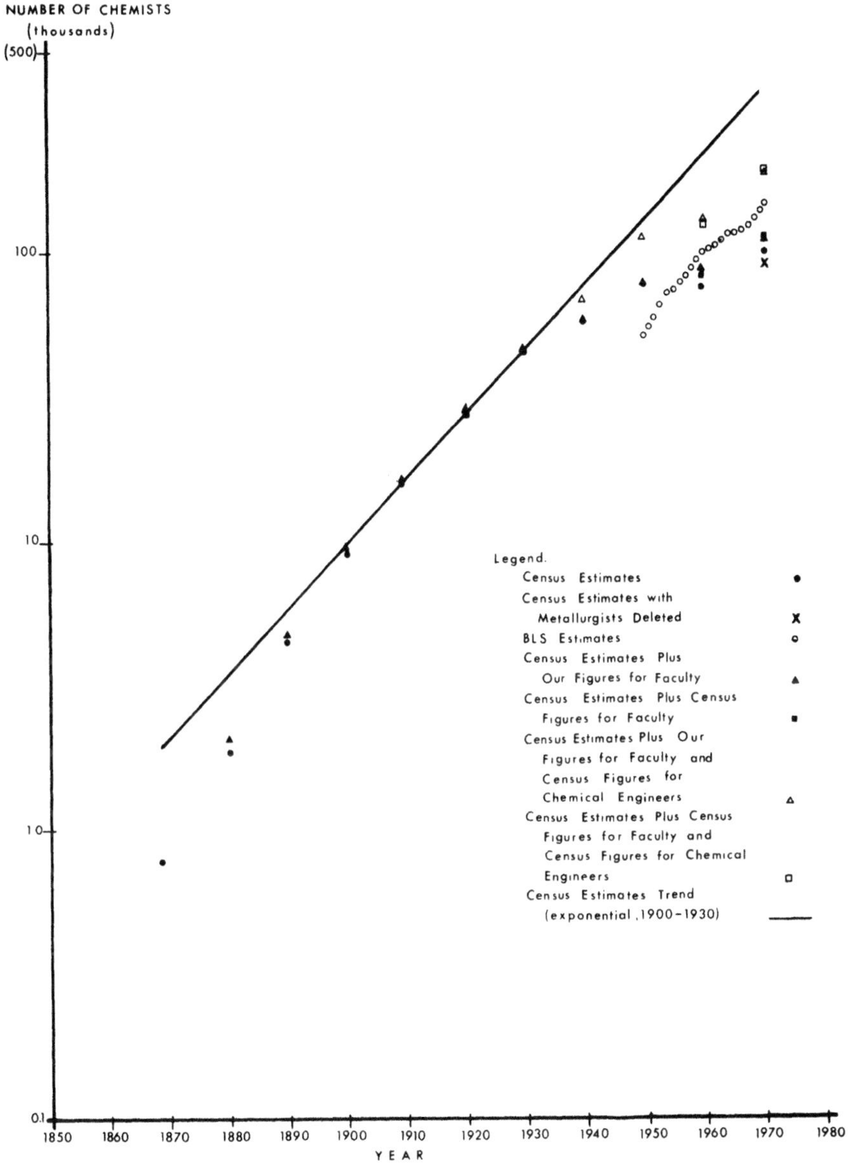

Fig. A.2–2. Trends in the number of chemists, with possible adjustments, 1870–1970. (See Table 2.2, column 1; Table 3.11, column 2; Table 5.1, column 2; Table 5.16, column 4; Table 5.17, column 2; and Table A.1, column 2.)

A similar result obtains in the case of chemists per thousand professional, technical, and kindred workers (see Figure 2.2–9). Figure A.2–3 reproduces this graph, along with the adjustments for chemistry faculty and chemical engineers, as in Figure A.2–1. Again, the addition of chemistry faculty has a minor effect upon the time series, and the inclusion of chemical engineers shifts the curve upward. Neither adjustment affects the *trend* of steady increase to 1950 and relative decline thereafter. Thus, the Census data as reported are adequate for the present analytical purpose of evaluating long-run trends in the relative occupational standing of American chemists.[6]

A.3. Degree Recipients in Chemistry and Related Fields

One of our most significant findings concerning chemistry degree conferrals in the United States over the last hundred years is that there has been an enduring trend of relative decline on all levels when compared with degree conferrals in other fields (see pp. 45, 49–54 above). This decline was illustrated in Figure 3.1–8. The skeptical reader might conclude that the apparent decline could be explained by the continuing exfoliation of chemical sub-disciplines. Such an objection implies that in recent years the growth of allied fields such as biochemistry and chemical engineering has more than compensated for the relative decline in chemistry degree conferrals *per se*. By this reasoning, chemistry in *all* its forms is just as prominent on American campuses today as it ever was, and the trend in Figure 3.1–8 is an artifact of our categories.

To test this alternative formulation, we adjusted the time series in Figure 3.1–8 to incorporate degree conferrals in biochemistry and in chemical-materials engineering. The results are illustrated in Figure A.3–1. As in the case of Census chemists, inclusion of data concerning allied fields shifts the time series on chemistry degree conferrals upward slightly, but the relative decline remains. When similar adjustments are made for the comparison of history degree conferrals to chemistry degree conferrals (see pp. 60–63), we arrive at an analogous conclusion. The divergent trend of chemistry and history as academic fields (shown in Figure 3.1–15) is still evident when one adopts a broader definition of chemistry. Figure A.3–2 displays the adjusted indicator.

[6] This is not meant to suggest that changes in occupational classification have *no* effects on studies designed to ask different questions of the data. For an example of how undetected biases inherent in early Census categories have vitiated recent studies of social mobility in the American occupational structure, see II, Conk, 1978b.

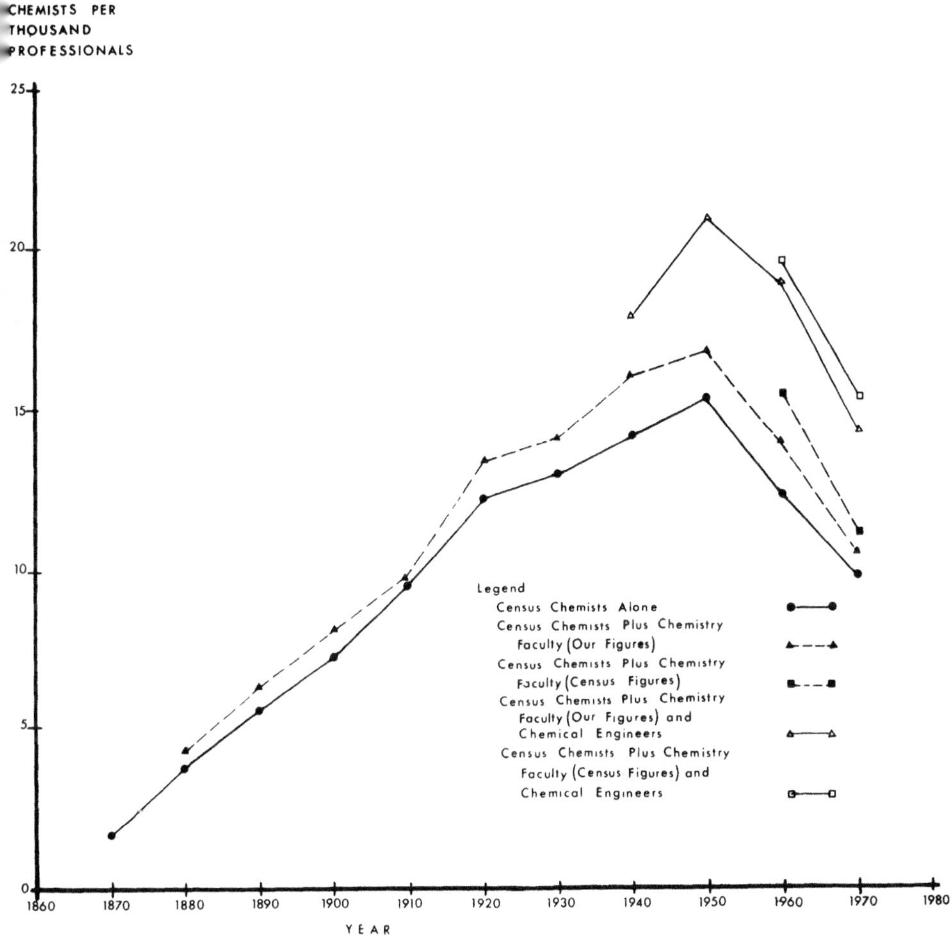

Fig. A.2–3. Chemists per thousand professional, technical, and kindred workers, with possible adjustments, 1870–1970. (See Table 2.7, column 3; Table 3.11, column 2; Table 5.16, column 4; and Table 5.17, column 2.)

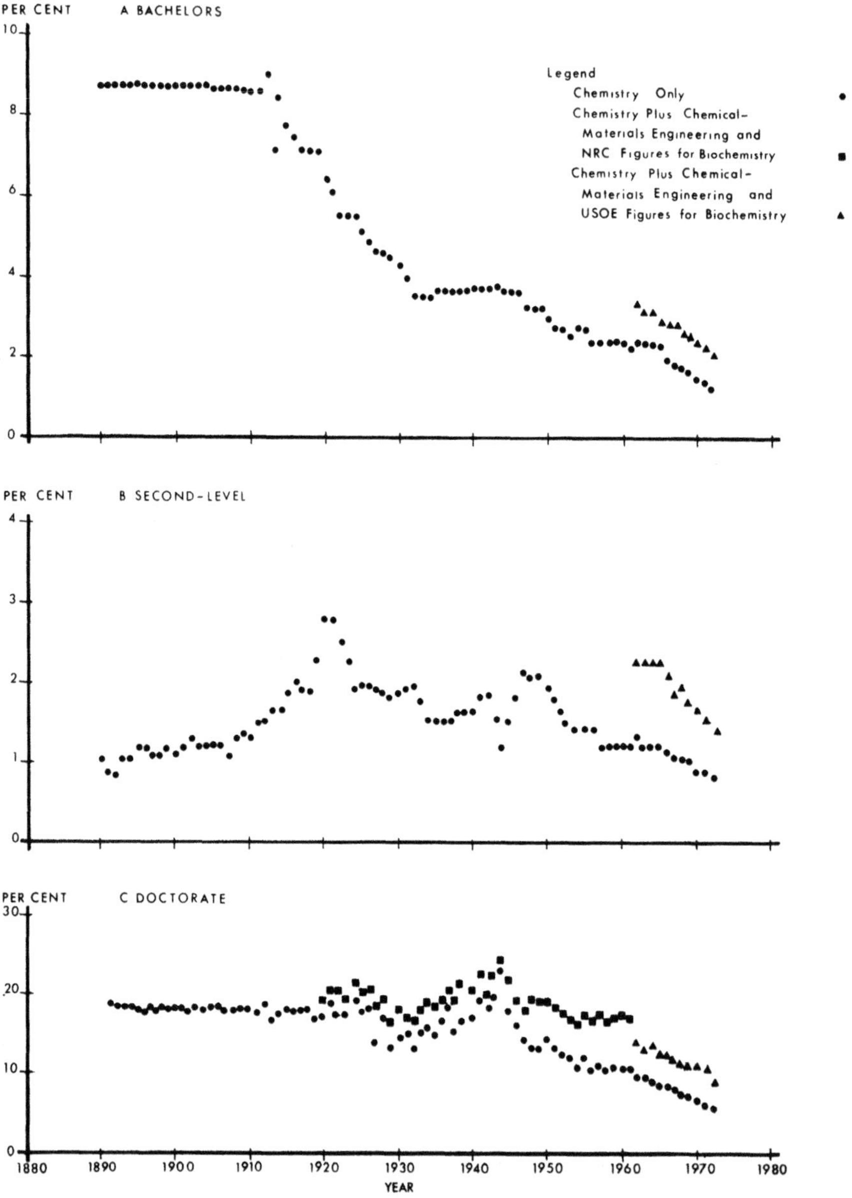

Fig. A.3–1. Chemistry as percentage of all degree conferrals, by level, with possible adjustments, 1890–1972. (See Table 3.1, column 5; Table 3.2, column 5; Table 3.3, column 5; Table 3.9, column 1; Table A.2; and Table A.3.)

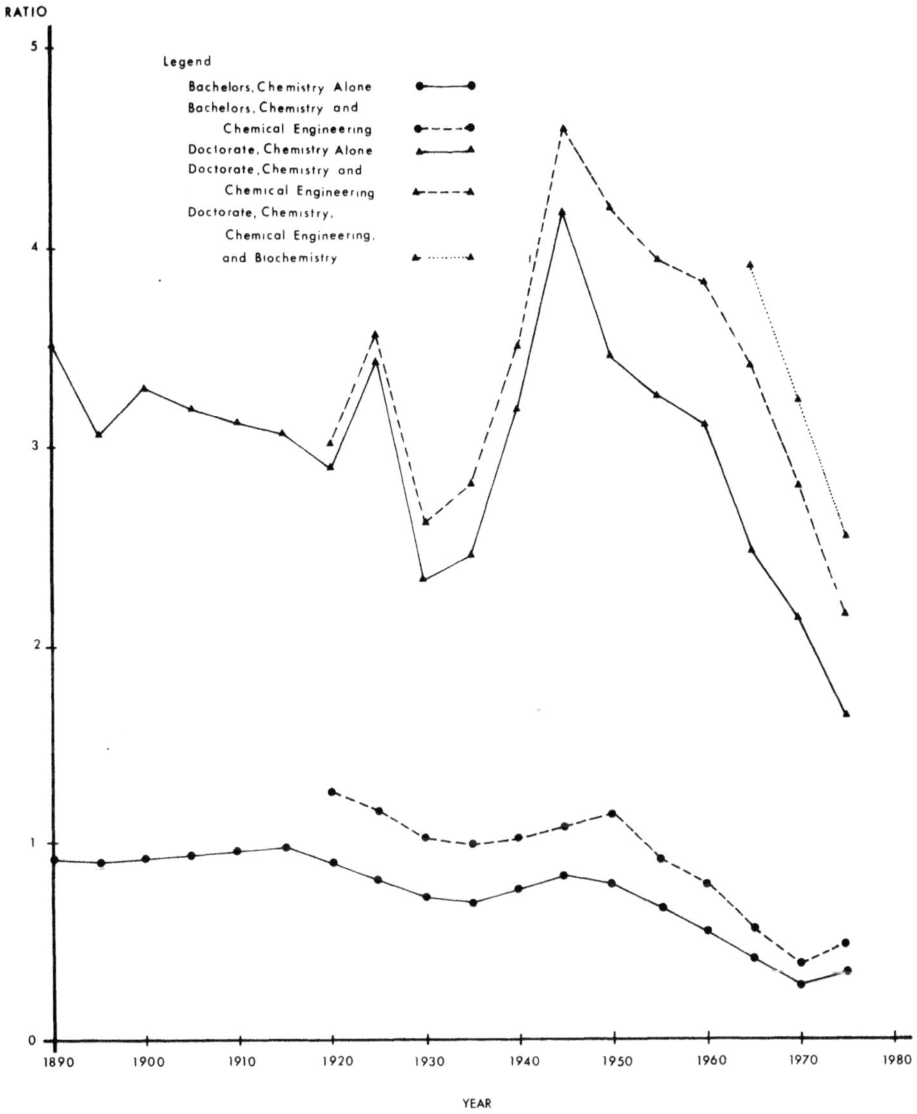

Fig. A.3–2. Ratio of chemistry degree conferrals to history degree conferrals, by level, with possible adjustments, 1890–1975. (See Table 3.9, column 2; Table 3.14, columns 2, 4, 5, and 6; Table A.2, column 2; Table A.3, columns 2 and 5; and Table A.6, column 2.)

A.4. Decoupling

In comparing the number of credentialled vs practicing chemists (pp. 54–60 above), a number of assumptions were made to produce Figure 3.1–13. First, consistent data on persons credentialled and practicing in *all* chemistry-related fields were not available. Accordingly, information on persons in chemistry and chemical engineering was used instead, on the premise that these two groups comprise the vast majority of persons in all chemistry-related fields.[7] Second, since there are no estimates of the number of persons holding *non*-terminal degrees in chemistry-related fields in a given year, a crude equivalent was created by counting *all* degrees conferred in these subjects over a 41-yr period ending in the sample year. (This is the duration used by Adkins to construct his estimates of *terminal* degree-holders.)

The defects of these time series are obvious. They exclude degree-holders in related fields such as biochemistry. They also underestimate slightly the number of chemistry-related degrees because the assumption that *every* holder of a higher degree obtained a bachelors in the same field is not correct. More important, these estimates are too high because they do not factor in mortality. For example, *all* persons who received a chemistry-related degree 40 yr earlier would be included in the 1971 estimate, although a significant fraction had already died by then. (The time series for practicing chemists and for *terminal* degree holders have been corrected for mortailty, rendering tentative the comparisons between these and the time series for all chemistry degree holders.)

Figure A.4–1 is analogous to Figure 3.1–13, except that all *higher* degrees are eliminated when counting persons credentialled in chemistry. Certain subtle and significant differences are revealed by focusing on baccalaureate chemists alone. The divergence between the number of practicing chemists (dashed line) and credentialled chemists with terminal bachelors degrees (solid dots) is obvious. The magnitude of that divergence and its rate of increase are more subdued on the bachelors level than overall (compare with Figure 3.1–13A). This suggests that, at least since the 1930s, the decoupling of chemical education from chemistry as occupation has been less pronounced at the undergraduate level than overall. One interpretation is that there has always been a certain amount of decoupling between undergraduate

[7] In 1972, 1413 degrees were conferred in biochemistry, probably the largest of all excluded fields. For chemistry and chemical-materials engineering the total was just under 22 000 (IA, NCES, 1975a, Table 5, 12–16). For the specialties included in these fields, see Table 3.9, notes a and b.

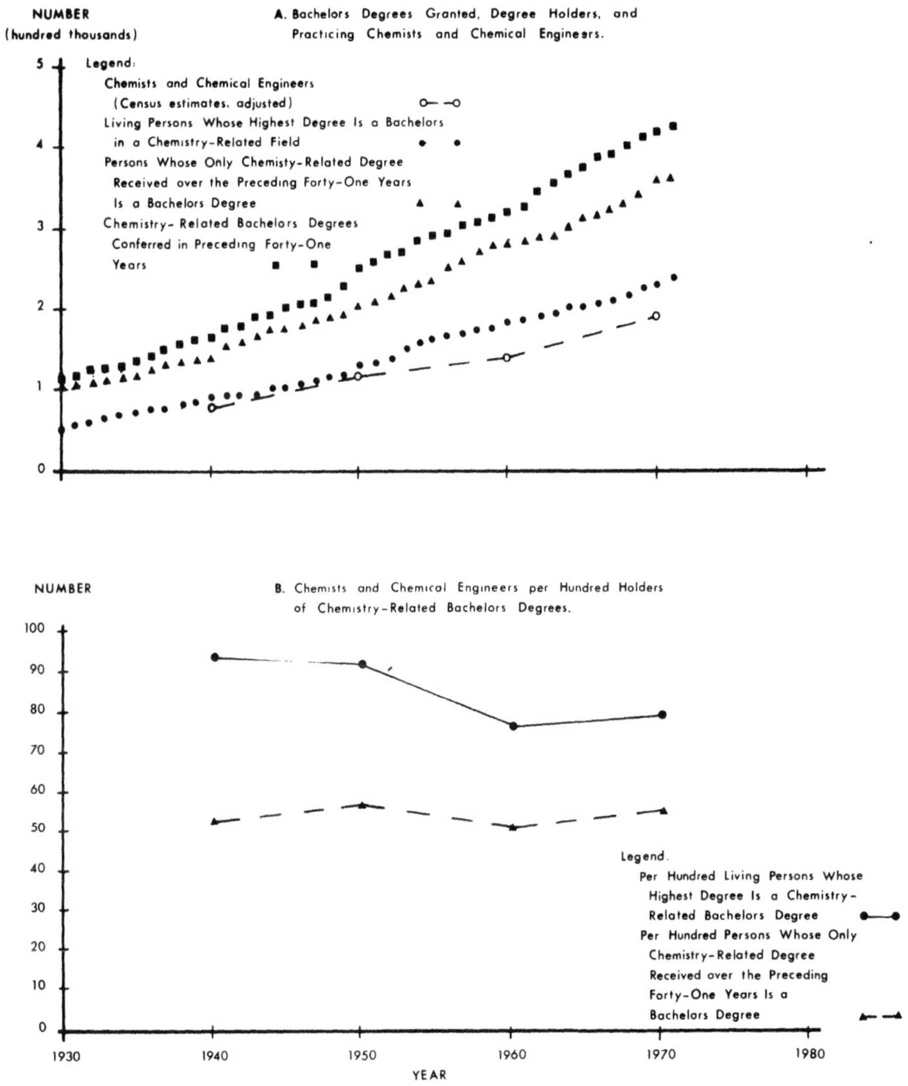

Fig. A.4–1. Chemists and chemical engineers versus holders of chemistry-related *bachelors* degrees, 1930–1971. (See Table 3.10, column 3; Table 3.11, columns 2 and 4; and Table A.5, columns 2 and 4.)

chemical education and chemistry as occupation, which has increased slowly over the years. A new development within the past four decades has been the steady decoupling of *graduate* education from chemistry as occupation. The evidence in Figures 3.1–13A and A.4–1A points in this direction. The very gradual decline illustrated by the solid line in Figure A.4–1B corroborates this tentative conclusion. It shows that the number of persons practicing chemistry and related occupations dropped only slightly relative to the number of persons with terminal bachelors degrees in those fields. Thus, the decline in the analogous indicator for terminally-credentialled chemists at all levels (see Figure 3.1–13B) is due mainly to a growing tendency for individuals to pursue graduate study in chemistry or related fields before moving into nonchemical careers, as suggested on p. 60.

Similar technical adjustments may be made to the indicators of the relationship between chemical education and the chemical profession displayed in Figure 2.2–8. A review of these modifications – which confirm the findings discussed on p. 33 – is informative, especially when juxtaposed to our discussion of the decoupling of chemical education and chemistry as occupation.

An obvious first adjustment is to incorporate chemistry-related fields into the time series used to construct Figure 2.2–8A. The next step is to include holders of non-terminal degrees among the credentialled chemists in the same graph. Both of these aims may be achieved by using the surrogate series given in Table 3.10, columns 2 and 3, rather than Adkins' figures for the stock of terminally-credentialled chemists (Table 2.6, column 3). These revised time series were used to construct Figure A.4–2, an adjusted version of Figure 2.2–8A. The top curve (solid dots) compares ACS membership with a measure of all persons with terminal degrees in chemistry-related fields. The bottom curve (solid triangles) is an approximation of the ratio of ACS membership to *all* holders of chemistry-related degrees, whether terminal or non-terminal. While each modification shifts the curve lower than it was in Figure 2.2–8, neither alters the shape of that curve significantly.

A final adjustment involves consideration of doctorates only. The number of ACS members has long exceeded the number of persons with a doctorate in chemistry and related fields. As Figure A.4–3 shows, this is the case whether terminal degree holders are counted alone or together with our approximation for non-terminal degree holders.[8] The ratio of ACS members

[8] As one might expect, most holders of doctorates possess *terminal* doctorates, as the near coincidence of the two bottom lines in Figure A.4–3 indicates.

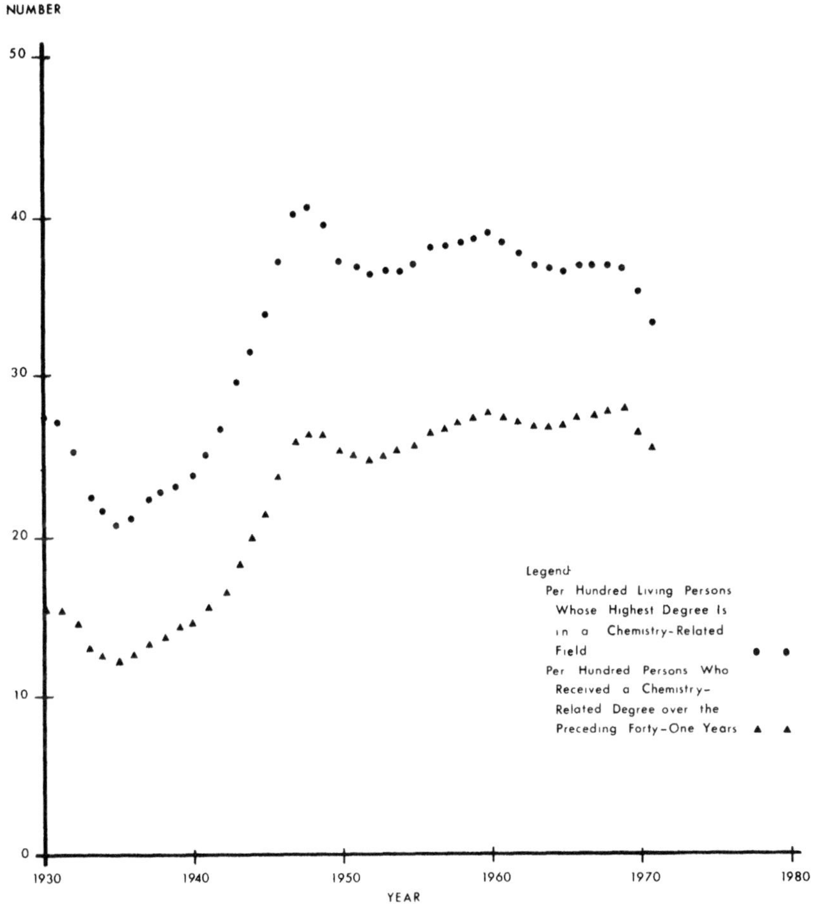

Fig. A.4–2. ACS members per hundred holders of chemistry-related degrees, 1930–1971. (See Table 2.4, column 1; and Table 3.10, columns 2 and 3.)

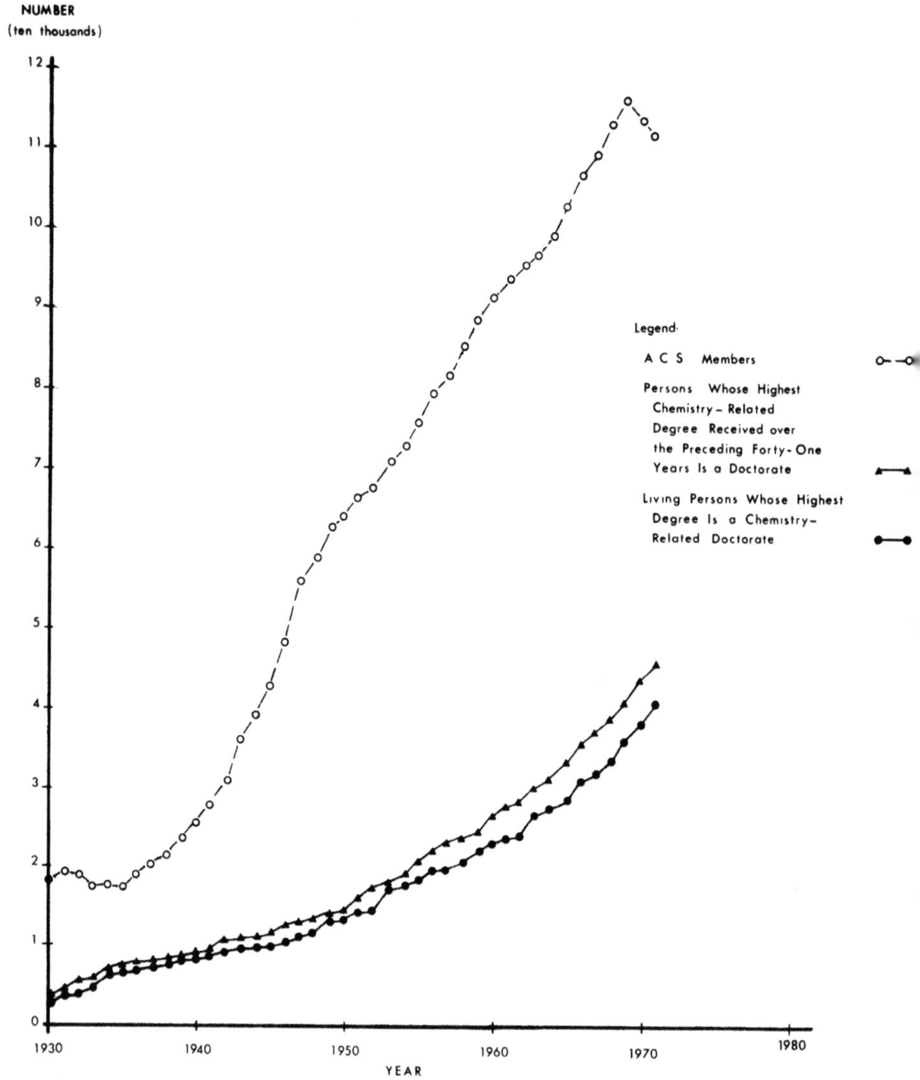

Fig. A.4–3. ACS members versus holders of doctorates in chemistry-related fields, 1930–1971. (See Table 2.4, column 1; and Table A.7, columns 3 and 5.)

to doctorates in chemistry, however, shows considerable change in recent decades (Figure A.4–4). A strong trend toward parity between ACS membership and holders of chemistry doctorates (a value of 100 on the vertical scale indicates parity) is evident after World War II. This is true whether non-terminal doctorate holders are included (top graph) or excluded (bottom graph) or whether chemistry is considered alone (dots) or in conjunction with chemical-materials engineering (triangles). Through all these changes the 1971 endpoint remains almost the same: there are about three ACS members for every one person with a relevant doctorate. The downward trend continues, as graduate education infiltrates the chemistry profession.

This finding seems paradoxical in light of our earlier assertion that chemical education has become increasingly decoupled from chemistry as occupation. Now the claim is that the chemistry profession is being overrun by chemistry PhDs. These two conclusions are compatible. For some time, American educational institutions have produced more people with chemical training than there have been niches within chemistry in which to place them. The "excess chemists" have had to find jobs outside of chemistry. Meanwhile, the American Chemical Society, as the bastion of the profession, has pursued the interests of those chemists with professional certification. It has expanded to the point where it represents virtually the entire *occupational* group of chemists, since nearly every chemist in the nation is now fully credentialled.[9] As the ACS has approached this saturation point, the production of persons credentialled in the science via higher degrees has continued unabated. The result is that these graduate-trained chemists are infiltrating both the occupation of chemistry and its contiguous profession, even as an increasing minority of them spill over into extrachemical careers.

A survey of persons in engineering and scientific occupations in 1972 places our findings about decoupled-but-credentialled chemists in perspective. This study shows that of an estimated 1.6 million scientists and engineers in the nation whose field of highest degree can be determined, about 86 000 (or just over 5%) have their highest degree in chemistry.[10] Fully

[9] In fact (as we have noted elsewhere), there are many nonchemists credentialled in the science, but since they do not consider themselves chemists they are not sought after by the ACS, at least at present. Just how far it may go in future to accomodate the "service occupations" more marginal than those listed in Table 2.1 is a matter of speculation.

[10] The information in this paragraph and the next two is from IA, USBC, 1974, Table 11, 92–109. For similar data from another survey, see IB, ACS, 1976, 8–13.

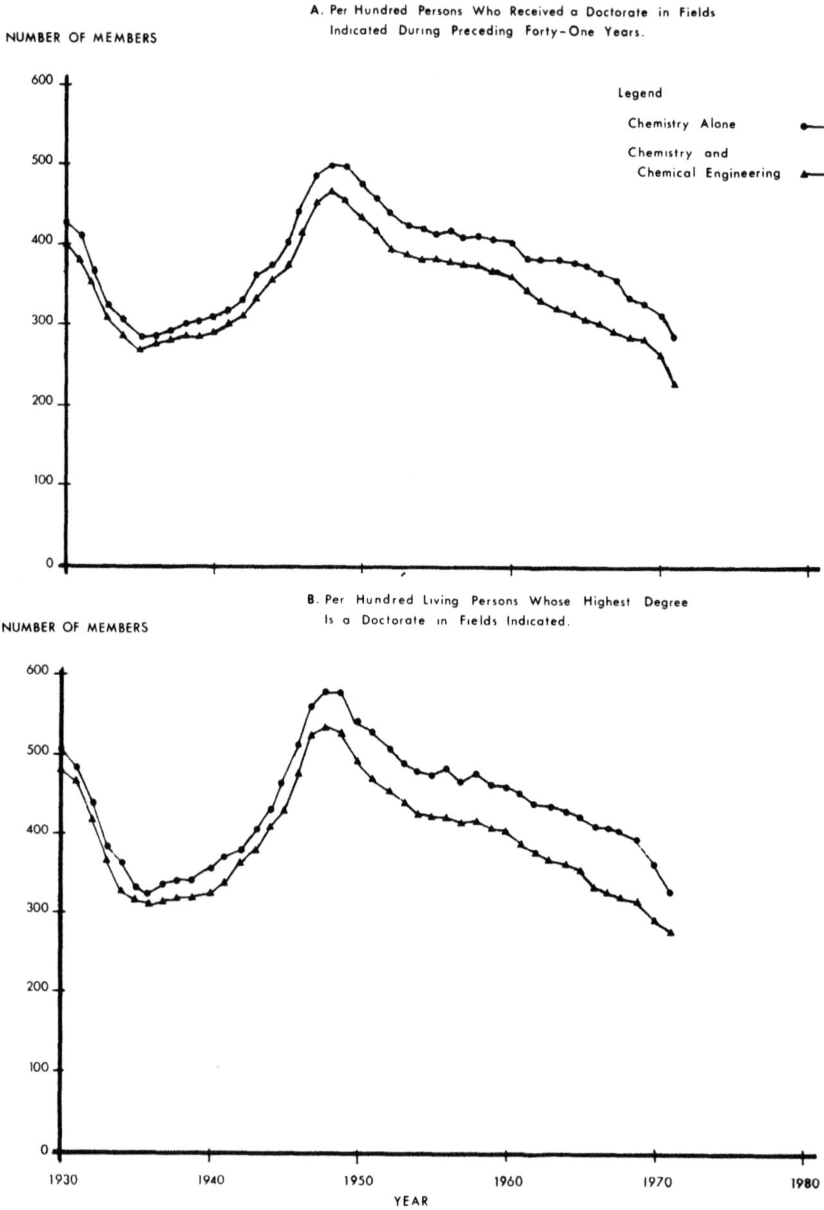

Fig. A.4–4. ACS members per hundred holders of chemistry and chemistry-related doctorates, 1930–1971. (See Table 2.4, column 1; and Table A.7, columns 1, 3, 4, and 5.)

three-fourths of these persons list themselves as practicing chemists. Another tenth (about 9000) are found among "other engineers", a category which includes chemical engineers. The occupations of chemistry and chemical engineering thus absorb nearly 85% – a fairly high capture ratio; the occupation of physics, by comparison, absorbed only 45% of those with highest degrees in physics.

We should re-emphasize that there are twice as many terminally-credentialled chemists as there are persons practicing in the field. This means that the small group of decoupled-but-credentialled chemists among the ranks of scientists and engineers must be compensated for by a relatively large group of such people with occupations entirely outside of science and engineering. These individuals would most likely be in managerial careers, medicine, patent law, or similar professional occupations.[11]

A closer look at the decoupled-but-credentialled chemists in this Census survey is also revealing. Of almost 13 000 people in this group, some 6500 are engineers in a specialty other than chemical engineering. About 2400 are employed as operations and computer specialists; some 1500 are in physical sciences other than chemistry; and there are about 1000 each in the life sciences and in the social sciences. This illustrates the diversity of extrachemical niches – even within science and engineering – into which decoupled-but-credentialled chemists have settled. Even more striking, almost 44% of these people have chemical training *beyond* the bachelors degree, an indication of just how far the decoupling phenomenon has infiltrated graduate education in chemistry.

[11] There is no recent comprehensive survey of the educational backgrounds of such non-scientific professionals.

B. CHEMICAL INDUSTRY: ALTERNATIVE DEFINITIONS

B.1. Chemical Process Industries versus the Chemicals and Allied Products Industry

Gauging the precise extent of the chemical industry is a difficult task. Our approach has been governed by a pragmatic concern to ascertain the economic power of chemical manufacturing in the aggregate relative to other manufacturing industries. In addition to trends in the chemicals and allied products industry itself, we were interested in measuring the more diffuse economic impact of chemistry in the chemical process industries, defined as those based mainly on the chemical conversion of raw materials. The industries included in this set vary. This report adopts the grouping used in the American Chemical Society study of *Chemistry in the Economy* (1973): food and kindred products; paper and allied products; chemicals and allied products; petroleum refining and related industries; rubber and miscellaneous plastics products; stone, clay, glass, and concrete products; and primary metal industries.[1]

In compiling the data on the chemical process industries reported in Chapter 4, we relied upon the categories of the U.S. Standard Industrial Classification (SIC), a scheme promulgated by the Statistical Policy Division of the Office of Management and Budget. The SIC, first published in 1939, grew out of the classification adopted by the U.S. Bureau of the Census in its surveys of manufacturing. It is designed to facilitate the collection of uniform and comparable data pertaining to economic activities in the United States.[2] The SIC assigns two-, three-, or four-digit codes to industries in a hierarchical system according to the degree of detail required in particular analyses. Thus, Chemicals and Allied Products is Major Group 28 of the SIC, industrial inorganic chemicals (SIC 281) is one industry group within Chemicals and Allied Products, and establishments producing alkalies and chlorine (SIC 2812) comprise one industry within the set of industrial inorganic chemicals manufacturers. The material in Chapter 4 involves comparisons among two-digit and, occasionally, three-digit industries.

[1] See IB, ACS, 1973, xi, 489. Part I of *Chemistry in the Economy* discusses the contributions of chemistry to manufacturing processes in the chemical process industries and other areas of the economy (e.g., electronics and automotive manufacturing).

[2] The purpose, scope, and structure of the SIC are explained in II, OMB, 1972, 9–13.

Table B.1 enumerates those industries included in our definition of chemical process industries, according to SIC codes. Table B.2 provides a detailed breakdown of the 28 classes of chemical producers included in SIC Major Group 28, Chemicals and Allied Products. The SIC defines this industrial group to include:

establishments producing basic chemicals, and establishments manufacturing products by predominantly chemical processes. Establishments classified in this major group manufacture three general classes of products: (1) basic chemicals such as acids, alkalies, salts, and organic chemicals; (2) chemical products to be used in further manufacture such as synthetic fibers, plastics materials, dry colors, and pigments; (3) finished chemical products to be used for ultimate consumption such as drugs, cosmetics, and soaps; or to be used as materials or supplies in other industries such as paints, fertilizers, and explosives .... Establishments primarily engaged in packaging, repackaging, and bottling of purchased chemical products, but not engaged in manufacturing chemicals and allied products, are classified in trade industries.[3]

B.2. Changing Definitions of the Chemicals and Allied Products Industry

The task of fixing boundaries for the chemical industry has perplexed analysts for as long as attempts to collect statistics on the enterprise have been made.[4] The working definition of the chemical industry adopted by the U.S. Bureau of the Census has changed periodically. During the past century the number of constituent industries in the chemicals and allied products group expanded along with the variety and number of chemical products manufactured in the United States. Older industries combined, new ones arose, and still others disappeared from the scene.

From the nine industries included in the chemicals and allied products group at the turn of the century (general chemicals; sulphuric, nitric, and mixed acids; wood distillation; paints and varnishes; fertilizers; explosives; dyestuffs and extracts; essential oils; and bone, carbon, and lamp black), some 31 branches of chemical industry had evolved by the late 1950s.[5] As

[3] *Ibid.*, 111.

[4] For instance, William Rowland, the consultant who compiled the data for the 1880 Census report on chemical products, noted that "manufacturing chemistry is so intimately associated with other branches of manufacture that it becomes difficult to locate the dividing line between them" (IA, Rowland, 1883, 989). See also IA, Munroe, 1908, 397.

[5] On changes in the definition of chemicals and allied products prior to World War I, see the summary discussion in IA, USBC, 1913, 529; cf. IA, USBC, 1919, 452. For chemicals and allied products as defined in the 1957 SIC, see II, BOB, 1957, 76–82.

a result of this differentiation, and the reallocation to other industries of some subgroups within the chemical industry, data on chemicals and allied products *as reported* by the Census Bureau since 1899 are not strictly comparable.

It is possible to construct adjusted time series for the chemical industry by recalculating aggregate figures on the basis of a constant set of four-digit industries. This was the course taken by Jules Backman in his investigation of *The Economics of the Chemical Industry* (1970), and his adjustments are instructive. By comparing the chemicals and allied products industry as defined in the 1957 SIC with earlier data in the Census of Manufactures, Backman obtained a set of eight four-digit industries which were no longer classified as chemical, and one (fireworks and pyrotechnics) which now was included. Armed with this information, he constructed homogeneous time series based on the 1957 SIC for the number of establishments, value added by manufacture, number of production workers, and wages in the chemicals and allied products industry from 1899 to 1967.[6] The result of this recalculation for production workers in the chemical industry is displayed in Figure B.2–1. (Similar curves could be drawn for value added by manufacture, or the number of establishments in the chemical industry.)

The only significant changes in Major Group 28 in the 1967 and 1972 SIC revisions involved the realignment of certain four-digit industries. Cyclic (coal tar) crudes; putty, caulking compounds, and allied products; agricultural pesticides; and fatty acids were incorporated in other four-digit industries in 1967, decreasing to 28 the number of constituent industries in chemicals and allied products. (Compare II, BOB, 1957, 76–77, 80–82, with II, BOB, 1967, 95–98, 105–107, 109.) This number rose to the present 28 when the 1967 category of "fertilizers" was split into two in the 1972 SIC: nitrogenous fertilizers and phosphatic fertilizers. Also, gum and wood chemicals (SIC 286) was renamed "industrial organic chemicals" in 1972, incorporating cyclic intermediates, dyes, organic pigments (lakes and toners), and cyclic (coal tar) crudes (SIC 2815) and industrial organic chemicals, not elsewhere classified (SIC 2818) from the old category of industrial inorganic and organic chemicals (SIC 281). The latter became "industrial inorganic chemicals" in the revised SIC. (Compare II, BOB, 1967, 94–98, 106–107, with II, OMB, 1972, 111–113, 120–124).

[6] IB, Backman, 1970, 307–316. The eight industries (and the years for which Backman subtracted the relevant data from the chemicals and allied products total) were cottonseed oil mills (1899–1954); linseed oil mills (1899–1954); grease and tallow (1899–1954); oils, not elsewhere classified (1899–1937); soybean oil mills (1939–1954); vegetable oil mills (1939–1954); animal oil mills (1939–1954); and ferroalloys (1921–1937). (*Ibid.*, 308.)

For a detailed analysis of the comparability of Census categories pertaining to chemicals and allied products from 1929 to 1958, see II, Goldstein, 1959, 29–31.

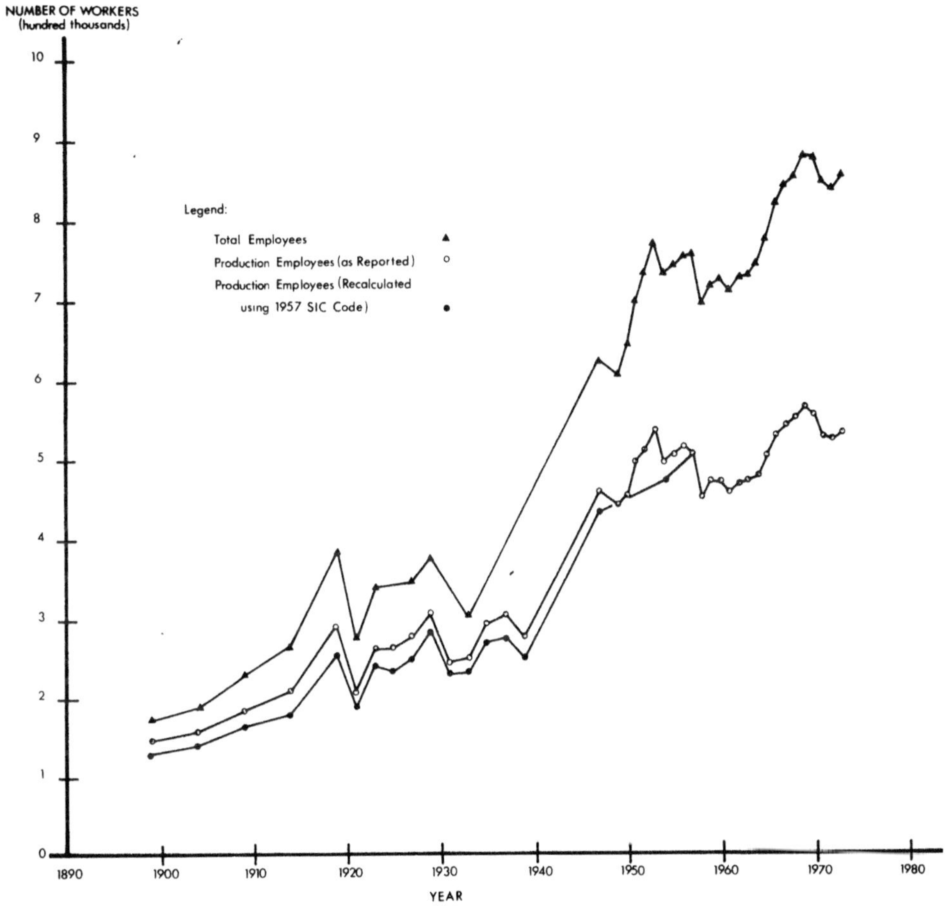

Fig. B.2–1. Employment levels in the chemicals and allied products industries, with possible adjustments, 1899–1973. (See Table 5.3, columns 2–4.)

Two observations are pertinent. First, while the recalculated time series is more homogeneous, and thus facilitates comparison over time, the discrepancies between the adjusted and the reported data are not great. For example, the 1958 figure on the number of employees in the chemicals and allied products industry (based on the 1957 SIC) was only 4.2% lower than it would have been using the older classification.[7] Second, as Figure B.2–1 shows, the *trend* in the growth of employment in the chemical industry is substantially the same, whichever definition of the industry is used. (Similar results for trends in degree conferrals in chemistry were discussed in Appendix A.) This suggests that if long-run trends – rather than absolute figures – are the major concern, Census data on chemicals and allied products may be used without laborious adjustments. We have adopted this approach in the main body of this report.

B.3. Industrial Research Personnel

The compilers of the National Research Council surveys of industrial research laboratories were well aware of the difficulty of defining "research". Their practical solution derived from an interest in disseminating information on the *variety* of research areas in American industry, rather than a concern to ascertain the exact *number* of persons engaged in a strictly-defined activity. From the outset, the NRC surveys adopted a broad classification: "Industrial research is the endeavor to learn how to apply scientific facts to the service of mankind. Many laboratories are engaged in both industrial research and industrial development. These two classes of investigations commonly merge so that no sharp boundary can be traced between them. Indeed, the term *research* is frequently applied to work which is nothing else than development of industrial processes, methods, equipment, production, or by-products."[8]

Decisions on what and who should be counted were left to the reporting companies, though the compilers sought to exclude product control and commercial testing activities: "This bulletin is not intended to list laboratories which are concerned merely with routine testing of raw materials and products, but it is desired to include all those which definitely devote some time to *research* looking towards the improvement and development of products. We believe that you are in [a] position to judge whether or

[7] IB, Backman, 1970, 311, Table A–1.
[8] IB, Hull, 1946, [iv].

not your company's laboratory falls within this class."[9] The criteria by which industrial research personnel were classified as chemists, physicists, or members of other professional groups were also left up to the companies to decide. Their categories were based on practice (i.e., those individuals who worked as scientists were classified as such), although by 1950 there was somewhat more concern with determining professional qualifications.[10]

College and university laboratories were excluded from the NRC surveys, but an appendix of those institutions offering research service to local industry was initiated in the 1946 edition. Government laboratories were also excluded, except for the National Bureau of Standards and the research service of the Department of Agriculture. The former was first included in 1938 and the latter in 1946. By 1950 an appendix had appeared covering some federal government laboratories.[11] Trade association laboratories, research institutes (such as the Mellon Institute), and consulting laboratories (such as A. D. Little, Inc.) were included in the directories.

The extent of coverage in the NRC surveys is treated in IB, Perazich and Field, 1940, 52–63, along with other technical issues in the analysis of these data. Perazich and Field conclude that coverage was quite complete by 1927, but that the 1921 results should be adjusted by 23%. They also discuss the problems in providing disciplinary breakdowns of industrial research personnel, due to the incomplete reporting of these categories by the laboratories surveyed; this explains why there is only an omnibus category for engineers in Table 5.5.[12]

Table 5.5 probably represents a lower bound for the number of persons employed in industrial research from 1921 to 1950. It reveals trends in activities *perceived as* research by the industrial community. Systematic bias may well exist in the disciplinary breakdowns, but the trends in employment of various professional groupings are clear.

[9] IB, Hull, 1938, [iii]; cf. IA, FSA, 1952, 2.

[10] IB, Rand, 1951b, 3604.

[11] IB, Hull, 1946, 349–355; IB, Rand, 1950, 399–405.

[12] IB, Perazich and Field, 1940, 54–58. For a discussion of the 1921 adjustment, see *ibid.*, 52–53, 59–60.

C. PROCEDURES USED IN ANALYSIS OF CITATIONS

C.1. Chemische Berichte

Although the international distribution of citations in *Chemische Berichte* has been studied by others, our report is distinguished by the length of the period examined, the frequency of sampling, and the comparison of the number of references to American research with the number of non-German references.[1] Issues of *Chemische Berichte* from as early as 1876 and 1880 were surveyed, but citations to American work were so scarce that an analysis of samples from these years was inappropriate. For this reason, the first data reported in Table 6.7 are for 1890. Until 1920 American publications were cited so infrequently in *Chemische Berichte* that only a decennial census was warranted. The shift to a very rapid rate of growth in the 1920s prompted a more detailed monitoring of the trend every two or three years. In order to minimize the perturbations in the long-run trend due to wartime interruption of scientific communications, 1938 was substituted for 1940, and 1945 was skipped. Sampling at five-year intervals was maintained from 1950 to 1975.

Since the modest aim of our analysis was to illuminate the long-run trend in the perception of American chemical research in Germany, rather than to offer a detailed analysis of *Chemische Berichte*, an elaborately constructed sample of citations would have yielded spurious accuracy. The citations to be examined were selected in the following manner:

1. Only footnote citations were counted, although some references were given in the text in the early years.
2. About 1000 citations were selected from the earliest issues of each year. Footnotes in all of the articles appearing in the first issue of the year onward were scrutinized, closing the sample at the end of the issue with which the number of citations counted was closest to 1000.[2]

[1] "German" in this context refers to publications written in the German language. In this study publications of Swiss and Austrian origin are grouped with German publications if written in German.

[2] In this way, bias resulting from the distribution of papers within individual issues was avoided. Any systematic error that may have resulted from the distribution of papers over the year remains, but inspection suggested that this was small.

As shown in Figure 6.2–1, the trend of citations to American publications followed a pattern of logistic growth. Observations from other studies support the values predicted by our long-run trend. Narin and co-workers analyzed citations in major chemical journals (including *Chemische Berichte*) published in 1939, 1949, 1959, and 1969.[3] Figures for citations of American publications as a percentage of all citations to non-German publications may be calculated from their data, and compared to the values predicted by the logistic trend. Apart from 1939 – an anomalous year for German science – the observed values are close to those predicted (see Table C.1). A similar study conducted for the 1965 Westheimer Report found that 42% of the citations to foreign articles from a sample of 1160 citations in German chemical research papers were to American publications.[4] This agrees well with the 1965 value of 44.0% predicted by our logistic curve.

C.2. Chemical Abstracts

As noted on p. 155, Figure 6.2–1 includes a graph of the relative volume of American chemical publications. This allows one to gauge the extent to which increased citation of American research was a function of the perception of its significance by German chemists, rather than merely the consequence of there being more and more publications by American authors. This measure was constructed in a relatively straightforward manner.

Information on the number of American articles, the number of German articles, and the total number of articles abstracted in *Chemical Abstracts* is available for selected years since 1909 (see Table 6.10).[5] Using these data, one can plot a graph of the number of American articles (y) versus year (t). Since the total number of abstracted articles has grown exponentially, and

[3] IB, Computer Horizons, 1972, 71–72, Table 68.

[4] IB, NAS, 1965, 32, Table 2.

[5] Our citation data are based on the country in which research is published, while the *Chemical Abstracts* figures are based on the country in which the research was conducted (IB, Baker, 1971, 38). Since an overwhelming majority of research papers are published in the country of the author, the figures are roughly comparable. In the study by Narin and his co-workers cited above, 80% of the articles published in American research journals since 1939 also originated in the United States. Similar results were found for *Chemische Berichte* and the *Journal of the Chemical Society*, which were dominated by German and British authors, respectively (IB, Computer Horizons, 1972, 68–69, Table 66).

the proportion of American articles has been roughly constant, we may assume that the number of American articles has grown exponentially between the years for which there are data points.[6] A curve of linked exponential segments may then be drawn on the graph. The area under each segment can then be obtained by integration, as follows:

$$y = a\,[\exp(kt)]$$

$$\int_{t_1}^{t_2} y\,\mathrm{d}t = a \int_{t_1}^{t_2} \exp(kt)\,\mathrm{d}t$$

$$\begin{array}{l}\text{total number of}\\ \text{American articles} = [(y_2 - y_1)/\ln(y_2/y_1)] \times (t_2 - t_1).\\ \text{from } t_1 \text{ to } t_2\end{array}$$

The areas under successive segments were summed to obtain cumulative totals of American publications. Cumulative totals of non-German articles were obtained in a similar manner (see Table 6.8, columns 2 and 5).

Since *Chemical Abstracts* only began publication in 1907, and its coverage during the first few years was limited, we needed an estimate of how much chemical literature had accumulated before the journal first appeared. A series of estimates was constructed by assuming that both American and non-German chemical publications had grown exponentially from their inception until 1913, the first year for which useful *Chemical Abstracts* figures were available. We assumed arbitrarily that American publications began in 1879, the year in which both the *American Chemical Journal* and the *Journal of the American Chemical Society* were founded. Non-German articles certainly began to appear earlier: for the purpose of this exercise, 1800 was chosen as the year of the first non-German chemical article and the beginning of continuous exponential growth.[7] The total

[6] On the exponential growth of the total number of articles abstracted, see IB, Baker, 1961.

[7] When the starting points for exponential growth in the number of American and non-German publications are calculated by extrapolation from the post-1914 *Chemical Abstracts* data in Table 6.8, the results are preposterous. Using log-linear regression on the data in columns 1 and 4 to obtain the date (x-intercept) when American and non-German publications would have begun to grow exponentially yields 1745 for American and 1751 for non-German publications. It would be absurd to claim that American

number of publications was then calculated by integrating the exponential curve from a value of 1 in the arbitrarily-chosen starting year to the value in 1913. Combining those pre-1914 estimates with the cumulative totals from *Chemical Abstracts* yielded the appropriate time series (Table 6.8, columns 3 and 6). Our indicator of American chemical articles as a proportion of the total number of non-German chemical articles published by a given year was then obtained by taking the ratio of these two estimates for that year (Table 6.8, column 7).

C.3. Annual Reports on the Progress of Chemistry

Annual Reports on the Progress of Chemistry is a series established by the Chemical Society (U.K.) in 1905, the first volume of which covered chemical research published in 1904. The objective of *Annual Reports* was

> to present an epitome of the principal definite steps in advance which have been accomplished in the preceding year for the benefit of all workers, students, or teachers of chemistry, or those chemists who are engaged in technical or manufacturing applications of chemistry, in order that specialists in any one department of the science may obtain without further difficulty information as to the nature and extent of progress in other branches of the subject to which they have not paid special attention.[8]

To achieve this goal, the journal devoted a chapter to each of the major branches of chemistry reviewing recent advances in that area. By the mid-1960s, it was acknowledged that "the enormous increase in the rate of accumulation of knowledge witnessed during the last few years now makes any attempt to comprehensiveness unattainable within the compass of an annual volume of reasonable size and cost." Accordingly, Chemical Society policy changed, and beginning in 1968 (covering 1967 literature) more selective criteria were instituted for deciding the coverage of *Annual Reports*. The journal continued to maintain a wide scope, covering major specialties in pure chemistry.[9]

---

chemical publication was initiated before other non-German publication. In any case, the number of American and non-German publications has probably not exhibited *continuous* exponential growth. This prompted our decision to utilize the arbitrary dates of 1879 and 1800 given above. (Choosing different takeoff points would only shift the curve up or down without affecting the slope.) There is some evidence that American chemical publications were indeed few before the 1880's; see III, Caldwell, 1892, and IB, Siegfried, 1952.

[8] IB, ARPC, 1905, [v].

[9] IB, ARPC, 1967, vii–ix. Quotation from vii.

The volumes of *Annual Reports* to be examined were selected by the same criteria used in our analysis of *Chemische Berichte*. However, since there were systematic differences among the various parts of each volume, a new method of sampling the citations themselves was employed. Instead of examining only early pages in each volume, a random sample of citations was obtained by choosing every *n*th page for scrutiny (with *n* set so that roughly 1000 citations from each year were included). Every other page was examined for volumes from 1905 to 1930; every fourth page for 1935, 1950 to 1960, and 1975; and every eighth page for 1965 and 1970. Both the number of citations to American work and the total number of citations in each volume were counted (see Table 6.9).[10]

Since *Annual Reports* cites only recent research, the citation pattern in a given year may be compared to the international distribution of chemical publications in the same year. For this reason, in Figure 6.2–2 *Chemical Abstracts* figures for *annual* production, rather than cumulative totals, were compared to the proportion of citations to American research in *Annual Reports.*

C.4. Distribution of Citations in American Journals

The international distribution of citations in leading American chemical journals was investigated in IB, Fussler, 1949, and IB, Computer Horizons, 1972. These studies covered complementary periods – Fussler concentrating on 1899 to 1946, and Narin and his colleagues examining 1939 to 1969 – and similar sets of journals. Fussler examined the following journals in the years indicated:

American Chemical Journal (1899)
Chemical Reviews (1939)
Journal of Chemical Physics (1939)
Journal of Organic Chemistry (1939)
Journal of Physical Chemistry (1899, 1919, 1939)
Journal of the American Chemical Society (1899, 1919, 1939, 1946)
Organic Syntheses (1939)

[10] We compared citations to American publications with the total set of references in *Annual Reports* since the British references did not favor British publications excessively; see IB, Narin and Carpenter, 1975, 91, Table 9.

Narin and his colleagues examined *Analytical Chemistry, Chemical Reviews, Journal of Chemical Physics, Journal of Organic Chemistry, Journal of Physical Chemistry*, and *Journal of the American Chemical Society* for the years 1939, 1949, 1959, and 1969. We amalgamated the observations of these two studies to derive time series roughly comparable to those obtained from our own analyses of *Annual Reports* and *Chemische Berichte* (see Figure 6.2–3 and Tables 6.11 and 6.12).[11]

[11] Our data on *Chemische Berichte* included Swiss and Austrian publications within the set of German chemical publications, while the data collected by Fussler and Narin and his colleagues did not. The number of such publications was small, however, and the effect of including them was negligible.

D. A NOTE ON THE TREATMENT OF ERRORS

Getting the numbers right is a difficult business. Statistical counts routinely suffer from measurement errors; for instance, the techniques used may fail to locate with sufficient precision the feature requiring scrutiny, and there may be inaccuracies in the reading of whatever *is* measured. Readings may also be subject to errors of recording, coding, and processing. If only sample measurements are made, further error is possible because the samples may not be representative. We would be remiss if we nourished the myth of magic in numbers by failing to point to the uncertainties which we suspect in our own findings. While it is more desirable to be correct than to be precise, there is no consensus on how to achieve this end. Although the derivation of standard errors of sample estimates (and the confidence intervals derived therefrom) is straightforward, the evaluation of nonsampling errors varies considerably with the methodology employed. And, to quote William Kruskal, "careful examination of measurement bias may indeed be expensive," even in the best of circumstances.[1]

The task is particularly daunting in the case of historical statistics, where most of the measuring has been done by a host of other parties. Our unwitting collaborators have not usually garnished their quantitative fare with many details of method, let alone firm estimates of error. As compilers we can evaluate the accuracy of their findings only through that proven qualitative technique of creative self-doubt, which the historian always needs in the evaluation of evidence. Because each of the surveys from which we have gathered data follows a unique blueprint, reflecting its own objectives but not necessarily ours, comparison of the results of different surveys is necessarily a precarious operation. One can rarely be positive the same thing has been measured in each case, even if the same agent (e.g., the Census Bureau) has done all the measuring. It is thus impractical to expect our analysis of errors to be uniformly rigorous in any formal sense.[2]

The obligation to convey an order-of-magnitude sense of the accuracy of

[1] This appendix follows from Kruskal's further admonition to launch at least "a reasonable first attack at error" (II, Kruskal, 1978, 159). Problems with the accuracy of statistical indicators are also discussed in II, Fienberg and Goodman, 1974.

[2] For a further discussion and an apologia for the attempt, see II, McCloskey, 1976, esp. 445.

our data remains. We have taken several steps to discharge that obligation. To gauge inaccuracies derived from problems of definition, we have attempted a number of possible adjustments to our series, as we discussed in Appendixes A and B. We invite the industrious critical reader to continue this endeavor. If we have discovered differing estimates of the same quantity, we have accompanied them with plausible reasons for the discrepancy and a rationale for our preference. In the tables and elsewhere, we have tried to give sufficient attention to rounding and to the number of significant digits. Whenever we have felt particularly suspicious of the accuracy of a given figure, we have aired our doubts in the text. For our own tabulations, we have tried to provide (usually in notes to tables) ample documentation of our methods, with an emphasis upon probable sources of bias, so that readers can evaluate the reliability of our findings for themselves and even attempt to reproduce the results, if desired. In those cases where we based our tabulations upon samples, we made no effort to calculate standard sampling errors, since we considered our resources better employed in checking nonsampling errors, which we suspect are far the more serious problem. For sampling errors in the findings of other observers, readers should consult the sources that generated the data; for this reason, among others, these sources are documented carefully in our tables.[3]

Despite any and all failings, our data seem accurate enough for present purposes. Certainly most of the data for earlier years are less reliable than for recent years, as we showed for one case in Table 5.19. Forced to provide an educated guess, we would say that a given estimate is probably accurate to within about 20%.[4] This is enough precision to substantiate most of the inferences we make in this report. It may not be precise enough for all possible inferences our readers may make: *caveat lector*!

[3] Percentages in the tables are usually given to the nearest tenth. A useful guide to good rounding practice is II, Croxton *et al.*, 1967, 738–742. Standard errors *are* provided for the estimates of the parameters of nonlinear regression curves (see Table E.3). Errors in the degree-conferral and degree-holders data are discussed in IB, Adkins, 1975, 19 and 49–59. A useful bibliography of official sources concerning errors in data generated by federal agencies appears in II, Bowman, 1964, 19–20. For errors in the 1970 Census of Populations, see IA, USBC-COP, 1973, App-61 to App-77. An excellent guide to the treatment of errors in survey and census data and their presentation in reports like ours is II, Gonzalez *et al.*, 1975.

[4] We offer this often generous outer bound in the spirit of II, Morgenstern, 1963a, 254–259, but admittedly without the rigor with which the estimates discussed by Morgenstern were derived. See also II, Morgenstern, 1963b; II, Bowman, 1964; and II, Morgenstern, 1964. A similar "expert guess" was offered and defended in III, Forman, Heilbron, and Weart, 1975, 10.

E. TREND ANALYSES: TECHNICAL DETAILS

The rudimentary state of the movement to compile reliable, quantitative, historical indicators of American science renders inappropriate to our work most of the precision instruments deployed by statisticians. A surgeon's scalpel is not suited to hacking paths through dense jungles.

Graphic representation of serial data is one indispensable means of conveying a rough, qualitative sense of trend, which is one reason why so many illustrations appear in this report. Regression analysis, the common technique of fitting curves to data by the method of least squares, augments this undertaking. We have employed it where appropriate. In deference to the imprecision of the data, though, we fitted very simple, classic growth curves to the time series. Only rarely did we engage in further fine tuning, such as breaking long-run series into segments and fitting separate curves to each piece.[1]

Brief explanations of the nature of the four types of curve we employed are provided below. Next, the actual equations for the curves and the estimated parameters for each calculated trend are discussed. The final section gives formulas for calculating compound interest rates and their equivalent doubling times.

E.1. Kinds of growth

Each of the curves we used implies something different about the process by which growth occurs. In linear growth, a series increases by a constant annual increment — the same amount every year. Such a phenomenon usually results when the agent producing the growth is both unchanging and independent of the entity which is growing. Since relatively few social systems work in this way, we have postulated a model of linear growth in only six cases, where such a pattern is at least plausible or where, because of the softness of the observed data, it did not seem worthwhile to do anything more sophisticated.

[1] Analysis of trend is but one of a collection of techniques known as time series analysis; general introductions to the subject (in ascending order of complexity) include: II, Floud, 1973, 85–124; II, Harnett, 1975, 439–465; II, Croxton *et al.*, 1967, 214–342, esp. 249–284; and II, Olinick, 1978, esp. 59–65. On the merits of graphic presentation, see II, Ehrenberg, 1978; II, Spear, 1969; and II, Tufte, 1983.

Exponential growth, more commonly encountered than linear growth, implies annual increase by a constant percentage of the amount from the preceding year. Since the addition to the total gets higher as the total itself climbs, exponential growth is much more rapid than linear growth. Such a process usually implies that the agent producing the growth is somehow related to, or incorporated within, the entity which grows exponentially. In human populations, for instance, fertile women are the critical (though demonstrably not the only!) agents of increase; their numbers, and consequently the annual number of births, depend heavily upon the total size of the population. Exponential growth is encountered often in such fields as demography and epidemiology. We postulated its presence on 20 occasions.

Little reflection is necessary to discover limitations to proposing exponential growth everywhere: it cannot be sustained forever. Left unchecked, the exponential increase in the number of certified chemists – to give an apposite example – would soon lead to a situation in which everyone on earth is a chemist. Obviously, saturation must occur.[2]

In forms appropriate for our purposes, the logistic and Gompertz equations, which comprise the two other types of growth we proposed, incorporate the notion of a ceiling. Systems growing according to these formulas increase neither by a constant amount nor a constant percentage; instead, their rates of increase climb for a period, passing through a more-or-less exponential phase in the middle, then fall again as saturation sets in. We employed logistic curves on seven occasions and Gompertz on only one. What is more, most of the 20 cases of exponential growth we found doubtless represent logistic or Gompertz growth in their most explosive phases of expansion.[3]

Choosing the "best" curve to fit to the data is as much art as science. Using rather complicated mathematics, one could produce an elaborate polynomial that comes as close to the observed values as desired, but such an equation rarely conveys any theoretical or qualitative understanding of the processes that cause growth. Nor is the well-known Pearson product-moment correlation coefficient of much use in choosing among the candidate

[2] Derek de Solla Price deserves applause for his lucid discussion of this notion in *Science since Babylon* (II, Price, 1961, 92–124). But see also II, Rose, 1967, and II, Gilbert and Woolgar, 1974, esp. 279–283.

[3] For the record, there *are* constancies in these equations. The Gompertz curve implies a constant rate of decline in the *increments* to the *logarithm* of the observed value. Logistic growth connotes a constant rate of decline in the *decrements* to the *reciprocal* of the observed value. These features become evident when the equations are converted to the form of a modified exponential curve.

equations, since the choice is usually between linear and nonlinear growth, or between different sorts of nonlinear growth, where Pearson's parameter is inapplicable.[4] The choice must therefore be primarily subjective. As a first step, we plotted each series on both arithmetic and semilogarithmic graph paper. If the resulting plot appeared linear on arithmetic axes, we fitted a linear equation. If it was linear on the semilogarithmic axes, we fitted a log-linear (exponential) curve. If it was nonlinear on both, but looked approximately sigmoid on the arithmetic axes, we fitted either a Gompertz or a logistic curve.

In cases of a sigmoid pattern, we employed mathematical and graphical techniques to choose between the Gompertz and the logistic equations. For each series, we calculated the first differences of the logarithms and of the reciprocals of the observed values. If the annual percentage change of the former of these was approximately constant, we fitted a Gompertz curve; if the latter, we fitted a logistic curve. To corroborate our choice, or if these mathematical prestidigitations proved inconclusive, we plotted the reciprocals of the observed values on arithmetic axes, then compared the shape of this trend with that of the original series plotted on semilogarithmic axes. If the former more nearly resembled a modified exponential curve, we fitted a logistic equation; if the latter, we fitted the Gompertz equation. When none of these tests provided a clear choice, we settled for a logistic curve.[5]

E.2. Calculating the trends

Once the type of curve had been chosen, trends were calculated by the method of least-squares regression. This was done directly in the case of linear growth, using the equation

$$Y = K + MT, \tag{1}$$

where Y is the variable or series to which a trend is fitted, T is the year in which Y is measured, K is the constant base value that Y assumes

[4] We calculated Pearson's r for all linear and log-linear curves nonetheless. It was invariably very high, and usually above 0.9; this is not surprising in time series analysis, especially where curves are fitted to the logarithms of Y rather than to the Y values themselves. For many other examples, see II, Hamblin *et al.*, 1973.

[5] For details, see II, Croxton *et al.*, 1967, 318–319. These techniques exploit the constancies in the two sigmoid equations, as explained in note 3 above.

in year zero (the y-intercept), and M is the constant increment of increase, called the slope. Plotting Y against T on arithmetic axes yields a straight line, and estimates for the values of K and M can be calculated by straightforward linear regression of the observed values of Y against T. Table E.1 gives the calculated estimates for the six linear curves we fitted.[6]

The same technique of linear least-squares regression was used to calculate exponential trends, but only after transformation of the observed data. Obviously, the basic equation for exponential growth,

$$Y = KB^T, \qquad (2)$$

does not express a linear relationship between Y and T. (In this standard accounting formula for compound interest, Y, K, and T are the same as in the linear equation, while $B = (1 + R/100)$, with R being the annual percentage growth rate.) To obtain a linear form of this equation, take the logarithm of both sides, yielding

$$\ln Y = \ln K + (\ln B)\, T. \qquad (3)$$

Comparison of Equations (1) and (3) shows the parallels between Y, K, and M in Equation (1), and $\ln Y$, $\ln K$, and $\ln B$, respectively, in Equation (3). Since it is the *logarithms* of Y that bear a linear relationship with T in cases of exponential growth, one must plot Y against T on *semilogarithmic* graph paper to obtain a straight line. And because the logarithmic form of the equation is linear in nature, estimates for $\ln K$ and $\ln B$ can be calculated by linear regression of the logarithms of the observed values of Y against T. Table E.2 gives our calculated estimates of $\ln K$, $\ln B$, and R for the 20 exponential curves derived in this way.

No convenient modifications of the basic logistic and Gompertz equations will yield a linear form, so it was necessary to resort to the relatively new technique of nonlinear least-squares regression to fit these curves to the data. We fitted the Gompertz equation in its usual form:

$$Y = KA^{B^T}. \qquad (4)$$

[6] Hand-held Texas Instruments calculators with a built-in linear regression subroutine were employed.

For logsitic growth, a somewhat less common formula was used:[7]

$$Y = \frac{K}{1 + e^{A + BT}}. \tag{5}$$

For both equations, in the uses which interest us here, K is the ceiling value that the curves approach at saturation, e is the base for natural logarithms (2.71828), A and B are parameters that influence the shape and placement of the curves, Y is the observed value, and T is the year of observation, minus 1870. (The year less 1870 was used instead of the year itself to avoid computations too large for the computer to handle during the regression procedure.)

The nonlinear regressions were performed on an IBM 370 computer, using the NLIN procedure in the SAS79 statistical package.[8] Although a derivative-free method of performing the regressions is now available, we used the methods employing partial derivatives instead, because they seemed to produce better results in those cases where we tried both. Of the three methods using derivatives, we used the Gauss-Newton because, on comparison, it seemed to work best in our cases. To force the maximum refinement out of the procedure, we decreased the convergence criterion to 10^{-14}. To prevent the computer from making nonsensical adjustments to the parameters, we inserted BOUNDS statements. These kept the parameter A above zero at all times; they kept the parameter B below zero for the logistic regressions, but above zero for the Gompertz. To provide initial values of the parameters, we calculated rough approximations using three equally-spaced points in the observed series.[9] If these did not work especially well (usually because the three points chosen were not particularly representative of the trend), we arbitrarily adjusted the Y values for one or more points to get better initial values, or we let the computer program search a range of values in the neighborhood of the initial values we had calculated. In all cases, the nonlinear regression was of the observed values (Y) against

[7] The standard representation for logistic growth is $1/Y = C + MP^T$. To derive Equation (5) from this formula, take its reciprocal, substitute $1/K$ for C, e^B for P, and e^A/K for M, then manipulate algebraically. We used the form in Equation (5) because it easily yields partial derivatives with respect to the estimated parameters K, A, and B, and because K is the ceiling value for the curve. The partial derivatives are necessary for the nonlinear least-squares regression procedure used. On Equation (5), see II, Croxton *et al.*, 1967, 274.

[8] See II, SAS Institute Inc., 1979, 317–329.

[9] In the case of the Gompertz equation, the exact computations employed here can be found in II, Harnett, 1975, 448–451; for the logistic equation, the computations used are from II, Olinick, 1978, 62–63.

year less 1870 (T). The computer produced estimates of K, A, and B, as well as calculated trend values, for the one Gompertz and the seven logistic curves in our report (see Table E.3).

E.3. Calculating growth rates

For an exponential trend, the growth rate is the most critical parameter, the fingerprint by which that curve is usually distinguished from others of the same nature. For those series to which we fitted an exponential curve by regression, the growth rate, R, is given in Table E.2. In many other cases in the text, however, we reported a compound percentage rate of growth on the *assumption* of exponential growth between two observed dates, but without resorting to regression analysis. Readers may obtain such rates by employing the following formula, derived directly from Equation (2):[10]

$$R\ [\%] = [\exp\left(\frac{\ln Y_2/Y_1}{T_2 - T_1}\right) - 1] \times 100, \tag{6}$$

where Y_1 is the value observed in year T_1 and Y_2 the value in the later T_2.

An alternative and perhaps more easily-grasped way to convey the same information is to compare the *doubling times* of the two trends, that is, the times the two respective trends take to double in size. As along as exponential growth continues unabated, this value remains constant and can be computed from the growth rate R (given as a percentage) as follows:[11]

$$\text{doubling time [years]} = \frac{\ln 2}{\ln (1 + R/100)}. \tag{7}$$

For purposes of comparison, an R of 10% implies a doubling time of about 7.3 yr, while 5% yields a doubling every 14.2 yr.

[10] To obtain this formula, write Equation (2) twice (once with each of the two data points T_i, Y_i), divide one of the equations by the other, and solve for R.

[11] This formula also can be obtained by writing Equation (2) twice (this time with data points T_1, Y and T_2, $2Y$), then solving simultaneously for $(T_2 - T_1)$, which is the doubling time.

TABLES

"Tonight, we're going to let the statistics speak for themselves."

TABLE 2.1.
Categories of chemist in the 1970 Census, arranged according to specialization

1. Analysts and assayers	
Analyst	Gold assayer
Analytical chemist	Spectrograph operator
Assayer	Spectrographer
Chemical analyst	Spectroscopist
2. General and research chemists	
Chemist	Organic chemist
Inorganic chemist	Physical chemist
Laboratory chemist	
3. Biological chemists	
Biochemist	Physical biochemist
Biological chemist	Physiological chemist
Nutritional chemist	
4. Medical and pharmaceutical chemists	
Medical chemist	Pharmaceutical chemist
Pharmaceutical analyst	Pharmacognosist
5. Agricultural and food chemists	
Agricultural chemist	Food scientist
Cereal chemist	Food technologist
Dairy chemist	Food-processing chemist
Fermentologist	Juice tester
Food analyst	Juice standardizer
Food chemist	
6. Industrial and manufacturing chemists	
Ceramic chemist	Formulator
Ceramist	Glass technologist
Coagulating-drying supervisor	Industrial chemist
Coal chemist	Mix chemist
Color consultant	Oil expert
Color maker	Paint formulator
Color maker, formulator	Pesticide chemist
Color matcher	Powder expert
Colorist	Quality-control chemist
Colorist, formulator	Rubber chemist

Table 2.1. (continued)

Compounder, formulator	Rubber compounder, formulator
Control chemist	Textile chemist
Dye colorist, formulator	Textile colorist, formulator
Dye expert	Textile technologist
Electrochemist	Tower-control man
7. Metallurgists	*8. Environmental chemists*
Metallographer	Atmospheric chemist
Metallurgical specialist	Soil chemist
Metallurgist	Water chemist
9. Service occupations	
Chemical economist	Inspector, chemical
Chemical educator	Patent chemist
Chemical librarian	Teacher, chemistry

See Figure 2.1–1.

Note: These seventy-five titles are distinct subcategories of the occupational code for chemists (045). That code, in turn, is but one of the entries included in the broader occupational category of "professional, technical, and kindred workers". Since industry classifications are used in conjunction with occupational titles in allocating individuals to particular occupational codes, not all persons with one of the job titles listed here are enumerated as chemists by the U.S. Bureau of the Census. A striking example is provided by the title of "teacher, chemistry", which includes as chemists only those chemistry teachers *not* connected with institutions of primary, secondary, or higher education; the Census Bureau allocates these individuals to subcategories of educational occupations. For a discussion of the relationship between occupational titles and industry designations, see II, USBC, 1971, iv-vi.

Source: II, USBC, 1971, O–5 to O–6. The groupings according to specialization are our own.

TABLE 2.2.
The number of chemists, 1870–1970

	Chemists (thousands)		
	Census estimates	Calculated trend (exponential)	Calculated trend (Gompertz)
Year	(1)	(2)	(3)
1870	0.774	1.022	0.750
1880	1.969	2.407	2.080
1890	4.503	3.897	4.827
1900	9	6	10
1910	16	10	17
1920	28	17	28
1930	45	27	41
1940	57	43	56
1950	77	70	73
1960	84	114	91
1970	110	184	109

See Figures 2.1–2, 2.1–3, 2.1–4, 2.1–5, 2.2–5, 2.2–10, 5.1–8, A.2–1, and A.2–2.

Note: For 1870–1890, estimates are to nearest chemist; for 1900–1970, to nearest thousand.

Sources: *Column 1*:

1870 to 1890 – IA, Edwards, 1943, Table 8, 104–112.
1900 to 1970 – IA, USBC, 1975b, 140, Series D245.

Columns 2 and 3 – calculated by regression on column 1; see Appendix E.

TABLE 2.3.
Chemists per ten thousand working population, 1870–1970

Year	Chemists[a]	Working population[b] (ten thousands)	Chemists per ten thousand working population	
			Observed	Calculated trend (logistic)
	(1)	(2)	(3)	(4)
1870	774	1292.5	0.6	0.6
1880	1 969	1739.2	1.1	1.1
1890	4 503	2331.8	1.9	1.9
1900	9 000	2903.0	3.1	3.1
1910	16 000	3729.1	4.3	4.8
1920	28 000	4220.6	6.6	6.8
1930	45 000	4868.6	9.2	9.0
1940	57 000	5174.2	11.0	10.8
1950	77 000	5899.9	13.1	12.2
1960	84 000	6799.0	12.4	13.2
1970	110 000	7980.2	13.8	13.8

See Figure 2.1–6.

[a] For 1870–1890, estimates are to nearest chemist; for 1900–1970, to nearest thousand.
[b] For 1870–1930, estimates are to nearest thousand persons ten years old or older in the "economically active population"; for 1940–1960, 14 yr old and older; for 1970, 16 yr and older. For other inconsistencies, see IA, USBC, 1975b, 125–126 and 145.

Sources: *Column 1* – see Table 2.2, column 1.

Column 2:

1870 to 1890 – IA, Edwards, 1943, Table 8, 104–112.
1900 to 1970 – IA, USBC, 1975b, 140, Series D233.

Column 4 – calculated by regression on column 3; see Appendix E.

TABLE 2.4.
Annual membership in the American Chemical Society, 1876–1982

	ACS members				
		Calculated trends			
			Logistic		
	Observed	Exponential 1876–1979	1889–1979	1889–1935	1935–1979
Year	(1)	(2)	(3)	(4)	(5)
1876	230	300	–	–	–
1877	265	321	–	–	–
1878	256	344	–	–	–
1879	289	368	–	–	–
1880	303	394	–	–	–
1881	314	421	–	–	–
1882	293	451	–	–	–
1883	306	483	–	–	–
1884	323	517	–	–	–
1885	255	553	–	–	–
1886	241	592	–	–	–
1887	235	634	–	–	–
1888	227	678	–	–	–
1889	204	726	379	226	–
1890	238	777	416	266	–
1891	302	832	457	313	–
1892	351	890	502	368	–
1893	460	953	551	433	–
1894	722	1 020	605	508	–
1895	903	1 091	664	597	–
1896	1 011	1 168	729	700	–
1897	1 156	1 250	800	820	–
1898	1 415	1 338	878	960	–
1899	1 569	1 432	963	1 122	–
1900	1 715	1 533	1 057	1 309	–
1901	1 933	1 641	1 160	1 525	–
1902	2 188	1 756	1 272	1 773	–
1903	2 428	1 880	1 396	2 057	–
1904	2 675	2 012	1 531	2 379	–
1905	2 919	2 154	1 680	2 745	–

Table 2.4. (continued)

	ACS members				
		Calculated trends			
			Logistic		
	Observed	Exponential 1876–1979	1889–1979	1889–1935	1935–1979
Year	(1)	(2)	(3)	(4)	(5)
1906	3 079	2 305	1 842	3 155	–
1907	3 389	2 467	2 020	3 613	–
1908	4 004	2 641	2 215	4 121	–
1909	4 502	2 826	2 428	4 678	–
1910	5 081	3 025	2 662	5 283	–
1911	5 603	3 238	2 917	5 934	–
1912	6 219	3 466	3 196	6 627	–
1913	6 673	3 709	3 502	7 355	–
1914	7 170	3 970	3 835	8 110	–
1915	7 417	4 250	4 199	8 884	–
1916	8 355	4 548	4 597	9 666	–
1917	10 603	4 868	5 030	10 445	–
1918	12 203	5 211	5 503	11 212	–
1919	13 686	5 577	6 018	11 956	–
1920	15 582	5 969	6 578	12 668	–
1921	14 318	6 389	7 188	13 343	–
1922	14 400	6 839	7 850	13 973	–
1923	14 346	7 320	8 570	14 557	–
1924	14 515	7 834	9 350	15 091	–
1925	14 381	8 385	10 195	15 576	–
1926	14 704	8 975	11 109	16 012	–
1927	15 188	9 606	12 097	16 401	–
1928	16 240	10 282	13 164	16 746	–
1929	17 426	11 005	14 312	17 051	–
1930	18 206	11 779	15 547	17 318	–
1931	18 963	12 608	16 874	17 550	–
1932	18 572	13 494	18 295	17 753	–
1933	17 465	14 444	19 815	17 928	–
1934	17 561	15 459	21 436	18 080	–
1935	17 541	16 547	23 162	18 210	18 677

Table 2.4. (continued)

	ACS members				
		Calculated trends			
			Logistic		
	Observed	Exponential 1876–1979	1889–1979	1889–1935	1935–1979
Year	(1)	(2)	(3)	(4)	(5)
1936	18 727	17 710	24 994	–	20 531
1937	20 677	18 956	26 935	–	22 530
1938	22 185	20 289	28 984	–	24 676
1939	23 519	21 716	31 142	–	26 971
1940	25 414	23 243	33 407	–	29 415
1941	28 738	24 878	35 775	–	32 006
1942	31 717	26 628	38 245	–	34 739
1943	36 001	28 501	40 811	–	37 608
1944	39 438	30 505	43 466	–	40 604
1945	43 075	32 651	46 203	–	43 715
1946	48 755	34 947	49 013	–	46 927
1947	55 100	37 405	51 888	–	50 224
1948	58 782	40 036	54 815	–	53 587
1949	62 211	42 851	57 784	–	56 996
1950	63 349	45 865	60 781	–	60 430
1951	66 009	49 091	63 794	–	63 869
1952	67 730	52 544	66 811	–	67 289
1953	70 155	56 239	69 817	–	70 669
1954	72 287	60 194	72 800	–	73 990
1955	75 223	64 428	75 747	–	77 232
1956	79 224	68 959	78 645	–	80 378
1957	81 927	73 809	81 485	–	83 414
1958	85 815	79 000	84 256	–	86 327
1959	88 806	84 557	86 948	–	89 106
1960	92 193	90 504	89 553	–	91 745
1961	93 637	96 869	92 066	–	94 238
1962	95 210	103 682	94 479	–	96 583
1963	96 749	110 974	96 790	–	98 779
1964	99 475	118 779	98 995	–	100 826
1965	102 525	127 133	101 092	–	102 729

Table 2.4. (continued)

	ACS members				
		Calculated trends			
			Logistic		
	Observed	Exponential 1876–1979	1889–1979	1889–1935	1935–1979
Year	(1)	(2)	(3)	(4)	(5)
1966	106 271	136 074	103 080	–	104 490
1967	109 528	145 644	104 960	–	106 115
1968	113 373	155 888	106 732	–	107 610
1969	116 816	166 851	108 399	–	108 982
1970	114 323	178 586	109 963	–	110 237
1971	112 016	191 147	111 427	–	111 383
1972	110 708	204 590	112 795	–	112 428
1973	110 285	218 979	114 069	–	113 377
1974	110 274	234 380	115 255	–	114 239
1975	110 820	250 865	116 357	–	115 020
1976	112 730	268 508	117 379	–	115 727
1977	115 141	287 393	118 325	–	116 366
1978	116 240	307 606	119 200	–	116 943
1979	118 214	329 240	120 008	–	117 463
1980	120 400				
1981	122 377				
1982	127 864				

See Figures 2.2–1, 2.2–2, 2.2–3, 2.2–4, 2.2–5, 2.2–7, and 2.2–10.

Note: Before 1974, the observed totals include members deceased during the calendar year. From 1974 on such deceased members are excluded. The discrepancy is miniscule; excluding such deceased members for 1974, for example, lowers the total from 110 799 to 110 274. Cf. IB, Skolnik and Reese, 1976, 456.

Sources: *Column 1*:

1876 to 1924, 1926 to 1975 – IB, Skolnik and Reese, 1976, 456.
1925 – III, Browne and Weeks, 1952, 127.
1976 to 1982 – IB, ACS, 1983a, 56.

Columns 2 to 5 – calculated by regression on column 1; see Appendix E.

TABLE 2.5.
American Chemical Society members versus Census chemists, 1870–1970

			ACS members per Census chemist	
	ACS members[a]	Chemists (Census estimates)	Observed	Calculated trend (logistic)
Year	(1)	(2)	(3)	(4)
1870	–	774	–	–
1880	303	1 969	0.15	–
1890	238	4 503	0.05	0.11
1900	1 715	9 000	0.19	0.17
1910	5 081	16 000	0.32	0.25
1920	15 582	28 000	0.56	0.36
1930	18 206	45 000	0.40	0.49
1940	25 414	57 000	0.45	0.65
1950	63 349	77 000	0.82	0.81
1960	92 193	84 000	1.10	0.96
1970	114 323	110 000	1.04	1.09

See Figure 2.2–6.

[a] Membership totals for all non-census years appear in Figure 2.2–5 and are included in Table 2.4, column 1.

Sources: *Column 1* – IB, Skolnik and Reese, 1976, 456.

Column 2:

1870 to 1890 – IA, Edwards, 1943, Table 8, 104–112.
1900 to 1970 – IA, USBC, 1975b, 140, Series D245.

Column 4 – calculated by regression on column 3; see Appendix E.

TABLE 2.6.
Indicators of the relationship between training and professionalization in chemistry, 1930–1971

Year	ACS members	Holders of terminal degrees in chemistry: Masters and doctorates only	Holders of terminal degrees in chemistry: All levels	ACS members per hundred holders of any terminal chemistry degree	Holders of terminal masters and doctorates in chemistry per hundred ACS members
	(1)	(2)	(3)	(4)	(5)
1930	18 206	11 254	42 477	42.9	61.8
1931	18 963	11 908	44 657	42.5	62.8
1932	18 572	12 580	46 688	39.8	67.7
1933	17 465	13 225	48 752	35.8	75.7
1934	17 561	13 791	50 851	34.5	78.5
1935	17 541	14 329	53 082	33.0	81.7
1936	18 727	14 893	55 482	33.8	79.5
1937	20 677	15 521	58 269	35.5	75.1
1938	22 185	16 140	61 363	36.2	72.8
1939	23 519	16 853	64 789	36.3	71.7
1940	25 414	17 645	68 634	37.0	69.4
1941	28 738	18 461	72 498	39.6	64.2
1942	31 717	19 212	76 425	41.5	60.6
1943	36 001	19 698	79 216	45.4	54.7
1944	39 438	20 064	81 105	48.6	50.9
1945	43 075	20 391	83 102	51.8	47.3
1946	48 755	20 814	85 147	57.3	42.7
1947	55 100	21 715	88 594	62.2	39.4
1948	58 782	23 059	93 091	63.1	39.2
1949	62 211	24 738	100 499	61.9	39.8
1950	63 349	26 612	109 205	58.0	42.0
1951	66 009	28 409	115 087	57.4	43.0
1952	67 730	30 095	119 511	56.7	44.4
1953	70 155	31 542	123 157	57.0	45.0
1954	72 287	32 851	127 066	56.9	45.4
1955	75 223	34 209	130 854	57.5	45.5
1956	79 224	35 526	134 313	59.0	48.8
1957	81 927	36 733	138 176	59.3	44.8

Table 2.6. (continued)

		Holders of terminal degrees in chemistry		ACS members per hundred holders of any terminal chemistry degree	Holders of terminal masters and doctorates in chemistry per hundred ACS members
	ACS members	Masters and doctorates only	All levels		
Year	(1)	(2)	(3)	(4)	(5)
1958	85 815	38 004	142 420	60.3	44.3
1959	88 806	39 342	146 924	60.4	44.3
1960	92 193	40 773	151 719	60.8	44.2
1961	93 637	42 304	156 479	59.8	45.2
1962	95 210	43 934	161 743	58.9	46.1
1963	96 749	45 616	167 719	57.7	47.1
1964	99 475	47 455	174 482	57.0	47.7
1965	102 525	49 536	181 484	56.5	48.3
1966	106 271	51 761	187 847	56.6	48.7
1967	109 528	54 042	194 099	56.4	49.3
1968	113 373	56 627	201 692	56.2	49.9
1969	116 816	59 335	210 201	55.6	50.8
1970	114 323	62 066	217 581	52.5	54.3
1971	112 016	64 852	224 620	49.9	57.9

See Figures 2.2–8, 2.2–10, A.4–2, A.4–3, and A.4–4.

Sources: *Column 1* – IB, Skolnik and Reese, 1976, 456.

Columns 2 and 3 – IB, Adkins, 1975, Table B–4.5, 474–475.

TABLE 2.7.
Chemists per thousand professional, technical, and kindred workers, 1870–1970

Year	Chemists (1)	Professional, technical, and kindred workers[a] (thousands) (2)	Chemists per thousand professional, technical, and kindred workers (3)
1870	774	342.107	2.26
1880	1 969	549.822	3.58
1890	4 503	876.299	5.14
1900	9 000	1 234	7.29
1910	16 000	1 758	9.10
1920	28 000	2 283	12.26
1930	45 000	3 311	13.59
1940	57 000	3 879	14.69
1950	77 000	5 081	15.15
1960	84 000	7 336	11.45
1970	110 000	11 561	9.51

See Figures 2.2–9 and A.2–3.

Note: For 1870–1890, estimates are to the nearest individual; for 1900–1970, to the nearest thousand.

[a] Described simply as "professional service" in the source for 1870–1890. See sources for other minor discrepancies.

Sources: *Columns 1 and 2*:

1870 to 1890 – IA, Edwards, 1943, Table 8, 111.
1900 to 1970 – IA, USBC, 1975b, 140, Series D234.

TABLE 3.1.
Bachelors degrees conferred in selected fields, 1890–1978

	Bachelors degrees conferred				
		Natural sciences[a]		Chemistry[b]	
	All fields	Number	As percentage of all fields	Number	As percentage of all fields
Year	(1)	(2)	(3)	(4)	(5)
1890	7 228	1 890	26.1	631	8.7
1891	7 693	2 013	26.2	671	8.7
1892	7 334	1 916	26.1	636	8.7
1893	9 090	2 378	26.2	793	8.7
1894	11 796	3 084	26.1	1 031	8.7
1895	13 529	3 537	26.1	1 184	8.8
1896	13 357	3 494	26.2	1 168	8.7
1897	13 244	3 464	26.2	1 156	8.7
1898	13 090	3 424	26.2	1 142	8.7
1899	14 140	3 700	26.2	1 235	8.7
1900	15 048	3 933	26.1	1 313	8.7
1901	15 838	4 143	26.2	1 381	8.7
1902	16 188	4 232	26.1	1 410	8.7
1903	17 000	4 445	26.1	1 481	8.7
1904	17 501	4 574	26.1	1 524	8.7
1905	17 380	4 381	25.2	1 495	8.6
1906	19 711	4 889	24.8	1 697	8.6
1907	19 918	4 760	23.9	1 698	8.5
1908	21 118	5 046	23.9	1 799	8.5
1909	25 690	6 144	23.9	2 194	8.5
1910	24 665	5 825	23.6	2 073	8.4
1911	25 192	5 909	23.5	2 098	8.3
1912	26 137	6 023	23.0	2 417	9.2
1913	29 571	6 846	23.2	2 136	7.2
1914	31 540	7 297	23.1	2 573	8.2
1915	31 186	7 095	22.8	2 397	7.7
1916	31 960	7 216	22.6	2 379	7.4
1917	31 048	6 943	22.4	2 196	7.1
1918	29 726	6 648	22.4	2 102	7.1
1919	35 235	7 881	22.4	2 497	7.1

Table 3.1. (continued)

	Bachelors degrees conferred				
		Natural sciences[a]		Chemistry[b]	
	All fields	Number	As percentage of all fields	Number	As percentage of all fields
Year	(1)	(2)	(3)	(4)	(5)
1920	40 741	8 312	20.4	2 623	6.4
1921	45 806	10 005	21.8	2 809	6.1
1922	51 304	8 995	17.5	2 827	5.5
1923	58 847	10 314	17.5	3 236	5.5
1924	67 325	11 798	17.5	3 694	5.5
1925	73 741	12 371	16.8	3 784	5.1
1926	80 664	13 229	16.4	3 988	4.9
1927	86 917	13 405	15.4	3 979	4.6
1928	93 712	14 655	15.6	4 271	4.6
1929	99 419	15 547	15.6	4 517	4.5
1930	105 295	15 297	14.5	4 392	4.2
1931	112 350	15 707	14.0	4 478	4.0
1932	119 825	15 451	12.9	4 350	3.6
1933	119 516	15 410	12.9	4 331	3.6
1934	119 205	15 369	12.9	4 313	3.6
1935	122 792	15 911	13.0	4 493	3.7
1936	126 574	16 440	13.0	4 653	3.7
1937	136 763	17 859	13.1	5 084	3.7
1938	147 652	17 279	11.7	5 475	3.7
1939	158 299	20 661	13.1	5 858	3.7
1940	169 582	22 976	13.5	6 366	3.8
1941	169 425	21 885	12.9	6 383	3.8
1942	169 401	21 733	12.8	6 455	3.8
1943	138 397	17 704	12.8	5 199	3.8
1944	113 673	14 669	12.9	4 207	3.7
1945	119 125	15 361	12.9	4 400	3.7
1946	123 926	15 969	12.9	4 564	3.7
1947	189 657	23 273	12.3	6 344	3.3
1948	247 772	29 535	11.9	7 833	3.2
1949	338 160	38 798	11.5	10 658	3.2
1950	403 783	49 184	12.2	12 272	3.0
1951	354 606	39 017	11.0	9 423	2.7

Table 3.1. (continued)

	Bachelors degrees conferred				
		Natural sciences[a]		Chemistry[b]	
	All fields	Number	As percentage of all fields	Number	As percentage of all fields
Year	(1)	(2)	(3)	(4)	(5)
1952	304 167	31 626	10.4	7 777	2.6
1953	277 652	28 064	10.1	6 906	2.5
1954	265 480	28 753	10.8	7 108	2.7
1955	261 030	27 269	10.4	7 038	2.7
1956	284 774	30 638	10.8	6 779	2.4
1957	312 891	34 337	11.0	7 253	2.3
1958	337 133	37 866	11.2	7 708	2.3
1959	354 867	41 867	11.8	7 989	2.3
1960	364 135	45 529	12.5	8 308	2.3
1961	370 600	47 139	12.7	8 289	2.2
1962	388 184	50 200	12.9	8 877	2.3
1963	417 178	54 618	13.1	9 719	2.3
1964	467 262	62 250	13.3	10 657	2.3
1965	502 044	66 580	13.3	11 093	2.2
1966	525 246	67 838	12.9	10 662	2.0
1967	562 952	71 386	12.7	10 707	1.9
1968	636 863	76 991	12.1	12 241	1.9
1969	734 002	86 645	11.8	13 318	1.8
1970	798 068	91 108	11.4	12 280	1.5
1971	846 108	88 563	10.5	12 049	1.4
1972	894 110	82 373	9.2	10 739	1.2
1973	930 272	86 704	9.3	10 246	1.1
1974	954 376	91 956	9.6	10 536	1.1
1975	931 663	91 478	9.8	10 667	1.1
1976	934 443	92 557	9.9	11 123	1.2
1977	928 228	91 114	9.8	11 350	1.2
1978	930 201	88 809	9.5	11 503	1.2

See Figures 3.1–1, 3.1–4, 3.1–6, 3.1–8, 3.1–9, 3.1–14, and A.3–1.

Note: For the exponential trends calculated for these series, see Appendix E.

Table 3.1. (*continued*)

[a] Includes the following subjects: mathematics and statistics (U.S. Office of Education specialty codes 1701 to 1703, plus 1799); chemistry (1905 to 1910, plus 1920); earth sciences (1913 to 1919, plus 1999); physics (1902 to 1904, 1911, and 1912); physical science not elsewhere classified (1901 and 1999–2); and biological sciences (0401 to 0427, plus 0499). For further information, see IB, Adkins, 1975, Table A–1, 181–190, and *passim*.

[b] Includes the following specialties: general chemistry (U.S. Office of Education specialty codes 1905); inorganic chemistry (1906); organic chemistry (1907); physical chemistry (1908); analytical chemistry (1909); pharmaceutical chemistry (1910); and metallurgy (1920). See IB, Adkins, 1975, Table A–1, 181–190.

Sources: *Columns 1, 2, and 4*:

1890 to 1971 – IB, Adkins, 1975. Totals for all fields are from Table A–2, 190–195; those for natural sciences are a combination from Tables A–4.1 and A–5.16, 210–215 and 314–317; those for chemistry are from Table A–5.5, 270–273. Typographical errors discovered in the source tables are corrected silently.

1972 – IA, NCES, 1975b, Table 5, 12 and 15. Notes a and b above indicate which codes to use in constructing the appropriate totals for this and subsequent years.

1973 – IA, NCES, 1976, 1972–73 Tables, Table 5, 12, 15–16. See comment for 1972 source.

1974 – IA, NCES, 1976, 1973–74 Tables, Table 5, 26 and 29. See comment for 1972 source.

1975 – IA, NCES, 1977, Table 5, 15 and 18. See comment for 1972 source.

1976 – IA, NCES, 1978b, Table 5, 16 and 19. See comment for 1972 source.

1977 – IA, NCES, 1980a, Table 5, 28, and 31. See comment for 1972 source.

1978 – IA, NCES, 1980b, Table 5. See comment for 1972 source.

TABLE 3.2
Second-level degrees conferred in selected fields, 1890–1978

	Second-level degrees conferred				
		Natural sciences[a]		Chemistry[b]	
	All fields	Number	As percentage of all fields	Number	As percentage of all fields
Year	(1)	(2)	(3)	(4)	(5)
1890	9 326	312	3.3	103	1.1
1891	9 923	236	2.4	77	0.8
1892	10 198	226	2.2	75	0.7
1893	10 681	350	3.3	116	1.1
1894	11 277	381	3.4	126	1.1
1895	11 911	420	3.5	139	1.2
1896	12 714	451	3.5	149	1.2
1897	13 400	425	3.2	141	1.1
1898	13 402	435	3.2	144	1.1
1899	13 382	465	3.5	154	1.2
1900	13 945	469	3.4	155	1.1
1901	14 587	519	3.6	171	1.2
1902	14 636	567	3.9	186	1.3
1903	14 625	530	3.6	174	1.2
1904	14 679	512	3.5	169	1.2
1905	16 064	578	3.6	190	1.2
1906	14 095	523	3.7	170	1.2
1907	13 935	463	3.3	155	1.1
1908	14 653	548	3.7	186	1.3
1909	14 390	594	4.1	207	1.4
1910	14 647	554	3.8	191	1.3
1911	14 745	638	4.3	221	1.5
1912	16 306	775	4.8	250	1.5
1913	15 850	733	4.6	268	1.7
1914	15 998	800	5.0	273	1.7
1915	16 303	899	5.5	312	1.9
1916	17 196	997	5.8	348	2.0
1917	14 057	808	5.7	269	1.9
1918	11 759	682	5.8	219	1.9
1919	11 578	838	7.2	263	2.3

Table 3.2. (continued)

	Second-level degrees conferred				
		Natural sciences[a]		Chemistry [b]	
	All fields	Number	As percentage of all fields	Number	As percentage of all fields
Year	(1)	(2)	(3)	(4)	(5)
1920	12 160	1 030	8.5	327	2.7
1921	14 006	1 209	8.6	385	2.7
1922	16 348	1 273	7.8	404	2.5
1923	19 604	1 375	7.0	432	2.2
1924	23 674	1 377	5.8	429	1.8
1925	25 033	1 514	6.0	471	1.9
1926	26 618	1 666	6.3	518	1.9
1927	28 010	1 645	5.9	499	1.8
1928	29 535	1 714	5.8	511	1.7
1929	30 558	1 708	5.6	494	1.6
1930	31 818	1 878	5.9	543	1.7
1931	34 517	2 195	6.4	631	1.8
1932	37 607	2 324	6.2	661	1.8
1933	36 413	2 150	5.9	605	1.7
1934	35 244	1 886	5.4	522	1.5
1935	35 101	1 892	5.4	519	1.5
1936	34 853	1 901	5.5	517	1.5
1937	36 787	2 041	5.5	559	1.5
1938	38 953	2 209	5.7	604	1.6
1939	41 153	2 436	5.9	671	1.6
1940	43 649	2 631	6.0	748	1.6
1941	42 073	2 592	6.2	702	1.7
1942	40 593	2 459	6.1	674	1.7
1943	31 274	1 692	5.4	457	1.5
1944	25 604	1 180	4.6	318	1.2
1945	27 795	1 545	5.6	414	1.5
1946	31 457	2 012	6.4	535	1.7
1947	44 837	3 634	8.1	982	2.2
1948	66 241	4 609	7.0	1 370	2.1
1949	78 529	6 022	7.7	1 614	2.1
1950	87 337	6 805	7.8	1 701	1.9
1951	94 080	7 154	7.6	1 577	1.7

Table 3.2. (continued)

	Second-level degrees conferred				
		Natural sciences[a]		Chemistry[b]	
	All fields	Number	As percentage of all fields	Number	As percentage of all fields
Year	(1)	(2)	(3)	(4)	(5)
1952	90 453	6 399	7.1	1 484	1.6
1953	87 433	4 488	5.1	1 272	1.5
1954	83 550	4 760	5.7	1 129	1.4
1955	83 848	4 994	6.0	1 210	1.4
1956	85 193	5 321	6.2	1 195	1.4
1957	88 578	5 504	6.2	1 086	1.2
1958	93 355	6 334	6.8	1 209	1.3
1959	98 901	7 021	7.1	1 247	1.3
1960	104 324	7 690	7.4	1 341	1.3
1961	108 383	8 871	8.2	1 440	1.3
1962	115 494	9 897	8.6	1 573	1.4
1963	123 704	10 881	8.8	1 576	1.3
1964	134 744	12 145	9.0	1 733	1.3
1965	147 835	13 711	9.3	1 941	1.3
1966	169 838	14 700	8.7	2 023	1.2
1967	188 431	16 515	8.8	1 998	1.1
1968	210 592	17 200	8.2	2 337	1.1
1969	228 715	19 018	8.3	2 374	1.0
1970	243 130	18 358	7.6	2 275	0.9
1971	268 133	17 994	6.7	2 390	0.9
1972	296 548	17 642	5.9	2 294	0.8
1973	314 960	17 601	5.6	2 265	0.7
1974	332 537	17 508	5.3	2 167	0.7
1975	349 910	16 759	4.8	2 035	0.6
1976	376 062	15 969	4.2	1 820	0.5
1977	385 021	16 197	4.2	1 820	0.5
1978	379 780	15 810	4.2	1 923	0.5

See Figures 3.1–2, 3.1–4, 3.1–7, 3.1–8, 3.1–9, and A.3–1.

Note: For the exponential trends calculated for these series, see Appendix E.

[a] For the definition of this aggregation of disciplines, see Table 3.1, note a.
[b] For the definition of "chemistry" as used here, see Table 3.1, note b.

Table 3.2. (continued)

Sources: *Columns 1, 2, and 4*:

1890 to 1971 – IB, Adkins, 1975. Totals for all fields are from Table A–2, 190–195; those for natural sciences are a combination from Tables A–4.1 and A–5.16, 210–215 and 314–317; those for chemistry are from Table A–5.5, 270–273. Typographical errors discovered in the source tables are corrected silently.

1972 – IA. NCES, 1975b, Table 5, 12 and 15, and Table 6, 17. Notes a and b above indicate where to find the specialty codes to use in constructing the appropriate totals for this and subsequent years.

1973 – IA, NCES, 1976, 1972–73 Tables, Table 5, 12, 15–16, and Table 6, 17. See comment for 1972 source.

1974 – IA, NCES, 1976, 1973–74 Tables, Table 5, 26 and 29, and Table 6, 31. See comment for 1972 source.

1975 – IA, NCES, 1977, Table 5, 15 and 18, and Table 6, 19. See comment for 1972 source.

1976 – IA, NCES, 1978b, Table 5, 16 and 19, and Table 6, 20. See comment for 1972 source.

1977 – IA, NCES, 1980a, Table 5, 28 and 31, and Table 6, 32. See comment for 1972 source.

1978 – IA, NCES, 1980b, Tables 5 and 6. See comment for 1972 source.

TABLE 3.3.
Doctorate degrees conferred in selected fields, 1890–1978

Year	Doctorate degrees conferred				
		Natural sciences[a]		Chemistry[b]	
	All fields	Number	As percentage of all fields	Number	As percentage of all fields
	(1)	(2)	(3)	(4)	(5)
1890	149	66	44.3	28	18.8
1891	187	80	42.8	35	18.7
1892	190	81	42.6	35	18.4
1893	218	94	43.1	40	18.3
1894	279	123	44.1	51	18.3
1895	272	119	43.8	49	18.0
1896	271	117	43.2	48	17.7
1897	319	128	40.1	58	18.2
1898	324	140	43.2	57	17.6
1899	345	150	43.5	63	18.3
1900	382	167	43.7	69	18.1
1901	365	158	43.3	66	18.1
1902	293	128	43.7	52	17.7
1903	337	147	43.6	61	18.1
1904	334	146	43.7	60	18.0
1905	369	162	43.9	67	18.2
1906	383	167	43.6	69	18.0
1907	349	153	43.8	63	18.1
1908	391	168	43.0	68	17.4
1909	451	195	43.2	79	17.5
1910	443	193	43.6	78	17.6
1911	497	217	43.7	87	17.5
1912	500	219	43.8	94	18.8
1913	538	235	43.7	87	16.2
1914	559	244	43.6	97	17.4
1915	611	268	43.9	107	17.5
1916	667	291	43.6	115	17.2
1917	699	308	44.1	121	17.3
1918	556	244	43.9	96	17.3
1919	394	171	43.4	65	16.5

Table 3.3. (continued)

Year	Doctorate degrees conferred				
		Natural sciences[a]		Chemistry[b]	
	All fields	Number	As percentage of all fields	Number	As percentage of all fields
	(1)	(2)	(3)	(4)	(5)
1920	615	268	43.6	104	16.9
1921	717	312	43.5	137	19.1
1922	836	382	45.7	147	17.6
1923	958	464	48.4	167	17.4
1924	1 098	515	46.9	218	19.9
1925	1 246	533	42.8	220	17.7
1926	1 415	622	44.0	248	17.5
1927	1 514	616	40.7	211	13.9
1928	1 481	635	42.9	232	15.7
1929	1 746	698	40.0	229	13.1
1930	2 299	945	41.1	332	14.4
1931	2 601	1054	40.5	368	14.1
1932	2 654	1044	39.3	362	13.6
1933	2 712	1150	42.4	420	15.5
1934	2 830	1212	42.8	436	15.4
1935	2 715	1140	42.0	402	14.8
1936	2 770	1212	43.8	461	16.6
1937	2 771	1238	44.7	510	18.4
1938	2 933	1284	43.8	439	15.0
1939	3 059	1382	45.2	502	16.4
1940	3 290	1459	44.3	532	16.2
1941	3 615	1615	44.7	658	18.2
1942	3 497	1538	44.0	606	17.3
1943	2 654	1232	46.4	519	19.6
1944	2 305	1063	46.1	548	23.8
1945	1 943	764	39.3	342	17.6
1946	1 966	731	37.2	317	16.1
1947	2 911	1154	39.6	418	14.4
1948	3 973	1403	35.3	569	14.3
1949	5 293	1830	34.6	757	14.3
1950	6 633	2351	35.4	967	14.6
1951	7 338	2682	36.5	1055	14.4

Table 3.3. (continued)

	Doctorate degrees conferred				
		Natural sciences[a]		Chemistry[b]	
	All fields	Number	As percentage of all fields	Number	As percentage of all fields
Year	(1)	(2)	(3)	(4)	(5)
1952	7683	2701	35.2	1038	13.5
1953	8306	3236	39.0	1005	12.1
1954	8996	3023	33.6	1032	11.5
1955	8840	2959	33.5	1009	11.4
1956	8903	2908	32.7	989	11.1
1957	8756	3023	34.5	1008	11.5
1958	8942	3020	33.8	947	10.6
1959	9360	3139	33.5	1027	11.0
1960	9829	3343	34.0	1062	10.8
1961	10575	3518	33.3	1145	10.8
1962	11622	3852	33.1	1141	9.8
1963	12822	4320	33.7	1256	9.8
1964	14490	4677	32.3	1303	9.0
1965	16467	5469	33.2	1428	8.7
1966	18239	5939	32.6	1583	8.7
1967	20621	6580	31.9	1764	8.6
1968	23091	7346	31.8	1785	7.7
1969	26189	8043	30.7	1980	7.6
1970	29870	8854	29.6	2224	7.4
1971	32111	9235	28.8	2174	6.8
1972	33369	8885	26.6	1992	6.0
1973	34790	8721	25.1	1900	5.5
1974	33826	8102	24.0	1842	5.4
1975	34086	7987	23.4	1834	5.4
1976	34076	7686	22.6	1639	4.8
1977	33244	7570	22.8	1593	4.8
1978	32156	7255	22.6	1544	4.8

See Figures 3.1–3, 3.1–4, 3.1–5, 3.1–8, 3.1–9, 3.1–14, and A.3–1.

Note: For the exponential trends calculated for these series, see Appendix E.

[a] For the definition of this aggregation of disciplines, see Table 3.1, note a.
[b] For the definition of "chemistry" as used here, see Table 3.1, note b.

Table 3.3. (continued)

Sources: *Columns 1, 2, and 4*:

1890 to 1971 – IB, Adkins, 1975. Totals for all fields are from Table A–2, 190–195; those for natural sciences are a combination from Tables A–4.1 and A–5.16, 210–215 and 314–317; those for chemistry are from Table A–5.5, 270–273. Typographical errors discovered in the source tables are corrected silently.
1972 – IA, NCES, 1975b, Table 5, 12–16. Notes a and b above indicate where to find the specialty codes to use in constructing the appropriate totals for this and subsequent years.
1973 – IA, NCES, 1976, 1972–73 Tables, Table 5, 12, 15–16. See comment for 1972 source.
1974 – IA, NCES, 1976, 1973–74 Tables, Table 5, 26 and 29. See comment for 1972 source.
1975 – IA, NCES, 1977, Table 5, 15 and 18. See comment for 1972 source.
1976 – IA, NCES, 1978b, Table 5, 16 and 19. See comment for 1972 source.
1977 – IA, NCES, 1980a, Table 5, 28 and 31. See comment for 1972 source.
1978 – IA, NCES, 1980b, Table 5. See comment for 1972 source.

TABLE 3.4.
Degree conferrals per unit of school-age population, 1890–1970

	Degrees conferred		Resident U.S. population aged 20 to 24 (millions)	Degrees conferred per million population aged 20 to 24 (hundreds)	
	All fields	Chemistry		All fields	Chemistry
Year	(1)	(2)	(3)	(4)	(5)
1890	16 703	762	6.20	26.9	1.2
1900	29 375	1 537	7.34	40.0	2.1
1910	39 755	2 342	9.06	43.9	2.6
1920	53 516	3 054	9.28	57.7	3.3
1930	139 412	5 267	10.87	128.3	4.8
1940	216 521	7 646	11.59	186.8	6.6
1950	497 753	14 940	11.48	433.6	13.0
1960	478 288	10 711	10.73	445.7	10.0
1970	1 071 068	16 779	16.37	654.3	10.3

Sources: *Columns 1 and 2* – IB, Adkins, 1975, Tables A–2 and A–5.5, 190–195 and 270–273.

Column 3 – IA, USBC, 1975b, 15, Series A124.

TABLE 3.5.
Holders of terminal degrees per unit of population, 1930–1970

	Holders of terminal degrees			Degree holders per million population (hundreds)	
	All fields	Chemistry	Resident U.S. population (millions)	All fields	Chemistry
Year	(1)	(2)	(3)	(4)	(5)
1930	1 450 124	42 477	123.08	117.8	3.5
1940	2 588 600	68 634	131.95	196.2	5.2
1950	4 305 208	109 205	151.24	284.7	7.2
1960	6 963 856	151 719	179.98	386.9	8.4
1970	11 603 444	217 581	203.81	569.3	10.7

Sources: *Columns 1 and 2* – IB, Adkins, 1975, Tables B–1 and B–4.5, 432–433 and 474–475.

Column 3 – IA, USBC, 1975b, 8, Series A7.

TABLE 3.6

Bachelors degrees conferred in selected non-scientific fields, 1890–1975

	Bachelors degrees conferred				
Year	Administration[a] (1)	Education[b] (2)	Engineering[c] (3)	Social sciences and related fields[d] (4)	All fields (5)
A. Number					
1890	64	54	642	1 161	7 228
1895	117	104	1 195	2 209	13 529
1900	132	112	1 333	2 418	15 048
1905	180	97	1 598	2 817	17 380
1910	367	216	2 610	4 199	24 665
1915	898	804	4 115	5 172	31 186
1920	2 549	2 363	6 099	6 688	40 741
1925	6 525	8 511	9 711	12 395	73 741
1930	9 631	19 227	11 009	18 154	103 295
1935	13 311	28 308	12 199	20 363	122 792
1940	19 863	41 068	15 653	27 844	169 582
1945	13 151	29 117	10 273	18 825	119 125
1950	77 346	62 508	54 260	74 839	403 783
1955	44 570	53 197	24 228	45 599	261 030
1960	55 339	90 147	39 327	62 947	364 135
1965	65 866	119 179	39 454	106 892	502 044
1970	113 737	125 060	53 642	203 683	798 068
1975	144 113	166 891	60 339	238 355	931 663

Table 3.6. (continued)

	Bachelors degrees conferred				
	Administration[a]	Education[b]	Engineering[c]	Social sciences and related fields[d]	All fields
Year	(1)	(2)	(3)	(4)	(5)
B. As percentage of all fields					
1890	0.9	0.7	8.9	16.1	–
1895	0.9	0.8	8.8	16.3	–
1900	0.9	0.7	8.9	16.1	–
1905	1.0	0.6	9.2	16.2	–
1910	1.5	0.9	10.6	17.0	–
1915	2.9	2.6	13.2	16.6	–
1920	6.3	5.8	15.0	16.4	–
1925	8.8	11.5	13.2	16.8	–
1930	9.3	18.6	10.7	17.6	–
1935	10.8	23.1	9.9	16.6	–
1940	11.7	24.2	9.2	16.4	–
1945	11.0	24.4	8.6	15.8	–
1950	19.2	15.5	13.4	18.5	–
1955	17.1	20.4	9.3	17.5	–
1960	15.2	24.8	10.8	17.3	–
1965	13.1	23.7	7.9	21.3	–
1970	14.3	15.7	6.7	25.5	–
1975	15.5	17.9	6.5	25.6	–

See Figure 3.1–10, part A.

Table 3.6. (continued)

[a] Includes business administration (U.S. Office of Education specialty codes 0501 to 0516, plus 0599) and administration other than business administration (0110, 0112, 0809, 0827, 0913, 1202, 1215, 1307, 1801 to 1803, 1899, 2102, and 2103). For further information about the definition of this and the other fields listed in this table, see IB, Adkins, 1975, Tables A–1 and A–4, 181–190 and 208–211, and *passim*.

[b] Includes the following subjects: education of primary-age, preprimary-age, and exceptional children (U.S. Office of Education specialty codes 0802, 0808, 0810 to 0820, 0823, and 0899–2); secondary education, adult education, and specialized teaching fields (0803 to 0807, 0830 to 0834, 0836, 0838, 0839, 0899–1, 0899–3, and 0899–4); physical education (0835); and education not elsewhere classified (0801, 0821, 0824 to 0826, 0828, 0829, and 0837).

[c] Includes the following subjects: computer and information science (U.S. Office of Education specialty codes 0701 to 0705, plus 0799); architecture (0201 to 0205, plus 0299); chemical-materials engineering (0906 and 0914 to 0916); civil and other heavy engineering (0903, 0904, 0908, and 0922); electrical-electronic engineering (0909); geological-mining engineering (0907, 0911, 0912, 0918, and 0924); mechanical-equipment engineering (0902, 0910, and 0923); and engineering and other technical specialties not elsewhere classified (0901, 0905, 0917, 0919 to 0921, 0925, 0999, and 1303).

[d] Includes the following subjects: anthropology and archaeology (U.S. Office of Education specialty codes 2202 and 2203); economics and agricultural economics (0111, 0517, and 2204); history (2205); political science (2207); sociology and social psychology (2005, 2208, and 2215); social science not elsewhere classified (0301 to 0314, 0399, 1505, 2201, 2206, 2210 to 2214, and 2299); journalism and communications (0601 to 0605, plus 0699); law (1401 and 1499); social work (1222, 2101, and 2104); social science professions not elsewhere classified (0206, 2105, 2106, 2199, and 2209); and psychology (0822, 2001 to 2004, 2006 to 2010, and 2099).

Sources: 1890 to 1970 – IB, Adkins, 1975. Totals for administration are from Table A–4.6, 232–235; those for education are from Table A–4.8, 240–243; those for engineering are from Table A–4.2, 214–219; those from social sciences and related fields are a combination from Tables A–4.5 and A–4.7, 228–231 and 236–239; those for all fields are from Table A–2, 190–195.
1975 – IA, NCES, 1977, Table 5, 15–19. Notes a through d above indicate which codes to use in constructing the appropriate totals.

TABLE 3.7.
Doctorate degrees conferred in selected non-scientific fields, 1890–1975

	Doctorate degrees conferred				
	Administration[a]	Education[b]	Engineering[c]	Social sciences and related fields[d]	All fields
Year	(1)	(2)	(3)	(4)	(5)
A. Number					
1890	4	2	1	43	149
1895	6	5	3	76	272
1900	9	6	3	110	382
1905	11	8	3	103	369
1910	15	12	5	118	443
1915	23	20	10	155	611
1920	26	22	11	153	615
1925	70	64	17	305	1 246
1930	162	152	71	516	2 299
1935	136	151	116	565	2 715
1940	204	288	106	629	3 290
1945	128	240	77	314	1 943
1950	391	732	427	1336	6 633
1955	566	984	606	1910	8 840
1960	459	1228	806	2075	9 829
1965	1084	1904	2101	3145	16 467
1970	1813	4585	3697	6063	29 870
1975	2499	5363	3233	8113	34 086

Table 3.7. (continued)

	Doctorate degrees conferred				
	Administration[a]	Education[b]	Engineering[c]	Social sciences and related fields[d]	All fields
Year	(1)	(2)	(3)	(4)	(5)
B. As percentage of all fields					
1890	2.7	1.3	0.7	28.9	–
1895	2.2	1.8	1.1	27.9	–
1900	2.4	1.6	0.8	28.8	–
1905	3.0	2.2	0.8	27.9	–
1910	3.4	2.7	1.1	26.6	–
1915	3.8	3.3	1.6	25.4	–
1920	4.2	3.6	1.8	24.9	–
1925	5.6	5.1	1.4	24.5	–
1930	7.0	6.6	3.1	22.4	–
1935	5.0	5.6	4.3	20.8	–
1940	6.2	8.8	3.2	19.1	–
1945	6.6	12.4	4.0	16.2	–
1950	5.9	11.0	6.4	20.1	–
1955	6.4	11.1	6.9	21.6	–
1960	4.7	12.5	8.2	21.1	–
1965	6.6	11.6	12.8	19.1	–
1970	6.1	15.3	12.4	20.3	–
1975	7.3	15.7	9.5	23.8	–

See Figure 3.1–10, part B.

Table 3.7. (continued)

[a] For the definition of this aggregation of disciplines, see Table 3.6, note a.
[b] For the definition of this aggregation of disciplines, see Table 3.6, note b.
[c] For the definition of this aggregation of disciplines, see Table 3.6, note c.
[d] For the definition of this aggregation of disciplines, see Table 3.6, note d.

Sources: 1890 to 1970 – IB, Adkins, 1975. Totals for administration are from Table A–4.6, 232–235; those for education are from Table A–4.8, 240–243; those for engineering are from Table A–4.2, 214–219; those for social sciences and related fields are a combination from Tables A–4.5 and A–4.7, 228–231 and 236–239; those for all fields are from Table A–2, 190–195.
1975 – IA, NCES, 1977, Table 5, 15–19. Notes a through d above indicate where to find the specialty codes to use in constructing the appropriate totals.

TABLE 3.8.
Practicing versus credentialled chemists, 1930–1970

Year	Practicing chemists[a] (1)	Holders of terminal chemistry degrees (hundreds) (2)	Chemists per hundred holders of terminal chemistry degrees (3)
1930	45 000	424.77	106
1940	57 000	686.34	83
1950	77 000	1 092.05	71
1960	84 000	1 517.19	55
1970	110 000	2 175.81	51

See Figures 3.1–11 and 3.1–12.

[a] To nearest thousand.

Sources: *Column 1* – IA, USBC, 1975b, 140, Series D245.
Column 2 – IB, Adkins, 1975, Table B–4.5, 474–475.

TABLE 3.9.
Chemistry-related degrees conferred, 1890–1971

Year	Degrees conferred			
	Bachelors		All levels	
	Chemical-materials engineering[a] (1)	Chemistry-related fields[b] (2)	Chemical-materials engineering[a] (3)	Chemistry-related fields[b] (4)
1890	102	733	117	879
1891	108	779	119	902
1892	104	740	115	861
1893	127	920	144	1 093
1894	164	1 195	185	1 393
1895	187	1 371	208	1 580
1896	186	1 354	208	1 573
1897	185	1 341	207	1 562
1898	183	1 325	205	1 548
1899	197	1 432	221	1 673
1900	210	1 523	234	1 771
1901	222	1 603	248	1 866
1902	227	1 637	255	1 903
1903	238	1 719	265	1 981
1904	245	1 769	271	2 024
1905	255	1 750	284	2 036
1906	293	1 990	320	2 256

Table 3.9. (continued)

	Degrees conferred			
	Bachelors		All levels	
	Chemical-materials engineering[a]	Chemistry-related fields[b]	Chemical-materials engineering[a]	Chemistry-related fields[b]
Year	(1)	(2)	(3)	(4)
1907	310	2 008	336	2 252
1908	329	2 128	360	2 413
1909	398	2 592	434	2 914
1910	424	2 497	458	2 800
1911	455	2 553	494	2 900
1912	515	2 932	566	3 327
1913	583	2 719	633	3 124
1914	622	3 195	685	3 628
1915	692	3 089	765	3 581
1916	745	3 124	828	3 670
1917	790	2 986	868	3 454
1918	757	2 859	826	3 243
1919	898	3 395	990	3 815
1920	1039	3 662	1158	4 212
1921	1170	3 979	1308	4 639
1922	1314	4 141	1479	4 857
1923	1505	4 741	1683	5 518
1924	1719	5 413	1917	6 258

Table 3.9. (continued)

Year	Degrees conferred			
	Bachelors		All levels	
	Chemical-materials engineering[a]	Chemistry-related fields[b]	Chemical-materials engineering[a]	Chemistry-related fields[b]
	(1)	(2)	(3)	(4)
1925	1656	5 440	1874	6 349
1926	1682	5 670	1926	6 680
1927	1525	5 504	1752	6 441
1928	1641	5 912	1874	6 888
1929	1703	6 220	1907	7 147
1930	1825	6 217	2068	7 335
1931	1937	6 415	2213	7 690
1932	2088	6 438	2406	7 779
1933	2037	6 368	2354	7 710
1934	1985	6 298	2305	7 576
1935	1900	6 393	2208	7 622
1936	1863	6 516	2142	7 773
1937	1862	6 946	2150	8 303
1938	1958	7 433	2242	8 760
1939	2045	7 903	2332	9 363
1940	2252	8 618	2583	10 229
1941	2238	8 621	2550	10 293
1942	2273	8 728	2562	10 297

Table 3.9. (continued)

	Degrees conferred			
	Bachelors		All levels	
Year	Chemical-materials engineering[a] (1)	Chemistry-related fields[b] (2)	Chemical-materials engineering[a] (3)	Chemistry-related fields[b] (4)
1943	1752	6 951	1939	8 114
1944	1333	5 540	1477	6 550
1945	1347	5 747	1536	6 692
1946	1344	5 908	1610	7 026
1947	2872	9 216	3352	11 096
1948	4199	12 032	5387	15 159
1949	4994	15 652	6078	19 107
1950	5494	17 766	6553	21 493
1951	4688	14 111	5879	17 934
1952	3597	11 374	4546	14 845
1953	2914	9 820	3641	12 824
1954	2662	9 770	3491	12 760
1955	2562	9 600	3414	12 671
1956	3045	9 824	3981	12 944
1957	3523	10 776	4507	13 854
1958	3837	11 545	4808	14 672
1959	4032	12 021	5080	15 343

Table 3.9. (continued)

	Degrees conferred			
	Bachelors		All levels	
Year	Chemical-materials engineering[a] (1)	Chemistry-related fields[b] (2)	Chemical-materials engineering[a] (3)	Chemistry-related fields[b] (4)
1960	3817	12 125	4921	15 632
1961	3751	12 040	4913	15 787
1962	3505	12 382	4766	16 357
1963	3619	13 338	5039	17 590
1964	3892	14 549	5381	19 074
1965	3951	15 044	5669	20 131
1966	3629	14 291	5513	19 781
1967	3696	14 403	5557	20 026
1968	4028	16 269	6159	22 522
1969	4465	17 783	6637	24 309
1970	4670	16 950	6802	23 581
1971	4499	16 548	6694	23 307

See Figures A.3–1 and A.3–2.

[a] Includes the following specialties: chemical engineering (U.S. Office of Education specialty code 0906); metallurgical engineering (0914); materials engineering (0915); and ceramic engineering (0916). See IB, Adkins, 1975, Table A–1, 181–190.

[b] Includes degrees in the various specialties of chemistry *per se* and in chemical-materials engineering, as defined above in note a. Chemistry *per se* includes the following specialties: general chemistry (U.S. Office of Education specialty code 1905); inorganic chemistry (1906);

Table 3.9. (continued)

organic chemistry (1907); physical chemistry (1908); analytical chemistry (1909); pharmaceutical chemistry (1910); and metallurgy (1920). See IB, Adkins, 1975, Table A–1, 181–190.

Source: IB, Adkins, 1975. Totals for chemical-materials engineering are from Table A–5.10, 290–293. Totals for bachelors degrees in chemistry-related fields are a combination of column 1 of this Table and column 4 of our Table 3.1. Totals for all degrees in chemistry-related fields are a combination of column 3 of this Table, column 4 of our Table 3.1, column 4 of our Table 3.2, and column 4 of our Table 3.3. Typographical errors discovered in the source tables are corrected silently.

TABLE 3.10.
Holders of chemistry-related degrees, 1930–1971

	Holders of terminal degrees		Chemistry-related degrees conferred in preceding forty-one years[b]	
	Chemical-materials engineering[a]	Chemistry-related fields[b]	Bachelors[c]	All levels[d]
Year	(1)	(2)	(3)	(4)
1930	23 267	65 744	116 157	134 346
1931	24 785	69 442	121 839	141 157
1932	26 434	73 122	127 498	148 034
1933	28 042	76 794	133 126	154 883
1934	29 594	80 445	138 504	161 366
1935	31 059	84 140	143 702	167 595
1936	32 480	87 962	148 847	173 788
1937	33 884	92 189	154 439	180 518
1938	35 365	96 728	160 531	187 716
1939	36 916	101 705	167 109	195 531
1940	38 657	107 291	174 295	204 087
1941	40 382	112 880	181 393	212 609
1942	42 136	118 561	188 518	221 040
1943	43 410	122 626	193 832	227 251
1944	44 288	125 393	197 653	231 820
1945	45 162	128 264	201 631	236 488
1946	46 018	131 165	205 789	241 478

Table 3.10. (continued)

	Holders of terminal degrees		Chemistry-related degrees conferred in preceding forty-one years[b]	
Year	Chemical-materials engineering[a]	Chemistry-related fields[b]	Bachelors[c]	All levels[d]
	(1)	(2)	(3)	(4)
1947	48 275	136 869	213 015	250 318
1948	51 759	144 850	223 039	263 225
1949	55 787	156 286	236 563	279 919
1950	60 217	169 422	251 737	298 498
1951	63 911	178 998	263 351	313 632
1952	66 544	186 055	272 172	325 577
1953	68 474	191 631	279 060	335 074
1954	70 213	197 279	286 111	344 710
1955	71 832	202 686	292 516	353 753
1956	73 917	208 230	299 251	363 116
1957	76 419	214 595	306 903	373 300
1958	79 127	221 547	315 462	384 518
1959	81 956	228 880	324 624	396 618
1960	84 565	235 384	333 354	408 435
1961	87 037	243 516	341 732	420 010
1962	89 218	250 961	350 135	431 728
1963	91 488	259 217	359 332	444 461
1964	93 933	268 415	369 140	458 017

Table 3.10. (continued)

	Holders of terminal degrees		Chemistry-related degrees conferred in preceding forty-one years[b]	
	Chemical-materials engineering[a]	Chemistry-related fields[b]	Bachelors[c]	All levels[d]
Year	(1)	(2)	(3)	(4)
1965	96 371	277 855	378 771	471 890
1966	98 435	286 282	387 622	485 322
1967	100 485	294 584	396 355	498 668
1968	102 818	304 510	407 120	514 749
1969	105 484	315 685	418 991	532 170
1970	108 263	325 844	429 721	548 604
1971	110 797	335 417	440 052	564 576

See Figure 3.1–13, part A, A.4–1, part A, and A.4–2.

[a] For the definition of this aggregate subject, see Table 3.9, note a.
[b] For the definition of this aggregate subject, see Table 3.9, note b.
[c] This series is employed as a substitute for the number of *persons* who received a chemistry-related degree in the preceding forty-one years. Such a substitution assumes that each person who received a higher degree in a chemistry-related field had first received a bachelors degree in one of these areas; while this is not exactly correct, it is probably approximately accurate.
[d] See Appendix A.4 for explanation.

Sources: *Column 1* – IB, Adkins, 1975, Table B–4.10, 484–485.
Column 2 – add column 1 and Table 2.6, column 3.
Column 3 – sum over the appropriate forty-one-year intervals of Table 3.9, column 2.
Column 4 – sum over the appropriate forty-one-year intervals of Table 3.9, column 4.

TABLE 3.11.
Practicing chemists and chemical engineers, 1940–1970

Year	Chemists (1)	Chemical engineers (2)	High school chemistry teachers (estimated) (3)	Chemistry faculty (estimated) (4)	Total[a] (5)
1940	57 000	13 000	7 000	2000	79 000
1950	77 000	34 000	7 000	3000	121 000
1960	84 000	41 000	12 000	5000	142 000
1970	110 000	53 000	20 000	7000	190 000

See Figures 3.1–13, A.2–1, A.2–2, A.2–3, and A.4–1.

Note: All totals are to nearest thousand.

[a] Represents an estimated upper limit for practicing chemists and chemical engineers.

Sources: *Column 1* – IA, USBC, 1975b, 140, Series D245.
Column 2 – IA, USBC, 1975b, 140, Series D257.
Column 3 – see Table 3.18, column 2.
Column 4 – see Table 5.17, column 2.

TABLE 3.12.
Chemists and chemical engineers per hundred holders of chemistry-related degrees, 1940–1970

	Chemists and chemical engineers		
	Total (estimated upper limit)	Per hundred holders of a terminal degree in any chemistry-related field	Per hundred bachelors degrees conferred in chemistry-related fields over the preceding forty-one years[a]
Year	(1)	(2)	(3)
1940	79 000	73.6	45.3
1950	121 000	71.4	48.1
1960	142 000	60.3	42.6
1970	190 000	58.3	44.2

See Figure 3.1–13, part B.

[a] This series is used as a substitute for chemists and chemical engineers per hundred *persons* who received a chemistry-related degree in the preceding forty-one years. For a discussion, see Appendix A.4, as well as Table 3.10, note c.

Sources: *Column 1* – see Table 3.11, column 5.
Column 2 – divide column 1 by Table 3.10, column 2, then multiply by 100.
Column 3 – divide column 1 by Table 3.10, column 3, then multiply by 100.

TABLE 3.13.
Bachelors and doctorate degrees conferred in history, 1890–1978

Year	History degrees conferred	
	Bachelors (1)	Doctorates (2)
1890	681	8
1891	719	10
1892	676	10
1893	854	12
1894	1 131	15
1895	1 303	16
1896	1 274	16
1897	1 252	17
1898	1 232	19
1899	1 337	19
1900	1 416	21
1901	1 485	20
1902	1 514	17
1903	1 590	20
1904	1 632	19
1905	1 575	21
1906	1 810	22
1907	1 792	20
1908	1 897	23
1909	2 339	27
1910	2 154	25
1911	2 144	28
1912	2 120	29
1913	2 398	31
1914	2 544	32
1915	2 438	35
1916	2 455	39
1917	2 351	40
1918	2 251	32
1919	2 677	24
1920	2 893	36
1921	3 152	41
1922	3 290	60
1923	3 761	55
1924	4 286	58

Table 3.13. (continued)

Year	History degrees conferred	
	Bachelors (1)	Doctorates (2)
1925	4 623	64
1926	5 008	70
1927	5 311	87
1928	5 691	84
1929	6 069	98
1930	6 154	142
1931	6 447	130
1932	6 573	136
1933	6 597	164
1934	6 621	156
1935	6 494	164
1936	6 540	137
1937	6 689	146
1938	7 239	172
1939	7 791	181
1940	8 371	167
1941	8 307	188
1942	8 240	172
1943	6 540	125
1944	5 061	72
1945	5 275	82
1946	5 439	115
1947	8 166	165
1948	10 558	173
1949	12 461	250
1950	15 372	280
1951	13 843	327
1952	11 374	317
1953	10 520	302
1954	10 381	361
1955	10 477	311
1956	11 311	267
1957	12 471	323
1958	13 412	300
1959	14 231	324

Table 3.13. (*continued*)

Year	History degrees conferred	
	Bachelors (1)	Doctorates (2)
1960	15 207	343
1961	16 262	371
1962	17 915	343
1963	20 684	379
1964	24 538	507
1965	26 922	580
1966	29 859	604
1967	32 845	662
1968	36 645	698
1969	42 558	838
1970	44 688	1044
1971	46 380	995
1972	43 975	1133
1973	41 223	1140
1974	37 381	1114
1975	31 768	1117
1976	28 636	1014
1977	25 608	921
1978	23 145	813

See Figure 3.1–14.

Note: History comprises U.S. Office of Education specialty code 2205.

Sources: 1890 to 1971 – IB, Adkins, 1975, Table A–5.25, 350–353.
1972 – IA, NCES, 1975b, Table 5, 15.
1973 – IA, NCES, 1976, 1972–73 Tables, Table 5, 16.
1974 – IA, NCES, 1976, 1973–74 Tables, Table 5, 30.
1975 – IA, NCES, 1977, Table 5, 18.
1976 – IA, NCES, 1978b, Table 5, 19.
1977 – IA, NCES, 1980a, Table 5, 31.
1978 – IA, NCES, 1980b, Table 5.

TABLE 3.14.
Ratio of chemistry degree conferrals to history degree conferrals, by level, 1890–1975

	Degrees conferred				Ratio of chemistry degree conferrals to history degree conferrals	
	Bachelors		Doctorates			
	Chemistry [a]	History [b]	Chemistry [a]	History [b]	Bachelors	Doctorates
Year	(1)	(2)	(3)	(4)	(5)	(6)
1890	631	681	28	8	0.93	3.5
1895	1 184	1 303	49	16	0.91	3.1
1900	1 313	1 416	69	21	0.93	3.3
1905	1 495	1 575	67	21	0.95	3.2
1910	2 073	2 154	78	25	0.96	3.1
1915	2 397	2 438	107	35	0.98	3.1
1920	2 623	2 893	104	36	0.91	2.9
1925	3 784	4 623	220	64	0.82	3.4
1930	4 392	6 154	332	142	0.71	2.3
1935	4 493	6 494	402	164	0.69	2.5
1940	6 366	8 371	532	167	0.76	3.2
1945	4 400	5 275	342	82	0.83	4.2
1950	12 272	15 372	967	280	0.80	3.5
1955	7 038	10 477	1009	311	0.67	3.2
1960	8 308	15 207	1062	343	0.55	3.1
1965	11 093	26 922	1428	580	0.41	2.5
1970	12 280	44 688	2224	1044	0.27	2.1
1975	10 667	31 768	1834	1117	0.34	1.6

See Figures 3.1–15 and A.3–2.

[a] Includes the following specialties: general chemistry (U.S. Office of Education specialty code 1905); inorganic chemistry (1906); organic chemistry (1907); physical chemistry (1908); analytical chemistry (1909); pharmaceutical chemistry (1910); and metallurgy (1920). See IB, Adkins, 1975, Table A–1, 181–190. Annual series of the same data appear in our Table 3.1, column 4, and Table 3.3, column 4.

[b] Comprises U.S. Office of Education specialty code 2205. Annual series of the same data appear in our Table 3.13.

Sources: *Columns 1 to 4*:

1890 to 1970 – IB, Adkins, 1975, Tables A–5.5 and A–5.25, 270–273 and 350–353.

1975 – IA, NCES, 1977, Table 5, 18. Notes a and b above list the specialty codes included in these totals.

TABLE 3.15.
Secondary school enrollment, 1890–1973

	Secondary school enrollment (thousands)		
	Public	Private	Both
Year	(1)	(2)	(3)
1890	203	95	298
1900	519	111	630
1910	915	117	1 032
1915	1 329	166	1 495
1922	2 873	246	3 119
1928	3 911	341	4 252
1934	5 669	360	6 029
1949	5 658	637	6 295
1955	6 574	785	7 359
1959	8 258	983	9 241
1961	8 821	1078	9 899
1963	10 372	1204	11 576
1965	11 628	1308	12 936
1970	13 022	1400	14 422
1973	13 909	1300	15 209

Note: Includes only regular schools. Data for years between surveys are interpolated, based on a crude assumption of simple linear growth between any two sequential surveys.

Sources: *Column 1*:

1890 to 1934 – IA, USBC, 1975b, 368–369, Series H424.
1949 to 1965 – IB, Ferriss, 1969, 376–377, Series A–7.
1970, 1973 – IA, USOE, 1974, 7, Table 3 (for 1970 figure), and 6, Table 1 (for 1973 figure).

Column 2:

1890 to 1970 – IA, USBC, 1975b, 368–369, Series H429.
1973 – IA, USOE, 1974, 6, Table 1.

TABLE 3.16.
High school chemistry enrollment, 1890–1973

Year	Total high school enrollment (thousands) (1)	High school chemistry enrollment (as percentage of public high school enrollment) (2)	High school chemistry enrollment (thousands, estimated) (3)
1890	298	10.1	30
1900	630	7.7	49
1910	1 032	6.9	71
1915	1 495	7.4	111
1922	3 119	7.4	231
1928	4 252	7.1	302
1934	6 029	7.6	458
1949	6 295	7.6	478
1955	7 359	7.5	552
1959	9 241	8.1	749
1961	9 899	9.1	901
1963	11 576	8.3	961
1965	12 936	9.3	1203
1970	14 422	9.3	1341
1973	15 209	7.7	1171

See Figure 3.2–1.

Sources: *Column 1* – see Table 3.15, column 3.

Column 2:

1890 to 1959, 1963, and 1965 – IA, USBC, 1975b, 377, Series H548.
1961 and 1973 – IA, NCES, 1975c, Table C, 18. Percentages were calculated by dividing the chemistry enrollment by the total enrollment.
1970 – IA, NSF, 1971a. Percentage was calculated by dividing the chemistry enrollment (2, Table 2) by the total enrollment (1, Table 1). Note that percentages in 3, Table 3, agree with ours.

TABLE 3.17.
High school chemistry enrollment as a percentage of population aged sixteen, 1890–1973

Year	Population aged sixteen (thousands, estimated) (1)	High school chemistry enrollment: Number (thousands, estimated) (2)	High school chemistry enrollment: As percentage of population aged sixteen (3)
1890	1388	30	2.2
1900	1519	49	3.2
1910	1813	71	3.9
1915	1889	111	5.9
1922	2041	231	11.3
1928	2298	302	13.1
1934	2358	458	19.4
1949	2106	478	22.7
1955	2255	552	24.5
1959	2920	749	25.7
1961	2774	901	32.5
1963	3728	961	25.8
1965	3517	1203	34.2
1970	3893	1341	34.4
1973	4196	1171	27.9

See Figure 3.2–2.

Sources: *Column 1*:

1890 to 1965 – IB, Ferriss, 1969, Series I–16, 414–415.
1970 and 1973 – IA, USBC, 1975a, Table 4, 29 (for 1970 figure) and Table 3, 25 (for 1973 figure).

Column 2 – see Table 3.16, column 3.

TABLE 3.18.
Rough estimates of the number of high school chemistry teachers, 1920–1974

	High school teachers (thousands)	
	Total	Chemistry (estimated)
Year	(1)	(2)
1920	102[a]	2
1930	213[b]	4
1940	330	7
1950	366	7
1960	580	12
1970	1024	20
1974	1093	22

[a] Public schools only, a minor deviation.
[b] Public schools only, a minor deviation. Includes junior high school teachers.

Sources: *Column 1*:

1920 and 1930 – IB, Ferriss, 1969, 383, Series B–6.
1940 and 1950 – IB, Ferriss, 1969, 383, combination of Series B–6 and B–10.
1960 and 1970 – IA, USOE, 1972, 6, Table 8. Slightly different totals for 1950, 1960, and 1970 appear in IA, USBC-COP, 1964, Table 201, 522, and IA, USBC-COP, 1973, Table 221, 719.
1974 – IA, NCES, 1975a, 11, Table 11.

Column 2 – multiply column 1 by 0.02. See page 67, note 15.

TABLE 3.19.
Baccalaureates conferred in chemistry per hundred students of the same age cohort who had enrolled in high school chemistry, 1910–1973

Year[a]	High school chemistry enrollment (thousands, estimated) (1)	Baccalaureates conferred in chemistry[b] five years later: Number (thousands) (2)	Per hundred students of the same age cohort[c] who had enrolled in high school chemistry[d] (3)
1910	71	2.397	3.4
1915	111	2.623	2.4
1922	231	3.979	1.7
1928	302	4.331	1.4
1934	458	5.858	1.3
1949	478	7.108	1.5
1955	552	8.308	1.5
1959	749	10.657	1.4
1961	901	10.662	1.2
1963	961	12.241	1.3
1965	1203	12.280	1.0
1970	1341	10.667	0.8
1973	1171	11.503	1.0

See Figure 3.2–3.

[a] In which students enrolled in chemistry.
[b] For the definition of chemistry used here, see Table 3.1, note b.
[c] This assertion rests upon the convenient assumption that only eleventh graders enroll in high school chemistry. Such a simplification, while reasonable (see IA, NSF, 1971a, 3, Table 4), undoubtedly introduces slight distortions.
[d] This series is identical to baccalaureates expressed as a *percentage* of cohorts exposed to high school chemistry, but it is labeled differently to emphasize that not all baccalaureates in chemistry necessarily took the subject in high school.

Sources: *Column 1* – see Table 3.16, column 3.

Column 2:

1915 to 1970 – IB, Adkins, 1975, Table A–5.5, 270–273.
1975 – IA, NCES, 1977, Table 5, 15–19. See note a above for the specialty codes included in this total.
1978 – IA, NCES, 1980b, Table 5. See comment for 1975 source.

TABLE 3.20.
Total enrollment in eleventh grade, 1910–1970

Year	Ratio of total high school enrollment to public high school enrollment	Eleventh grade enrollment (thousands)		
		Public high schools[a]	Private high schools	Both[a]
	(1)	(2)	(3)	(4)
1910	–	163.176	22.693	185.869
1915	–	287.326	30.224	317.550
1922	–	515.542	36.723	552.265
1928	–	767.706	59.208	826.914
1934	1.064	1209.180	–	1286.568[b]
1949	1.113	1267.483	–	1410.709[b]
1955	1.119	1520	–	1701[b]
1959	1.119	1955	–	2188[b]
1961	–	2024	243.695	2268
1963	1.116	2348	–	2620[b]
1965	1.112	2778	–	3089[b]
1970	1.108	3171	–	3513[b]

[a] For 1910 to 1949, estimates are to nearest student; after 1949, to nearest thousand.
[b] Estimated.

Sources: *Column 1* – divide Table 3.15, column 3, by Table 3.15, column 1.

Column 2:

1910 – IA, USBE, 1912, II, 1187.
1915 – IA, USBE, 1925, II, Chapter 2, Table 2, 34.
1922 and 1928 – IA, USOE, 1930, Chapter 20, Table 2, 454.
1934 – IA, USOE, 1937, Chapter 2, Table 2, 8.
1949 – IA, USOE, 1954, Chapter 1, Table 14, 18.
1955 to 1961 – IA, USOE, 1963a, Table 3, 8.
1963 to 1970 – IA, NCES, 1975a, Table 30, 31.

Column 3:

1910 to 1922 – IA, USBE, 1925, II, Chapter 7, Table 3, 605.
1928 – IA, USOE, 1937, Chapter 6, 20, Table 9.
1961 – IA, Gertler, 1963, Table 6, 39.

Column 4:

1910 to 1928, 1961 – add columns 2 and 3.
1934 to 1959, 1963 to 1970 – multiply column 1 by column 2, to achieve a rough estimate.

TABLE 3.21.
Total enrollment in twelfth grade, 1910–1970

Year	Ratio of total high school enrollment to public high school enrollment	Twelfth grade enrollment (thousands)		
		Public high schools[a]	Private high schools	Both[a]
	(1)	(2)	(3)	(4)
1910	–	111.444	17.674	129.118
1915	–	218.618	27.108	245.726
1922	–	362.201	29.192	391.393
1928	–	622.091	51.141	673.232
1934	1.064	1005.375	–	1069.719[b]
1949	1.113	1126.022	–	1253.262[b]
1955	1.119	1246	–	1394[b]
1959	1.119	1538	–	1721[b]
1961	–	1851	235.641	2087
1963	1.116	1866	–	2082[b]
1965	1.112	2560	–	2847[b]
1970	1.108	2841	–	3148[b]

[a] For 1910 to 1949, estimates are to nearest student; after 1949, to nearest thousand.
[b] Estimated.

Sources: see sources for Table 3.20.

TABLE 3.22.
Propensity of eleventh graders to graduate from college, 1910–1970

Year[a]	Total enrollment in eleventh grade[b] (thousands, estimated) (1)	Baccalaureates five years later (thousands) (2)	Propensity of eleventh graders to graduate from college[c] (3)
1910	185.869	31.186	0.168
1915	317.550	40.741	0.128
1922	552.265	86.917	0.157
1928	826.914	119.516	0.145
1934	1286.568	158.299	0.123
1949	1410.709	265.480	0.188
1955	1701	364.135	0.214
1959	2188	467.262	0.214
1961	2268	525.246	0.232
1963	2620	636.863	0.243
1965	3089	798.068	0.258
1970	3513	931.663	0.265

[a] In which students were enrolled in eleventh grade.
[b] For 1910 to 1949, estimates are to nearest student; after 1949, to nearest thousand.
[c] Assumes that all students who graduate from college get only one bachelors degree and had been in eleventh grade five years before graduation. Both of these simplifications distort the series, but probably do not alter the trends significantly.

Sources: *Column 1* – see Table 3.20, column 4.

Column 2:

1915 to 1970 – IB, Adkins, 1975, Table A–2, 190–194.
1975 – IA, NCES, 1977, Table 5, 15–19.

Column 3 – divide column 2 by column 1.

TABLE 3.23.
Propensity of twelfth graders to graduate from college, 1910–1970

Year[a]	Total enrollment in twelfth grade[b] (thousands, estimated) (1)	Baccalaureates four years later (thousands) (2)	Propensity of twelfth graders to graduate from college[c] (3)
1910	129.118	31.540	0.244
1915	245.726	35.235	0.143
1922	391.393	80.664	0.206
1928	673.232	119.825	0.178
1934	1069.719	147.652	0.138
1949	1253.262	277.652	0.222
1955	1394	354.867	0.255
1959	1721	417 178	0.242
1961	2087	502.044	0.241
1963	2082	562.952	0.270
1965	2847	734.002	0.258
1970	3148	954.376	0.303

[a] In which students were enrolled in twelfth grade.
[b] For 1910 to 1949, estimates are to nearest student; after 1949, to nearest thousand.
[c] Assumes, conveniently, that all students who graduate from college get only one bachelors degree and had been in twelfth grade four years before graduation. Both of these simplifications distort the series slightly, but probably do not alter the trends significantly.

Sources: *Column 1* – see Table 3.21, column 4.

Column 2:

1914 to 1969 – IB, Adkins, 1975, Table A–2, 190–194.
1974 – IA, NCES, 1976, 1973–74 Tables, Table 1, 21.

Column 3 – divide column 2 by column 1.

TABLE 3.24.
Mean propensity of eleventh and twelfth graders to graduate from college, 1910–1970

	Propensity to graduate from college		
	Eleventh graders	Twelfth graders	Eleventh and twelfth graders
Year[a]	(1)	(2)	(3)
1910	0.168	0.244	0.206
1915	0.128	0.143	0.136
1922	0.157	0.206	0.182
1928	0.145	0.178	0.162
1934	0.123	0.138	0.131
1949	0.188	0.222	0.205
1955	0.214	0.255	0.235
1959	0.214	0.242	0.228
1961	0.232	0.241	0.237
1963	0.243	0.270	0.257
1965	0.258	0.258	0.258
1970	0.265	0.303	0.284

[a] In which student was enrolled in stated year of secondary school.

Sources: *Column 1* – see Table 3.22, column 3.
Column 2 – see Table 3.23, column 3.

TABLE 3.25.
Baccalaureates in chemistry per hundred college-bound students of the same age cohort who had enrolled in high school chemistry, 1910–1970

				Baccalaureates in chemistry five years later	
	High school chemistry enrollment (thousands, estimated)	Mean propensity of these eleventh and twelfth graders to graduate from college[b]	Students in high school chemistry who continue through college (thousands, estimated)	Number (thousands)	Per hundred college-bound students of the same age cohort[c] who had enrolled in high school chemistry (estimated)
Year[a]	(1)	(2)	(3)	(4)	(5)
1910	71	0.206	14.626	2.397	16.4
1915	111	0.136	15.096	2.623	17.4
1922	231	0.182	42.042	3.979	9.5
1928	302	0.162	48.924	4.331	8.9
1934	458	0.131	59.998	5.858	9.8
1949	478	0.205	97.990	7.108	7.3
1955	552	0.235	129.720	8.308	6.4
1959	749	0.228	170.772	10.657	6.2
1961	901	0.237	213.537	10.662	5.0
1963	961	0.257	246.977	12.241	5.0
1965	1203	0.258	310.374	12.280	4.0
1970	1341	0.284	380.844	10.667	2.8

See Figure 3.2–4.

Table 3.25. (continued)

[a] In which students completed the grade that included high school training in chemistry.
[b] Weights eleventh and twelfth graders equally, even though most high school chemistry students are eleventh graders.
[c] Assumes that all baccalaureates in chemistry were in eleventh grade five years earlier – another slight distortion.

Sources: *Column 1* – see Table 3.16, column 3.

Column 2 – see Table 3.24, column 3.

Column 3 – multiply column 1 by column 2.

Column 4 – see Table 3.19, column 2.

Column 5 – divide column 4 by column 3 and convert to percentage.

TABLE 3.26.
Public high school enrollment in selected special science subjects, 1890–1973

Year	Total public high school enrollment[a] (thousands)	Percentage enrolled in				Chemistry enrollment as percentage of enrollment in most popular special science subject[b]
		Biology	Earth sciences	Physics	Chemistry	
	(1)	(2)	(3)	(4)	(5)	(6)
1890	203	–	–	22.8	10.1	–
1900	519	–	29.8	19.0	7.7	26
1910	739	1.1	21.0	14.6	6.9	33
1915	1 165	6.9	15.3	14.2	7.4	48
1922	2 155	8.8	4.5	8.9	7.4	83
1928	2 897	13.6	2.8	6.8	7.1	52
1934	4 497	14.6	1.7	6.3	7.6	52
1949	5 399	18.4	0.4	5.4	7.6	41
1955	6 480	20.0	–	4.7	7.5	38
1959	8077	20.8	–	4.7	8.1	39
1961	8 219	21.6	0.9	4.9	9.1	42
1963	10 372	24.0	–	3.8	8.3	35
1965	11 628	23.2	–	4.5	9.3	40
1970	12 442	25.7	–	3.9	9.3	36
1973	13 438	21.3	4.2	4.3	7.7	36

See Figures 3.2–5 and 3.2–6.

Table 3.26. (continued)

Note: Physiology and all the mathematical subjects are excluded. As far as can be seen from the available data, the former has never been the most popular special science subject, although it was surprisingly widespread at the turn of the century. Like general science courses, high school training in mathematics subjects has been ubiquitous.

[a] For 1910 to 1934, represents only pupils enrolled in schools *that returned usable questionnaires*. For other years, represents an estimate of *all* pupils, but derived from a partial survey. Cf. Table 3.15, column 1.
[b] The most popular special science subjects were: 1900–1915, earth sciences; 1922, physics; 1928–1973, biology.

Sources: *Columns 1 to 5*:

1890 to 1959, 1963, and 1965 – IA, USBC, 1975b, 377, Series H545, H547, H548, H549, and H551.
1961 and 1973 – IA, NCES, 1975c, Table C, 18. Percentages calculated by dividing the special subject enrollment by the total enrollment.
1970 – IA, NSF, 1971a. Percentages calculated by dividing the special subject enrollment (2, Table 2) by the total enrollment (1, Table 1).

Column 6:

1900 to 1915 – divide column 5 by column 3.
1922 – divide column 5 by column 4.
1928 to 1973 – divide column 5 by column 2.

TABLE 3.27
Prestige rankings of selected professionals, 1947 and 1963

Profession	Rank	
	1947 (1)	1963 (2)
Physician	2.5	2
College professor	8	8
Scientist	8	3.5
Chemist	18	11
Nuclear physicist	18	3.5
Sociologist	26.5	26
Biologist	29	24.5

Source: IB, Hodge *et al.*, 1964, Table 1, 290.

TABLE 3.28.
Percentages of encyclopedia yearbooks devoted to chemistry, 1922–1979

Year covered[a]	Americana Annual				Britannica Book of the Year			
	Total pages[d]	Pages on chemistry[b]			Total pages[d]	Pages on chemistry[c]		
		Number[e]	As percentage of total			Number[e]	As percentage of total	
			Annual number[f]	Five-year moving average[g]			Annual number[f]	Five-year moving average[g]
	(1)	(2)	(3)	(4)	(5)	(6)	(7)	(8)
1922	927	6.54	0.71	–	–	–	–	–
1923	865	5.80	0.67	–	–	–	–	–
1924	791	6.51	0.82	0.73	–	–	–	–
1925	846	5.71	0.68	0.72	–	–	–	–
1926	909	7.13	0.78	0.78	–	–	–	–
1927	842	5.62	0.67	0.78	–	–	–	–
1928	789	7.43	0.94	0.80	–	–	–	–
1929	834	6.78	0.81	0.68	–	–	–	–
1930	839	6.69	0.80	0.61	–	–	–	–
1931	785	1.27	0.16	0.48	–	–	–	–
1932	822	2.74	0.33	0.39	–	–	–	–
1933	656	1.86	0.28	0.32	–	–	–	–

Table 3.28. (continued)

Year covered[a]	Americana Annual				Britannica Book of the Year			
	Total pages[d]	Pages on chemistry[b]			Total pages[d]	Pages on chemistry[c]		
		Number[e]	As percentage of total			Number[e]	As percentage of total	
			Annual number[f]	Five-year moving average[g]			Annual number[f]	Five-year moving average[g]
	(1)	(2)	(3)	(4)	(5)	(6)	(7)	(8)
1934	791	2.93	0.37	0.35	–	–	–	–
1935	806	3.53	0.44	0.34	–	–	–	–
1936	771	2.55	0.33	0.36	–	–	–	–
1937	771	2.00	0.26	0.42	720	2.56	0.35	–
1938	844	3.56	0.42	0.46	720	3.42	0.48	–
1939	853	5.55	0.65	0.66	714	3.83	0.54	0.44
1940	774	4.91	0.63	0.81	714	3.50	0.49	0.45
1941	799	10.74	1.34	0.89	720	2.62	0.36	0.43
1942	864	8.92	1.03	0.94	776	3.05	0.39	0.39
1943	813	6.63	0.82	1.02	776	2.89	0.37	0.37
1944	819	7.15	0.87	0.88	776	2.65	0.34	0.37
1945	839	8.69	1.04	0.79	840	3.35	0.40	0.38
1946	785	4.88	0.62	0.76	840	2.98	0.36	0.42

Table 3.28. (continued)

Year covered[a]	Americana Annual				Britannica Book of the Year			
	Total pages[d]	Pages on chemistry[b]			Total pages[d]	Pages on chemistry[c]		
		Number[e]	As percentage of total			Number[e]	As percentage of total	
			Annual number[f]	Five-year moving average[g]			Annual number[f]	Five-year moving average[g]
	(1)	(2)	(3)	(4)	(5)	(6)	(7)	(8)
1947	755	4.67	0.62	0.72	806	3.60	0.45	0.47
1948	746	4.78	0.64	0.64	776	4.33	0.56	0.48
1949	749	5.23	0.70	0.65	732	4.07	0.56	0.51
1950	742	4.78	0.64	0.65	734	3.59	0.49	0.51
1951	785	5.01	0.64	0.64	734	3.54	0.48	0.49
1952	777	4.91	0.63	0.61	744	3.58	0.48	0.47
1953	796	4.87	0.61	0.58	744	3.33	0.45	0.46
1954	815	4.33	0.53	0.55	744	3.31	0.44	0.45
1955	818	4.04	0.49	0.52	744	3.36	0.45	0.43
1956	854	4.07	0.48	0.48	742	3.01	0.41	0.40
1957	846	4.24	0.50	0.46	744	2.95	0.40	0.38
1958	843	3.53	0.42	0.44	744	2.31	0.31	0.36
1959	861	3.40	0.39	0.43	744	2.51	0.34	0.37

Table 3.28. (continued)

Year covered[a]	Americana Annual				Britannica Book of the Year			
	Total pages[d]	Pages on chemistry[b]			Total pages[d]	Pages on chemistry[c]		
		Number[e]	As percentage of total			Number[e]	As percentage of total	
			Annual number[f]	Five-year moving average[g]			Annual number[f]	Five-year moving average[g]
	(1)	(2)	(3)	(4)	(5)	(6)	(7)	(8)
1960	852	3.52	0.41	0.42	744	2.57	0.35	0.37
1961	860	3.57	0.41	0.43	750	3.32	0.44	0.43
1962	744	3.49	0.47	0.46	760	3.24	0.43	0.45
1963	724	3.55	0.49	0.45	784	4.55	0.58	0.47
1964	748	3.71	0.50	0.45	777	3.58	0.46	0.47
1965	748	2.98	0.40	0.42	748	3.45	0.46	0.46
1966	748	2.96	0.40	0.38	745	3.07	0.41	0.43
1967	748	2.45	0.33	0.33	745	3.03	0.41	0.44
1968	735	1.95	0.27	0.30	741	3.18	0.43	0.45
1969	729	1.96	0.27	0.27	738	3.55	0.48	0.44
1970	707	1.63	0.23	0.25	719	3.69	0.51	0.43
1971	703	1.67	0.24	0.25	680	2.49	0.37	0.42
1972	709	1.65	0.23	0.27	682	2.37	0.35	0.40

Table 3.28. **(continued)**

Year covered[a]	Americana Annual				Britannica Book of the Year			
	Total pages[d]	Pages on chemistry[b]			Total pages[d]	Pages on chemistry[c]		
		Number[e]	As percentage of total			Number[e]	As percentage of total	
			Annual number[f]	Five-year moving average[g]			Annual number[f]	Five-year moving average[g]
	(1)	(2)	(3)	(4)	(5)	(6)	(7)	(8)
1973	591	1.65	0.28	0.29	687	2.70	0.39	0.37
1974	536	2.00	0.37	0.28	695	2.57	0.37	0.37
1975	540	1.68	0.31	0.27	655	2.49	0.38	0.41
1976	490	1.00	0.20	0.26	656	2.36	0.36	0.46
1977	486	1.00	0.21	0.23	598	3.38	0.57	0.49
1978	490	1.00	0.20	–	607	3.70	0.61	–
1979	487	1.00	0.21	–	612	3.24	0.53	–

See Figure 3.3–1.

[a] Each volume covers the events of the year preceding the date of publication. For example, *Britannica Book of the Year 1977* covers the events of 1976 and is entered in this table under the earlier year.

[b] Our intent was to include the major article or articles on the discipline of chemistry and the chemical industry. The yearbook sometimes lumped discussion of such subjects under a single heading and sometimes published equivalent treatment in two or more articles. Our totals include articles with the following names: "chemical analysis, optical methods of" (1942 only); "chemical and fertilizer industries" (1939);

Table 3.28. (continued)

"chemical manufacturing" (1922–1930); "chemicals" (1922 and 1940); "chemicals and allied products" (1939); and "chemistry" (1922–1975). They excluded articles dealing with chemical specialties and institutions, such as "biochemistry", "Chemistry, Federal Bureau of", "chemotherapy", "chemurgy", "industrial review – foods and chemicals" (1974 only) and "Mellon Institute". For 1938–1940, the "chemistry" article is entitled "chemistry in [year]".

[c] Following the general guidelines given in note b, includes articles with the following titles: "chemical industry" (1957–1964); "chemistry" (1937–1956, 1961–1975); "chemistry, applied" (1937–1940); "industrial production and technology – chemicals" (1974 only); and "industrial review – chemicals" (1965–1973, 1975). Excludes, *inter alia*, the following: "chemical warfare"; "Chemistry and Soils, U.S. Bureau of"; "chemotherapy"; and "chemurgy".

[d] Refers to total space devoted to alphabetically-ordered articles. Also includes necrologies if they appear separately. Excludes introductions, indexes, and the like. Feature articles are included only if they appear in the alphabetized portion.

[e] Fractional parts of articles were measured to the nearest column-millimeter, and this measurement was divided by the height of a column times the number of columns per page to yield a measurement in fractional pages.

[f] Uses number of pages on chemistry calculated to the thirteen-digit accuracy of a hand-held calculator. Using the rounded series from the table may produce slightly different results.

[g] Arithmetic means of series in preceding column, recorded in middle year of averaged period.

Sources: *Americana Annual*, 1923–1980, and *Britannica Book of the Year*, 1938–1980. Our tabulations.

TABLE 4.1.
Production of the chemical and of all manufacturing industries, 1899–1957

	Index of output (1929 = 100)		Index of output per unit of capital input (1929 = 100)	
	Chemicals and allied products	Total manufacturing	Chemicals and allied products	Total manufacturing
Year	(1)	(2)	(3)	(4)
1899	18.6	27.5	72.4	93.9
1909	31.4	43.4	66.2	79.8
1919	51.5	61.0	55.8	65.7
1929	100.0	100.0	100.0	100.0
1937	123.9	103.3	126.9	121.0
1948	309.8	184.2	183.2	152.5
1953	470.9	243.4	194.4	158.4
1957	612.8	264.6	–	–

See Figures 4.2–1 and 4.2–3.

Source: IB, Kendrick, 1961, Tables D–I, 464, and D–IV, 471.

TABLE 4.2.
Capital in the chemical and the chemical process industries, 1879–1957
(billions of 1929 dollars)

			Chemicals and allied products[b]	
	Total manufacturing	Chemical process industries[a]	Amount	As percentage of total manufacturing
Year	(1)	(2)	(3)	(4)
1879	4.821	1.984	0.206	4.3
1889	11.157	4.531	0.478	4.3
1899	17.452	8.107	0.869	5.0
1904	23.295	11.635	1.134	4.9
1909	31.563	15.779	1.531	4.9
1914	36.737	19.124	2.078	5.7
1919	46.094	24.197	2.777	6.0
1929	63.022	34.896	4.221	6.7
1937	55.319	33.558	3.965	7.2
1948	78.067	45.675	6.487	8.3
1953	97.843	55.764	8.845	9.0
1957	110.455	65.634	10.564	9.6

See Figures 4.2–2 and 4.2–4.

Note: For 1879–1889, includes custom and neighborhood shops; for 1899–1919, covers factories having annual production of $500 or more; for 1929–1957, of $5000 or more. See source for other minor discrepancies.

[a] For the names of the subgroups that constitute this heading, see Table 4.3.
[b] Listed in the source as "chemicals and allied substances".

Source: IA, USBC, 1975b, 685. Total manufacturing appears as Series P123. Chemical process industries is a combination of Series P126, P146, P152, P154, P158, P159, P160, and P164. Chemicals and allied products appears as Series P154.

TABLE 4.3.
Distribution of capital within the chemical process industries, 1879–1957
(billions of 1929 dollars)

Year	Food and kindred products (1)	Rubber products (2)	Paper, pulp, and products (3)	Chemicals and allied products[a] (4)	Petroleum refining (5)	Stone, clay, and glass products (6)	Primary metals[b] (7)	Total (8)
1879	0.897	0.010	0.090	0.206	0.037	0.156	0.588	1.984
1889	1.839	0.036	0.200	0.478	0.151	0.408	1.419	4.531
1899	3.598	0.074	0.453	0.869	0.195	0.709	2.209	8.107
1904	4.656	0.093	0.670	1.134	0.254	1.138	3.690	11.635
1909	5.517	0.139	1.002	1.531	0.327	1.755	5.508	15.779
1914	6.515	0.265	1.246	2.078	0.552	1.937	6.531	19.124
1919	7.593	0.704	1.524	2.777	1.380	1.676	8.543	24.197
1929	9.591	1.131	2.239	4.221	6.092	2.592	9.030	34.896
1937	9.180	0.816	2.062	3.965	6.503	1.975	9.057	33.558
1948	10.488	1.422	2.476	6.487	11.188	2.128	11.486	45.675
1953	12.878	1.660	3.086	8.845	12.455	2.631	14.209	55.764
1957	13.361	1.842	4.039	10.564	16.134	3.375	16.319	65.634

Note: For 1879–1889, includes custom and neighborhood shops; for 1899–1919, covers factories having annual production of $500 or more; for 1929–1957, of $5000 or more. See source for other minor discrepancies. See Table 4.4 for a *percentage* distribution.

[a] Listed in the source as "chemicals and allied substances".
[b] Listed in the source as two subgroups: "iron and steel and products" and "nonferrous metals and products".

Source: IA, USBC, 1975b, 685, Series P126, P146, P152, P154, P158, P159, P160, and P164.

TABLE 4.4.
Percentage distribution of capital within the chemical process industries, 1879–1957

	Percentage in							
	Food and kindred products	Rubber products	Paper, pulp, and products	Chemicals and allied products[a]	Petroleum refining	Stone, clay, and glass products	Primary metals[b]	Total
Year	(1)	(2)	(3)	(4)	(5)	(6)	(7)	(8)
1879	45.2	0.5	4.5	10.4	1.9	7.9	29.6	100.0
1889	40.6	0.8	4.4	10.5	3.3	9.0	31.3	99.9
1899	44.4	0.9	5.6	10.7	2.4	8.7	27.2	99.9
1904	40.0	0.8	5.8	9.7	2.2	9.8	31.7	100.0
1909	35.0	0.9	6.4	9.7	2.1	11.1	34.9	100.1
1914	34.1	1.4	6.5	10.9	2.9	10.1	34.2	100.1
1919	31.4	2.9	6.3	11.5	5.7	6.9	35.3	100.0
1929	27.5	3.2	6.4	12.1	17.5	7.4	25.9	100.0
1937	27.4	2.4	6.1	11.8	19.4	5.9	27.0	100.0
1948	23.0	3.1	5.4	14.2	24.5	4.7	25.1	100.0
1953	23.1	3.0	5.5	15.9	22.3	4.7	25.5	100.0
1957	20.4	2.8	6.2	16.1	24.6	5.1	24.9	100.1

Note: For 1879–1889, includes custom and neighborhood shops; for 1899–1919, covers factories having annual production of $500 or more; for 1929–1957, of $5000 or more. See source for other minor discrepancies.

[a] Listed in the primary source as "chemicals and allied substances".
[b] I.e. the two subgroups: "iron and steel and products" and "nonferrous metals and products".

Source: Table 4.3.

TABLE 4.5.
Manufacturing income originating in the chemical process industries, 1929–1978

	Amount (billions of current dollars)			As percentage of total manufacturing[a]	
	Total manufacturing	Chemical process industries[a]	Chemicals and allied products	Chemical process industries[b]	Chemicals and allied products
Year	(1)	(2)	(3)	(4)	(5)
1929	21.9	9.7	1.1	44.3	5.0
1930	18.3	8.5	1.0	46.4	5.5
1931	12.5	5.6	0.8	44.8	6.4
1932	7.3	3.3	0.6	45.2	8.2
1933	7.7	3.5	0.6	45.5	7.8
1934	11.1	4.9	0.7	44.1	6.3
1935	13.4	6.0	0.8	44.8	6.0
1936	16.3	7.3	1.0	44.8	6.1
1937	19.5	9.0	1.2	46.2	6.2
1938	15.2	7.2	1.0	47.4	6.6
1939	18.1	8.4	1.2	46.4	6.6
1940	22.5	10.3	1.5	45.8	6.7
1941	33.2	14.5	1.9	43.7	5.7
1942	45.4	18.9	2.8	41.6	6.2
1943	58.3	23.6	3.4	40.5	5.8
1944	60.3	24.3	3.4	40.3	5.6
1945	52.2	22.1	3.2	42.3	6.1
1946	49.2	22.4	3.3	45.5	6.7
1947	59.6	26.9	3.8	45.1	6.4
1948	68.8	27.7	4.1	40.3	6.0
1949	64.8	25.9	4.1	40.0	6.3
1950	76.3	30.4	4.9	39.8	6.4
1951	90.3	36.2	6.0	40.1	6.6
1952	92.6	34.9	5.8	37.7	6.3
1953	100.5	38.3	6.1	38.1	6.1
1954	94.7	36.7	6.2	38.8	6.5
1955	108.0	43.0	7.4	39.8	6.9
1956	113.2	45.3	7.6	40.0	6.7
1957	116.3	45.7	8.0	39.3	6.9
1958	107.9	43.1	7.8	39.9	7.2
1959	124.2	48.8	9.1	39.3	7.3
1960	125.4	48.8	9.1	38.9	7.3

Table 4.5. (continued)

Year	Amount (billions of current dollars) Total manufacturing (1)	Chemical process industries[a] (2)	Chemicals and allied products (3)	As percentage of total manufacturing[a] Chemical process industries[b] (4)	Chemicals and allied products (5)
1961	124.7	48.5	9.4	38.9	7.5
1962	136.0	50.5	9.8	37.1	7.2
1963	143.2	53.0	10.5	37.0	7.3
1964	154.1	57.2	11.3	37.1	7.3
1965	170.4	61.5	12.4	36.1	7.3
1966	189.4	67.7	13.6	35.7	7.2
1967	193.2	68.3	13.7	35.4	7.1
1968	210.8	73.5	15.4	34.9	7.3
1969	221.4	77.2	16.0	34.9	7.2
1970	215.4	78.7	16.0	36.5	7.4
1971	224.7	81.6	16.8	36.3	7.5
1972	251.8	89.7	18.3	35.6	7.3
1973	283.5	100.8	20.3	35.6	7.2
1974	298.2	116.5	21.7	39.1	7.3
1975	312.5	121.4	23.8	38.8	7.6
1976	363.1	139.4	28.1	38.4	7.7
1977	409.4	151.3	30.1	37.0	7.4
1978	459.5	167.2	32.7	36.4	7.1

[a] Calculated using amounts rounded to the nearest hundred million dollars. Use of the more precise data in the sources would yield only slightly different results.

[b] Includes the following groups: food and kindred products; paper and allied products; chemicals and allied products; products of petroleum and coal; rubber products; stone, clay, and glass products; iron and steel and their products, including ordnance; non-ferrous metals and their products.

Sources: 1929 to 1972 – IA, BEA, 1977, Table 6.3, 186–189.
1973 and 1974 – IA, "U.S. National Income", 1976, 50, Table 6.3.
1975 to 1978 – IA, "U.S. National Income", 1979, 53, Table 6.3.

TABLE 4.6.
Chemical firms among the top hundred corporations, 1909–1975

Industrial group (or chemical firm)	Rank		Companies in top hundred	Assets		
					Industrial groups	
	Among industrial groups	Among individual firms		Chemical firms (millions of current dollars)	Amount (billions of current dollars)	As percentage of total
	(1)	(2)	(3)	(4)	(5)	(6)
	1909					
Iron and steel	1		13		2.497	28.6
Petroleum	2		8		1.103	12.6
Nonferrous metals	3		13		0.763	8.7
Food products	4		10		0.692	7.9
Transporation and equipment	5		7		0.579	6.6
Private transport	6		5		0.448	5.1
Chemicals and allied products	7		8		0.360	4.1
Du Pont		30		75		
Virginia-Carolina Chemical Co.		33		72		
American Agricultural Chemical Co.		42		55		
American Cotton Oil Co.		65		38		
Eastman Kodak Co.		72		35		

Table 4.6. (continued)

Industrial group (or chemical firm)	Rank		Companies in top hundred	Assets		
					Industrial groups	
	Among industrial groups	Among individual firms		Chemical firms (millions of current dollars)	Amount (billions of current dollars)	As percentage of total
	(1)	(2)	(3)	(4)	(5)	(6)
American Linseed Co.		79		34		
International Salt Co.		94		26		
General Chemical Co.		99		25		
Coal mining	8		6		0.331	3.8
Tobacco	8		2		0.312	3.6
Miscellaneous machinery	10		5		0.301	3.4
All others	–		23		1.360	15.5
Total for top hundred	–		100		8.746	99.9
			1919			
Iron and steel	1		12		3.962	23.1
Petroleum	2		19		3.211	18.7
Food products	3		10		1.844	10.8
Transportation equipment	4		10		1.742	10.2

Table 4.6. (continued)

Industrial group (or chemical firm)	Rank		Companies in top hundred	Assets		
					Industrial groups	
	Among industrial groups	Among individual firms		Chemical firms (millions of current dollars)	Amount (billions of current dollars)	As percentage of total
	(1)	(2)	(3)	(4)	(5)	(6)
Nonferrous metals	5		12		1.529	8.9
Chemicals and allied products	6		8		0.946	5.5
Du Pont		18		214		
Union Carbide Corp.		20		200		
Virginia-Carolina Chemical Corp.		50		121		
American Agricultural Chemical Corp.		52		111		
Procter & Gamble Co.		63		94		
Eastman Kodak Co.		69		89		
American Cotton Oil Co.		94		63		
General Chemical Co.		100		54		
Rubber	7		4		0.668	3.9
Electrical equipment and machinery	8		4		0.606	3.5
Tobacco	9		4		0.548	3.2
Coal mining	10		3		0.381	2.2
All others	–		14		1.701	9.9
Total for top hundred	–		100		17.138	99.9

Table 4.6. (continued)

Industrial group (or chemical firm)	Rank		Companies in top hundred	Assets		
				Chemical firms (millions of current dollars)	Industrial groups	
	Among industrial groups	Among individual firms			Amount (billions of current dollars)	As percentage of total
	(1)	(2)	(3)	(4)	(5)	(6)
	1929					
Petroleum	1		20		7.986	28.0
Iron and steel	2		11		4.556	16.0
Transportation equipment	3		8		3.260	11.4
Nonferrous metals	4		7		1.910	6.7
Food products	5		8		1.720	6.0
Chemicals and allied products	6		5		1.353	4.7
Du Pont		12		497		
Union Carbide Corp.		25		307		
Allied Chemical & Dye Corp.		27		277		
Eastman Kodak Co.		60		163		
Procter & Gamble Co.		91		109		
Retail distribution	7		8		1.319	4.6
Electrical equipment and machinery	8		4		1.238	4.3

Table 4.6. (continued)

Industrial group (or chemical firm)	Rank		Companies in top hundred	Assets		
				Chemical firms (millions of current dollars)	Industrial groups	
	Among industrial groups	Among individual firms			Amount (billions of current dollars)	As percentage of total
	(1)	(2)	(3)	(4)	(5)	(6)
Motion picture production and distribution	9		4		0.888	3.1
Rubber	10		4		0.877	3.1
All others	–		21		3.380	11.9
Total for top hundred	–		100		28.487	99.8
	1935					
Petroleum	1		16		6.719	26.8
Iron and steel	2		10		3.828	15.3
Transportation equipment	3		7		3.182	12.7
Nonferrous metals	4		8		1.880	7.5
Food products	5		9		1.467	5.8
Chemicals and allied products	6		5		1.399	5.6
Du Pont		9		581		
Union Carbide Corp.		22		271		

Table 4.6. (continued)

Industrial group (or chemical firm)	Rank		Companies in top hundred	Assets		
				Chemical firms (millions of current dollars)	Industrial groups	
	Among industrial groups	Among individual firms			Amount (billions of current dollars)	As percentage of total
	(1)	(2)	(3)	(4)	(5)	(6)
Allied Chemical & Dye Corp.		25		252		
Eastman Kodak Co.		49		168		
Procter & Gamble Co.		63		127		
Retail distribution	7		9		1.247	5.0
Electrical equipment and machinery	8		5		0.967	3.9
Coal mining and coke	9		5		0.799	3.2
Rubber	10		4		0.614	2.4
All others	–		22		2.986	11.9
Total for top hundred	–		100		25.088	100.1
1948						
Petroleum	1		18		14.402	29.1
Iron and steel	2		9		5.932	12.0

Table 4.6. (continued)

Industrial group (or chemical firm)	Rank		Companies in top hundred	Assets		
					Industrial groups	
	Among industrial groups	Among individual firms		Chemical firms (millions of current dollars)	Amount (billions of current dollars)	As percentage of total
	(1)	(2)	(3)	(4)	(5)	(6)
Transportation equipment	3		6		5.761	11.6
Chemicals and allied products	4		10		4.333	8.7
Du Pont		7		1304		
Union Carbide Corp.		14		723		
Eastman Kodak Co.		33		412		
Procter & Gamble Co.		38		356		
Allied Chemical & Dye Corp.		44		339		
Dow Chemical Co.		53		294		
Celanese Corp. of America		59		283		
American Viscose Corp.		72		227		
American Cyanamid Co.		78		212		
Monsanto Chemical Co.		87		183		
Electrical equipment and machinery	5		6		3.265	6.6
Retail distribution	6		8		2.887	5.8
Nonferrous metals	7		7		2.809	5.7

Table 4.6. (continued)

Industrial group (or chemical firm)	Rank		Companies in top hundred	Assets		
					Industrial groups	
	Among industrial groups	Among individual firms		Chemical firms (millions of current dollars)	Amount (billions of current dollars)	As percentage of total
	(1)	(2)	(3)	(4)	(5)	(6)
Food products	8		7		2.146	4.3
Tobacco	9		3		1.643	3.3
Rubber	10		4		1.420	2.9
All others	–		22		4.930	10.0
Total for top hundred	–		100		49.528	100.0
			1960			
Petroleum	1		19		39.251	31.3
Transportation equipment	2		9		18.264	14.6
Iron and steel	3		9		12.885	10.3
Chemicals and allied products	4		11		12.315	9.8
Du Pont		8		3134		
Union Carbide Corp.		16		1713		
Monsanto Chemical Co.		27		1090		

Table 4.6. (continued)

Industrial group (or chemical firm)	Rank		Companies in top hundred	Assets		
					Industrial groups	
	Among industrial groups	Among individual firms		Chemical firms (millions of current dollars)	Amount (billions of current dollars)	As percentage of total
	(1)	(2)	(3)	(4)	(5)	(6)
Dow Chemical Co.		29		1040		
Eastman Kodak Co.		33		959		
Procter & Gamble Co.		36		931		
Olin Mathieson Chemical Corp.		41		860		
Allied Chemical & Dye Corp.		51		801		
American Cyanamid Co.		64		641		
W. R. Grace and Co.		69		606		
National Distillers & Chemical Corp.[a]		79		540		
Electrical equipment and machinery	5		7		10.881	8.7
Nonferrous metals	6		8		6.559	5.2
Retail distribution	7		7		5.665	4.5
Miscellaneous	8		5		3.177	2.5
Rubber	9		4		3.145	2.5
Lumber and paper	10		4		2.747	2.2
All others	–		17		10.629	8.5
Total for top hundred	–		100		125.518	100.1

Table 4.6. (continued)

Industrial group (or chemical firm)	Rank		Companies in top hundred	Assets		
				Chemical firms (millions of current dollars)	Industrial groups	
	Among industrial groups	Among individual firms			Amount (billions of current dollars)	As percentage of total
	(1)	(2)	(3)	(4)	(5)	(6)
	1975					
Petroleum	1		19		156.236	35.5
Motor vehicles[b]	2		5		45.505	10.4
Chemicals and allied products	3		13		40.776	9.3
Du Pont		16		6425		
Dow Chemical Co.		18		5847		
Union Carbide Corp.		19		5741		
Procter & Gamble Co.		32		3653		
Monsanto Chemical Co.		35		3451		
W. R. Grace and Co.		52		2524		
Allied Chemical & Dye Corp.		61		2236		
Pfizer Co.		70		2019		
Celanese Co.		78		1908		
PPG Industries		80		1869		

Table 4.6. (continued)

Industrial group (or chemical firm)	Rank		Companies in top hundred	Assets		
					Industrial groups	
	Among industrial groups	Among individual firms		Chemical firms (millions of current dollars)	Amount (billions of current dollars)	As percentage of total
	(1)	(2)	(3)	(4)	(5)	(6)
Warner-Lambert Co.		86		1808		
American Cyanamid Co.		89		1722		
Merck		100		1574		
Metal manufacturing[c]	4		14		39.176	8.9
Electronics and appliances[d]	5		6		35.563	8.1
Office equipment (including computers)[e]	6		7		29.276	6.7
Industrial and farm equipment[f]	7		5		12.855	2.9
Measuring, scientific, and photographic equipment[g]	8		3		12.529	2.9
Paper, fiber, and wood products	9		4		10.962	2.5
Rubber and plastic products	10		4		10.557	2.4
All others	–		20		46.062	10.5
Total for top hundred	–		100		439.497	100.1

Note: Detail in column 4 may not add to totals in column 5 because of rounding errors. Percentages may not add to 100% because of rounding errors. Assets rounded to nearest million dollars. For 1909–1960, includes retail and transportation companies as well as

Table 4.6. (continued)

industrials; for 1975, includes industrials only. Cf. IB, Navin, 1970, for the top five hundred industrials in 1917, and IB, "Top 100", 1977, for the top hundred industrials for selected years from 1917 to 1977.

[a] Grouped with distillers until the early 1950s, when it diversified into industrial chemicals; see III, Berenson, 1963, 134–141.
[b] Included with transportation equipment in earlier years.
[c] Includes both iron and steel and nonferrous metals.
[d] Incorporates electrical equipment and machinery from earlier years. Also includes International Telephone and Telegraph Corp., which was grouped with miscellaneous in earlier years, and Singer Manufacturing Co., which was grouped with miscellaneous machinery in earlier years.
[e] Included with miscellaneous in earlier years.
[f] Includes Caterpillar Tractor Co. and Deere and Co., which were grouped with miscellaneous machinery in earlier years, and International Harvester Co., which was grouped with transportation equipment in earlier years.
[g] Includes Eastman Kodak Co., which was grouped with chemicals in earlier years, and Minnesota Mining & Manufacturing Co., which was grouped with miscellaneous in earlier years.

Sources: 1909 to 1960 – IB, Kaplan, 1964, Tables 7–2 through 7–7, 140–153.
1975 – IB, "Fortune 500", 1976, 4–15, 24. Data rearranged by assets, not sales, to correspond with data for other years.

TABLE 4.7.
Extent of oligopoly in chemicals and allied products industrial group, 1909–1972

	Industries in group			Product value of group (millions of current dollars)		
		Oligopolistic			Controlled by oligopolies	
	Total	Number	As percentage of total	Total	Amount	As percentage of total
Year	(1)	(2)	(3)	(4)	(5)	(6)
1909	25	8	32	803.1	70.2	9
1919	30	8	27	2 651.4	238.9	9
1929	27	8	30	3 276.5	550.1	17
1939	27	9	33	3 288.6	592.6	18
1947	41	28	68	12 080	5 321	44
1958	41	25	61	22 141	7 420	31
1963	28	15	54	31 773	9 862	31
1967[a]	27	16	59	42 148	13 974[b]	13
1972[a]	28	11	39	57 350	10 386[c]	18

See Figure 4.3–1.

Note: "Industries" as used here refers to the four-digit classification of subcategories within this industrial group employed by the Bureau of the Budget in 1957 (II, BOB, 1957, 76–82); see Appendix B. An industry is oligopolistic if "six or fewer firms manufacture 50 per cent

Table 4.7. (continued)

or more of the total product value [of that industry], or [if] twelve or fewer firms manufacture 75 per cent or more of total industry product, or some number of firms between six and twelve manufactures a proportionate percentage of total product, e.g., eight firms with 58 per cent . . . " (IB, Porter and Livesay, 1969, 282).

[a] Based on value of shipments, rather than on total product value. For industries in group, either the four-company or the eight-company concentration criterion was used. For product value (i.e., value of shipments), only the eight-company criterion was used.
[b] For industry code 2823, a percentage control of "99+" was given; this was interpreted as 99.5%. For code 2895, the mean of the four-company and the twenty-company percentages was used.
[c] For industry code 2823, the mean of the four-company and the twenty-company percentages was used. For code 2895, a percentage control of "99+" was given; this was interpreted as 99.5%.

Sources: 1909 to 1963 – IB, Porter and Livesay, 1969, Table 1, 283–289.
1967 – IA, USBC-COM, 1970, Table 5, SR2–19 to SR2–21.
1972 – IA, USBC-COM, 1975, Table 5, SR2–21 to SR2–24.

TABLE 4.8.
Distribution of U.S. patents for inventions, by patentee, 1901–1978

	Corporations		Individuals[b]	U.S. Government[c]	Total[d]
	Domestic	Foreign[a]			
Year	(1)	(2)	(3)	(4)	(5)
1901	4 370	280	20 896	–	25 546
1906	6 040	380	24 750	–	31 170
1911	7 580	520	24 756	–	32 856
1916	11 540	610	31 742	–	43 892
1921	9 860	840	27 098	–	37 798
1922	10 300	700	27 369	–	38 369
1923	10 800	800	27 016	–	38 616
1924	12 400	1 000	29 174	–	42 574
1925	14 800	1 300	30 332	–	46 432
1926	15 200	900	28 633	–	44 733
1927	15 100	1 200	25 417	–	41 717
1928	17 800	1 200	23 357	–	42 357
1929	18 500	1 400	25 367	–	45 267
1930	19 700	1 800	23 726	–	45 226
1931	23 149	1 961	26 618	28	51 756
1932	24 822	2 325	26 274	37	53 458
1933	23 667	2 343	22 713	51	48 774
1934	22 529	2 131	19 731	29	44 420
1935	20 821	2 018	17 757	22	40 618
1936	21 207	1 903	16 639	33	39 782
1937	19 831	1 824	15 995	33	37 683
1938	19 635	2 063	16 304	59	38 061
1939	21 800	2 640	18 583	50	43 073
1940	22 165	2 406	17 627	40	42 238
1941	22 632	2 112	16 322	43	41 109
1942	22 019	1 286	14 534	62	38 449[e]
1943	18 022	524	11 654	48	31 054[e]
1944	16 769	645	9 636	106	28 053[e]
1945	15 665	580	8 981	87	25 695[e]
1946	13 486	585	7 444	147	21 803[e]
1947	11 448	669	7 784	155	20 139[e]
1948	13 124	628	9 812	352	23 963[e]

Table 4.8. (continued)

Year	Corporations: Domestic (1)	Corporations: Foreign[a] (2)	Individuals[b] (3)	U.S. Government[c] (4)	Total[d] (5)
1949	18 536	1 127	14 957	485	35 131[e]
1950	21 782	1 660	18 960	622	43 040[e]
1951	22 305	2 163	19 192	659	44 326[e]
1952	22 340	2 035	18 538	695	43 616[e]
1953	21 230	2 294	16 284	658	40 468[e]
1954	18 319	2 301	12 531	658	33 809
1955	16 084	1 744	11 914	689	30 432[e]
1956	25 502	3 690	16 643	982	46 817
1957	23 255	3 372	15 154	963	42 744
1958	27 116	4 230	15 706	1278	48 330
1959	29 888	5 081	16 017	1422	52 408
1960	28 187	4 670	13 069	1244	47 170
1961	28 351	5 161	13 383	1473	48 368
1962	32 560	6 380	15 470	1281	55 691
1963	26 632	5 501	12 525	1021	45 679
1964	27 836	5 854	12 504	1182	47 376
1965	37 158	8 096	16 063	1540	62 857
1966	41 634	9 222	16 018	1532	68 406
1967	38 353	9 895	15 647	1757	65 652
1968	34 886	9 172	13 555	1489	59 102
1969	38 847	12 188	14 772	1750	67 557
1970	36 896	12 294	13 511	1726	64 427
1971	43 022	16 048	17 099	2147	78 316
1972	38 890	16 414	17 729	1775	74 808
1973	38 615	16 513	16 929	2082	74 139
1974	37 807	18 686	18 083	1699	76 275
1975	34 577	18 344	17 192	1881	71 994
1976	34 391	19 934	14 084	1827	70 236
1977	31 531	18 220	14 027	1491	65 269
1978	31 309	19 286	14 259	1248	66 102

Note: Some of these data are estimates based on samples, while others are actual counts. See IA, USBC, 1975b, 955, for particulars. See Table 4.9 for a *percentage* distribution.

Table 4.8. (continued)

---

[a] Includes patents assigned to foreign governments.
[b] According to the source, this series is derived by subtraction of the other components from the total, less patents issued to the Alien Property Custodian during and after World War II. It includes domestic and foreign individuals; the latter received 12.6% of total invention patents during 1951–1957, and 22.4% during 1964–1970 (IA, USBC, 1975b, 955; see also note e below).
[c] Excludes patents issued to the Alien Property Custodian; see note e.
[d] Since 1942, includes patents issued to the Alien Property Custodian, not shown separately; see note e.
[e] Exceeds the sum of the components shown, apparently because patents issued to the Alien Property Custodian are included. The numbers of such patents may be derived by subtracting the sum of columns 1 to 4 from column 5; they never exceed 900 in any given year.

Sources: 1901 to 1970 – IA, USBC, 1975b, 957–958, Series W99–W103.
1971 to 1972 – IA, USBC, 1977, Table 908, 557.
1973 to 1978 – IA, USBC, 1979, Table 954, 573.

TABLE 4.9.
Percentage distribution of U.S. patents for inventions, by patentee, 1901–1978

	Corporations		Individuals[b]	U.S. Government[c]	Total[d]
	Domestic	Foreign[a]			
Year	(1)	(2)	(3)	(4)	(5)
1901	17.1	1.1	81.8	–	100.0
1906	19.4	1.2	79.4	–	100.0
1911	23.1	1.6	75.3	–	100.0
1916	26.3	1.4	72.3	–	100.0
1921	26.1	2.2	71.7	–	100.0
1922	26.8	1.8	71.3	–	99.9
1923	28.0	2.1	70.0	–	100.1
1924	29.1	2.3	68.5	–	99.9
1925	31.9	2.8	65.3	–	100.0
1926	34.0	2.0	64.0	–	100.0
1927	36.2	2.9	60.9	–	100.0
1928	42.0	2.8	55.1	–	99.9
1929	40.9	3.1	56.0	–	100.0
1930	43.6	4.0	52.5	–	100.1
1931	44.7	3.8	51.4	0.1	100.0
1932	46.4	4.3	49.1	0.1	99.9
1933	48.5	4.8	46.6	0.1	100.0
1934	50.7	4.8	44.4	0.1	100.0
1935	51.3	5.0	43.7	0.1	100.1
1936	53.3	4.8	41.8	0.1	100.0
1937	52.6	4.8	42.4	0.1	99.9
1938	51.6	5.4	42.8	0.2	100.0
1939	50.6	6.1	43.1	0.1	99.9
1940	52.5	5.7	41.7	0.1	100.0
1941	55.1	5.1	39.7	0.1	100.0
1942	57.3	3.3	37.8	0.2	98.6[e]
1943	58.0	1.7	37.5	0.2	97.4[e]
1944	59.8	2.3	34.3	0.4	96.8[e]
1945	61.0	2.3	35.0	0.3	98.6[e]
1946	61.9	2.7	34.1	0.7	99.4[e]
1947	56.8	3.3	38.7	0.8	99.6[e]
1948	54.8	2.6	40.9	1.5	99.8[e]

Table 4.9. (continued)

	Corporations		Individual[b]	U.S. Government[c]	Total[d]
	Domestic	Foreign[a]			
Year	(1)	(2)	(3)	(4)	(5)
1949	52.8	3.2	42.6	1.4	100.0[e]
1950	50.6	3.9	44.1	1.4	100.0[e]
1951	50.3	4.9	43.3	1.5	100.0[e]
1952	51.2	4.7	42.5	1.6	100.0[e]
1953	52.5	5.7	40.2	1.6	100.0[e]
1954	54.2	6.8	37.1	1.9	100.0
1955	52.9	5.7	39.1	2.3	100.0[e]
1956	54.5	7.9	35.5	2.1	100.0
1957	54.4	7.9	35.5	2.3	100.1
1958	56.1	8.8	32.5	2.6	100.0
1959	57.0	9.7	30.6	2.7	100.0
1960	59.8	9.9	27.7	2.6	100.0
1961	58.6	10.7	27.7	3.0	100.0
1962	58.5	11.5	27.8	2.3	100.1
1963	58.3	12.0	27.4	2.2	99.9
1964	58.8	12.4	26.4	2.5	100.1
1965	59.1	12.9	25.6	2.5	100.1
1966	60.9	13.5	23.4	2.2	100.0
1967	58.4	15.1	23.8	2.7	100.0
1968	59.0	15.5	22.9	2.5	99.9
1969	57.5	18.0	21.9	2.6	100.0
1970	57.3	19.1	21.0	2.7	100.1
1971	54.9	20.5	21.8	2.7	99.9
1972	52.0	21.9	23.7	2.4	100.0
1973	52.1	22.3	22.8	2.8	100.0
1974	49.6	24.5	23.7	2.2	100.0
1975	48.0	25.5	23.9	2.6	100.0
1976	49.0	28.4	20.1	2.6	100.1
1977	48.3	27.9	21.5	2.3	100.0
1978	47.4	29.2	21.6	1.9	100.1

See Figure 4.3–2.

Table 4.9. (continued)

[a] Includes patents assigned to foreign governments.
[b] Estimated; includes foreign residents. See Table 4.8, note b.
[c] Excludes patents issued to the Alien Property Custodian.
[d] May not equal 100% because of rounding errors. See also note e.
[e] May not equal 100% because a total including more than the sum of the component parts shown was used in the calculations of percentage. This minor discrepancy is not detailed in Figure 4.3–2. See Table 4.8, note e.

Source: Table 4.8.

TABLE 4.10.
Chemical patent issues as percentage of total U.S. patent issues, by pentade, 1891–1965

Pentade	Total patents issued	Chemical patents issued	
		Number	As percentage of total
	(1)	(2)	(3)
1891–1895	108 515	3 194	2.9
1896–1900	112 325	4 001	3.6
1901–1905	143 791	5 586	3.9
1906–1910	171 560	5 959	3.5
1911–1915	186 241	7 251	3.9
1916–1920	197 644	8 520	4.3
1921–1925	203 977	10 562	5.2
1926–1930	219 384	13 613	6.2
1931–1935	239 092	23 423	9.8
1936–1940	200 902	24 921	12.4
1941–1945	164 438	23 736	14.4
1946–1950	144 160	22 577	15.7
1951–1955	192 897	28 107	14.6
1956–1960	237 768	42 681	18.0
1961–1965	260 058	52 411	20.2

See Figure 4.3–3.

Note: The series in the source begin with data for 1836–1840. Between that pentade and the ones shown, the percentage for chemistry fluctuated only between 2.4 and 3.9%.

Source: IB, Hurd, 1970, 169, Table I.

TABLE 5.1.
Rough estimates of the number of chemists in industry, 1917–1970 (thousands)

	Chemists in the labor force		Chemists in industry (BLS estimates)		
	Census estimates	BLS estimates	Chemicals and allied products	Chemical process industries[a]	Total
Year	(1)	(2)	(3)	(4)	(5)
1910	16	–	–	–	–
1917	–	–	–	–	13.7
1920	28	–	–	–	–
1930	45	–	–	–	–
1940	57	–	–	–	–
1944	–	–	–	–	22.7
1950	77	51.9	13.4	24.6	36.6
1951	–	56.8	15.8	28.1	40.9
1952	–	62.9	18.8	32.0	46.4
1953	–	67.9	21.0	35.2	50.5
1954	–	71.6	23.3	37.9	53.9
1955	–	73.9	23.7	38.4	55.1
1956	–	79.2	25.1	40.4	58.3
1957	–	84.5	27.0	43.0	62.6
1958	–	90.6	29.2	45.8	66.3
1959	–	95.4	30.1	46.9	68.7
1960	84	99.7	32.0	49.5	72.1
1961	–	102.8	33.1	51.4	74.7
1962	–	106.8	35.3	53.3	77.7
1963	–	110.0	35.3	53.6	79.7
1964	–	115.0	36.3	54.4	83.4
1965	–	116.7	37.8	55.1	84.5
1966	–	119.6	40.3	57.6	87.3
1967	–	122.8	40.5	57.4	86.7
1968	–	127.3	40.9	58.6	89.5
1969	–	131.0	41.6	59.6	92.0
1970	110	132.9	42.8	61.0	93.5

See Figures 2.2–10, 5.1–1, A.2–1, and A.2–2.

[a] Includes the following groups: food; paper; chemicals; petroleum refining; rubber, stone, clay, and glass; and primary metals. See Table 5.2 for further details.

Table 5.1. (*continued*)

Sources: *Column 1* – IA, USBC, 1975b, 140, Series D245.

Columns 2–5:

1917 – IA, Fay, 1917, 10–11. The total used equals the total number of chemists (15 000), less professors and instructors (1285); this adjustment was made to render the estimate compatible with Census data. See Sections 2.1 and 5.4.

1944 – IB, "Census", 1944, 1005.

1950 to 1970 – IA, BLS, 1973, Table A–16, 50–51.

TABLE 5.2.
Chemists in the chemical process industries, 1950–1970

Year	Food and kindred products (1)	Paper and allied products (2)	Chemicals and allied products (3)	Petroleum refining (4)	Rubber and products (5)	Stone, clay and glass (6)	Primary metals (7)	Total[a] (8)
A. Number (thousands)								
1950	2.9	1.5	13.4	2.3	1.6	0.9	2.0	24.6
1955	3.6	2.1	23.7	3.5	2.0	1.2	2.3	38.4
1960	4.4	2.8	32.0	3.9	2.2	1.5	2.7	49.5
1965	4.5	2.9	37.8	3.4	2.8	1.5	2.2	55.1
1970	4.5	3.6	42.8	3.2	3.0	1.5	2.4	61.0
B. As percentage of total chemists in industry								
1950	7.9	4.1	36.6	6.3	4.4	2.5	5.5	67.2
1955	6.5	3.8	43.0	6.4	3.6	2.2	4.2	69.7
1960	6.1	3.9	44.4	5.4	3.1	2.1	3.7	68.7
1965	5.3	3.4	44.7	4.0	3.3	1.8	2.6	65.2
1970	4.8	3.9	45.8	3.4	3.2	1.6	2.6	65.2

[a] Detail may not add to total because of rounding errors.

Sources: *Part A* – IA, BLS, 1973, Table A–16, 50–51.
Part B – divide columns in part A by Table 5.1, column 5.

TABLE 5.3.
Employment in the chemicals and allied products industries, 1899–1977
(hundred thousands)

		Production workers		
Year	Chemists	As reported	Recalculated[a]	Total[b]
	(1)	(2)	(3)	(4)
1899	–	1.44	1.29	1.70
1904	–	1.58	1.38	1.91
1909	–	1.85	1.62	2.35
1914	–	2.08	1.79	2.69
1919	–	2.94	2.53	3.88
1921	–	2.12	1.87	2.79
1923	–	2.64	2.41	3.40
1925	–	2.61	2.34	–
1927	–	2.78	2.50	3.48
1929	–	3.07	2.85	3.82
1931	–	2.48	2.31	–
1933	–	2.54	2.35	3.02
1935	–	2.94	2.73	–
1937	–	3.03	2.77	–
1939	–	2.76	2.50	–
1947	–	4.64	4.32	6.26
1949	–	4.40	–	6.12
1950	0.134	4.57	–	6.43
1951	0.158	4.98	–	7.03
1952	0.188	5.13	–	7.39
1953	0.210	5.36	–	7.68
1954	0.233	4.99	4.71	7.34
1955	0.237	5.08	–	7.41
1956	0.251	5.15	–	7.55
1957	0.270	5.06	–	7.57
1958	0.292	4.53	–	6.98
1959	0.301	4.71	–	7.18
1960	0.320	4.70	–	7.26
1961	0.331	4.60	–	7.13
1962	0.353	4.70	–	7.27
1963	0.353	4.74	–	7.37
1964	0.363	4.80	–	7.49
1965	0.378	5.02	–	7.80

Table 5.3. (*continued*)

Year	Chemists	Production workers		Total[b]
		As reported	Recalculated[a]	
	(1)	(2)	(3)	(4)
1966	0.403	5.29	–	8.22
1967	0.405	5.41	–	8.41
1968	0.409	5.51	–	8.56
1969	0.416	5.66	–	8.83
1970	0.428	5.56	–	8.81
1971	–	5.29	–	8.49
1972	–	5.25	–	8.36
1973	–	5.35	–	8.53
1974	–	5.41	–	8.65
1975	–	5.10	–	8.42
1976	–	5.20	–	8.51
1977	–	5.41	–	8.78

See Figures 5.1–2 and B.2–1.

[a] Using the 1957 SIC code. See Appendix B and IB, Backman, 1970, Appendix A, 307–312.
[b] Includes employees in administration, construction, and distribution, as well as in production.

Sources: *Column 1* – IA, BLS, 1973, Table A–16, 50–51.

Columns 2 and 4:

1899 to 1971 – IA, USBC-COM, 1976, 28–2, Table 1. Includes only *operating* manufacturing establishments; see notes to source table for other qualifications.
1972 to 1976 – IA, USBC-ASM, 1978, 129, Table SIC: 28.
1977 – IA, USBC-COM, 1979, Table 2, 10.

Column 3 – IB, Backman, 1970, Appendix Table 2, 315. Note that Backman's totals for 1967, which differ from those given in column 2, are preliminary estimates.

TABLE 5.4.
The formation of industrial research laboratories, 1890–1940

Year	Industrial research laboratories formed: Annual number (1)	Mean for pentade ending on date (2)	Cumulative to date (3)
1890	4	–	4
1891	2	–	6
1892	5	–	11
1893	5	–	16
1894	2	3.6	18
1895	4	3.6	22
1896	2	3.6	24
1897	3	3.2	27
1898	8	3.8	35
1899	4	4.2	39
1900	10	5.4	49
1901	5	6.0	54
1902	15	8.4	69
1903	10	8.8	79
1904	10	10.0	89
1905	11	10.2	100
1906	11	11.4	111
1907	7	9.8	118
1908	15	10.8	133
1909	3	9.4	136
1910	44	16.0	180
1911	20	17.8	200
1912	27	21.8	227
1913	22	23.2	249
1914	28	28.2	277
1915	33	26.0	310
1916	37	29.4	347
1917	28	29.6	375
1918	26	30.4	401
1919	40	32.8	441
1920	78	41.8	519
1921	34	41.2	553

Table 5.4. (continued)

Year	Industrial research laboratories formed: Annual number (1)	Mean for pentade ending on date (2)	Cumulative to date (3)
1922	39	43.4	592
1923	42	46.6	634
1924	41	46.8	675
1925	53	41.8	728
1926	62	47.4	790
1927	45	48.6	835
1928	43	48.8	878
1929	63	53.2	941
1930	89	60.4	1030
1931	44	56.8	1074
1932	42	56.2	1116
1933	27	53.0	1143
1934	29	46.2	1172
1935	41	36.6	1213
1936	51	38.0	1264
1937	22	34.0	1286
1938	18	32.2	1304
1939	21	30.6	1325
1940	13	25.0	1338

See Figures 5.1–3 and 5.1–4.

Note: Date assigned for initiation of research in each corporation is the year in which research activity was introduced as a recognized, permanent function within the organization. Excludes testing, consulting, and trade association laboratories. Represents a 59.1% sample of the 2264 corporate units reporting laboratories in 1940 (Table 5.5, column 1). See source for further details.

Source: *Columns 1 and 3* – IB, Cooper, 1941. Data for 1890–1939 obtained by inspection from Figure 46, 176. Data for 1940 inferred from 177, note 12.

Column 2 – five-year moving averages of data in column 1, recorded in final years of averaged periods.

TABLE 5.5.
Industrial research personnel, by selected field, 1921–1960

Year	Corporate units surveyed[a]	Research personnel			Chemists	
		Total	Physicists	Engineers	Number	As percentage of total
	(1)	(2)	(3)	(4)	(5)	(6)
1921[b]	568	11 500	150	1 898	3 830	33.3
1927	926	18 982	437	3 018	5 163	27.2
1931	1520	32 830	689	6 993	8 470	25.8
1933	1462	27 567	414	5 541	7 526	27.3
1938	1722	44 292	1 550	10 276	12 623	28.5
1940	2210	70 033	2 031	14 987	15 687	22.4
1946	2443	133 515	2 660	20 637	21 095	15.8
1950	2795	165 032	2 969	35 601	23 159	14.0
1954	–	440 440	4 800	105 100	26 800	6.1
1957	–	618 600	6 800	151 300	32 200	5.2
1960	–	–	13 800	224 300	42 800	–

See Figures 5.1–7 and 5.1–8.

Note: See Appendix B.3 for a discussion of the limitations of the surveys from which these data are derived.

[a] A "corporate unit" is a single corporation and all its subsidiaries, regardless of the number of laboratories maintained by the entire entity. Totals provided do not necessarily equal the number of corporate units reporting laboratories to the NRC, because some of the survey questionnaires contained inadequate information about the research personnel employed. Column 1 therefore gives the number of corporate units that employed the personnel listed in the other columns. For the companies or corporate units reporting, see the original NRC survey reports: IB, Flinn and Cobb, 1921; IB, West and Risher, 1927; IB, West and Hull, 1931a; IB, West and Hull, 1933; IB, Hull, 1938; IB, Hull, 1940; IB, Hull, 1946; and IB, Rand, 1950.
[b] The totals for 1921 in columns 1–5 have been adjusted upwards by 23% [(11 500–9350) × 100/9350] to compensate for underreporting (see IB, Perazich and Field, 1940, 59–60). This is an approximate correction, especially in the case of column 1.

Table 5.5. (continued)

Sources: *Column 1*:

1921 to 1938 – IB, Perazich and Field, 1940, 65, Table A–3.
1940 – IB, Cooper, 1941, 177.
1946 – IB, Hull, 1946, [iii-vi]. Also in IB, "Industrial Research", 1946, 2905.
1950 – IA, FSA, 1952, 2. Also printed as IB, Rand, 1951b; see 3604.

Columns 2 to 6:

1921 to 1938 – IB, Perazich and Field, 1940. Column 2 is from 65, Table A–3. Columns 3 to 5 were obtained by multiplying column 2 by the appropriate percentages from 78, Table A–19. Column 6 is from 78, Table A–19.
1940 – IB, Cooper, 1941, 176, Table 1. The absolute numbers in this source are rounded. Ours – obtained by multiplying the percentages in the source by the total number of personnel as reported in the source – are not.
1946 – IB, Hull, 1946, [iii-vi]. Also in IB, "Industrial Research", 1946, 2905.
1950 – IA, FSA, 1952, 9, Table 1. Also printed as IB, Rand, 1951b; see 3605, Table I.
1954 – IA, NSF, 1956, 72, Table A–18. For column 2, multiply the total in Table A–18 by 2.8 to adjust for support personnel; see ibid., 78, Table A–24.
1957 – IA, NSF, 1959, 58, Table A–19, and 62, Table A–24.
1960 – IA, NSF, 1961b, 28, Table A–10.

TABLE 5.6.
Distribution of research personnel in the chemical process industries, 1927 and 1938

Industrial group[a]	Companies				Employees			
	1927		1938		1927		1938	
	Number	As percentage of total	Number	As percentage of total	Number	As percentage of total	Number	As percentage of total
	(1)	(2)	(3)	(4)	(5)	(6)	(7)	(8)
Food and kindred products	62	6.7	108	6.3	401	2.1	1 424	3.2
Paper and allied products	28	3.0	57	3.3	271	1.4	7 52	1.7
Chemicals and allied products	231	24.9	395	22.9	3 463	18.2	9 542	21.5
Petroleum and its products	28	3.0	53	3.1	788	4.2	5 033	11.4
Rubber products	20	2.2	35	2.0	1 115	5.9	2 250	5.1
Stone, clay, and glass products	45	4.9	99	5.7	527	2.8	1 404	3.2
Primary metals[b]	97	10.5	144	8.4	1 222	6.4	2 728	6.2
Total, chemical process industries[c]	511	55.2	891	51.7	7 787	41.0	23 133	52.2
Total, all industries	926	100.0	1722	100.0	18 982	100.0	44 292	100.0

Table 5.6. (continued)

[a] For the definition of these groups, see Appendix B and IB, Perazich and Field, 1940, 61–62.
[b] Combines the two groups: iron and steel and their products, not including machinery; and nonferrous metals and their products.
[c] Detail may not add to totals in percentage columns because of rounding errors.

Source: IB, Perazich and Field, 1940, 73, Table A–13.

TABLE 5.7.
Companies maintaining research laboratories in selected industrial groups, 1920–1938

Year	Industrial chemicals[a] (1)	Petroleum and petroleum products (2)	Rubber products (3)
1920	20	5	9
1921	30	10	8
1927	41	28	20
1931	50	47	39
1933	49	48	32
1938	54	53	35

See Figure 5.1–5.

Note: All data as reported. Making the recommended adjustments to the totals for 1920 and 1921 (see IB, Perazich and Field, 1940, 59–60) would not counteract significantly the overall pattern of growth.

[a] Not to be confused with chemicals and allied products, of which this is but one sector. Cf. Table 5.6, row 3, columns 1 to 4.

Source: IB, Perazich and Field, 1940, 77, Table A–17.

TABLE 5.8.
Research personnel in selected industrial groups, 1920–1938

Year	Industrial chemicals[a] (1)	Petroleum and its products (2)	Rubber products (3)
1920	1690	145	590
1921	1178	167	495
1927	1918	788	1115
1931	3257	2957	1561
1933	2929	2724	1939
1938	5248	5033	2250

See Figure 5.1–6.

Note: All data as reported. Making the recommended adjustments to the totals for 1920 and 1921 (see IB, Perazich and Field, 1940, 59–60) would not counteract significantly the overall pattern of growth.

[a] Not to be confused with chemicals and allied products, of which this is but one sector. Cf. Table 5.6, row 3, columns 5–8.

Source: IB, Perazich and Field, 1940, 77, Table A–17.

TABLE 5.9.
Employment of respondents to 1941 ACS survey of membership, selected occupations, 1926–1941

Year	Total respondents reporting an occupation	Respondents in industrial research		Respondents in teaching			
						Both	
		Number	As percentage of total	College and university	Secondary school	Number	As percentage of total
	(1)	(2)	(3)	(4)	(5)	(6)	(7)
1926	8 009	1280	16.0	1616	269	1885	23.5
1929	9 982	1929	19.3	1861	239	2100	21.0
1932	12 010	2250	18.7	2168	242	2410	20.1
1934	13 671	2695	19.7	2250	263	2513	18.4
1937	16 484	3818	23.2	2502	235	2737	16.6
1938	17 136	3996	23.3	2600	219	2819	16.5
1939	17 801	4175	23.5	2654	219	2873	16.1
1940	18 366	4487	24.4	2690	224	2914	15.9
1941	18 721	4673	25.0	2626	192	2818	15.1

See Figure 5.1–9.

Note: Questionnaires were mailed to the 24 990 regular and junior members of the ACS in 1941. Of these, 19 009 returned usable responses. The 18 721 reporting an occupation in 1941 thus represent a 74.9% sample of ACS membership in 1941. See IB, Fraser, 1942a, 1289–1290.

Source: IB, Fraser, 1942c, 1565, Table 12. Percentages in source table contain many minor rounding errors, so all percentages have been recalculated.

TABLE 5.10.
Distribution of scientists in industry, by field, 1950–1970
(thousands)

Year	Physicial scientists			Mathematicians	Life scientists	Total
	Chemists	Physicists	Other			
	(1)	(2)	(3)	(4)	(5)	(6)
1950	36.6	5.8	16.6	8.4	8.5	76.0
1951	40.9	6.7	17.3	9.0	10.0	83.9
1952	46.4	7.7	19.0	10.2	10.9	94.2
1953	50.5	8.8	21.6	11.7	11.6	104.2
1954	53.9	9.6	22.8	12.6	12.1	111.0
1955	55.1	9.8	23.3	13.3	12.4	113.9
1956	58.3	10.4	24.7	14.3	13.2	120.9
1957	62.6	11.7	27.2	15.6	14.5	131.5
1958	66.3	12.2	27.6	16.2	15.7	138.1
1959	68.7	13.1	28.5	17.7	16.1	144.2
1960	72.1	13.2	28.4	18.9	16.8	149.4
1961	74.7	13.7	29.4	19.8	17.9	155.3
1962	77.7	14.7	29.3	22.2	18.5	162.2
1963	79.7	15.5	29.7	23.9	19.9	168.7
1964	83.4	16.4	29.5	25.5	20.2	175.0
1965	84.5	16.1	30.9	27.1	20.2	178.9
1966	87.3	17.1	32.3	29.4	22.0	188.1
1967	86.7	17.4	34.0	33.2	21.9	193.4
1968	89.5	17.9	35.3	34.9	22.9	200.5
1969	92.0	19.3	36.3	37.9	24.0	209.5
1970	93.5	19.8	37.0	38.7	24.1	213.1

See Figure 5.1–10.

Note: Detail may not add to total because of rounding.

Source: IA, BLS, 1973. Column 1 is from Table A–16, 50–51. Column 2 is from Table A–18, 54–55. Column 3 combines geologists and geophysicists (Table A–20, 58–59) and other physical scientists (Table A–22, 62–63). Column 4 is from Table A–24, 66–67. Column 5 combines agricultural scientists (Table A–26, 70–71), biological scientists (Table A–28, 74–75), and medical scientists (Table A–30, 78–79). Column 6 is from Table A–11, 40–41.

TABLE 5.11.
Percentage distribution of scientists in industry, by field, 1950–1970

Year	Physicial scientists			Mathematicians	Life scientists	Total[a]
	Chemists	Physicists	Other			
	(1)	(2)	(3)	(4)	(5)	(6)
1950	48.2	7.6	21.8	11.1	11.2	99.9
1951	48.7	8.0	20.6	10.7	11.9	99.9
1952	49.3	8.2	20.2	10.8	11.6	100.1
1953	48.5	8.4	20.7	11.2	11.1	99.9
1954	48.6	8.6	20.5	11.4	10.9	100.0
1955	48.4	8.6	20.5	11.7	10.9	100.1
1956	48.2	8.6	20.4	11.8	10.9	99.9
1957	47.6	8.9	20.7	11.9	11.0	100.1
1958	48.0	8.8	20.0	11.7	11.4	99.9
1959	47.6	9.1	19.8	12.3	11.2	100.0
1960	48.3	8.8	19.0	12.7	11.2	100.0
1961	48.1	8.8	18.9	12.7	11.5	100.0
1962	47.9	9.1	18.1	13.7	11.4	100.2
1963	47.2	9.2	17.6	14.2	11.8	100.0
1964	47.7	9.4	16.9	14.6	11.5	100.1
1965	47.2	9.0	17.3	15.1	11.3	99.9
1966	46.4	9.1	17.2	15.6	11.7	100.0
1967	44.8	9.0	17.6	17.2	11.3	99.9
1968	44.6	8.9	17.6	17.4	11.4	99.9
1969	43.9	9.2	17.3	18.1	11.5	100.0
1970	43.9	9.3	17.4	18.2	11.3	100.1

[a] May not add to 100% because of rounding errors.

Source: Table 5.10.

TABLE 5.12.
Leading private industrial employers of chemists, 1950–1970

Year and industrial group[a]	Chemists employed	
	Number (thousands)	As percentage of chemists in private industry[b]
	(1)	(2)
1950		
Chemicals and allied products	13.4	36.6
Food and kindred products	2.9	7.9
Miscellaneous business services	2.8	7.7
Petroleum refining	2.3	6.3
Primary metals	2.0	5.5
Other private industry	13.2	36.1
Total	36.6	100.1
1955		
Chemicals and allied products	23.7	43.0
Food and kindred products	3.6	6.5
Miscellaneous business services	3.6	6.5
Petroleum refining	3.5	6.4
Primary metals	2.3	4.2
Other private industry	18.4	33.4
Total	55.1	100.0
1960		
Chemicals and allied products	32.0	44.4
Miscellaneous business services	5.6	7.8
Food and kindred products	4.4	6.1
Petroleum refining	3.9	5.4
Paper and allied products	2.8	3.9
Other private industry	23.4	32.5
Total	72.1	100.1

Table 5.12. (continued)

Year and industrial group[a]	Chemists employed: Number (thousands) (1)	Chemists employed: As percentage of chemists in private industry[b] (2)
1965		
Chemicals and allied products	37.8	44.7
Miscellaneous business services	7.3	8.6
Food and kindred products	4.5	5.3
Other nonmanufacturing	3.9	4.6
Petroleum refining	3.4	4.0
Other private industry	27.6	32.7
Total	84.5	99.9
1970		
Chemicals and allied products	42.8	45.8
Miscellaneous business services	8.6	9.2
Food and kindred products	4.5	4.8
Other nonmanufacturing	4.5	4.8
Paper and allied products	3.6	3.9
Other private industry	29.5	31.6
Total	93.5	100.1

[a] Most of the industrial groups listed here are among the chemical process industries; see Appendix B. "Miscellaneous business services" includes, among other things, research and development laboratories and commercial testing laboratories. "Other nonmanufacturing" as used here includes: agriculture, forestry, and fisheries; wholesale and retail sales; finance, insurance, etc.; and other services. See the source and II, OMB, 1972, for details.

[b] May not add to 100% for each year because of rounding errors.

Source: IA, BLS, 1973, Table A–16, 50–51.

TABLE 5.13.
Civilian chemists in the federal government, 1879–1978
(thousands)

Year	Observed (1)	Calculated trend (logistic) 1911–1978 (2)
1879	0.002	–
1883	0.013	–
1887	0.014	–
1891	0.022	–
1897	0.018	–
1901	0.022	–
1905	0.032	–
1911	0.292	0.224
1914	0.474[a]	0.289
1916	0.716	0.343
1925	0.805	0.721
1928	0.882[b]	0.916
1931	0.958	1.158
1938	1.293	1.940
1947	2.9[c]	3.448
1950	3.2	4.057
1951	4.346	4.268
1952	4.7	4.481
1953	5.0	4.697
1954	4.671	4.913
1955	4.9	5.129
1956	5.2	5.345
1957	5.243	5.559
1958	5.606	5.772
1959	5.730	5.981
1960	5.845	6.186
1961	6.211	6.388
1962	6.789	6.584
1963	7.1	6.775
1964	7.716	6.960
1965	8.2	7.138
1966	8.135	7.310
1967	8.302	7.475
1968	8.474	7.633
1969	8.248	7.784

Table 5.13. (continued)

Year	Observed (1)	Calculated trend (logistic) 1911–1978 (2)
1970	8.138	7.928
1971	8.313	8.064
1972	8.377	8.193
1973	7.966	8.315
1974	7.799	8.430
1975	7.945	8.538
1976	8.080	8.640
1977	8.025	8.735
1978	8.256	8.825

See Figures 5.2–1 and 5.2–2.

Note: Calculated trends, while estimated to nearest person, are probably only accurate to nearest hundred or so. See Appendixes D and E.

[a] Excludes 30 "junior aides in chemistry" with annual salaries ranging between $720 and $1199. The qualifications of people in this capacity (viz., high school chemistry, high school physics, high school diploma, "accuracy and carefulness") are insufficient for their inclusion in our totals (IB, Kebler, 1920, 1207 and 1211).
[b] Approximate only. The Navy Department is listed in the source as employing "about 60 chemists" (IA, USCSC, 1929, 21; see also the disclaimer in italics on p. 15).
[c] The source gives exactly 2855 chemists, but the accompanying explanatory text emphasizes that the figures are "rough estimates".

Sources: *Column 1*:

1879 – IA, ORUS, 1879. Our tabulations, counting only persons designated as "chemist" in Volume I.
1883 – IA, ORUS, 1883. See 1879 for selection criterion.
1887 – IA, ORUS, 1887. See 1879 for selection criterion.
1891 – IA, ORUS, 1892. See 1879 for selection criterion.
1897 – IA, ORUS, 1897. See 1879 for selection criterion.
1901 – IA, ORUS, 1901. See 1879 for selection criterion.
1905 – IA, ORUS, 1905. See 1879 for selection criterion.
1911 – IA, ORUS, 1911. See 1879 for selection criterion.
1914 – IB, Kebler, 1920, 1209, Table IV.
1916 – IB, Breithut, 1917, 66, Table 2.
1925 – IA, USCSC, 1926, 9–15. Our tabulations. Source mentions that its data are "not intended to be exhaustive" (p. 9).

Table 5.13. (continued)

1928 – IA, USCSC, 1929, 5, 8, 10, 12, 14, 16–25. Our tabulations.
1931 – IB, Baruch, 1932, 60. Our tabulations. Baruch's totals for chemists in field centers date from 1928 – a minor distortion (p. 52).
1938 – IA, Smith, 1941, 83, Table 8. Smith gives 1455 chemists and metallurgists. Since the latter constituted 11.1% of civilian chemists and metallurgists in the federal government in 1931 (see IB, Baruch, 1932, 60–62), an identical percentage has been deducted from the combined total for 1938.
1947 – IA, BLS, 1951b, Table 10.
1950, 1952–1953, 1955–1956, 1963, and 1965 – IA, BLS, 1973, Table A–16, 50–51.
1951 – IA, BLS, 1951a, Table B, 20–21.
1954 – IA, NSF, 1957b, 75, Table III–5. See Appendix F of this source for a discussion of the limitations of the data from 1931 to 1954.
1957 and 1958 – IA, NSF, 1961c, Table A, 17. The total of 5171 for 1957 in IA, USCSC, 1958, Table B, 46, appears to be preliminary.
1959 and 1960 – IA, NSF, 1962, Table B, 34 and 38.
1961 and 1962 – IA, NSF, 1965, Table A–2, 32, and Table A–1, 28.
1964 – IA, NSF, 1967, Appendix Table B–1, 20.
1966 – IA, NSF, 1968, Table A, 14.
1967 – IA, NSF, 1969, Table B–1, 32.
1968 – IA, NSF, 1970, Table B–1, 26.
1969 – IA, NSF, 1971b, Table B–1, 30.
1970 – IA, USCSC, 1972, Table E, 194–195.
1971 – IA, NSF, 1973, Table A, 6.
1972 – IA, USCSC, 1975, Table E, 224–225.
1973 – IA, NSF, 1976a, Table A, 14.
1974 and 1975 – IA, USCSC, 1977, Table E–1, 1974, 198–199, and Table E–1, 1975, 90–91.
1976 – IA, USCSC, 1978, Table A–2, 27, and Table E–1, 110–111.
1977 – IA, USOPM, 1979, Table A–2, 38, and Table E–1, 124–125.
1978 – IA, USOPM, 1980, Table A–2, 40, and Table E–1, 126–127.

Column 2 – calculated by regression on column 1; see Appendix E.

TABLE 5.14.
Distribution of civilian chemists in the federal government, by selected department, 1901–1978

Year	All agencies	Department of Agriculture		Department of Defense[a]		Department of Health, Education, and Welfare[b]	
		Number	As percentage of all agencies	Number	As percentage of all agencies	Number	As percentage of all agencies
	(1)	(2)	(3)	(4)	(5)	(6)	(7)
1901	22	12	55	4	18	–	–
1905	32	15	47	8	25	–	–
1911	292	204	70	20	7	–	–
1916	716	397	55	93	13	–	–
1925	805	360	45	160	20	–	–
1928	882	421[c]	48	156[d]	18	–	–
1951	4346	760	17	1893	44	465	11
1954	4671	725	16	1895	41	549	12
1957	5243	829	16	2037	39	762	15
1958	5606	911	16	2091	37	930	17
1959	5730	934	16	2067	36	967	17
1960	5845	935	16	2067	35	1008	17
1961	6211	985	16	2156	35	1197	19
1962	6789	1060	16	2228	33	1440	21
1964	7716	1110	14	2471	32	1882	24
1966	8135	1204	15	2642	32	1869	23
1967	8302	1255	15	2755	33	1812	22

Table 5.14. (continued)

Year	All agencies	Department of Agriculture		Department of Defense[a]		Department of Health, Education, and Welfare[b]	
		Number	As percentage of all agencies	Number	As percentage of all agencies	Number	As percentage of all agencies
	(1)	(2)	(3)	(4)	(5)	(6)	(7)
1968	8474	1256	15	2781	33	1868	22
1969	8248	1254	15	2766	34	1755	21
1970	8138	1232	15	2542	31	1864	23
1971	8313	1205	14	2489	30	1681	20
1972	8377	1192	14	2450	29	1731	21
1973	7966	1127	14	2376	30	1610	20
1974	7799	1082	14	2187	28	1627	21
1975	7945	1064	13	2212	28	1692	21
1976	8080	1073	13	2078	26	1916	24
1977	8025	1064	13	1973	25	1905	24
1978	8256	1055	13	1939	23	2098	25

See Figure 5.2–3.

[a] Includes predecessors of the Department of Defense, viz., Departments of War, Army, Air Force, and Navy.
[b] Includes predecessor of the Department of Health, Education, and Welfare, viz., the Federal Security Agency.
[c] Approximate only. The source states that the "number of employees in the Bureau of Plant Industry classed as chemists fails to give an accurate picture of the number of men with chemical training employed" (p. 15).
[d] Approximate only. The Navy Department is listed in the source as employing "about 60 chemists" (p. 21).

Sources: see sources for Table 5.13, column 1. For 1957 and 1958, see IA, NSF, 1961c, Table B, 20–21.

TABLE 5.15.
Chemists in state and local governments, 1950–1970
(hundreds)

Year	State government (1)	Local government (2)
1950	8	4
1951	8	5
1952	9	5
1953	9	6
1954	9	7
1955	9	7
1956	10	8
1957	11	9
1958	11	9
1959	12	10
1960	12	10
1961	13	11
1962	14	11
1963	14	12
1964	14	12
1965	15	12
1966	16	13
1967	17	14
1968	18	15
1969	18	15
1970	19	16

See Figure 5.2–4.

Note: See Appendix E for the linear and exponential trends calculated for these series.

Source: IA, BLS, 1973, Table A–16, 50–51.

TABLE 5.16.
Faculty in higher education, 1870–1978

	College and university faculty			Chemists in higher education	
		Chemistry			
Year	Total	Total	Graduate departments only	Teachers (Census estimates)	Total (BLS estimates, thousands)
	(1)	(2)	(3)	(4)	(5)
1870	5 553[a]	–	–	–	–
1876	–	113	–	–	–
1880	11 522[a]	–	–	–	–
1890	15 809[a]	–	–	–	–
1900	23 868	–	–	–	–
1901	–	569	–	–	–
1910	36 480	–	–	–	–
1920	48 615	–	–	–	–
1926	–	1107[b]	–	–	–
1927	–	1003[b]	–	–	–
1928	–	1089[b]	–	–	–
1929	–	1291[b]	–	–	–
1930	82 386	1404[b]	–	–	–
1931	–	1392[b]	–	–	–
1932	100 789	1486[b]	–	–	–
1933	–	1432[b]	–	–	–
1934	108 873	1534[b]	–	–	–
1935	–	1607[b]	–	–	–
1936	121 036	–	–	–	–
1938	135 989	–	–	–	–
1940	146 929	–	–	–	–
1942	151 066	–	–	–	–
1944	150 980	–	–	–	–
1946	165 324	–	–	–	–
1948	223 660	–	–	–	–
1950	246 722	–	–	–	10.2
1951	–	–	–	–	9.9
1952	244 488	–	–	–	9.5
1953	–	–	–	–	9.8
1954	265 911	–	–	–	10.1

Table 5.16. (continued)

	College and university faculty			Chemists in higher education	
		Chemistry			
Year	Total	Total	Graduate departments only	Teachers (Census estimates)	Total (BLS estimates, thousands)
	(1)	(2)	(3)	(4)	(5)
1955	–	–	1614	–	11.1
1956	298 910	–	–	–	12.6
1957	–	–	–	–	13.2
1958	344 525	–	–	–	15.4
1959	–	–	–	–	17.3
1960	380 554	–	–	8 551	18.1
1961	–	–	2157	–	18.4
1962	424 862	–	–	–	18.6
1963	–	5238[c]	2516	–	19.0
1964	494 514	–	–	–	19.7
1965	–	–	2989	–	19.7
1966	596 400	–	–	–	19.3
1967	–	–	3347	–	22.8
1968	674 000[d]	–	–	–	23.9
1969	–	–	3757	–	25.1
1970	825 000[d]	–	–	15 720	25.3
1972	–	–	4255	–	–
1973	881 665[e]	–	–	–	–
1974	–	–	4087	–	–
1976	–	–	4108	–	–
1978	–	–	3907	–	–

See Figures 5.3–1, A.2–1, A.2–2, and A.2–3.

Note: See Appendix E for the exponential trends calculated for these series. See Tables 5.17 and 5.18 for the calculated values of column 2 in selected years.

[a] Estimated.
[b] Includes only faculty engaged at least part time in research.
[c] Includes only faculty engaged at least part time in teaching.
[d] Estimated, apparently to nearest thousand.
[e] Fall 1972.

Table 5.16. (*continued*)

Sources: *Column 1*:

Census years and 1973 – IA, NCES, 1978a, Table 97, 94. Other years – IA, USBC, 1975b, 382–383, Series H696. This source gives erroneous or apparently preliminary totals for 1880 and 1970.

Column 2:

1876 and 1901 – IB, Baskerville *et al.*, 1902, 99–137.
1926 – IB, West and Hull, 1927b, 909, Table II.
1927 – IB, West and Hull, 1928, 883, Table III. The total of 936 faculty given in Tables II and III seems erroneous; while summation of the faculty column in Table II yields 936, summation of the same columns in Tables I and III yields 1003.
1928 – IB, West and Hull, 1929, 1338, Table II.
1929 – IB, West and Hull, 1930, 1674, Table II.
1930 – IB, West and Hull, 1931b, 1374, Table II.
1931 – IB, Hull and West, 1932, 1472, Table II.
1932 – IB, Hull and West, 1933, 499, Table II.
1933 – IB, Hull and West, 1934, 471, Table 2, plus correction on 542.
1934 – IB, Hull and West, 1935, 339, Table 2.
1935 – IB, Hull and West, 1936, 339, Table 2.
1963 – IA, Dunham *et al.*, 1966, 77, Table 8.

Column 3:

1955 – IB, ACS-CPT, 1955, 429–437. Our tabulations from index.
1961 – IB, ACS-CPT, 1961, x–xi.
1963 – IB, ACS-CPT, 1963, xii.
1965 – IB, ACS-CPT, 1965, xii.
1967 – IB, ACS-CPT, 1967, xviii.
1969 – IB, ACS-CPT, 1969, xvi.
1972 – IB, ACS-CPT, 1974, xviii.
1974 – IB, ACS-CPT, 1975, xiv.
1976 – IB, ACS-CPT, 1977b, xiv.
1978 – IB, ACS-CPT, 1979, xiv.

Column 4 – IA, USBC-COP, 1973, Table 221, 718.

Column 5 – IA, BLS, 1973, Table A–16, 50–51.

TABLE 5.17.
Chemistry faculty as percentage of total faculty, 1870–1970

Year	Total	Chemistry	
		Number (estimated)	As percentage of total
	(1)	(2)	(3)
1870	5 553[a]	107	1.9
1880	11 522[a]	162	1.4
1890	15 809[a]	246	1.6
1900	23 868	375	1.6
1910	36 480	569	1.6
1920	48 615	866	1.8
1930	82 386	1316	1.6
1940	146 929	2001	1.4
1950	246 722	3041	1.2
1960	380 554	4623	1.2
1970	825 000[b]	7028	0.9

See Figures 5.3–2, A.2–1, A.2–2, and A.2–3.

[a] Estimated.
[b] Estimated, apparently to nearest thousand.

Sources: *Column 1* – IA, NCES 1978a, Table 97, 94.

Column 2 – calculated by regression on Table 5.16, column 2; see Appendix E. Exponential growth assumed.

TABLE 5.18.
Chemistry faculty versus American Chemical Society membership, 1876–1979

Year	Chemistry faculty (estimated) (1)	ACS members (2)	Chemistry faculty per ACS member (3)
1876	137	230	0.60
1877	143	265	0.54
1878	149	256	0.58
1879	155	289	0.54
1880	162	303	0.53
1881	169	314	0.54
1882	176	293	0.60
1883	184	306	0.60
1884	192	323	0.59
1885	200	255	0.78
1886	208	241	0.86
1887	217	235	0.92
1888	227	227	1.00
1889	236	204	1.16
1890	246	238	1.03
1891	257	302	0.85
1892	268	351	0.76
1893	279	460	0.61
1894	291	722	0.40
1895	304	903	0.34
1896	317	1 011	0.31
1897	330	1 156	0.29
1898	344	1 415	0.24
1899	359	1 569	0.23
1900	375	1 715	0.22
1901	391	1 933	0.20
1902	407	2 188	0.19
1903	425	2 428	0.18
1904	443	2 675	0.17
1905	462	2 919	0.16
1906	482	3 079	0.16
1907	502	3 389	0.15
1908	524	4 004	0.13
1909	546	4 502	0.12
1910	569	5 081	0.11
1911	594	5 603	0.11
1912	619	6 219	0.10

Table 5.18. (*continued*)

Year	Chemistry faculty (estimated) (1)	ACS members (2)	Chemistry faculty per ACS member (3)
1913	646	6 673	0.10
1914	673	7 170	0.09
1915	702	7 417	0.09
1916	732	8 355	0.09
1917	763	10 603	0.07
1918	796	12 203	0.07
1919	830	13 686	0.06
1920	866	15 582	0.06
1921	903	14 318	0.06
1922	941	14 400	0.07
1923	982	14 346	0.07
1924	1 024	14 515	0.07
1925	1 067	14 381	0.07
1926	1 113	14 704	0.08
1927	1 161	15 188	0.08
1928	1 210	16 240	0.07
1929	1 262	17 426	0.07
1930	1 316	18 206	0.07
1931	1 372	18 963	0.07
1932	1 431	18 572	0.08
1933	1 492	17 465	0.09
1934	1 556	17 561	0.09
1935	1 623	17 541	0.09
1936	1 692	18 727	0.09
1937	1 764	20 677	0.09
1938	1 840	22 185	0.08
1939	1 918	23 519	0.08
1940	2 001	25 414	0.08
1941	2 086	18 738	0.07
1942	2 175	31 717	0.07
1943	2 268	36 001	0.06
1944	2 365	39 438	0.06
1945	2 467	43 075	0.06
1946	2 572	48 755	0.05
1947	2 682	55 100	0.05
1948	2 797	58 782	0.05
1949	2 916	62 211	0.05
1950	3 041	63 349	0.05

Table 5.18. (continued)

Year	Chemistry faculty (estimated) (1)	ACS members (2)	Chemistry faculty per ACS member (3)
1951	3 171	66 009	0.05
1952	3 307	67 730	0.05
1953	3 448	70 155	0.05
1954	3 596	72 287	0.05
1955	3 750	75 223	0.05
1956	3 910	79 224	0.05
1957	4 077	81 927	0.05
1958	4 252	85 815	0.05
1959	4 434	88 806	0.05
1960	4 623	92 193	0.05
1961	4 821	93 637	0.05
1962	5 027	95 210	0.05
1963	5 242	96 749	0.05
1964	5 467	99 475	0.05
1965	5 700	102 525	0.06
1966	5 944	106 271	0.06
1967	6 199	109 528	0.06
1968	6 464	113 373	0.06
1969	6 740	116 816	0.06
1970	7 028	114 323	0.06
1971	7 329	112 016	0.07
1972	7 643	110 708	0.07
1973	7 970	110 285	0.07
1974	8 310	110 799	0.08
1975	8 666	110 820	0.08
1976	9 037	112 730	0.08
1977	9 423	115 141	0.08
1978	9 826	116 240	0.08
1979	10 246	118 214	0.09

See Figure 5.3–3.

Sources: *Column 1* – calculated by regression on Table 5.16, column 2; see Appendix E. Exponential growth assumed.
Column 2 – Table 2.4, column 1.

TABLE 5.19.
Comparison of Census estimates for chemists with disaggregated data on employment of chemists from other sources, 1920 and 1970
(thousands)

Item	1920 (1)	1970 (2)
Census estimate		
Chemists in the labor force	28	110
Disaggregated data		
Chemists in industry	13.7[a]	93.5
Chemists in government		
Federal	0.7	8.9[b]
State and local	1.7[a]	3.5
Total		
Number	16.1	105.9
As percentage of Census estimate	57.5%	96.3%
Remainder		
Number	11.9	4.1
As percentage of Census estimate	42.5%	3.7%

Note: All numbers approximate, intended to convey only a sense of the terrain. Academic chemists and persons in chemistry-related fields (e.g., chemical engineering) are excluded from all estimates shown.

[a] Estimate is for 1917.
[b] This Bureau of Labor Statistics estimate is higher than the U.S. Civil Service Commission total (Table 5.13). The higher figure is employed to make the best possible case for the disaggregated data.

Sources: *Census estimates* – Table 2.2, column 1.

Disaggregated data, 1920:

Industry – Table 5.1, column 5.
Federal government – Table 5.13, column 1.
State and local government – arithmetic mean of the estimates in IB, Wiley, 1917, 83, and IB, Breithut, 1917, 78. See p. 136.

Disaggregated data, 1970 – IA, BLS, 1973, Table A–16, 51. Cf. note b above.

TABLE 6.1.
American doctoral dissertations, by field of study and by decade, 1861–1899

Field	Period					
					1861–1899	
	1860s	1870s	1880s	1890s	Number	As percentage of total
	(1)	(2)	(3)	(4)	(5)	(6)
I. Humanities	5	55	220	1084	1365	59.9
A. Communications and the arts	–	–	1	2	3	0.1
1. Fine arts	–	–	1	1	2	0.1
2. Theater	–	–	–	1	1	0.1
B. Education	–	–	4	75	79	3.5
1. General	–	–	2	22	24	1.1
2. Administration	–	–	–	1	1	0.0
3. Curriculum and instruction	–	–	–	7	7	0.3
4. History	–	–	–	8	8	0.4
5. Industrial	–	–	–	1	1	0.0
6. Music	–	–	–	1	1	0.0
7. Philosophy	–	–	–	5	5	0.2
8. Physical	–	–	–	1	1	0.0
9. Preschool	–	–	–	1	1	0.0
10. Psychology	–	–	–	6	6	0.3
11. Religion	–	–	1	1	2	0.1
12. Sciences	–	–	–	5	5	0.2

Table 6.1. (continued)

Field	Period					
					1861–1899	
	1860s (1)	1870s (2)	1880s (3)	1890s (4)	Number (5)	As percentage of total (6)
13. Secondary	–	–	–	2	2	0.1
14. Special	–	–	–	2	2	0.1
15. Teacher training	–	–	–	5	5	0.2
16. Theory and practice	–	–	1	5	6	0.3
17. Vocational	–	–	–	2	2	0.1
C. Language, literature, and linguistics	4	36	92	442	574	25.2
1. Language and literature	4	33	91	426	554	24.3
a. General	4	28	56	303	391	17.2
b. Classical	–	5	31	101	137	6.0
c. Medieval	–	–	1	5	6	0.3
d. Modern	–	–	–	12	12	0.5
e. Linguistics	–	–	3	5	8	0.4
2. Literature	–	–	–	11	11	0.5
a. Anglo-Saxon	–	–	–	5	5	0.2
b. English	–	–	–	5	5	0.2
c. Religious	–	–	–	1	1	0.0
3. Oriental literature and linguistics	–	1	–	2	3	0.1
4. Philology	–	2	1	3	6	0.3

Table 6.1. (continued)

Field	Period					
					1861–1899	
	1860s	1870s	1880s	1890s	Number	As percentage of total
	(1)	(2)	(3)	(4)	(5)	(6)
D. Philosophy, religion, and theology	2	4	61	215	282	12.4
1. Philosophy	2	4	38	157	201	8.8
2. Religion	–	–	17	44	61	2.7
3. Theology	–	–	6	14	20	0.9
E. Social sciences	–	15	62	350	427	18.8
1. Anthropology	–	–	–	2	2	0.1
2. Biography	–	–	–	3	3	0.1
3. Business administration	–	–	1	2	3	0.1
4. Economics	–	2	7	74	83	3.6
a. General	–	2	7	70	79	3.5
b. Commerce-business	–	–	–	1	1	0.0
c. History	–	–	–	1	1	0.0
d. Theory	–	–	–	2	2	0.1
5. Folklore	–	–	–	1	1	0.0
6. History	–	8	43	189	240	10.5
a. General	–	8	41	172	221	9.7
b. Archaeology	–	–	1	3	4	0.2
c. Ancient	–	–	–	5	5	0.2

Table 6.1. (continued)

Field	Period					
					1861–1899	
	1860s	1870s	1880s	1890s	Number	As percentage of total
	(1)	(2)	(3)	(4)	(5)	(6)
d. Medieval	–	–	–	1	1	0.0
e. American	–	–	1	8	9	0.4
7. Law	–	–	5	15	20	0.9
8. Political science	–	5	4	39	48	2.1
a. General	–	5	4	38	47	2.1
b. Public administration	–	–	–	1	1	0.0
9. Sociology	–	–	2	25	27	1.2
a. General	–	–	1	22	23	1.0
b. Demography	–	–	–	1	1	0.0
c. Labor relations	–	–	–	1	1	0.0
d. Socialism, communism, anarchism	–	–	1	1	2	0.1
II. Sciences	10	54	163	675	902	39.6
A. Biological sciences	1	12	32	174	219	9.6
1. Agriculture	–	–	4	1	5	0.2
a. General	–	–	–	1	1	0.0
b. Agronomy	–	–	2	–	2	0.1
c. Plant pathology	–	–	1	–	1	0.0
d. Plant physiology	–	–	1	–	1	0.0

Table 6.1. (continued)

Field	Period					
					1861–1899	
	1860s	1870s	1880s	1890s	Number	As percentage of total
	(1)	(2)	(3)	(4)	(5)	(6)
2. Anatomy	–	–	–	3	3	0.1
3. Bacteriology and microbiology[a]	–	–	–	4	4	0.2
4. Biology	–	–	1	16	17	0.7
5. Botany	–	2	4	46	52	2.3
6. Physiology	–	1	12	20	33	1.4
7. Veterinary science	–	1	–	8	9	0.4
8. Zoology	1	8	11	76	96	4.2
B. Earth sciences	2	11	19	70	102	4.5
1. Geography	–	–	1	2	3	0.1
2. Geology	–	11	12	60	83	3.6
3. Geophysics	–	–	–	1	1	0.0
4. Mineralogy	–	–	2	1	3	0.1
5. Paleontology	1	–	–	4	5	0.2
6. Paleozoology	1	–	4	2	7	0.3
C. Health sciences	–	1	1	3	5	0.2
1. General	–	–	1	3	4	0.2
2. Public health	–	1	–	–	1	0.0

Table 6.1. (continued)

Field	Period					
					1861–1899	
	1860s	1870s	1880s	1890s	Number	As percentage of total
	(1)	(2)	(3)	(4)	(5)	(6)
D. Physical sciences	7	30	108	366	511	22.4
1. Pure sciences	7	27	98	354	486	21.3
a. Astronomy	–	1	6	15	22	1.0
b. Chemistry	2	12	46	191	251	11.0
1. General	2	12	45	184	243	10.7
2. Agricultural and biological	–	–	–	1	1	0.0
3. Biological-biochemistry [b]	–	–	1	6	7	0.3
c. Mathematics	2	8	26	75	111	4.9
d. Physics	3	6	20	73	102	4.5
1. General	3	6	19	71	99	4.3
2. Atmospheric science	–	–	1	1	2	0.1
3. Spectroscopy	–	–	–	1	1	0.0
2. Applied sciences	–	3	10	12	25	1.1
a. Engineering	–	3	10	12	25	1.1
1. General	–	–	8	6	14	0.6
2. Civil	–	–	–	1	1	0.0
3. Electronics and electrical	–	–	–	1	1	0.0
4. Metallurgy	–	1	1	3	5	0.2
5. Mining	–	2	–	1	3	0.1
6. Sanitary and municipal	–	–	1	–	1	0.0

Table 6.1. (continued)

Field	Period					
					1861–1899	
	1860s	1870s	1880s	1890s	Number	As percentage of total
	(1)	(2)	(3)	(4)	(5)	(6)
E. Psychology	–	–	3	62	65	2.9
1. General	–	–	3	60	63	2.8
2. Physiological	–	–	–	2	2	0.1
III. Field unknown	–	1	3	6	10	0.4
Total[c]	16	110	386	1765	2277	....

See Figure 6.1–1.

Note: A dash ("–") indicates "zero", not "unknown".

[a] Contains one in "microbiology" alone, and one in "bacteriology" alone.
[b] One dissertation in the 1880s is in "biochemistry"; one in the 1890s is in "chemistry, biological", while the other five are in "biochemistry".
[c] Indicates total number of dissertations. Columns do not add to these sums because the row entries are hierarchical; summation over any one level of the hierarchy (e.g., adding only the estimates for "Humanities", "Sciences", and "Field unknown") will yield totals.

Source: IB, Xerox University Microfilms, 1973+. The machine-readable version of this data base was searched through the Online Search Services of Van Pelt Library, University of Pennsylvania, using the data files of Bibliographic Retrieval Services, Inc., Schenectady, New York.

TABLE 6.2.
American doctoral dissertations in chemistry, by granting institution and by decade, 1861–1899

Granting institution	Period					
					1861–1899	
	1860s (1)	1870s (2)	1880s (3)	1890s (4)	Number (5)	As percentage of total[a] (6)
Johns Hopkins University	–	1	24	75	100	39.8
Yale University[b]	2	7	2	25	36	14.3
Harvard University	–	1	8	19	28	11.2
University of Pennsylvania	–	–	–	26	26	10.4
Columbia University[c]	–	–	6	11	17	6.8
University of Chicago	–	–	–	13	13	5.2
Cornell University	–	–	–	8	8	3.2
University of Michigan[d]	–	1	2	4	7	2.8
Clark University	–	–	–	3	3	1.2
George Washington University	–	–	–	2	2	0.8
Stevens Institute of Technology	–	1	1	–	2	0.8
Vanderbilt University	–	1	1	–	2	0.8
Other[e]	–	–	2	5	7	2.8
Total	2	12	46	191	251	100.1

See Figure 6.1–2.

Table 6.2. (continued)

Note: A dash ("–") indicates "zero", not "unknown". With exceptions noted below, all dissertations are listed under "chemistry, general".

[a] Does not add to 100% because of rounding errors.
[b] Includes one "biochemistry" each in 1880, 1893, 1894, and 1897, and two in 1898.
[c] Includes one "chemistry, agricultural and biological" in 1894.
[d] Includes one "chemistry, biological" in 1898.
[e] Includes one dissertation each from Boston University, University of California, Lehigh University, Northwestern University, Ohio State University, University of Virginia, and University of Wisconsin.

Source: IB, Xerox University Microfilms, 1973+. The machine-readable version of this data base was searched through the Online Search Services of Van Pelt Library, University of Pennsylvania, using the data files of Lockheed Information Systems, Palo Alto, California, and Bibliographic Retrieval Services, Inc., Schenectady, New York.

TABLE 6.3.

American doctoral dissertations in chemistry, by granting institution, 1976

Granting institution[a]	Doctorates granted		
		As percentage of total	
	Number	Individual[b]	Cumulative[c]
	(1)	(2)	(3)
1. University of Illinois, Urbana	58	3.5	3.5
2. University of California, Berkeley	50	3.0	6.6
3. University of Wisconsin, Madison	41	2.5	9.1
4. Massachusetts Institute of Technology	35	2.1	11.2
5. Purdue University	32	1.9	13.1
Total, top five	216	13.0	–
6. Harvard University	31	1.9	15.0
7. Ohio State University	30	1.8	16.8
8. University of California, Los Angeles	25	1.5	18.4
9. University of Colorado	25	1.5	19.9
10. University of Maryland	25	1.5	21.4
11. Michigan State University	25	1.5	22.9
12. California Institute of Technology	24	1.5	24.4
13. Cornell University	22	1.3	25.7
14. University of Georgia	22	1.3	27.1
15. University of Texas, Austin	22	1.3	28.4

Table 6.3. (continued)

	Doctorates granted		
		As percentage of total	
Granting institution[a]	Number	Individual[b]	Cumulative[c]
	(1)	(2)	(3)
16. Indiana University	21	1.3	29.7
17. Iowa State University	21	1.3	31.0
18. University of Minnesota	21	1.3	32.2
19. University of Florida	20	1.2	33.5
20. Northwestern University	20	1.2	34.7
21. Wayne State University	20	1.2	35.9
22. Columbia University	19	1.2	37.0
23. University of North Carolina, Chapel Hill	19	1.2	38.2
24. Stanford University	19	1.2	39.4
25. University of California, San Diego	18	1.1	40.5
26. University of Cincinnati	18	1.1	41.5
27. New York University	18	1.1	42.6
28. Texas A&M University	18	1.1	43.7
29. University of Iowa	17	1.0	44.8
30. Pennsylvania State University	17	1.0	45.8
31. Yale University	17	1.0	46.8
32. Case Western Reserve University	16	1.0	47.8

Table 6.3. (continued)

	Doctorates granted		
		As percentage of total	
Granting institution[a]	Number	Individual[b]	Cumulative[c]
	(1)	(2)	(3)
33. University of Michigan	16	1.0	48.8
34. University of Pittsburgh	16	1.0	49.8
35. City University of New York	15	0.9	50.7
36. University of Pennsylvania	15	0.9	51.6
37. Princeton University	15	0.9	52.5
38. Duke University	14	0.9	53.3
39. University of Washington	14	0.9	54.2
40. Boston University	13	0.8	55.0
41. University of Houston	13	0.8	55.8
42. Louisiana State University, Baton Rouge	13	0.8	56.6
43. University of Nebraska	13	0.8	57.4
44. Oregon State University	13	0.8	58.2
45. Southern Illinois University, Carbondale	13	0.8	58.9
46. University of Arizona	12	0.7	59.7
47. Brown University	12	0.7	60.4
48. University of California, Santa Barbara	12	0.7	61.1
49. University of Chicago	12	0.7	61.9

Table 6.3. (continued)

Granting institution[a]	Doctorates granted		
		As percentage of total	
	Number	Individual[b]	Cumulative[c]
	(1)	(2)	(3)
50. University of Oregon	12	0.7	62.6
51. State University of New York, Buffalo	12	0.7	63.3
52. University of Alabama	11	0.7	64.0
53. Carnegie-Mellon University	11	0.7	64.7
54. Emory University	11	0.7	65.3
55. Howard University	11	0.7	66.0
56. University of Massachusetts	11	0.7	66.7
57. Rensselaer Polytechnic Institute	11	0.7	67.3
58. Rutgers University, New Brunswick	11	0.7	68.0
59. Virginia Polytechnic Institute	11	0.7	68.7
60. Florida State University	10	0.6	69.3
61. Georgia Institute of Technology	10	0.6	69.9
62. University of South Carolina	10	0.6	70.5
Total, sixth through sixty-second	943	57.4	–
63. Johns Hopkins University	9	0.5	71.0
64. University of Oklahoma	9	0.5	71.6
65. Polytechnic Institute of New York	9	0.5	72.1

Table 6.3. (continued)

Granting institution[a]	Doctorates granted		
		As percentage of total	
	Number (1)	Individual[b] (2)	Cumulative[c] (3)
66. Arizona State University	8	0.5	72.6
67. Brandeis University	8	0.5	73.1
68. University of California, Davis	8	0.5	73.6
69. University of Idaho	8	0.5	74.1
70. Institute of Paper Chemistry	8	0.5	74.6
71. Oklahoma State University	8	0.5	75.1
72. University of Rochester	8	0.5	75.5
73. Rutgers University, Newark	8	0.5	76.0
74. Temple University	8	0.5	76.5
75. Auburn University	7	0.4	76.9
76. University of California, Irvine	7	0.4	77.4
77. University of California, Riverside	7	0.4	77.8
78. Clarkson College of Technology	7	0.4	78.2
79. University of Illinois, Chicago Circle	7	0.4	78.6
80. University of Kansas	7	0.4	79.1
81. Kent State University	7	0.4	79.5
82. University of New Mexico	7	0.4	79.9
83. University of Notre Dame	7	0.4	80.4

Table 6.3. (continued)

Granting institution[a]	Doctorates granted	As percentage of total	
	Number	Individual[b]	Cumulative[c]
	(1)	(2)	(3)
84. University of Southern Mississippi	7	0.4	80.8
85. State University of New York, Albany	7	0.4	81.2
86. University of Tennessee	7	0.4	81.6
87. Texas Tech University	7	0.4	82.1
88. University of Utah	7	0.4	82.5
89. Washington University	7	0.4	82.9
90. University of Arkansas	6	0.4	83.3
91. Catholic University of America	6	0.4	83.6
92. University of Delaware	6	0.4	84.0
93. Fordham University	6	0.4	84.4
94. Georgetown University	6	0.4	84.7
95. North Texas State University	6	0.4	85.1
96. Rice University	6	0.4	85.5
97. University of Southern California	6	0.4	85.8
98. State University of New York, Stony Brook	6	0.4	86.2
99. Vanderbilt University	6	0.4	86.6
100. Villanova University	6	0.4	86.9
101. University of Virginia	6	0.4	87.3
102. Worcester Polytechnic Institute	6	0.4	87.7

Table 6.3. (continued)

Granting institution[a]	Doctorates granted		
		As percentage of total	
	Number (1)	Individual[b] (2)	Cumulative[c] (3)
103. University of Akron	5	0.3	88.0
104. American University	5	0.3	88.3
105. University of Detroit	5	0.3	88.6
106. University of Hawaii	5	0.3	88.9
107. Lehigh University	5	0.3	89.2
108. University of Louisville	5	0.3	89.5
109. University of New Orleans	5	0.3	89.8
110. North Carolina State University	5	0.3	90.1
111. University of South Florida	5	0.3	90.4
112. Syracuse University	5	0.3	90.7
113. University of Vermont	5	0.3	91.0
114. Washington State University	5	0.3	91.3
115. Colorado State University	4	0.2	91.5
116. Kansas State University	4	0.2	91.8
117. University of Kentucky	4	0.2	92.0
118. University of Miami	4	0.2	92.3
119. Montana State University	4	0.2	92.5
120. North Dakota State University	4	0.2	92.8

Table 6.3. (continued)

Granting institution[a]	Doctorates granted		
		As percentage of total	
	Number	Individual[b]	Cumulative[c]
	(1)	(2)	(3)
121. Northern Illinois University	4	0.2	93.0
122. Ohio University	4	0.2	93.2
123. University of Rhode Island	4	0.2	93.5
124. State University of New York, College of Environmental Sciences and Forestry	4	0.2	93.7
125. University of Wyoming	4	0.2	94.0
126. Baylor University	3	0.2	94.2
127. Boston College	3	0.2	94.3
128. Bryn Mawr College	3	0.2	94.5
129. Clemson University	3	0.2	94.7
130. Dartmouth College	3	0.2	94.9
131. George Washington University	3	0.2	95.1
132. University of Lowell	3	0.2	95.3
133. University of Missouri, Kansas City	3	0.2	95.4
134. University of New Hampshire	3	0.2	95.6
135. New Mexico State University	3	0.2	95.8
136. University of North Dakota	3	0.2	96.0
137. Northeastern University	3	0.2	96.2

Table 6.3. (continued)

Granting institution[a]	Doctorates granted		
		As percentage of total	
	Number	Individual[b]	Cumulative[c]
	(1)	(2)	(3)
138. University of Puerto Rico	3	0.2	96.4
139. San Diego State University	3	0.2	96.5
140. State University of New York, Binghamton	3	0.2	96.7
141. Utah State University	3	0.2	96.9
142. Virginia Commonwealth University	3	0.2	97.1
143. West Virginia University	3	0.2	97.3
144. Western Michigan University	3	0.2	97.4
145. University of Wisconsin, Milwaukee	3	0.2	97.6
146. Brigham Young University	2	0.1	97.7
147. University of California, Santa Cruz	2	0.1	97.9
148. University of Connecticut	2	0.1	98.0
149. Drexel University	2	0.1	98.1
150. Illinois Institute of Technology	2	0.1	98.2
151. Marquette University	2	0.1	98.4
152. Memphis State University	2	0.1	98.5
153. Miami University	2	0.1	98.6
154. University of Missouri, Columbia	2	0.1	98.7
155. University of Nevada, Reno	2	0.1	98.8

Table 6.3. (continued)

Granting institution[a]	Doctorates granted		
		As percentage of total	
	Number	Individual[b]	Cumulative[c]
	(1)	(2)	(3)
156. Philadelphia College of Pharmacy and Science	2	0.1	99.0
157. Rockefeller University	2	0.1	99.1
158. Texas Christian University	2	0.1	99.2
159. Clark University	1	0.1	99.3
160. Clemson University, Textile Chemistry	1	0.1	99.3
161. University of Denver	1	0.1	99.4
162. Duquesne University	1	0.1	99.5
163. Mississippi State University	1	0.1	99.5
164. University of Mississippi	1	0.1	99.6
165. University of Missouri, Rolla	1	0.1	99.6
166. North Carolina State University, Textile Chemistry	1	0.1	99.7
167. Oregon Graduate Center for Study and Research	1	0.1	99.8
168. State University of New York, Buffalo, Roswell Park Division	1	0.1	99.8
169. Tulane University	1	0.1	99.9
170. Wesleyan University	1	0.1	99.9
171. Wichita State University	1	0.1	100.0
Total, sixty-third through end	485	29.5	–
Grand total	1644	100.1	–

See Figure 6.1–2.

Table 6.3. (continued)

Note: Includes only degrees from those programs approved for doctorate study by the ACS Committee on Professional Training; this group confers the overwhelming majority of doctorates in chemistry today. Also, such a limited list is probably the best to compare with that in Table 6.2, since the latter includes only those doctorates which required a dissertation. Excludes degrees from programs in biochemistry, chemical engineering, and pharmaceutical or medicinal chemistry. For the 1979 tabulations, see IB, "Number of Grads," 1980.

[a] The following schools, while approved by the ACS to confer the doctorate in chemistry, awarded no such degrees in 1976: Cleveland State University; Loyola University of Chicago; University of Maine; University of Maryland, Baltimore County; Michigan Technological University; University of Missouri, St. Louis; University of Montana; University of the Pacific; Providence College; Seton Hall University; Stevens Institute of Technology; Texas Woman's University; and Wake Forest University.

[b] Does not add to 100% because of rounding errors.

[c] Computed by dividing the cumulative number by the total rather than by summing column 2.

Source: IB, ACS-CPT, 1977b, xi–xiv.

TABLE 6.4.
Doctoral dissertations at the Johns Hopkins University, by field of study and by decade, 1878–1899

Field	Period				
				1878–1899	
	1870s (1)	1880s (2)	1890s (3)	Number (4)	As percentage of total (5)
I. Humanities	4	67	195	266	51.8
A. Language, literature, and linguistics	2	36	117	155	30.2
1. Language and literature	1	36	110	147	28.6
a. General	–	20	55	75	14.6
b. Classical	1	16	52	69	13.4
c. Medieval	–	–	3	3	0.6
2. Literature, Anglo-Saxon	–	–	5	5	1.0
3. Oriental literature and linguistics	1	–	2	3	0.6
B. Philosophy	1	8	–	9	1.8
C. History	1	23	78	102	19.8
II. Sciences	6	74	168	248	48.2
A. Biological sciences	2	22	28	52	10.1
1. Botany	–	–	1	1	0.2
2. Microbiology	–	–	1	1	0.2
3. Physiology	1	12	8	21	4.1
4. Zoology	1	10	18	29	5.6
B. Earth sciences	–	4	17	21	4.1
1. Geology	–	4	16	20	3.9
2. Paleontology	–	–	1	1	0.2

Table 6.4. (continued)

Field	Period				
				1878–1899	
	1870s	1880s	1890s	Number	As percentage of total
	(1)	(2)	(3)	(4)	(5)
C. Physical sciences	4	46	123	173	33.7
1. Astronomy	–	–	5	5	1.0
2. Chemistry	1	24	75	100	19.5
3. Mathematics	2	12	16	30	5.8
4. Physics	1	10	27	38	7.4
a. General	1	10	26	37	7.2
b. Atmospheric science	–	–	1	1	0.2
D. Psychology	–	2	–	2	0.4
Total[a]	10	141	363	514	. . . .

See Figure 6.1–3.

Note: A dash ("–") indicates "zero", not "unknown".

[a] Indicates total number of dissertations. Columns do not add to these sums because the row entries are hierarchical; summation over any one level of heirarchy will yield totals.

Source: IB, Xerox University Microfilms, 1973+. The machine-readable version of this data base was searched through the Online Search Services of Van Pelt Library, University of Pennsylvania, using the data files of Bibliographic Retrieval Services, Inc., Schenectady, New York.

TABLE 6.5.
Chemists appointed to college presidencies, 1821–1959

Name	Institution	Tenure[b]
Samuel Avery	University of Nebraska	1909–1927
W. G. Ballantine	Oberlin College	1891–1896
A. Charles Baugher	Elizabethtown College	1941–1960+
J. L. Beeson	Georgia State College for Women	1928–1935
Katherine Blunt	Connecticut College	1929–1943, 1945–1946
Carl W. Borgmann	Vermont College	1952–1958
C. E. Brewer	Meridith College	1915–1935+
M. F. Colbaugh	Colorado School of Mines	1925–1935+
James S. Coles	Bowdoin College	1952–1960+
James B. Conant	Harvard University	1933–1953
Thomas Cooper	University of South Carolina	1821–1833
James M. Crafts	Massachusetts Institute of Technology	1897–1900
Charles W. Dabney	University of Tennessee	1887–1904
James E. Danieley	Elon College	1957–1960+
Arthur L. Dean	University of Hawaii	1914–1927
Thomas M. Drown	Lehigh University	1895–1904
Charles W. Eliot	Harvard University	1869–1909
E. C. Elliott	University of Montana	1916–1922
	Purdue University	1922–1935+
Conrad A. Elvehjem	University of Wisconsin	1958–1960+
Paul H. Fall	Hiram College	1940–1957
Charles C. French	Washington State College	1952–1960+
Edward C. Fuller	Bard College	1946–1950
Clifford C. Furnas	Buffalo College	1954–1960+

Table 6.5. (continued)

Name	Institution	Tenure[b]
Paul M. Glasoe	Spokane College	1907–1910
	Augustana College (South Dakota)	1916–1918
H. C. Graham	New Mexico State Teachers College, Silver City	1932–1935+
Reuben G. Gustavson	University of Colorado	1943–1945
	University of Nebraska	1946–1953
J. C. Hessler	James Millikin University	1935+
S. L. Hornbeak	Trinity University	1908–1920
Frederick L. Hovde	Purdue University	1946–1960+
R. M. Hughes	Miami University	1913–1927
	Iowa State University	1927–1935+
J. L. Jarman	Virginia State Teachers College, Farmville	1902–1935+
Frank S. Kedzie	Michigan State University	1915–1921
Henry J. Long	Greenville College	1936–1960+
Herbert E. Longenecker	Tulane University	1960+
Nels Minne	Winona State Teachers College	1944–1960+
Howard C. Parmelee	Colorado School of Mines	1916–1917
H. J. Patterson	Maryland Agricultural College	1913–1917
Evan Pugh	Pennsylvania State University	1859–1864
Raymond B. Purdum	Davis and Elkins College	1944–1953
Ira Remsen	Johns Hopkins University	1901–1913
V. L. Roy	Louisiana State Normal College	1911–1929
H. L. Sawyer	Colby Junior College	1928–1935+
J. A. Schaeffer	Franklin and Marshall College	1935+
Glenn T. Seaborg	University of California, Berkeley	1958–1960+
H. E. Simmons	University of Akron	1933–1951

Table 6.5. (continued)

Name	Institution	Tenure[b]
Edgar Fahs Smith	University of Pennsylvania	1911–1920
P. R. Stewart	Waynesburg College	1921–1935+
Winthrop E. Stone	Purdue University	1900–1921
Roscoe W. Thatcher	Massachusetts State College	1927–1932
C. H. Trowbridge	Weaver College	1923–1935+
G. H. Vande Bogart	Northern Montana College	1929–1935+
Francis P. Venable	University of North Carolina	1900–1914
John C. Warner	Carnegie Institute of Technology	1950–1960+
J. A. Widtsoe	Agricultural College of Utah	1907–1916
	University of Utah	1916–1921
A. C. Willard	University of Illinois	1934–1935+

[a] A plus sign ("+") following 1935 or 1960 indicates that the tenure continued beyond the time that the article listing the individual's name went to press.

Sources: IB, Hamor, 1935; IB, DeLoach and Jeanes, 1963. On Drown, see III, Bowen, 1924, 99–103, and III, Billinger, 1930. Tenures as acting president are excluded.

TABLE 6.6.
Appointments of chemists to permanent college presidencies, by decade of appointment, 1870–1959

Decade	Chemists appointed
1870–1879	0
1880–1889	1
1890–1899	3
1900–1909	8
1910–1919	11
1920–1929	10
1930–1939	7
1940–1949	9
1950–1959	8

See Figure 6.1–4.

Source: Table 6.5. *Appointments*, not people, are tabulated, so some chemists are counted more than once.

TABLE 6.7.
Citation of American publications in *Chemische Berichte*, 1890–1975

Year	Issues examined	Citations in examined issues				
				To American publications		
					As percentage of non-German publications	
		Total	To non-German publications[a]	Number	Observed	Calculated trend (logistic)
	(1)	(2)	(3)	(4)	(5)	(6)
1890	January-April	992	61	7	11.5	7.3
1900	January	832	77	9	11.7	10.4
1910	January	764	106	11	10.4	14.3
1920	January-March	818	76	8	10.5	19.3
1923	January-February	990	120	32	26.7	21.0
1925	January-February	897	103	20	19.4	22.1
1927	January-February	1139	149	34	22.8	23.2
1930	January-February	1001	248	74	29.8	25.0
1935	January-February	865	161	34	21.1	27.9
1938	January-February	955	260	85	32.7	29.7
1950	January-July	810	192	76	39.6	36.7
1955	*Hefte* 1–2	951	379	163	43.0	39.3
1960	*Heft* 1	721	330	160	48.5	41.8
1965	*Hefte* 1–2	1353	741	286	38.6	44.0

Table 6.7. (continued)

		Citations in examined issues				
				To American publications		
					As percentage of non-German publications	
Year	Issues examined	Total	To non-German publications[a]	Number	Observed	Calculated trend (logistic)
	(1)	(2)	(3)	(4)	(5)	(6)
1970	*Hefte* 1–2	1207	697	303	43.5	46.0
1975	*Heft* 1	804	439	204	46.5	47.8

See Figure 6.2–1.

Note: For details of our citation-analysis methodology, see Appendix C.

a Excludes all citations to publications written in the German language, whether from Austria, Germany, or Switzerland.

Sources: *Chemische Berichte* (*Berichte der Deutschen Chemischen Gesellschaft* for 1890–1938 entries), noted issues. Our tabulations. For details concerning the calculation of the logistic trend, see Appendix E.

TABLE 6.8.
American and non-German publications in chemistry, 1913–1975

Year	Non-German			American			
		Cumulative (calculated estimates)			Cumulative (calculated estimates)		
						Since 1879	
	Annual number	Since last observed year	Since 1800	Annual number	Since last observed year	Number	As percentage of non-German since 1800
	(1)	(2)	(3)	(4)	(5)	(6)	(7)
1913	12 486	–	149 571	3 940	–	16 177	10.8
1917	8 856	42 269	191 840	4 604	17 054	33 231	17.3
1918	8 025	8 025	199 865	4 136	4 136	37 367	18.7
1923	14 443	54 608	254 473	6 014	25 083	62 450	24.5
1929	21 241	105 744	360 217	7 498	40 372	102 822	28.5
1939	36 911	283 581	643 798	12 615	98 357	201 179	31.2
1940	35 605	35 605	679 403	14 054	14 054	215 233	31.7
1943	23 320[a]	87 092	766 495	9 340[a]	34 611	249 844	32.6
1951	48 537	275 214	1 041 709	19 288	109 744	359 588	34.5
1956	73 280	300 307	1 342 016	22 720	104 786	464 374	34.6
1960	98 285	340 686	1 682 702	28 889	102 725	567 099	33.7
1965	153 629	619 519	2 302 221	47 852	187 882	754 981	32.8
1970	218 416	920 634	3 222 855	64 006	277 690	1 032 671	32.0
1975	301 782	1 289 284	4 512 139	83 540	366 700	1 399 371	31.0
1980	368 969	1 671 254	6 183 393	106 552	472 899[b]	1 872 270	30.3

Table 6.8. (continued)

See Figure 6.2–1.

Note: Excludes patents. Although given to nearest publication, most estimates are accurate to only three significant digits at best. Exceptions are annual numbers of American publications (column 4) from 1913 to 1940, which the source gives to nearest publication; even these precise totals are subject to measurement errors of unknown magnitude (see Appendix D). Calculated estimates are undoubtedly approximate; for methodology, see Appendix C.

[a] May be approximate. See Table 6.10, and also under sources for columns 1 and 4, below.
[b] Actual cumulative total for 1980 was 476 235, according to Dale B. Baker, Director, Chemical Abstracts Service.

Sources: Columns 1 and 4:

1913 to 1943 – IB, Crane, 1944. Data for 1913 to 1940 taken directly from 1480, Table II. Data for 1943 calculated by multiplying totals in 1478, Table I by percentage in 1480, Table III. These percentages are not necessarily accurate, however; see Table 6.10.
1951 to 1975 – IB, Baker, 1976. Multiply number of abstracts of papers and books on page 23 by percentages on page 24.
1980 – IB, Baker, 1981.

Columns 2, 3, 5, and 6 – calculated from columns 1 and 4; see Appendix C. Entries for 1918 and 1940 created directly from the totals in columns 1 and 4 for those years.

TABLE 6.9.

Citations of American publications in *Annual Reports on the Progress of Chemistry*, 1904–1974

		Citations of American publications		
			As percentage of total in volume	
Year covered[a]	Total citations examined	Number	Observed	Calculated trend (linear)
	(1)	(2)	(3)	(4)
1904	801	57	7.1	7.1
1912	574	43	7.5	11.5
1919	522	95	18.2	15.3
1924	1185	235	19.8	18.1
1929	1308	243	18.6	20.8
1934	916	198	21.6	23.5
1937	1337	306	22.9	25.2
1949	1354	595	43.9	31.7
1954	1250	429	34.3	34.5
1959	1442	501	34.7	37.2
1964	749	308	41.1	39.9
1969	1083	434	40.1	42.7
1974	1015	436	43.0	45.4

See Figures 6.2–2 and 6.2–3.

Note: For details of our citation-analysis methodology, see Appendix C.

[a] Each volume covers the events of the year preceding the date of publication.

Sources: Chemical Society, London, *Annual Reports on the Progress of Chemistry*, noted issues. Our tabulations. For details concerning the calculation of the linear trend, see Appendix E.

TABLE 6.10.
American publications in chemistry as percentage of all publications in *Chemical Abstracts*,
1909–1980

Year	Total[a] (thousands) (1)	American: Number[b] (2)	American: As percentage of total[c] (3)
1909	11.455	2 294	20.0
1913	19.025	3 940	20.7
1917	10.921	4 604	42.2
1918	9.283	4 136	44.6
1923	19.507	6 014	30.8
1929	29.082	7 498	25.8
1939	45.414	12 615	27.8
1940	40.624	14 054	34.6
1943	30.523	9 340	30.6
1947	–	–	41.8
1951	52.7	19 288	36.6
1956	80.0	22 720	28.4
1960	106.6	28 889	27.1
1965	167.9	47 852	28.5
1970	233.6	64 006	27.4
1975	323.8	83 540	25.8
1980	407.2	106 552	26.2

See Figure 6.2–2.

Note: Excludes patents. For details of our citation-analysis methodology, see Appendix C.

[a] For 1909–1943, given in source to nearest publication; for 1951–1980, to nearest hundred.
[b] For 1909–1940, represents actual totals, given in source to nearest publication. For 1943–1980, estimates calculated by multiplying Columns 1 and 3; results are given to nearest publication, but are accurate, at best, to three significant digits. Estimate for 1943 is particularly suspect; see note c.
[c] For 1909–1940, the source gives slightly different percentages than those shown (IB, Crane, 1944, 1480, Table III). Publishing before the age of the pocket calculator, Crane apparently resorted to a slide rule or a table of logarithms to produce his percentages, which are accordingly approximate. For 1943, Crane's percentage is the best available estimate, so data for that year should be used with caution.

Table 6.10. (continued)

Sources: *Column 1*:

1909 to 1943 – IB, Crane, 1944, 1478, Table I.
1951 to 1975 – IB, Baker, 1976, 23.
1980 – IB, Baker, 1981, 29.

Column 2:

1909 to 1943 – IB, Crane, 1944. Data for 1909 to 1940 taken directly from 1480, Table II. Entry for 1943 calculated by multiplying total in 1478, Table I by percentage in 1480, Table III.
1951 to 1975 – multiply column 1 by column 3.
1980 – IB, Baker, 1981, 29.

Column 3:

1909 to 1940 – divide column 2 by column 1 and convert to percentage.
1943 – IB, Crane, 1944, 1480, Table III.
1947 – IB, Baker, 1961, 80. According to this source, the percentages for 1943, 1951, 1956, and 1960 are based only upon a statistically valid sample; presumably the percentages for more recent years were derived similarly.
1951 to 1975 – IB, Baker, 1976, 24.
1980 – IB, Baker, 1981, 29.

TABLE 6.11.
Citations of American publications in American and German chemical journals, 1899–1975
(percentages of total citations during year)

Year	In American journals[a]			In German journal[b]	
	Observed				
	By Computer Horizons	By Fussler	Calculated trend (linear)	Observed	Calculated trend (linear)
	(1)	(2)	(3)	(4)	(5)
1899	–	25.1	24.4	–	–
1919	–	38.2	36.4	–	–
1920	–	–	–	1.0	0.3
1923	–	–	–	3.2	1.7
1925	–	–	–	2.2	2.6
1927	–	–	–	3.0	3.5
1930	–	–	–	7.4	4.9
1935	–	–	–	3.9	7.3
1938	–	–	–	8.9	8.7
1939	44.0	48.5	48.5	–	–
1946	–	53.1	52.7	–	–
1949	51.4	–	54.5	–	–
1950	–	–	–	9.4	14.3
1955	–	–	–	17.1	16.7
1959	64.7	–	60.5	–	–
1960	–	–	–	22.2	19.0
1965	–	–	–	21.1	21.3
1969	66.8	–	66.5	–	–
1970	–	–	–	25.1	23.7
1975	–	–	–	25.4	26.0

See Figure 6.2–3.

Note: For comparable data from British journal (also plotted in Figure 6.2–3), see Table 6.9, columns 3 and 4. The slope of the trend in column 3 is approximately 0.60; in column 5, 0.47; in Table 6.9, column 4, 0.55. Cf. Table 6.12.

[a] Computer Horizons examined *Analytical Chemistry, Chemical Reviews, Journal of the American Chemical Society, Journal of Chemical Physics, Journal of Organic Chemistry*, and *Journal of Physical Chemistry* (IB, Computer Horizons, 1972, 70). Fussler examined the *Journal of the American Chemical Society* in all of his sample

Table 6.11. (continued)

years, the *Journal of Physical Chemistry* for all but 1946, the *American Chemical Journal* for 1899, and the following for 1939: *Chemical Reviews, Journal of Chemical Physics, Journal of Organic Chemistry*, and *Organic Syntheses* (IB, Fussler, 1949, 22–24). Fussler gives the wrong name for the last journal.

[b] *Chemische Berichte* (formerly *Berichte der Deutschen Chemischen Gesellschaft*).

Sources: *Column 1* – IB, Computer Horizons, 1972, Table 68, 71–72. For each year, divide the sum of citations to American papers in the six American journals by the sum of all citations per year in these journals, then convert to percentages.

Column 2 – IB, Fussler, 1949, 32, Table 7.

Column 3 – calculated by regression on the data in Columns 1 *and* 2; see Appendix E. Both estimates for 1939 were entered into the calculations. Series in column 1 not rounded to tenths of a % before computing regression curve.

Column 4 – divide Table 6.7, column 4, by Table 6.7, column 2, and convert to percentage.

Column 5 – calculated by regression on the data in column 4; see Appendix E. Series in column 4 not rounded to tenths of a % before computing regression curve.

TABLE 6.12.
Citations of German publications in American and German chemical journals, 1890–1975
(percentages of total citations during year)

Year	In American journals[a]			In German journals[b]	
	Observed				
	By Computer Horizons	By Fussler	Calculated trend (linear)	Observed	Calculated trend (linear)
	(1)	(2)	(3)	(4)	(5)
1890	–	–	–	93.9	104.3
1899	–	49.1	51.7	–	–
1900	–	–	–	90.7	97.6
1910	–	–	–	86.1	90.9
1919	–	40.5	39.1	–	–
1920	–	–	–	90.7	84.2
1923	–	–	–	87.9	
1925	–	–	–	88.5	
1927	–	–	–	86.9	
1930	–	–	–	75.2	
1935	–	–	–	81.4	74.2
1938	–	–	–	72.8	72.1
1939	30.0	25.0	26.4	–	–
1946	–	24.0	22.0	–	–
1949	21.5	–	20.1	–	–
1950	–	–	–	76.3	64.1
1955	–	–	–	60.1	60.7
1959	11.2	–	13.8	–	–
1960	–	–	–	54.2	57.4
1965	–	–	–	45.2	54.0
1969	5.6	–	7.4	–	–
1970	–	–	–	42.3	50.7
1975	–	–	–	45.4	47.3

See Figure 6.2–3.

Note: The slope of the trend in column 3 is approximately –0.63; in column 5, –0.67. Cf. Table 6.11.

[a] For the list of which journals were examined, see Table 6.11, note a.
[b] *Chemische Berichte* (formerly *Berichte der Deutschen Chemischen Gesellschaft*).

Table 6.12. (continued)

Sources: *Column 1* – IB, Computer Horizons, 1972, Table 68, 71–72. For each year, divide the sum of citations to German papers in the six American journals by the sum of all citations per year in these journals, then convert to percentages.

Column 2 – IB, Fussler, 1949, 32, Table 7.

Column 3 – calculated by regression on the data in columns 1 *and* 2; see Appendix E. Both estimates for 1939 were entered into the calculations. Series in column 1 not rounded to tenths of a % before computing regression curve.

Column 4 – subtract Table 6.7, column 3, from Table 6.7, column 2, divide the result by Table 6.7, column 2, and convert to percentage.

Column 5 – calculated by regression on the data in column 4; see Appendix E. Series in column 4 not rounded to tenths of a % before computing regression curve.

TABLE 6.13.

Nobel Prizes awarded in chemistry, selected countries, 1901–1976

Period	Country					
	France	Germany[a]	Sweden	Switzerland	United Kingdom	United States[b]
	(1)	(2)	(3)	(4)	(5)	(6)
A. Nobel Prizes per ten million population						
1901–1919	1.03	1.15	1.70	2.58	0.46	0.09
1920–1938	0.49	1.31	1.59	2.39	0.84	0.15
1939–1957	–	0.70	1.36	1.95	0.97	0.47
1958–1976	–	0.38	–	1.56	1.61	0.47
B. Nobel Prizes awarded						
1901–1919	4	7	1	1	2	1
1920–1938	2	9	1	1	4	2
1939–1957	–	5	1	1	5	8
1958–1976	–	3	–	1	9	10
Total	6	24	3	4	20	21
C. Total population[c] (millions, estimated)						
1920	38.750	60.679	5.876	3.877	43.718	106.466
1938	41.100	68.558	6.297	4.192	47.494	129.825
1957	44.31	71.17	7.37	5.13	51.43	171.98
1975	52.91	78.68[d]	8.20[d]	6.40	55.96[d]	213.63

Table 6.13. (continued)

See Figure 6.2–4.

Note: Includes all countries whose residents received more than two Nobels in chemistry through 1976. A dash ("–") indicates "zero", not "unknown".

[a] Population estimates for 1957 and 1975 include totals for the Federal Republic of Germany, the German Democratic Republic, East Berlin, and West Berlin.

[b] Excludes armed forces abroad.

[c] These population estimates are either mid-year estimates or mean averages of official end-of-year estimates. A single-year estimate of population at the end of each interval is used, since the construction of representative nineteen-year averages from available annual estimates presented too complex and time-consuming a task for present purposes. Also, it seemed advisable to use comparable population data, so United Nations estimates alone – the most reliable single set of series – are given in this table. Counts for 1919 and 1976 are unavailable from UN sources at present, so 1920 and 1975 estimates have been substituted; this should not affect international comparisons seriously.

[d] Provisional.

Sources: *Nobel Prizes*:

1901 to 1970 – IB, Nobel Foundation, 1972a, 638–645.
1971 to 1976 – IB, Nobel Foundation, 1972b–.

Population:

1920 and 1938 – IB, UN, 1952, Table 3, 122–131.
1957 – IB, UN, 1971, Table 4, 126–135.
1975 – IB, UN, 1976, Table 5, 160–166.

TABLE 6.14.
Appointments of chemists to deanships of American graduate schools, 1893–1970

Name of appointee	Institution	Year of appointment
Charles E. Munroe	George Washington University	1893
Henry W. Harper	University of Texas. Austin	1910
Edward C. Franklin	Stanford University	1910
William McPherson	Ohio State University	1911
Edward C. Franklin	Stanford University	1913
Edward C. Franklin	Stanford University	1917
John D. Clark	University of New Mexico	1919
Clare Chrisman Todd	Washington State University	1922
George Abbott	University of North Dakota	1925
Julius W. Sturmer	Philadelphia College of Pharmacy	1926
James Kendall	New York University	1927
T. Palmer Nash, Jr.	University of Tennessee	1928
F. W. Upson	University of Nebraska	1929
A. M. Pardee	University of South Dakota	1929
Stephen P. Burke	West Virginia University	1930
Willis H. Clark	Texas Women's University	1930
Gilbert H. Boggs	Georgia Institute of Technology	1932
Charles Watkins	Carnegie Institute of Technology	1933
Harry F. Lewis	Lawrence University	1933
Harry B. Weiser	Rice University	1933
Wilber Dwight Engle	University of Denver	1933
Victor J. Chambers	University of Rochester	1934
Frederick W. Sohon	Georgetown University	1934

Table 6.14. (continued)

Name of appointee	Institution	Year of appointment
Martin A. Rosanoff	Duquesne University	1936
Friend A. Clark	West Virginia University	1936
Thomas G. Chapman	University of Arizona	1937
Alfred C. Nelson	University of Denver	1937
Henry E. Bent	University of Missouri	1938
Robert C. Goodwin	Texas Technical College	1938
C. S. Hamilton	University of Nebraska	1939
Arthur B. Lamb	Harvard University	1940
C. S. Hamilton	University of Nebraska	1940
John L. Daniel	Georgia Institute of Technology	1940
Robert L. Nugent	University of Arizona	1941
Reuben G. Gustavson	University of Colorado	1942
Ivor Griffith	Philadelphia College of Pharmacy	1943
Chester M. Alter	Boston University	1944
James A. Bradley	Newark College of Engineering	1944
Raymond E. Kirk	Polytechnic Institute of Brooklyn	1944
Ralph Huston	Michigan State University	1944
Hugh S. Taylor	Princeton University	1945
John C. Warner	Carnegie Institute of Technology	1946
Herbert E. Longenecker	University of Pittsburgh	1946
Henry Eyring	University of Utah	1946
Joseph S. McGrath	University of Portland	1946
Hoke S. Greene	University of Cincinnati	1947
Paul M. Gross	Duke University	1947

Table 6.14. (continued)

Name of appointee	Institution	Year of appointment
W. T. Gooch	Baylor University	1947
S. E. Hazlet	Washington State University	1947
Ralph M. Hixon	Iowa State University	1948
Albert F. Daggett	University of New Hampshire	1949
Harvey A. Neville	Lehigh University	1949
C. Clement French	Virginia Polytechnic Institute	1949
Harold J. Tormey	St. Bonaventure University	1950
George Holmes Richter	Rice University	1950
George S. Parks	Stanford University	1950
Chapman Harris	Lowell Technical Institute	1951
Frederick T. Wall	University of Illinois	1951
Donald R. Clippinger	Ohio University	1951
Charles Ernest Braun	University of Vermont	1952
W. Albert Noyes, Jr.	University of Rochester	1952
Clark B. Carpenter	Colorado School of Mines	1952
Donald F. Hornig	Brown University	1953
Henry M. Woodburn	State University of New York, Buffalo	1953
Louis C. Cady	University of Idaho	1953
Marcus E. Hobbs	Duke University	1954
Joe Eugene Moose	University of Nevada	1954
Harry B. Feldman	Worcester Polytechnic Institute	1955
Warren C. Johnson	University of Chicago	1955
Frederick T. Wall	University of Illinois	1955
Charles T. Lester	Emory University	1957

Table 6.14. (continued)

Name of appointee	Institution	Year of appointment
Herbert D. Rhodes	University of Arizona	1957
Arthur A. Vernon	Northeastern University	1958
Melvin J. Astle	Case Western Reserve University	1958
John E. Willard	University of Wisconsin	1958
Lloyd E. Swearingen	University of Oklahoma	1958
Milton C. Kloetzel	University of Southern California	1958
Arthur Osol	Philadelphia College of Pharmacy	1959
Eric A. Arnold	Case Western Reserve University	1959
Walter J. Peterson	North Carolina State University	1959
Charles Ernest Braun	University of Vermont	1960
Bryce L. Crawford, Jr.	University of Minnesota	1960
W. Knowlton Hall	Medical College of Georgia	1960
William Lewis Nobles	University of Mississippi	1960
Alfred C. Nelson	University of Denver	1960
Thomas D. O'Brien	University of Nevada	1960
John T. Norton	Massachusetts Institute of Technology	1961
Ralph S. Halford	Columbia University	1961
Louis Gordon	Case Western Reserve University	1961
Andrew D. Suttle, Jr.	Mississippi State University	1961
Hilton A. Smith	University of Tennessee	1961
Norris W. Rakestraw	University of California, San Diego	1961
George M. Murphy	New York University	1962
Louis A. Reber	Philadelphia College of Pharmacy	1963
Herbert E. Carter	University of Illinois	1963

Table 6.14. (continued)

Name of appointee	Institution	Year of appointment
Robert A. Alberty	University of Wisconsin	1963
Warren W. Brandt	Virginia Polytechnic Institute	1963
Donald G. Davis	Louisiana State University, New Orleans	1963
William J. Burke	Arizona State University	1963
James F. Hornig	Dartmouth College	1964
W. Donald Cooke	Cornell University	1964
S. E. Wiberley	Rensselaer Polytechnic Institute	1964
Robert H. Baker	Northwestern University	1964
Wouter Bosch	University of Missouri, Rolla	1964
C. Ernest Birchenall	University of Delaware	1964
Armin H. Gropp	University of Miami	1964
Francis J. Behal	Medical College of Georgia	1964
Harrison Shull	Indiana University	1965
Harold S. Bailey	South Dakota State University	1965
James R. Arnold	University of California, San Diego	1965
Martin J. Kamen	University of California, San Diego	1965
J. Reid Shelton	Case Western Reserve University	1966
Wesley J. Dale	University of Missouri, Kansas City	1966
F. I. Brownley, Jr.	Clemson University	1966
William K. Easley	Northeast Louisiana State College	1966
Frederick T. Wall	University of California, San Diego	1966
Edward L. Alexander	Lowell Technical Institute	1967
Frank H. Hurley	Case Western Reserve University	1967
R. F. Kruh	Kansas State University	1967

Table 6.14. (continued)

Name of appointee	Institution	Year of appointment
A. William Johnson	University of North Dakota	1967
George K. Fraenkel	Columbia University	1968
Bernard T. Gillis	Duquesne University	1968
John N. Hobstetter	University of Pennsylvania	1968
William Rostoker	University of Illinois, Chicago Circle	1968
Alfred F. Foster	University of Toledo	1968
Thomas H. Whitehead	University of Georgia	1968
H. Willard Davis	University of South Carolina	1968
Joseph Sam	University of Mississippi	1968
John M. Stewart	University of Montana	1968
C. J. Nyman	Washington State University	1968
Daniel H. Murray	State University of New York, Buffalo	1969
Raymond P. Mariella	Loyola University	1969
Jan Rocek	University of Illinois, Chicago Circle	1969
Normal H. Cromwell	University of Nebraska	1970
Stanley M. Williamson	University of California, Santa Cruz	1970

Source: IA, Chase, 1970. Chemists extracted using computer assistance.

TABLE 6.15.
Chemists as cumulative percentage of appointments to graduate deanships, by pentade of appointment, 1870–1969

	Deanship appointments				
	Total		Chemists		
Pentade	Annual number (1)	Cumulative to date (2)	Annual number (3)	Cumulative to date (4)	As cumulative percentage of total (5)
1870–1874	2	2	0	0	0.0
1875–1879	0	2	0	0	0.0
1880–1884	2	4	0	0	0.0
1885–1889	4	8	0	0	0.0
1890–1894	8	16	1	1	6.3
1895–1899	11	27	0	1	3.7
1900–1904	15	42	0	1	2.4
1905–1909	23	65	0	1	1.5
1910–1914	33	98	4	5	5.1
1915–1919	27	125	2	7	5.6
1920–1924	44	169	1	8	4.7
1925–1929	66	235	6	14	6.0
1930–1934	79	314	9	23	7.3
1935–1939	79	393	7	30	7.6
1940–1944	94	487	10	40	8.2
1945–1949	149	636	13	53	8.3
1950–1954	128	764	14	67	8.8
1955–1959	135	899	13	80	8.9
1960–1964	217	1116	27	107	9.6
1965–1969	226	1342	26	133	9.9

See Figure 6.3–1.

Note: Table 6.14 lists the chemists who received deanship appointments.

Source: IA, Chase, 1970. Chemists extracted using computer assistance. Chase's summary data differ from our totals because he counts persons and we count appointments, and also because we exclude 1970.

TABLE 6.16.

A "graduate deans" indicator of the entrenchment of academic chemistry, 1910–1969

	Chemistry		Total		
Pentade	Deanship appointments (1)	Doctorates conferred (2)	Deanship appointments (3)	Doctorates conferred (4)	Entrenchment indicator[a] (5)
1910–1914	4	443	33	2 537	0.69
1915–1919	2	504	27	2 927	0.43
1920–1924	1	773	44	4 224	0.12
1925–1929	6	1140	66	7 402	0.59
1930–1934	9	1918	79	13 096	0.78
1935–1939	7	2314	79	14 248	0.55
1940–1944	10	2863	94	15 361	0.57
1945–1949	13	2403	149	16 086	0.58
1950–1954	14	5097	128	38 956	0.84
1955–1959	13	4980	135	44 801	0.87
1960–1964	27	5907	217	59 338	1.25
1965–1969	26	8540	226	104 607	1.41

See Figure 6.3–2.

[a] Assumes that deanships are distributed in proportion to doctorates conferred in each discipline.

Sources: *Columns 1 and 3* – Table 6.15, columns 3 and 1.
Columns 2 and 4 – sum over the appropriate pentades of Table 3.3, columns 4 and 1.
Column 5 – divide the ratio of column 1 to column 2 by the ratio of column 3 to column 4.

TABLE 6.17.
Appointments of chemists to deanships of American graduate schools, by region of appointing institution and by pentade of appointment, 1890–1969

Pentade when chemist started deanship[b]	Region of institution[a]				
	Northeast	North central	South	West	Total
	(1)	(2)	(3)	(4)	(5)
1890–1894	0	0	1	0	1
1910–1914	0	1	1	2	4
1915–1919	0	0	0	2	2
1920–1924	0	0	0	1	1
1925–1929	2	3	1	0	6
1930–1934	2	1	5	1	9
1935–1939	1	2	2	2	7
1940–1944	5	2	1	2	10
1945–1949	5	2	3	3	13
1950–1954	6	2	2	4	14
1955–1959	3	5	3	2	13
1960–1964	8	6	9	4	27
1965–1969	5	11	5	5	26
Total, 1890 through 1924	0	1	2	5	8
Total, 1925 through 1969	37	34	31	23	125
Total	37	35	33	28	133

[a] Following the geographical categories used by the U.S. Bureau of the Census, the *Northeast* region includes Maine, New Hampshire, Vermont, Massachusetts, Rhode Island, Connecticut, New York, New Jersey, and Pennsylvania. The *North Central* region includes Ohio, Indiana, Illinois, Michigan, Wisconsin, Minnesota, Iowa, Missouri, North Dakota, South Dakota, Nebraska, and Kansas. The *South* region includes Delaware, Maryland, the District of Columbia, Virginia, West Virginia, North Carolina, South Carolina, Georgia, Florida, Kentucky, Tennessee, Alabama, Mississippi, Arkansas, Louisiana, Oklahoma, and Texas. And the *West* region includes Montana, Idaho, Wyoming, Colorado, New Mexico, Arizona, Utah, Nevada, Washington, Oregon, California, Alaska, and Hawaii. No chemist has ever been appointed to a graduate deanship in an American territory or possession.

[b] No chemist was appointed to a deanship between 1895 and 1909.

Source: IA, Chase, 1970. Computer assistance used to produce tallies.

TABLE 6.18.
Presidents of the American Association for the Advancement of Science, 1848–1981

Name	Field[a]	Tenure[b]
William B. Rogers[c]	geology	1848
W. C. Redfield	geology	1848
Joseph Henry	physics	1849
A. D. Bache	geography	1850
Louis Agassiz	zoology	1851
Benjamin Peirce	physics	1853
James D. Dana	anthropology	1854
John Torrey	botany	1855
James Hall	geology	1856
J. W. Bailey	chemistry	1857
Jeffries Wyman	medicine	1858
Stephen Alexander	astronomy	1859
Isaac Lea	geology	1860
F. A. P. Barnard	astronomy	1866
J. S. Newberry	geology	1867
B. A. Gould	astronomy	1868
J. W. Foster	geography	1869
William Chauvenet	mathematics	1870
Asa Gray	botany	1871
J. Lawrence Smith	chemistry	1872
Joseph Lovering	physics	1873
J. L. LeConte	entomology	1874
J. E. Hilgard	geography	1875
William B. Rogers	geology	1876
Simon Newcomb	astronomy	1877
O. C. Marsh	geology	1878
G. F. Barker	chemistry	1879
Lewis H. Morgan	anthropology	1880
G. J. Brush	geology	1881
J. W. Dawson	geology	1882
C. A. Young	astronomy	1883
J. P. Lesley	geology	1884
H. A. Newton	mathematics	1885
Edward S. Morse	zoology	1886
S. P. Langley	physics	1887
J. W. Powell	geology	1888
T. C. Mendenhall	physics	1889
G. Lincoln Goodale	botany	1890
Albert B. Prescott	chemistry	1891

Table 6.18. (continued)

Name	Field[a]	Tenure[b]
Joseph Le Conte	geology	1892
William Harkness	astronomy	1893
Daniel G. Brinton	anthropology	1894
E. W. Morley	chemistry	1895
Edward D. Cope	zoology	1896
Theodore Gill	zoology	1896[d]
Wolcott Gibbs	chemistry	1897
F. W. Putnam	anthropology	1898
Edward Orton	geology	1899
R. S. Woodward	mathematics	1900
C. S. Minot	medicine	1901
Asaph Hall	astronomy	1902
Ira Remsen	chemistry	1902
Carroll D. Wright	economics	1903
W. G. Farlow	botany	1904
C. M. Woodward	mathematics	1905
William H. Welch	medicine	1906
E. L. Nichols	physics	1907
T. C. Chamberlin	geology	1908
David S. Jordan	biology	1909
A. A. Michelson	physics	1910
Charles E. Bessey	botany	1911
E. C. Pickering	astronomy	1912
Edmund B. Wilson	zoology	1913
Charles W. Eliot	education[e]	1914
W. W. Campbell	astronomy	1915
Charles Richard Van Hise	geology	1916
Theodore W. Richards	chemistry	1917
John Merle Coulter	botany	1918
Simon Flexner	medicine	1919
L. O. Howard	entomology	1920
E. H. Moore	mathematics	1921
J. Playfair McMurrich	anatomy	1922
Charles D. Walcott	paleontology	1923
J. McKeen Cattell	psychology	1924
Michael I. Pupin	engineering	1925
L. H. Bailey	horticulture	1926
Arthur A. Noyes	chemistry	1927
Henry Fairfield Osborn	paleontology	1928

Table 6.18. (*continued*)

Name	Field[a]	Tenure[b]
Robert A. Millikin	physics	1929
Thomas H. Morgan	zoology	1930
Franz Boas	anthropology	1931
John J. Abel	pharmacology	1932
Henry N. Russell	astronomy	1933
Edward L. Thorndike	psychology	1934
Karl F. Compton	physics	1935
Edwin G. Conklin	biology	1936
George D. Birkhoff	mathematics	1937
Wesley C. Mitchell	economics	1938
Walter B. Cannon	physiology	1939
Albert F. Blakeslee	genetics	1940
Irving Langmuir	chemistry	1941
Arthur H. Compton	physics	1942
Isaiah Bowman	geography	1943
Anton J. Carlson	physiology	1944
Charles F. Kettering	engineering	1945
James B. Conant	chemistry	1946
Harlow Shapley	astronomy	1947
Edmund W. Sinnott	botany	1948
Elvin C. Stakman	plant pathology	1949
Roger Adams	chemistry	1950
Kirtley F. Mather	geology	1951
Detlev W. Bronk	physiology[f]	1952
E. U. Condon	physics	1953
Warren Weaver	mathematics[g]	1954
George W. Beadle	biology	1955
Paul B. Sears	botany	1956
Laurence H. Snyder	genetics	1957
Wallace R. Brode	chemistry	1958
Paul C. Klopsteg	physics	1959
Chauncey D. Leake	pharmacology	1960
Thomas Parke	zoology	1961
Paul M. Gross	chemistry[h]	1962
Alan T. Waterman	electrical engineering	1963
Laurence M. Gould	geology	1964
Henry Eyring	chemistry	1965
Alfred S. Romer	zoology	1966
Don K. Price	public administration	1967
Walter Orr Roberts	astrophysics	1968

Table 6.18. (*continued*)

Name	Field[a]	Tenure[b]
H. Bentley Glass	genetics	1969
Athelstan F. Spilhaus	meteorology	1970
Mina S. Rees	mathematics	1971
Glenn T. Seaborg	chemistry[i]	1972
Leonard Rieser	physics	1973
Roger Revelle	population studies	1974
Margaret Mead	anthropology	1975
William D. McElroy	biology[j]	1976
Emilio Q. Daddario	politics and law	1977
Edward E. David, Jr.	engineering	1978
Kenneth E. Boulding	economics	1979
Frederick Mosteller	mathematics	1980
D. Allan Bromley	physics	1981

[a] For 1848–1947, provided by the source; thereafter, found in standard contemporary biographical dictionaries, most notably *American Men and Women of Science*.
[b] There were no meetings and no presidents of the AAAS in 1852 and in 1861–1865, and there were two in 1902; see the source.
[c] Acting president until the election of Redfield.
[d] Cope died in office; Gill completed his term.
[e] Trained as a chemist.
[f] Also listed widely as a biophysicist.
[g] Also well known as a foundation administrator.
[h] Physical chemistry.
[i] Nuclear chemistry.
[j] Also listed widely as a biochemist.

Sources: 1848 to 1947 – IB, AAAS, 1948, 18.
1948 to 1981 – *Science*, listings in tables of contents of appropriate issues.

TABLE 6.19.
Disciplinary affiliations of presidents of the American Association for the Advancement of Science, 1848–1981

Period	Disciplines having this many presidents			
	Four	Three	Two	One
1848–1857		geology	physics	anthropology botany *chemistry* geography zoology
1858–1867	–	–	astronomy geology	medicine
1868–1877	–	–	astronomy geography	botany *chemistry* entomology geology mathematics physics
1878–1887	geology	–	–	anthropology astronomy *chemistry* mathematics physics zoology
1888–1897		*chemistry*	geology zoology	anthropology astronomy botany physics
1898–1907		–	mathematics medicine	anthropology astronomy botany *chemistry* economics geology physics
1908–1917			astronomy geology	biology botany *chemistry* education physics zoology

Table 6.19. (continued)

Period	Disciplines having this many presidents			
	Four	Three	Two	One
1918–1927	–		–	anatomy
				botany
				chemistry
				engineering
				entomology
				horticulture
				mathematics
				medicine
				paleontology
				psychology
1928–1937	–	–	physics	anthropology
				astronomy
				biology
				mathematics
				paleontology
				pharmacology
				psychology
				zoology
1938–1947	–		*chemistry*	astronomy
			physiology	economics
				engineering
				genetics
				geography
				physics
1948–1957	–	–	botany	biology
				chemistry
				genetics
				geology
				mathematics
				physics
				physiology
				plant pathology
1958–1967	–	*chemistry*	zoology	electrical engineering
				geology
				pharmacology
				physics
				public administration

Table 6.19. (continued)

	Disciplines having this many presidents			
Period	Four	Three	Two	One
1968–1977	–	–	–	anthropology astrophysics biology *chemistry* genetics mathematics meteorology physics politics and law population studies
1978–1981	–	–	–	economics engineering mathematics physics

Note: Chemistry is italicized for easier reference. See also Table A in text.

Source: Table 6.18.

TABLE 6.20.
Selected American professional societies

Organization	Year established	Membership in 1979 (thousands)	Staff
American Medical Association	1847	213.9	900
American Association for the Advancement of Science	1848	133.0	171
American Society of Civil Engineers	1852	77.7	116
National Education Association	1857	1600.8	600
American Institute of Mining, Metallurgical, and Petroleum Engineers[a]	1871	69.0	–
American Chemical Society	1876	118.2	1500
American Bar Association	1878	250.0	480
American Society of Mechanical Engineers	1880	80.0	250
Modern Language Association of America	1883	30.0	60
American Historical Association	1884	14.5	20
Institute of Electrical and Electronics Engineers[b]	1884	180.0	250
National Science Teachers Association	1895	42.0	40
American Institute of Chemical Engineers	1908	45.0	75
American Association of University Professors	1915	68.1	55

a "Petroleum" added to name in 1957.
b Merger in 1963 of American Institute of Electrical Engineers (1884) and Institute of Radio Engineers (1912).

Source: IB, Yakes and Akey (eds.), 1980. This source gives 776 000 ACS members in 1979, an obvious error. We have substituted the correct number from Table 2.4, column 1.

TABLE 6.21.
Section affiliations of living members of the National Academy of Sciences, as of 1 July 1972

Section		Members in section[a]	
Number	Subject	Number	As percentage of total[b]
5	chemistry	127	13.3
3	physics	120	12.6
14	biochemistry	99	10.4
4	engineering	60	6.3
1	mathematics	58	6.1
8	zoology	57	6.0
17	medical sciences	49	5.1
6	geology	42	4.4
2	astronomy	41	4.3
13	geophysics	41	4.3
7	botany	40	4.2
18	genetics	38	4.0
12	psychology	37	3.9
9	physiology	35	3.7
16	applied physical and mathematical sciences	33	3.5
11	anthropology	26	2.7
10	microbiology	23	2.4
15	applied biology	20	2.1
19	social, economic, and political sciences	8	0.8
Total		954	100.1

[a] Includes the 939 regular members living in 1972, the 14 members emeriti, and the one living resigned member (Richard C. Lewontin). Excludes the 769 deceased members and all foreign associates.

[b] Does not add to 100% because of rounding errors.

Sources: IB, NAS, 1971–1972, 199–241. Our tallies. Lewontin's section affiliation is from IB, NAS, 1970–1971, 202.

TABLE 6.22.
Chemists and other members elected to the National Academy of Sciences, selected periods, 1870–1980

Period[a]	Members elected[b] Total	Chemists[c] Number	Chemists[c] As percentage of total
	(1)	(2)	(3)
1870–1872	23	4	17.4
1880–1883	10	3	30.0
1890–1895	11	1	9.1
1900–1902	14	1	7.1
1910–1911	17	3	17.6
1920	14	2	14.3
1930	15	3	20.0
1940	16	2	12.5
1950	30	3	10.0
1960	35	3	8.6
1970	50	6	12.0
1980	59	6	10.2
Total	294	37	12.6[d]

[a] Periods were created by starting with the eleven census years between 1870 and 1970 and, for each one, adding subsequent years as needed until a total of at least ten newly-elected members had been included.
[b] For the 110 members in this sample elected before 1980 and still living in 1972, each member's field was determined by that member's own choice of section (see Table 6.21), as listed in the source. For the 125 persons in this sample who had died by 1972, the section affiliations of 19 were found in similar listings in previous issues of the source, and those for the remaining 106 were assigned on the basis of information in biographies of the members. The Academy's own *Biographical Memoirs* were consulted for 103 of the latter, *World Who's Who in Science* for 2 more, and *Who Was Who in America* for the remaining individual.
[c] Excludes biochemists.
[d] Not a sum of column 3, but rather the total of column 2 as a percentage of the total of column 1.

Sources: 1870–1970 – IB, NAS, 1971–1972, 199–241 and 276–285. Our tallies. Earlier issues in this series and various standard biographical dictionaries, as mentioned in the notes, were also consulted to assign a specialty to each member.
1980 – IB, "Academy of Sciences Picks", 1980, and IB, "NAS Elects", 1980. Our tallies. *American Men and Women of Science* provided the necessary biographical detail.

TABLE 6.23.
Holders of chemistry and physics doctorates as percentages of holders of natural sciences doctorates, 1930–1970

Year	Stock of living doctorate holders (estimated)				
		Chemistry[a]		Physics[b]	
	Natural sciences[a]	Number	As percentage of natural sciences	Number	As percentage of natural sciences
	(1)	(2)	(3)	(4)	(5)
1930	9 049	3 596	39.7	1 117	12.3
1940	18 771	7 176	38.2	2 148	11.4
1950	28 866	11 581	40.1	3 342	11.6
1960	52 167	19 627	37.6	7 451	14.3
1970	98 437	31 465	32.0	16 278	16.5

[a] For the definitions of these subjects, see Table 3.1, notes a and b.
[b] Includes the following subjects: general physics (U.S. Office of Education specialty code 1902); molecular physics (1903); nuclear physics (1904); astronomy (1911); and astrophysics (1912).

Source: IB, Adkins, 1975. Totals for natural sciences are a combination from Table B–3.1 and B–4.16, 444–445 and 496–497. Those for chemistry are from Table B–4.5, 474–475. Those for physics are from Table B–4.7, 478–479.

TABLE 6.24.
Presidents of the National Academy of Sciences, 1863–1981

Name	Field[a]	Academy section[b]	Volume for obituary[c]	Tenure
A. D. Bache	*physics*	3	1	1863–1867
Joseph Henry	*physics*	3	5	1868–1878
William B. Rogers	geology	6	3	1879–1882
O. C. Marsh	paleontology	6	20	1883–1895
Wolcott Gibbs	*chemistry*	5	7	1895–1900
Alexander Agassiz	zoology	8	7	1901–1907
Ira Remsen	*chemistry*	5	14	1907–1913
W. H. Welch	medical sciences	17	22	1913–1917
Charles D. Walcott	geology	6	39	1917–1923
A. A. Michelson	*physics*	3	19	1923–1927
T. H. Morgan	genetics	18	33	1927–1931
W. W. Campbell	astronomy	2	25	1931–1935
Frank R. Lillie	zoology	8	30	1935–1939
Frank B. Jewett	engineering	4[d]	27	1939–1947
Alfred N. Richards	biochemistry	14[e]	42	1947–1950
Detlev W. Bronk	physiology	9[f]	–	1950–1962
Frederick M. Seitz	*physics*	3[g]	–	1962–1969
Philip M. Handler	biochemistry	14[g]	–	1969–1981

Note: Chemistry and physics are italicized for easier reference.

[a] When possible, derived from section affiliation (see notes b and d through g); otherwise, derived from Academy obituary notice (see note c).
[b] Follows the 1972 system of classification; see IB, NAS, 1971–1972, 199. Before 1940, sections are assigned according to the field(s) listed in each president's obituary. Thereafter, each president's own choice of section is used; see notes d through g.
[c] Indicates volume of the Academy's *Biographical Memoirs* which contains each president's obituary.
[d] See IB, NAS, 1948–1949, 144 and 149.
[e] See IB, NAS, 1953–1954, 132 and 142. Converted to 1972 classification.
[f] See IB, NAS, 1974–1975, 202 and 210. Converted to 1972 classification.
[g] See IB, NAS, 1975–1976, 188, 216 and 245. Converted to 1972 classification.

Sources: IB, NAS, 1975–1976, 187 and 303. Determination of field and section involved other references, as listed in the notes.

TABLE 6.25.
Census estimates of chemists, physicists, and natural scientists in the labor force, 1950–1970

Year	Natural scientists	Chemists		Physicists	
		Number	As percentage of natural scientists	Number	As percentage of natural scientists
	(1)	(2)	(3)	(4)	(5)
1950	118 874	75 958	63.9	7 541	6.3
1960 (1960 classification)	151 130	84 349	55.8	14 014	9.3
1960 (1970 classification)	157 842	95 759	60.7	14 014	8.9
1970	205 501	110 167	53.6	22 283	10.8

Sources: 1950 and 1960 (1960 classification) – IA, USBC-COP, 1964, Table 201, 522.
1960 (1970 classification) and 1970 – IA, USBC-COP, 1973, Table 221, 718.

TABLE 6.26.
National Research Fellows in the natural sciences, 1919–1950, by selected fields of employment in 1950

Field of employment in 1950	National Research Fellows, 1919–1950	
	Number	As percentage of total
	(1)	(2)
Chemistry	229	26.7
Physics	196	22.8
Zoology	164	19.1
Mathematics	126	14.7
Botany	112	13.1
Astronomy	16	1.9
Geology and geography	15	1.7
Total	858	100.0

Note: Because it treats only the natural sciences, this table excludes the 332 fellows with careers in the medical sciences, 93 in psychology, 41 in agriculture, 27 in anthropology, and 8 in forestry. For fellows deceased or unemployed in spring 1950, it uses field of latest employment to date.

Source: IB, Rand, 1951a, 78, Table 4.

TABLE 6.27.

Doctorates conferred in the natural sciences, selected years, 1920–1950

Year	Natural sciences[a] doctorates conferred				
		Chemistry[a]		Physics[b]	
	Total	Number	As percentage of total	Number	As percentage of total
	(1)	(2)	(3)	(4)	(5)
1920	268	104	38.8	32	11.9
1930	945	332	35.1	109	11.5
1940	1 459	532	36.5	132	9.0
1950	2 351	967	41.1	379	16.1
Total, 1919–1950	32 026	12 443	38.9	3653	11.4

a For the definitions of these subjects, see Table 3.1, notes a and b.
b For the definition of this subject, see Table 6.23, note b.

Sources: *Column 1* – Table 3.3, column 2.
Column 2 – Table 3.3, column 4.
Column 4 – IB, Adkins, 1975, Table A–5.7, 278–281.

TABLE 6.28.

Chemists as fellows of the John Simon Guggenheim Memorial Foundation, 1925–1974

	Fellows living in 1974					
	During period			Cumulative to period		
		Chemists[a]			Chemists[a]	
Period	Total (1)	Number (2)	As percentage of total (3)	Total (4)	Number (5)	As percentage of total (6)
1925–1929	106	9	8.5	106	9	8.5
1930–1934	146	7	4.8	252	16	6.3
1935–1939	202	14	6.9	454	30	6.6
1940–1944	321	13	4.0	775	43	5.5
1945–1949	602	33	5.5	1377	76	5.5
1950–1954	897	100	11.1	2274	176	7.7
1955–1959	1480	132	8.9	3754	308	8.2
1960–1964	1465	131	8.9	5219	439	8.4
1965–1969	1564	115	7.4	6783	554	8.2
1970–1974	1788	162	9.1	8571	716	8.4

Note: Because these totals include only fellows still alive in 1974, the comparability of the percentages over different time periods assumes, plausibly, that chemists die at the same rate as members of other disciplines. This procedure also implies that the absolute numbers in columns 1, 2, 4, and 5 are *not* comparable over time.

Table 6.28. (continued)

[a] Determined on the basis of 1974 job title. Includes fellows in: analytical chemistry, biochemistry, biological chemistry, chemical biodynamics, chemical biology, chemical engineering, chemical physics, chemistry, geochemistry, immunochemistry, inorganic chemistry, kinetics and thermochemistry, marine biochemistry, metallurgy, microbial biochemistry, molecular biology in chemistry, molecular biophysics, nuclear chemistry, organic chemistry, pharmaceutical biochemistry, pharmacology, physical biochemistry, physical chemistry, physical metallurgy, physiological chemistry, plant biochemistry, soil chemistry, theoretical chemistry, and theoretical physical chemistry.

Source: IB, Guggenheim Foundation, 1975. Our tallies from list of living members. Each of the 7403 living fellows was counted once for each Guggenheim award received, for the total of 8571 awards shown in the table.

TABLE 6.29.
Two "Guggenheim Fellows" indicators of the entrenchment of academic chemistry, 1930–1969

Pentade	Chemistry			All fields			Entrenchment indicators	
	Guggenheim Fellowships[a]	Faculty (estimated)[b]	Doctorate holders[c]	Guggenheim Fellowships[a]	Faculty (estimated)[b]	Doctorate holders[c]	Based upon estimated faculty[d]	Based upon doctorate holders[e]
	(1)	(2)	(3)	(4)	(5)	(6)	(7)	(8)
1930–1934	7	1433	4 224	146	103 876	24 946	3.48	0.28
1935–1939	14	1567	5 982	202	131 607	35 898	5.82	0.42
1940–1944	13	2179	8 149	321	166 741	47 930	3.10	0.24
1945–1949	33	2687	9 881	602	211 253	57 169	4.31	0.32
1950–1954	100	3313	13 291	897	267 649	81 342	9.01	0.68
1955–1959	132	4085	17 301	1480	339 100	116 129	7.40	0.60
1960–1964	131	5036	21 382	1465	429 625	156 140	7.63	0.65
1965–1969	115	6209	26 916	1564	544 316	223 496	6.45	0.61

See Figure 6.3–3.

[a] Counts the number of times persons holding chemical positions in 1974 had received Guggenheim Fellowships in each pentade listed.
[b] Provides the arithmetic mean of the estimated annual numbers of faculty members during the five years of each pentade.
[c] Provides the arithmetic mean of the estimated annual numbers of doctorate holders during the five years of each pentade.
[d] Assumes that, *ceteris paribus*, Guggenheim Fellowships are distributed among disciplines roughly in proportion to the number of faculty members in each discipline.

Table 6.29. (continued)

[e] Assumes that, *ceteris paribus*, Guggenheim Fellowships are distributed among disciplines roughly in proportion to the stock of terminal doctorate holders in each discipline.

Sources: *Columns 1 and 4* – Table 6.28, columns 2 and 1.

Columns 2 and 5 – calculated by log-linear regressions upon Table 5.16, columns 2 and 1; see Appendix E. For method of averaging over pentades, see note b above.

Columns 3 and 6 – IB, Adkins, 1975. Chemistry from Table B–4.5, 474–475. All fields from Table B–1, 432–433. For method of averaging over pentades, see note c above.

Column 7 – divide the ratio of column 1 to column 2 by the ratio of column 4 to column 5.

Column 8 – divide the ratio of column 1 to column 3 by the ratio of column 4 to column 6.

TABLE 6.30.
Current research and development expenditures in universities and colleges, 1954–1974

		Chemistry	
Year	Total (thousands of current dollars)	Amount (thousands of current dollars)	As percentage of total
	(1)	(2)	(3)
1954	205 500	14 700	7.2
1964	1 275 436	70 022	5.5
1966	1 714 684	87 955	5.1
1968	2 148 708	104 695	4.9
1970	2 334 859	102 002	4.4
1972	2 676 511	110 015	4.1
1973	2 936 707	114 293	3.9
1974	3 017 391[a]	117 479	3.9

See Figure 6.3–4.

[a] Excludes Draper Laboratory, with $55 million in such expenditures, because it is classified as part of the independent nonprofit sector in 1974.

Sources: 1954 – IA, NSF, 1957a, 3, Tables 2 and 3. Add entries from the two tables to combine federal and non-federal patronage.
1964 to 1974 – IA, NSF, 1976b, Table B–3, 26.

TABLE 6.31.
American Chemical Society journals

Journal	Year established	Circulation in 1979
Journal of the American Chemical Society	1879[a]	14 236
Journal of Physical Chemistry	1896[b]	4 502
Chemical Abstracts	1907[c]	– [d]
Journal of Industrial and Engineering Chemistry	1909[e]	– [f]
Chemical and Engineering News	1923[g]	128 133
Chemical Reviews	1924	5 473
Journal of Chemical Education[h]	1924	24 500[i]
SciQuest	1927[j]	47 456
Rubber and Chemistry Technology[k]	1928	5 300[l]
Analytical Chemistry	1929[m]	30 590
Journal of Organic Chemistry	1936[n]	9 997
Journal of Agricultural and Food Chemistry	1953	5 090
Journal of Chemical and Engineering Data	1959	1 962
Journal of Medicinal Chemistry	1959[o]	4 308
Journal of Chemical Information and Computer Sciences	1961[p]	2 209
Biochemistry	1962	6 845
Industrial and Engineering Chemistry –		
Fundamentals	1962[e]	4 967
Process Design and Development	1962[e]	5 204
Product Research and Development	1962[e]	4 853
Inorganic Chemistry	1962	4 721
Accounts of Chemical Research	1967[q]	9 316
Environmental Science and Technology	1967	17 354
Macromolecules	1968	2 373
Chemical Technology (CHEMTECH)	1971[e]	15 291
Journal of Physical and Chemical Reference Data	1972[r]	1 175

[a] The *Journal of the American Chemical Society* (JACS) was preceded by the Society's *Proceedings*, which were published in the *American Chemist* (with separate editions for ACS members) in 1876, and separately in 1877 and 1878. The *Proceedings* were incorporated into the new JACS in 1879. When Edward Hart (Lafayette) assumed the editorship of JACS in 1893, he merged his *Journal of Analytical and Applied Chemistry* with it. Ira Remsen's *American Chemical Journal*, published at Johns Hopkins from 1879, was merged with JACS in 1913.

[b] Begun by Wilder D. Bancroft (Cornell); transferred to American Chemical Society control in 1933; published under the title *Journal of Physical and Colloid Chemistry* from 1947–50.

Table 6.31. (*continued*)

[c] The predecessor of *Chemical Abstracts* was Arthur A. Noyes' *Review of American Chemical Research*, which first appeared as a supplement to the MIT *Technology Quarterly* in 1895. Two years later the *Review* was incorporated into the *Journal of the American Chemical Society*. Separate publication as *Chemical Abstracts* began in 1907, the culmination of a campaign led by William Noyes to institute an ACS-sponsored abstracting journal covering the world chemical literature.
[d] Unknown.
[e] Begun in 1909 as the *Journal of Industrial and Engineering Chemistry*. Under Harrison E. Howe's editorship, the *Journal* began publication of separate News and Analytical editions in 1923 and 1929, respectively. Three research-oriented supplements on theoretical studies in chemical engineering and process and product research and development first appeared in 1962; these achieved independent status in 1965. The original *Journal* became *Chemical Technology* in 1971.
[f] Now defunct. See note e.
[g] Begun as the *News Edition of Industrial and Engineering Chemistry* in 1923; renamed in 1942.
[h] Owned and published by the Division of Chemical Education, Inc.
[i] In January 1976.
[j] Published by Pennsylvania State University under the title *Chemistry Leaflet* beginning in 1927; name changed to *Science Leaflet* in 1933, and back to *Chemistry Leaflet* in 1941; acquired by Science Service in 1944 and renamed *Chemistry*; bought by the American Chemical Society in 1962; name changed to *SciQuest*, and editorial policy radically revised, in 1979. Now defunct.
[k] Owned and published by Rubber Division, Inc.
[l] In December 1975.
[m] Begun as the *Analytical Edition* of *Industrial and Engineering Chemistry* in 1929; name changed in 1948.
[n] Begun in this year under editorship of Morris Kharasch (University of Chicago); acquired by the American Chemical Society in 1955.
[o] Begun as *Journal of Medicinal and Pharmaceutical Chemistry*, published by Inter-Science; bought by the American Chemical Society in 1961; name changed in 1962.
[p] Begun as *Journal of Chemical Documentation*; name changed in 1975.
[q] Begun as supplement to *Chemical and Engineering News*; established separately in 1968.
[r] Published jointly with the U.S. National Bureau of Standards and the American Institute of Physics.

Sources: Names and origins of journals are from IB, Skolnik and Reese, 1976, 100–125, 399–401. Circulation data are from *ibid.*, 95, and IB, ACS, 1980, 41.

TABLE 6.32.
Specialist chemical societies in the United States

Society	Year established
American Water Works Association	1881
Association of Official Analytical Chemists[a]	1884
Society of Chemical Industry, American Section	1894
American Society for Testing Materials	1898
American Ceramic Society	1899
Electrochemical Society[b]	1902
American Leather Chemists Society	1903
American Society of Biological Chemists	1906
American Society for Pharmacology and Experimental Therapeutics	1908
American Institute of Chemical Engineers	1908
American Oil Chemists Society[c]	1909
American Association of Cereal Chemists	1915
Societé de Chimie Industrielle, American Section	1918
American Association of Textile Chemists and Colorists	1921
Association of Consulting Chemists and Chemical Engineers	1928
American Society of Brewing Chemists	1934
American Microchemical Society	1935
Association of Analytical Chemists	1941
Association of Vitamin Chemists	1943
Society of Cosmetic Chemists	1945
American Association for Clinical Chemistry	1948
Histochemical Society	1950
Coblentz Society[d]	1954
Geochemical Society	1955
Society of Flavor Chemists	1959
Phytochemical Society of North America	1960
Catalysis Society of North America[e]	1966
American Society for Neurochemistry	1969

Note: Includes some societies (such as the American Water Works Association) in which chemists have been only one interest group among many. Excludes chemical trade associations, of which the Chemical Manufacturers Association is the best-known example.

Table 6.32. (continued)

[a] Name changed from Association of Official Agricultural Chemists in 1965.
[b] Name changed from American Electrochemical Society in 1928; see III, Burns and Enck, 1977, 24.
[c] Originally named the Society of Cotton Products Analysts; present name adopted in 1920.
[d] Chemists and others interested in molecular spectroscopy and its applications.
[e] Originally named the Catalysis Society.

Sources: IB, Yakes and Akey, 1980; IB, NAS, 1971; and IB, West and Hull, 1927a.

TABLE 6.33.
Specialty divisions in the American Chemical Society

Division	Year established[a]	Membership in 1979
Industrial and Engineering Chemistry	1908[b]	3022
Agricultural and Food Chemistry	1908	1247
Fertilizer and Soil Chemistry	1908[c]	334
Organic Chemistry	1908	4715
Physical Chemistry	1908[d]	1960
Medicinal Chemistry	1909[e]	2946
Rubber	1909[f]	2146
Biological Chemistry	1913	2887
Environmental Chemistry	1913[g]	1580
Carbohydrate Chemistry	1919[h]	660
Cellulose, Paper, and Textile	1919[i]	641
Dye Chemistry	1919[j]	–
Leather and Gelatin Chemistry	1919[k]	–
Chemical Education	1921[l]	2643
History of Chemistry	1921[m]	270
Fuel Chemistry	1922[n]	874
Petroleum Chemistry	1922[o]	1714
Organic Coatings and Plastics Chemistry	1923[p]	2660
Colloid and Surface Chemistry	1926[q]	1291
Analytical Chemistry	1936[r]	4219
Chemical Information	1948[s]	848
Polymer Chemistry	1950	4195
Chemical Marketing and Economics	1952	525
Inorganic Chemistry	1956[t]	2205
Microbial and Biochemical Technology	1961[u]	686
Fluorine Chemistry	1963[v]	363
Nuclear Chemistry and Technology	1963[w]	757
Pesticide Chemistry	1969[x]	1104
Professional Relations	1972	516
Computers in Chemistry	1974	548
Chemical Health and Safety	1977	454
Geochemistry	1978	105
Small Chemical Businesses	1978	113

Table 6.33. (continued)

---

[a] Marks the beginning of a separate section with the Society. Some divisions began as a group within an established division, while others started as sections which were later authorized as new divisions.
[b] Originally Industrial Chemists and Chemical Engineers; present name adopted in 1919.
[c] Name changed in 1952 from Fertilizer Chemistry.
[d] Name changed in 1958 from Physical and Inorganic Chemistry, after establishment in 1956 of separate Divison of Inorganic Chemistry.
[e] Established in 1909 as Pharmaceutical Chemistry; name changed in 1920 to Chemistry of Medicinal Products; present name adopted in 1927.
[f] Originally named the Section of India Rubber Chemistry; authorized as a division in 1919, at which time the present name was adopted.
[g] Authorized in 1915 as Water, Sewage, and Sanitation Chemistry; name changed in 1959 to Water and Waste Chemistry, and in 1964 to Water, Air, and Waste Chemistry; present name adopted in 1973.
[h] Authorized in 1921 as Sugar Chemistry; name changed in 1939 to Sugar Chemistry and Technology; present name adopted in 1952.
[i] Authorized in 1922 as Cellulose Chemistry; name changed in 1961 to Cellulose, Wood, and Fiber Chemistry; present name adopted in 1974.
[j] Authorized as a division in 1920; merged with Organic Chemistry in 1935.
[k] Authorized in 1921 as Leather Chemistry; name changed to Leather and Gelatin in 1923; discontinued in 1938.
[l] A section of chemical education met as early as 1909, and papers were presented in this section at annual meetings of the American Chemical Society in 1910 and 1911. Revived in 1921 at the instigation of Neil Gordon, with the support of Edgar Fahs Smith (who was ACS President at the time). The Section of Chemical Education attained divisional status in 1924.
[m] Authorized as a division in 1927.
[n] Authorized in 1925 as Gas and Fuel Chemistry; present name adopted in 1960.
[o] Begun in 1921 as a section of Industrial and Engineering Chemistry.
[p] Authorized in 1927 as Paint and Varnish Chemistry; name changed in 1940 to Paint, Varnish, and Plastics Chemistry, then changed in 1952 to Paint, Plastics, and Printing Ink Chemistry; present name adopted in 1960.
[q] Name changed in 1960 from Colloid Chemistry.
[r] Authorized in 1938 as Microchemistry; name changed in 1940 to Analytical and Microchemistry; present name adopted in 1949.
[s] Name changed in 1975 from Chemical Literature.
[t] Split off in 1956 from Physical and Inorganic Chemistry.
[u] Chemists, microbiologists, and engineers interested in fermentation science and technology established a Fermentation Section (later Subdivision) within the Division of Agricultural and Food Chemistry in 1946. Attained divisional status in 1961. Name changed in 1976 from Microbial Chemistry and Technology.
[v] Established in 1950 as a subdivision of Industrial and Engineering Chemistry, after the stimulus of two symposia on fluorine chemistry in 1946 and 1950; attained separate divisional status in 1963.

Table 6.33. (continued)

---

[w] Existed for a few years previously as a subdivision within Industrial and Engineering Chemistry.

[x] Established as a subdivision of Agricultural and Food Chemistry in 1951; attained divisional status in 1969.

Sources: IB, Skolnik and Reese, 1976, 236–237 and 401–406; IB, ACS, 1977a, 42; and IB, ACS, 1979a, 40. Membership data, as of 31 December, 1979, from IB, ACS, 1980, 37.

TABLE 6.34.
American Chemical Society membership versus membership in selected specialist chemical societies, 1900–1980
(thousands)

Year	ACS	Electrochemical Society	American Institute of Chemical Engineers	American Institute of Chemists
	(1)	(2)	(3)	(4)
1900	1.72	–	–	–
1901	1.93	–	–	–
1902	2.19	0.37	–	–
1903	2.43	0.61	–	–
1904	2.68	0.63	–	–
1905	2.92	0.63	–	–
1906	3.08	0.62	–	–
1907	3.39	0.61	–	–
1908	4.00	0.78	0.04	–
1909	4.51	1.07	0.10	–
1910	5.08	1.21	0.12	–
1911	5.60	1.32	0.15	–
1912	6.22	1.35	0.17	–
1913	6.67	1.38	0.20	–
1914	7.17	1.38	0.21	–
1915	7.42	1.48	0.22	–
1916	8.36	1.54	0.23	–
1917	10.60	1.68	0.27	–
1918	12.20	1.90	0.28	–
1919	13.69	2.21	0.31	–
1920	15.58	2.30	0.34	–
1921	14.32	2.17	0.45	–
1922	14.40	1.96	0.53	–
1923	14.35	1.76	0.57	0.41
1924	14.52	1.69	0.61	0.45
1925	14.38	1.69	0.64	0.51
1926	14.70	1.67	0.68	0.53
1927	15.19	1.67	0.72	0.52
1928	16.24	–	0.79	0.53
1929	17.43	1.86	0.81	0.53
1930	18.21	–	0.87	0.59
1931	18.96	–	0.94	0.65

Table 6.34. (continued)

Year	ACS	Electrochemical Society	American Institute of Chemical Engineers	American Institute of Chemists
	(1)	(2)	(3)	(4)
1932	18.57	1.40	1.02	0.65
1933	17.47	1.19	1.07	0.73
1934	17.56	–	1.10	0.76
1935	17.54	1.16	1.16	0.67
1936	18.73	1.24	1.32	0.84
1937	20.68	–	1.44	1.12
1938	22.19	1.22	1.73	1.30
1939	23.52	1.22	1.98	1.46
1940	25.41	1.26	2.26	1.63
1941	28.74	1.29	2.53	1.66
1942	31.72	1.33	2.92	1.67
1943	36.00	1.53	3.35	1.79
1944	39.44	1.70	3.93	–
1945	43.08	1.91	4.86	1.99
1946	48.76	2.23	5.77	2.04
1947	55.10	2.37	6.73	2.00
1948	58.78	1.80	7.91	2.23
1949	62.21	1.88	8.94	2.39
1950	63.35	1.99	9.68	2.46
1951	66.01	–	10.41	2.49
1952	67.73	2.04	11.43	2.51
1953	70.16	2.09	12.59	2.51
1954	72.29	2.14	13.60	2.58
1955	75.22	2.27	14.45	2.72
1956	79.22	2.52	15.08	2.81
1957	81.93	2.74	15.87	2.90
1958	85.82	2.96	17.97	2.87
1959	88.81	3.23	18.66	2.84
1960	92.19	3.12	19.71	2.86
1961	93.64	3.55	21.42	2.90
1962	95.21	3.72	23.26	2.91
1963	96.75	3.80	24.27	2.99
1964	99.48	3.86	25.90	2.97

Table 6.34. (continued)

Year	ACS (1)	Electrochemical Society (2)	American Institute of Chemical Engineers (3)	American Institute of Chemists (4)
1965	102.53	3.87	27.32	3.02
1966	106.27	4.07	29.07	3.00
1967	109.53	4.08	31.47	2.99
1968	113.37	4.20	34.10	3.04
1969	116.82	4.30	36.56	5.68
1970	114.32	4.25	38.22	7.50
1971	112.02	4.16	38.62	7.93
1972	110.71	3.91	39.06	7.83
1973	110.29	3.85	39.03	7.83
1974	110.27	4.09	39.26	6.75
1975	110.82	4.15	38.97	5.74
1976	112.73	4.34	39.84	5.38
1977	115.14	4.62	40.01	5.22
1978	116.24	4.93	42.24	5.39
1979	118.21	5.26	45.25	5.21
1980	120.40	5.72	47.12	5.16

See Figure 6.4–1.

Sources: *Column 1:* Table 2.4, column 1.

Column 2: 1902–1976 – III, Burns and Enck, 1977, 151–160.
1977–1980 – IB, 'Membership Statistics', 1983, 475C.

Column 3: III, Reynolds, 1983, 200.

Column 4: 1923–1925, 1927–1928, 1944, and 1948 – IB, Van Doren, 1948, 190. Data for all years except 1923 and 1948 obtained by inspection from graph.
1926, 1959, and 1969 – III, Carmichael, 1974, October installment, 28.
1929–1943, 1945–1947, 1949–1958, 1960–1968, and 1970–1974 – Annual AIC Secretary's Reports in *The Chemist*. Number for 1970 is estimated.
1975 – Supplied by David A. H. Roethel, Executive Director of the AIC.
1976–1980 – 'Membership Report' for 1980 by AIC Secretary Roger F. Jones; copy supplied by David A. H. Roethel.

TABLE 6.35.
Graduate research in chemistry, by subject, 1924–1935 and 1967–1975

Year	Agricultural and food (1)	Analytical (2)	General and physical (3)	Inorganic (4)	Organic (5)	Pharmaceutical (6)	Other[a] (7)	Total[b] (8)
A. Graduate students engaged in research [number (row percentage)]								
1924	126(7.4)	71(4.2)	240(14.1)	101(5.9)	422(24.8)	20(1.2)	720(42.4)	1700(100.0)
1925	104(5.9)	44(2.5)	332(18.8)	86(4.9)	430(24.4)	39(2.2)	728(41.3)	1763(100.0)
1926	109(5.8)	54(2.9)	343(18.2)	109(5.8)	475(25.2)	21(1.1)	771(41.0)	1882(100.0)
1927	116(6.0)	75(3.9)	430(22.2)	116(6.0)	570(29.5)	14(0.7)	613(31.7)	1934(100.0)
1928	163(7.8)	96(4.6)	406(19.5)	124(6.0)	668(32.1)	16(0.8)	608(29.2)	2081(100.0)
1929	122(5.5)	95(4.3)	448(20.2)	116(5.2)	719(32.5)	27(1.2)	686(31.0)	2213(99.9)
1930	142(5.8)	102(4.1)	520(21.1)	129(5.2)	811(32.9)	34(1.4)	725(29.4)	2463(99.9)
1931	158(5.4)	149(5.1)	577(19.8)	159(5.5)	930(31.9)	54(1.9)	887(30.4)	2914(100.0)
1932	196(6.7)	147(5.0)	595(20.3)	162(5.5)	943(32.1)	47(1.6)	847(28.8)	2937(100.0)
1933	178(6.7)	142(5.3)	527(19.8)	139(5.2)	866(32.5)	26(1.0)	788(29.6)	2666(100.1)
1934	167(6.1)	156(5.7)	498(18.3)	147(5.4)	870(31.9)	34(1.2)	856(31.4)	2728(100.0)
1935	187(6.7)	150(5.4)	502(17.9)	150(5.4)	944(33.7)	26(0.9)	842(30.1)	2801(100.1)
B. Doctorates awarded [number (row percentage)]								
1967	58(2.6)	144(6.4)	495(21.9)	193(8.5)	717(31.7)	65(2.9)	587(26.0)	2259(100.0)
1968	44(1.9)	129(5.5)	506(21.5)	230(9.8)	723(30.7)	44(1.9)	679(28.8)	2355(100.1)
1969	50(2.0)	141(5.7)	547(21.9)	264(10.6)	764(30.6)	52(2.1)	676(27.1)	2494(100.0)
1970	62(2.2)	158(5.6)	640(22.8)	302(10.8)	838(29.9)	57(2.0)	746(26.6)	2803(99.9)
1971	58(2.0)	173(6.0)	599(21.0)	297(10.4)	808(28.3)	65(2.3)	857(30.0)	2857(100.0)

Table 6.35. (continued)

Year	Agricultural and food (1)	Analytical (2)	General and physical (3)	Inorganic (4)	Organic (5)	Pharmaceutical (6)	Other[a] (7)	Total[b] (8)
1972	22(0.8)	147(5.7)	598(23.1)	289(11.2)	713(27.5)	48(1.9)	774(29.9)	2591(100.1)
1973	9(0.4)	128(5.2)	593(23.9)	261(10.5)	643(25.9)	50(2.0)	801(32.2)	2485(100.1)
1974	13(0.5)	145(6.0)	571(23.8)	204(8.5)	623(26.0)	62(2.6)	780(32.5)	2398(99.9)
1975	8(0.3)	140(5.9)	555(23.3)	229(9.6)	600(25.2)	66(2.8)	784(32.9)	2382(100.0)
Total, graduate researchers, 1924–1935	1768(6.3)	1281(4.6)	5418(19.3)	1538(5.5)	8648(30.8)	358(1.3)	9071(32.3)	28 082(100.1)
Total, doctorates awarded, 1967–1975	324(1.4)	1305(5.8)	5104(22.6)	2269(10.0)	6429(28.4)	509(2.2)	6684(29.5)	22 624(99.9)

See Figure 6.4–2.

Note: Since slightly different things are counted in the earlier years than in the later ones, the ***absolute*** numbers are ***not*** directly comparable between the two periods; the row percentages given in parentheses, however, should be roughly comparable.

[a] For the 1924–1935 period, includes the following: colloid, catalysis, subatomic and radio, electro-inorganic, electro-organic, photochemistry and photography, metallurgical, physiological, pharmacological, sanitary, nutrition, and industrial and engineering (not to be confused with chemical engineering, which is excluded). For the 1967–1975 period, includes the following: nuclear, theoretical, polymer, biochemistry, and other (not specified further in source used).

[b] Percentages may not add to 100% because of rounding errors.

Table 6.35. (continued)

Sources: 1924 and 1925 – IB, Hull and West, 1934, 471, Table 2. Corrections for the 1933 totals, given on p. 542 of the source volume, were incorporated into our 1933 tallies.

1926 to 1935 – IB, Hull and West, 1936, 339, Table 2. Excluded from our tallies are the estimates for chemical engineering researchers from 1929 to 1935, the one chemical education researcher in 1933, and the 94 masters-level researchers included in the totals provided by the source but not classified by field within the body of the source table.

1967 to 1975 – IB, ACS, 1977b, 17, Table 15. Data compiled by the National Research Council. Included in our tallies are all totals listed under "chemistry", plus the totals for "biochemistry" listed under "biological sciences" at the bottom of the source table.

TABLE 6.36.

Subject sections in *Chemical Abstracts*, 1907 and 1977

Section number[a]	Name
	A. Volume 1, Number 1, 1907
1	Apparatus
2	General and Physical Chemistry
3	Radioactivity
4	Electrochemistry
5	Photography
6	Inorganic Chemistry
7	Analytical Chemistry
8	Mineralogical and Geological Chemistry
9	Metallurgy
10	Organic Chemistry
11	Biological Chemistry
12	Foods
13	Nutrition
14	Water, Sewage, and Sanitation
15	Soils and Fertilizer
16	Fermented and Distilled Liquors
17	Pharmaceutical Chemistry
18	Acids, Alkalies, and Salt
19	Glass and Pottery
20	Cements and Mortar
21	Fuel, Gas, and Coke
22	Petroleum, Asphalt, and Wood Products
23	Cellulose and Paper
24	Explosives
25	Dyes, Bleaching, and Textile Fabrics
26	Pigments, Resins, Varnishes, and India Rubber
27	Fats, Fatty Oils, and Soap
28	Sugar, Starch, and Gums
29	Leather
30	Patents
	B. Volume 86, Numbers 1 and 2, 1977
Biochemistry Sections	
1	Pharmacodynamics
2	Hormone Pharmacology

Table 6.36. (continued)

Section number[a]	Name
3	Biochemical Interactions
4	Toxicology
5	Agrochemicals
6	General Biochemistry
7	Enzymes
8	Radiation Biochemistry
9	Biochemical Methods
10	Microbial Biochemistry
11	Plant Biochemistry
12	Nonmammalian Biochemistry
13	Mammalian Biochemistry
14	Mammalian Pathological Biochemistry
15	Immunochemistry
16	Fermentations
17	Foods
18	Animal Nutrition
19	Fertilizers, Soils, and Plant Nutrition
20	History, Education, and Documentation
Organic Chemistry Sections	
21	General Organic Chemistry
22	Physical Organic Chemistry
23	Aliphatic Compounds
24	Alicyclic Compounds
25	Noncondensed Aromatic Compounds
26	Condensed Aromatic Compounds
27	Heterocyclic Compounds (One Hetero Atom)
28	Heterocyclic Compounds (More Than One Hetero Atom)
29	Organometallic and Organometalloidal Compounds
30	Terpenoids
31	Alkaloids
32	Steroids
33	Carbohydrates
34	Synthesis of Amino Acids, Peptides, and Proteins
Macromolecular Chemistry Sections	
35	Synthetic High Polymers
36	Plastics Manufacture and Processing
37	Plastics Fabrication and Uses

Table 6.36. (continued)

Section number[a]	Name
38	Elastomers, Including Natural Rubber
39	Textiles
40	Dyes, Fluorescent Whitening Agents, and Photosensitizers
41	Leather and Related Materials
42	Coatings, Inks, and Related Products
43	Cellulose, Lignin, Paper, and Other Wood Products
44	Industrial Carbohydrates
45	Fats and Waxes
46	Surface-Active Agents and Detergents
Applied Chemistry and Chemical Engineering Sections	
47	Apparatus and Plant Equipment
48	Unit Operations and Processes
49	Industrial Inorganic Chemicals
50	Propellants and Explosives
51	Fossil Fuels, Derivatives, and Related Products
52	Electrochemical, Radiational, and Thermal Energy Technology
53	Mineralogical and Geological Chemistry
54	Extractive Metallurgy
55	Ferrous Metals and Alloys
56	Nonferrous Metals and Alloys
57	Ceramics
58	Cement and Concrete Products
59	Air Pollution and Industrial Hygiene
60	Sewage and Wastes
61	Water
62	Essential Oils and Cosmetics
63	Pharmaceuticals
64	Pharmaceutical Analysis
Physical and Analytical Chemistry Sections	
65	General Physical Chemistry
66	Surface Chemistry and Colloids
67	Catalysis and Reaction Kinetics
68	Phase Equilibriums, Chemical Equilibriums and Solutions
69	Thermodynamics, Thermochemistry, and Thermal Properties
70	Nuclear Phenomena
71	Nuclear Technology

Table 6.36. (continued)

Section number[a]	Name
72	Electrochemistry
73	Spectra by Absorption, Emission, Reflection, or Magnetic Resonance, and Other Optical Properties
74	Radiation Chemistry, Photochemistry, and Photographic Processes
75	Crystallization and Crystal Structure
76	Electric Phenomena
77	Magnetic Phenomena
78	Inorganic Chemicals and Reactions
79	Inorganic Analytical Chemistry
80	Organic Analytic Chemistry

[a] The numbering of sections first appears in volume 5, number 16, published 20 August, 1911; the numbering which appears for the 1907 sections in the table is meant only to indicate the order of appearance of the sections in the first number for that year.

Sources: 1907 – *Chemical Abstracts*, 1 (1907).
1977 – *Chemical Abstracts*, 86, Nos. 1 and 2 (1977).

TABLE 6.37.

American Chemical Society presidents, 1876–1981

Name	Date of Birth	Date of Death	Tenure
John William Draper	1811	1882	1876
J. Lawrence Smith	1818	1883	1877
Samuel W. Johnson	1830	1909	1878
Thomas Sterry Hunt	1826	1892	1879, 1888
Frederick A. Genth	1820	1893	1880
Charles F. Chandler	1836	1925	1881, 1889
John William Mallet	1832	1912	1882
James Curtis Booth	1810	1888	1883–1885
Albert B. Prescott	1832	1905	1886
Charles A. Goessmann	1827	1910	1887
Henry B. Nason	1831	1895	1890
George F. Barker	1835	1910	1891
George C. Caldwell	1834	1907	1892
Harvey Washington Wiley	1844	1930	1893–1894
Edgar Fahs Smith	1854	1928	1895, 1921–1922
Charles B. Dudley	1842	1909	1896–1897
Charles E. Munroe	1849	1938	1898
Edward Williams Morley	1838	1923	1899
William McMurtrie	1851	1913	1900
Frank Wigglesworth Clarke	1847	1931	1901
Ira Remsen	1846	1927	1902
John H. Long	1856	1918	1903
Arthur A. Noyes	1866	1936	1904
Francis Preston Venable	1856	1934	1905
William F. Hillebrand	1853	1925	1906
Marston Taylor Bogert	1868	1954	1907–1908
Willis Rodney Whitney	1868	1958	1909
Wilder Dwight Bancroft	1867	1953	1910
Alexander Smith	1865	1922	1911
Arthur Dehon Little	1863	1935	1912–1913
Theodore William Richards	1868	1928	1914
Charles H. Herty	1867	1938	1915–1916
Julius Stieglitz	1867	1937	1917
William H. Nichols	1852	1930	1918–1919
William A. Noyes	1857	1941	1920

Table 6.37. (continued)

Name	Date of Birth	Date of Death	Tenure
Edward Curtis Franklin	1862	1937	1923
Leo Hendrik Baekeland	1863	1944	1924
James Flack Norris	1871	1940	1925–1926
George D. Rosengarten	1869	1936	1927
Samuel W. Parr	1957	1931	1928
Irving Langmuir	1881	1957	1929
William McPherson	1864	1951	1930
Moses Gomberg	1866	1947	1931
Lawrence V. Redman	1880	1946	1932
Arthur B. Lamb	1880	1952	1933
Charles L. Reese	1862	1940	1934
Roger Adams	1889	1971	1935
Edward Bartow	1870	1958	1936
Edward R. Weidlein	1887	1983	1937
Frank C. Whitmore	1887	1947	1938
Charles A. Kraus	1875	1967	1939
Samuel C. Lind	1879	1965	1940
William Lloyd Evans	1870	1954	1941
Harry N. Holmes	1879	1958	1942
Per K. Frolich	1899	1977	1943
Thomas Midgley, Jr.	1889	1944	1944
Carl S. Marvel	1894	–	1945
Bradley Dewey	1887	1974	1946
W. Albert Noyes, Jr.	1898	1980	1947
Charles Allen Thomas	1900	1982	1948
Linus Pauling	1901	–	1949
Ernest H. Volwiler	1893	–	1950
Nathaniel Howell Furman	1892	1965	1951
Edgar C. Britton	1891	1962	1952
Farrington Daniels	1889	1972	1953
Harry L. Fisher	1885	1961	1954
Joel H. Hildebrand	1881	1983	1955
John Christian Warner	1897	–	1956
Roger J. Williams	1893	–	1957
Clifford F. Rassweiler	1899	1976	1958
John C. Bailar, Jr.	1904	–	1959

Table 6.37. (continued)

Name	Date of Birth	Date of Death	Tenure
Albert L. Elder	1901	1976	1960
Arthur C. Cope	1909	1966	1961
Karl Folkers	1906	–	1962
Henry Eyring	1901	1981	1963
Maurice Henshaw Arveson	1902	1974	1964
Charles C. Price	1913	–	1965
William J. Sparks	1905	1976	1966
Charles G. Overberger	1920	–	1967
Robert W. Cairns	1909	–	1968
Wallace Reed Brode	1900	1974	1969
Byron Riegel	1906	1975	1970
Melvin Calvin	1911	–	1971
Max Tishler	1906	–	1972
Alan C. Nixon	1908	–	1973
Bernard S. Friedman	1907	–	1974
William J. Bailey	1921	–	1975
Glenn T. Seaborg	1912	–	1976
Henry A. Hill	1915	1979	1977
Anna J. Harrison	1912	–	1978
Gardner Stacy	1921	–	1979
James D'Ianni	1914	–	1980
Albert C. Zettlemoyer	1915	–	1981

Note: For sources of biographical information, see p. 189, note 28.

Sources: 1876–1976 – IB, Skolnik and Reese, 1976, 383–393.
1977 and 1978 – III, "Harrison Wins", 1976.
1979 – III, "Stacy Tops Crawford", 1977.
1980 – III, "D'Ianni Wins", 1978.
1981 – III, "Zettlemoyer Wins", 1979.

TABLE 6.38.

Average age of American Chemical Society presidents during tenure of office, by decade, 1876–1981

Period	Average age of presidents	
	Mean	Median
	(1)	(2)
1876–1885	60.2	59.5
1886–1895	53.2	53.5
1896–1905	51.2	51.5
1906–1915	45.5	46
1916–1925	60.6	62
1926–1935	58.6	56.5
1936–1945	59.6	61
1946–1955	58.8	59
1956–1965	58.0	59
1966–1975	61.2	62.5
1976–1981	63.7	65

See Figure 6.5–1.

Note: James Curtis Booth, at 75, has been the oldest president to date; Arthur Amos Noyes, at 38, the youngest.

Source: Table 6.37. Means and medians of beginning of tenure minus birth date.

TABLE 6.39.

Educational backgrounds of American Chemical Society presidents, 1876–1981

Period	Total	European[a]		American				Other[b]	
				MD		PhD			
		Number	As percentage of total	Number	As percentage of total	Number	As percentage of total	Number	As percentage of total
	(1)	(2)	(3)	(4)	(5)	(6)	(7)	(8)	(9)
1876–1885	8	5	62.5	2	25.0	0	0.0	1	12.5
1886–1895	9	5	55.6	3	33.3	0	0.0	1	11.1
1896–1905	9	3.5[c]	38.9	0.5[c]	5.6	2	22.2	3	33.3
1906–1915	8	4	50.0	0	0.0	3	37.5	1	12.5
1916–1925	8	3	37.5	0	0.0	4	50.0	1	12.5
1926–1935	10	3	30.0	0	0.0	5	50.0	2	20.0
1936–1945	10	2	20.0	0	0.0	6	60.0	2	20.0
1946–1955	10	0	0.0	0	0.0	8	80.0	2	20.0
1956–1965	10	0	0.0	0	0.0	9	90.0	1	10.0
1966–1975	10	0	0.0	0	0.0	10	100.0	0	0.0
1976–1981	6	0	0.0	0	0.0	6	100.0	0	0.0

See Figure 6.5–2.

Note: See also Table 6.41. Five chemists – Chandler, Herty, Hunt, Norris, and E. F. Smith – served as president in two of the periods shown; they are counted once in each.

[a] I.e., postgraduate study, even if no formal degree was conferred. An overwhelming majority of these individuals received German doctorates.

[b] Includes persons with no formal traning beyond the baccalaureate, plus those trained through industrial apprenticeships.

[c] Apportions Ira Remsen between two categories (German PhD and American MD).

Sources: Table 6.37. Biographical information from sources listed on p. 189, note 28.

TABLE 6.40.

Leading institutions at which American Chemical Society presidents received highest degree, 1876–1981

Institution	Presidents educated	
	Number	Names
Universität Göttingen	11	Bartow, Caldwell, Chandler, Goessmann, Langmuir, Mallet, Nason, Remsen, Richards, E. F. Smith, Venable
University of Illinois	11	Bailey, Brode, Elder, Friedman, Marvel, Overberger, Rassweiler, Riegel, Sparks, Stacy, Volwiler
Harvard University	10	Adams, Clarke, Daniels, Lamb, Munroe, Price, Richards, Tishler, Whitmore, Wiley
Massachusetts Institute of Technology	7	Dewey, Frolich, Hill, Kraus, Little, Thomas, Zettlemoyer
Universität Leipzig	6	Bancroft, Johnson, Lind, A. A. Noyes, Richards, Whitney
Johns Hopkins University	5	Franklin, Herty, Holmes, Norris, W. A. Noyes
University of Michigan	4	Bailar, Britton, Gomberg, Prescott
University of California	3	Eyring, Nixon, Seaborg
University of Chicago	3	Evans, McPherson, Williams
University of Wisconsin	3	Cope, D'Ianni, Folkers
Yale University	3	Barker, Dudley, Johnson

Note: A threshold of three presidents is used as the criterion of inclusion.

Sources: Table 6.37. Biographical information from sources listed on p. 189, note 28.

TABLE 6.41.

Institutions at which American Chemical Society presidents received highest level of training, 1876–1981

Institution[a]	Period of presidency						
	1876–1915		1916–1955		1956–1981		
	Number (1)	As percentage of total for period (2)	Number (3)	As percentage of total for period (4)	Number (5)	As percentage of total for period (6)	Total (7)
A. European							
Göttingen	8.00	25.0	3	8.1	0	0.0	11.00
Leipzig	3.33[b]	10.4	1	2.7	0	0.0	4.33
Heidelberg	1.00	3.1	1	2.7	0	0.0	2.00
Munich	1.33[b]	4.2	0	0.0	0	0.0	1.33
Other German	3.5[c]	10.9	2[d]	5.4	0	0.0	5.50
Other European	0.83[e]	2.6	1[f]	2.7	0	0.0	1.83
Total European	18	56.3	8	21.6	0	0.0	26
B. American							
Northeast Region:							
Harvard	4[g]	12.5	4	10.8	2	7.7	10
MIT	1	3.1	4	10.8	2	7.7	7
Yale	3[h]	9.4	0	0.0	0	0.0	3
Columbia	1	3.1	1	2.7	0	0.0	2
Cornell	0	0.0	2	5.4	0	0.0	2
Pennsylvania	1[i]	3.1	1	2.7	0	0.0	2
Other Northeast	2[j]	6.3	2[k]	5.4	0	0.0	4
Total	12	37.5	14	37.8	4	15.4	30

Table 6.41. (continued)

Institution[a]	Period of presidency						
	1876–1915		1916–1955		1956–1981		
	Number	As percentage of total for period	Number	As percentage of total for period	Number	As percentage of total for period	Total
	(1)	(2)	(3)	(4)	(5)	(6)	(7)
North Central Region:							
Illinois	0	0.0	2	5.4	9	34.6	11
Michigan	1	3.1	2	5.4	1	3.8	4
Chicago	0	0.0	2	5.4	1	3.8	3
Kansas	0	0.0	2	5.4	0	0.0	2
Wisconsin	0	0.0	0	0.0	3	11.5	3
Other North Central	0	0.0	1[l]	2.7	5[m]	19.2	6
Total	1	3.1	9	24.3	19	73.1	29
Other Regions:							
Johns Hopkins	1	3.1	5	13.5	0	0.0	6
California	0	0.0	0	0.0	3	11.5	3
Caltech	0	0.0	1	2.7	0	0.0	1
Total	1	3.1	6	16.2	3	11.5	10
Total American	14	43.8	29	78.4	26	100.0	69
Total	32[n]	–	37[n]	–	26	–	95[m]

Table 6.41. (continued)

Note: Percentage columns may not add to 100% because of rounding errors. Percentage detail may not add to percentage subtotals because of rounding errors. See also Table 6.39.

[a] For the definitions of the Northeast and North Central regions of the United States, see Table 6.17, note a.
[b] S. W. Johnson studied under Erdmann at Leipzig, Liebig at Munich, and Frankland at Manchester; he is apportioned accordingly.
[c] J. L. Smith studied under Liebig at Giessen and Dumas and Orfila at Paris; he is apportioned accordingly. Others in this total are Booth (Berlin and Cassell), Genth (Marburg), and Long (Tübingen).
[d] One each from Berlin and Jena.
[e] See notes b and c for Johnson at Manchester and Smith at Paris.
[f] Ghent.
[g] Lawrence Scientific School.
[h] Includes Sheffield Scientific School.
[i] Pennsylvania Medical School.
[j] One each from Lafayette and Williams College (Massachusetts).
[k] One each from NYU and Princeton.
[l] Grinnell.
[m] One each from Indiana, Lawrence College (Missouri), Minnesota, Missouri, and Ohio State.
[n] Two chemists – E. F. Smith (Göttingen) and Herty (Johns Hopkins) – served as president in both of the two earlier periods shown; they are counted once in each.

Sources: Table 6.37. Biographical information from sources listed on p. 189, note 28.

TABLE 6.42.

Types of institutions employing American Chemical Society presidents during their tenures of office, 1876–1981

Period	Total	Academic[a]		Government		Industrial: Entrepreneurs[b]		Industrial: Employees	
		Number	As percentage of total	Number	As percentage of total	Number	As percentage of total	Number	As percentage of total
	(1)	(2)	(3)	(4)	(5)	(6)	(7)	(8)	(9)
1876–1885	8	4[c]	50.0	0.5[c]	6.3	3.5[c]	43.8	0	0.0
1886–1895	9	7	77.8	1	11.1	1	11.1	0	0.0
1896–1905	9	6	66.7	1	11.1	0	0.0	2	22.2
1906–1915	8	5.5[d]	68.8	0.5[d]	6.3	1	12.5	1	12.5
1916–1925	8	6	75.0	0	0.0	2	25.0	0	0.0
1926–1935	10	6	60.0	0	0.0	2	20.0	2	20.0
1936–1945	10	8	80.0	0	0.0	0	0.0	2	20.0
1946–1955	10	5.5[e]	55.0	0	0.0	1	10.0	3.5[e]	35.0
1956–1965	10	6	60.0	0	0.0	0	0.0	4	40.0
1966–1975	10	4	40.0	0	0.0	4	40.0	2	20.0
1976–1981	6	4	66.7	0	0.0	1	16.7	1	16.7

Table 6.42. (continued)

See Figure 6.5–3.

Note: Row percentages may not add to 100% because of rounding errors. Five chemists – Chandler, Herty, Hunt, Norris, and E. F. Smith – served as president in two of the periods shown; they are counted once in each. See Table 6.43 for employment experiences both before and during tenures of office.

[a] Includes joint appointments in pharmacy schools (1881, 1886, 1889, and 1906); a medical school chemistry chait (1903); and directorship of the Mellon Institute for Industrial Research (1937).
[b] Includes proprietors and partners of private consulting laboratories or firms and chemist-entrepreneurs (e.g., Baekeland, Nichols, Redman, and Rosengarten) who worked in companies which they or their families had established.
[c] Johnson is apportioned between academic and government; Genth between academic and industrial entrepreneurs.
[d] Hillebrand is apportioned between academic and government.
[e] Fisher is apportioned between academic and industrial employees.

Sources: Table 6.37. Biographical information from sources listed on p. 189, note 28.

TABLE 6.43.

Employment experience of American Chemical Society presidents through tenures of office, 1876–1981

Type of employment	Period of presidency							
	1876–1885[a]		1886–1895[b]		1896–1905[c]		1906–1915[d]	
	Number	As percentage of total for period	Number	As percentage of total for period	Number	As percentage of total for period	Number	As percentage of total for period
Academic *only*	1	12.5	5	55.6	6	66.7	4	50.0
Government	5	62.5	3	33.3	1.5	16.7	1.5	18.8
State/local[e]	4	50.0	2	22.2	0	0.0	0	0.0
Federal[f]	1	12.5	1	11.1	1.5	16.7	1.5	18.8
Industrial	2	25.0	1	11.1	1.5	16.7	2.5	31.3
Entrepreneurs[g]	2	25.0	0.5	5.6	0	0.0	1	12.5
Employees	0	0.0	0	0.0	1.5	16.7	1.5	18.8
Chemical firms[h]	0	0.0	0.5	5.6	0.5	5.6	0.5	6.3
Other firms[i]	0	0.0	0	0.0	1	11.1	1	12.5
Total	8	–	9	–	9	–	8	–

	1916–1925[j]		1926–1935		1936–1945	
	Number	As percentage of total for period	Number	As percentage of total for period	Number	As percentage of total for period
Academic *only*	4	50.0	5	50.0	5	50.0
Government	1.5	18.8	0	0.0	2	20.0
State/local[e]	0	0.0	0	0.0	1	10.0
Federal[f]	1.5	18.8	0	0.0	1	10.0

Table 6.43. (continued)

Type of employment	Period of presidency					
	1916–1925[j]		1926–1935		1936–1945	
	Number	As percentage of total for period	Number	As percentage of total for period	Number	As percentage of total for period
Industrial	2.5	31.3	5	50.0	3	30.0
Entrepreneurs[g]	1.5	18.8	3	30.0	1[k]	10.0
Employees	1	12.5	2	20.0	2	20.0
Chemical firms[h]	1	12.5	1	10.0	2	20.0
Other firms[i]	0	0.0	1	10.0	0	0.0
Total	8	–	10	–	10	–

	1946–1955[l]		1956–1965[m]		1966–1975[n]		1976–1981[o]	
	Number	As percentage of total for period	Number	As percentage of total for period	Number	As percentage of total for period	Number	As percentage of total for period
Academic *only*	5	50.0	6	60.0	3	30.0	3	50.0
Government	0	0.0	0.5	5.0	0.5	5.0	1	16.7
State/local[e]	0	0.0	0.5	5.0	0	0.0	0	0.0
Federal[f]	0	0.0	0	0.0	0.5	5.0	1	16.7
Industrial	5	50.0	3.5	35.0	6.5	65.0	2	33.3
Entrepreneurs[g]	0.5	5.0	0	0.0	2	20.0	0.5	8.3
Employees	4.5	45.0	3.5	35.0	4.5	45.0	1.5	25.0
Chemical firms[h]	4	40.0	3	30.0	4.5	45.0	1.5	25.0
Other firms[i]	0.5	5.0	0.5	5.0	0	0.0	0	0.0
Total	10	–	10	–	10	–	6	–

Table 6.43. (continued)

Note: Five chemists – Chandler, Herty, Hunt, Norris, and E. F. Smith – served as president in two of the periods shown; they are counted once in each. Three years used as a threshold for inclusion in any category. Chemists listed in government or industrial positions may also have had academic posts; those listed as academic did not work three years or more in an off-campus capacity.

[a] Booth, Genth, Hunt, and J. L. Smith are apportioned between state/local government and industrial entrepreneurs.

[b] Goessmann is apportioned between state/local government and industrial employees in chemical firms. Hunt is apportioned between state/local government and industrial entrepreneurs.

[c] McMurtrie is apportioned between federal government and industrial employees in chemical firms.

[d] Herty is apportioned between federal government and industrial employees in chemical firms.

[e] Includes state assayers and chemists for geological surveys, agricultural experiment stations, water surveys, and public health departments.

[f] Includes Bureaus of Chemistry in the Department of Agriculture, the Forest Service, and the Geological Survey; the Ordnance Department of the Confederate Army; and the Atomic Energy Commission.

[g] Includes proprietors and partners of private consulting laboratories or firms and chemist-entrepreneurs who worked in companies which they or their families had established (e.g., A. D. Little, Inc.; Bakelite Corporation; Dewey & Almy Chemical Company; William H. Nichols Company; and Powers-Weightman-Rosengarten Chemical Company).

[h] For example, Onandaga Salt Company; New York Tartar Company; Du Pont; Dow; Monsanto; Esso Research and Engineering Co.; Abbott; Merck; Hercules, Inc.; and G. D. Searle & Co., Inc.

[i] For example, General Electric Company or Pennsylvania Railroad.

[j] Herty is apportioned between federal government and industrial employees in chemical firms. Baekeland is apportioned between industrial entrepreneurs and industrial employees in chemical firms.

[k] E. R. Weidlein, Director of Mellon Institute for Industrial Research.

[l] Dewey is apportioned between industrial entrepreneurs and industrial employees in non-chemical firms.

[m] Elder is apportioned between state/local government and industrial employees in chemical firms. Rassweiler is apportioned between industrial employees in chemical firms and those in non-chemical firms.

[n] Brode is apportioned between federal government and industrial entrepreneurs. Friedman, Nixon, and Riegel are apportioned between industrial entrepreneurs and industrial employees of chemical firms.

[o] Hill is apportioned between industrial entrepreneurs and industrial employees of chemical firms.

Sources: Table 6.37. Biographical information from sources listed in p. 189, note 28.

TABLE 6.44.

Leading institutions with which American Chemical Society presidents were affiliated, 1876–1981

Institution	Presidents affiliated	
	Number	Names
University of Illinois	18	Adams, Bailar, Bartow, Brode, Elder, Folkers, Friedman, McMurtrie, Marvel, W. A. Noyes, Overberger, Parr, Price, Rassweiler, Riegel, Sparks, Stacy, Volwiler
Harvard University	15	Adams, Bancroft, Caldwell, Chandler, Clarke, Cope, Daniels, Lamb, Munroe, Price, Richards, Riegel, Tishler, Whitmore, Wiley
Massachusetts Institute of Technology	12	Cope, Dewey, Frolich, Hill, Kraus, Lind, Little, Norris, A. A. Noyes, Thomas, Whitney, Zettlemoyer
Universität Göttingen	11	Bartow, Caldwell, Chandler, Goessmann, Langmuir, Mallet, Nason, Remsen, Richards, E. F. Smith, Venable
Johns Hopkins University	7	Franklin, Herty, Holmes, Norris, W. A. Noyes, Reese, Remsen
University of Pennsylvania	7	Barker, Booth, Genth, Hildebrand, Price, Rosengarten, E. F. Smith
University of Chicago	6	Evans, McPherson, W. A. Noyes, Jr., A. Smith, Stieglitz, Williams
Columbia University	6	Bogert, Chandler, Cope, Fisher, Langmuir, A. Smith
Cornell University	6	Bancroft, Caldwell, Hillebrand, Midgley, Parr, Stacy

Table 6.44. (continued)

	Presidents affiliated	
Institution	Number	Names
Federal government laboratories[a]	6	Clarke, Hillebrand, Lind, McMurtrie, W. A. Noyes, Wiley
Universität Leipzig	6	Bancroft, Johnson, Lind, A. A. Noyes, Richards, Whitney
University of California	5	Calvin, Eyring, Hildebrand, Nixon, Seaborg
University of Michigan	5	Bailar, Britton, Gomberg, Lind, Prescott
University of Wisconsin	5	Cope, Daniels, D'Ianni, Eyring, Folkers
Merck, Inc.	4	Folkers, Frolich, Rosengarten, Tishler
Ohio State University	4	Brode, Cairns, Evans, McPherson

Note: Four presidents used as a threshold for inclusion. A president is considered as having been affiliated with an institution if, at any time in his post-baccalaureate career, he was a student, post-doctoral fellow, research worker, or faculty member there.

a Includes Bureaus of Chemistry in the Department of Agriculture, the Forest Service, and the Geological Survey, plus the Bureau of Mines and the National Bureau of Standards. Excludes non-laboratory administration (e.g., Seaborg at Atomic Energy Commission).

Sources: Table 6.37. Biographical information from sources listed on p. 189, note 28.

TABLE A.1

Exponential trend in the number of chemists, 1900–1930
(thousands)

Year	Census estimates (1)	Calculated trend (exponential) (2)
1900	9	9
1910	16	16
1920	28	27
1930	45	46

See Figure A.2–2.

Note: Cf. Table 2.2, column 2.

Sources: *Column 1* – Table 2.2, column 1.
Column 2 – calculated by regression on column 1; see Appendix E.

TABLE A.2.

Series used to adjust the annual number of chemistry doctorates conferred, 1920–1961

Year	Doctorates in chemical-materials engineering[a] (1)	Doctorates in biochemistry[b] (2)
1920	6	9
1921	6	7
1922	9	18
1923	6	11
1924	8	20
1925	9	22
1926	14	32
1927	18	44
1928	25	37
1929	20	38
1930	38	41
1931	39	36
1932	38	37
1933	51	49
1934	63	59
1935	58	57
1936	35	11
1937	45	22
1938	35	108
1939	31	116
1940	50	127
1941	56	104
1942	43	137
1943	24	115
1944	34	97
1945	34	72
1946	44	30
1947	50	45
1948	102	91
1949	164	123
1950	204	125
1951	202	140

Tables A.2 (continued)

Year	Doctorates in chemical-materials engineering[a] (1)	Doctorates in biochemistry[b] (2)
1952	203	152
1953	177	179
1954	204	202
1955	217	210
1956	222	178
1957	213	237
1958	201	231
1959	214	208
1960	247	272
1961	262	285

See Figures A.3–1 and A.3–2.

[a] See Table 3.9, note a.
[b] U.S. Office of Education specialty code 0414.

Sources: *Column 1* – IB, Adkins, 1975, Table A–5.10, 290–293.
Column 2 – IB, Harmon and Soldz, 1963, Table 26, 50–53.

TABLE A.3.

Series used to adjust the annual number of chemistry degrees conferred, 1962–1975

Year	Chemical-materials engineering degrees conferred[a]		Biochemistry[b] degrees conferred		
	Second-level	Doctorate	Bachelors	Second-level	Doctorate
	(1)	(2)	(3)	(4)	(5)
1962	946	315	141	178	183
1963	1073	347	174	196	212
1964	1102	387	190	207	264
1965	1173	545	200	236	290
1966	1367	517	264	231	315
1967	1357	504	262	257	331
1968	1572	559	295	255	442
1969	1545	627	347	269	471
1970	1446	686	455	240	449
1971	1537	658	568	251	517
1972	1643	636	699	252	462
1975[c]	1461	556	1355	270	437

See Figures A.3–1 and A.3–2.

Note: The dots in Parts A–C of Figure A.3–1 consist of columns 5 in Tables 3.1, 3.2 and 3.3, respectively. For the triangles in Part A, add the sum of column 3 above and Table 3.9, column 1, to the numerator during the computation of Table 3.1, column 5. For the triangles in Part B, add columns 1 and 4 to the numerator during the computation of Table 3.2, column 5. For the triangles in Part C, add columns 2 and 5 to the numerator during the computation of Table 3.3, column 5. For the squares in Part C, add the two series in Table A.2 to the numerator during the computation of Table 3.3, column 5. For the dots and triangles connected by solid lines in Figure A.3–2, see Table 3.14, columns 5 and 6, respectively. For the dots connected by dashes, divide Table 3.9, column 2, by Table 3.14, column 2. For the triangles connected by dashes, divide Table A.6, column 2, by Table 3.14, column 4. For the dotted line, divide the sum of Table A.6, column 2, and either column 5 above or Table A.2, column 2, by Table 3.14, column 4.

[a] For the definition of this aggregate subject, see Table 3.9, note a.

[b] U.S. Office of Education specialty code 0414.

[c] In 1975, 3835 bachelors degrees were conferred in chemical-materials engineering. This additional data point was used in construction of Figure A.3–2. See source for 1975 data.

Table A.3. (continued)

Sources: *Columns 1 and 2*:

1962 to 1971 – IB, Adkins, 1975, Table A–5.10, 290–293.
1972 – IA, NCES, 1975b, Table 5, 13–14.
1975 – IA, NCES, 1977, Table 5, 16–17.

Columns 3 to 5:

1962 – IA, USOE, 1963b, Table 6, 12.
1963 – IA, USOE, 1965, Table 16, 45.
1964 – IA, USOE, 1966, Table 18, 79.
1965 – IA, USOE, 1967, Table 8, 49.
1966 – IA, USOE, 1968, Table 9, 11.
1967 – IA, NCES, 1968, Table 9, 12.
1968 – IA, NCES, 1969, Table 5, 11.
1969 – IA, NCES, 1971, Table 5, 9.
1970 – IA, NCES, 1972, Table 5, 11.
1971 – IA, NCES, 1973, Table 5, 13.
1972 – IA, NCES, 1975b, Table 5, 12.
1975 – IA, NCES, 1977, Table 5, 15.

TABLE A.4.

Second-level degrees conferred in chemical-materials engineering, 1890–1971

Year	Degrees conferred	Year	Degrees conferred	Year	Degrees conferred
1890	14	1920	113	1950	855
1891	10	1921	132	1951	989
1892	10	1922	156	1952	746
1893	16	1923	172	1953	550
1894	19	1924	190	1954	625
1895	19	1925	209	1955	635
1896	21	1926	230	1956	714
1897	20	1927	209	1957	771
1898	20	1928	208	1958	770
1899	22	1929	184	1959	834
1900	22	1930	205	1960	857
1901	24	1931	237	1961	900
1902	26	1932	280	1962	946
1903	25	1933	266	1963	1073
1904	24	1934	257	1964	1102
1905	27	1935	250	1965	1173
1906	24	1936	244	1966	1367
1907	24	1937	243	1967	1357
1908	28	1938	249	1968	1572
1909	33	1939	256	1969	1545
1910	31	1940	281	1970	1446
1911	35	1941	256	1971	1537
1912	47	1942	246		
1913	46	1943	163		
1914	58	1944	110		
1915	68	1945	155		
1916	77	1946	222		
1917	72	1947	430		
1918	64	1948	1086		
1919	89	1949	920		

Note: For the definition of this aggregate subject, see Table 3.9, note a.

Source: IB, Adkins, 1975, Table A–5.10, 290–293.

TABLE A.5.

Persons certified with a chemistry-related bachelors degree, 1930–1971

Year	Holders of terminal bachelors degrees: Chemical-materials engineering[a] (1)	Holders of terminal bachelors degrees: Chemistry-related fields[b] (2)	Second-level chemistry-related degrees conferred in preceding forty-one years (3)	Persons whose only chemistry-related degree received over preceding forty-one years is a bachelors degree (approximation)[c] (4)
1930	20 600	51 823	13 637	102 520
1931	21 916	54 665	14 388	107 451
1932	23 324	57 432	15 242	112 256
1933	24 700	60 227	16 028	117 098
1934	26 030	63 090	16 675	121 829
1935	27 281	66 034	17 299	126 403
1936	28 502	69 091	17 902	130 945
1937	29 705	72 453	18 534	135 905
1938	30 985	76 208	19 226	141 305
1939	32 333	80 269	19 989	147 120
1940	33 843	84 832	20 842	153 453
1941	35 362	89 399	21 623	159 770
1942	36 924	94 137	22 348	166 170
1943	38 091	97 609	22 756	171 076
1944	38 910	99 951	22 985	174 668
1945	39 685	102 396	23 361	178 270
1946	40 377	104 710	23 901	181 888
1947	42 278	109 157	25 119	187 896
1948	44 785	114 817	27 396	195 643
1949	47 983	123 744	29 716	206 847

Table A.5. (continued)

Year	Holders of terminal bachelors degrees		Second-level chemistry-related degrees conferred in preceding forty-one years	Persons whose only chemistry-related degree received over preceding forty-one years is a bachelors degree (approximation)[c]
	Chemical-materials engineering[a] (1)	Chemistry-related fields[b] (2)	(3)	(4)
1950	51 641	134 234	32 032	219 705
1951	54 443	141 121	34 376	228 975
1952	56 419	145 835	36 350	235 822
1953	57 887	149 502	37 875	241 185
1954	59 091	153 306	39 315	246 796
1955	60 165	156 810	40 829	251 687
1956	61 636	160 423	42 358	256 893
1957	63 479	164 922	43 790	263 113
1958	65 542	169 958	45 428	270 034
1959	67 673	175 255	47 226	277 398
1960	69 560	180 506	49 072	284 282
1961	71 282	185 457	50 972	290 760
1962	72 667	190 476	52 974	297 161
1963	74 025	196 128	55 063	304 269
1964	75 534	202 561	57 294	311 846
1965	76 934	208 882	59 789	318 982
1966	77 801	213 887	62 499	325 123
1967	78 683	218 740	65 106	331 249
1968	79 643	224 708	68 307	338 813
1969	80 957	231 823	71 507	347 484

Table A.5. (continued)

Year	Holders of terminal bachelors degrees		Second-level chemistry-related degrees conferred in preceding forty-one years	Persons whose only chemistry-related degree received over preceding forty-one years is a bachelors degree (approximation)[c]
	Chemical-materials engineering[a] (1)	Chemistry-related fields[b] (2)	(3)	(4)
1970	82471	237986	74550	355171
1971	83672	243440	77729	362323

See Figure A.4–1.

Note: Column 2 is plotted as solid dots in Figure A.4–1A. Column 4 appears as solid triangles. Table 3.11, column 5, provides the open dots, and Table 3.10, column 3, the solid squares. To obtain the dots and triangles in Figure A.4–1B, divide Table 3.11, column 5, by columns 2 and 4 above, respectively, and multiply by 100. Cf. Table 3.10.

[a] For the definition of this aggregate subject, see Table 3.9, note a.
[b] For the definition of this aggregate subject, see Table 3.9, note b.
[c] Assumes that all persons who receive a doctorate in a chemistry-related field have first received a second-level degree in one of these fields, and that all those receiving a second-level degree in this area have first received a bachelors in the same area.

Sources: *Column 1* – IB, Adkins, 1975, Table B–4.10, 484–485.
Column 2 – add column 1 and Table 2.6, column 3, then subtract Table 2.6, column 2.
Column 3 – sum over appropriate intervals of both series in Table A.4 and Table 3.2, column 4.
Column 4 – subtract column 3 from Table 3.10, column 3.

TABLE A.6.
Chemistry-related doctorate degrees conferred, 1890–1971

Year	Chemical-materials engineering[a] (1)	Chemistry-related fields[b] (2)
1890	1	29
1891	1	36
1892	1	36
1893	1	41
1894	2	53
1895	2	51
1896	1	49
1897	2	60
1898	2	59
1899	2	65
1900	2	71
1901	2	68
1902	2	54
1903	2	63
1904	2	62
1905	2	69
1906	3	72
1907	2	65
1908	3	71
1909	3	82
1910	3	81
1911	4	91
1912	4	98
1913	4	91
1914	5	102
1915	5	112
1916	6	121
1917	6	127
1918	5	101
1919	3	68
1920	6	110
1921	6	143
1922	9	156
1923	6	173
1924	8	226

Table A.6. (continued)

Year	Chemical-materials engineering[a] (1)	Chemistry-related fields[b] (2)
1925	9	229
1926	14	262
1927	18	229
1928	25	257
1929	20	249
1930	38	370
1931	39	407
1932	38	400
1933	51	471
1934	63	499
1935	58	460
1936	35	496
1937	45	555
1938	35	474
1939	31	533
1940	50	582
1941	56	714
1942	43	649
1943	24	543
1944	34	582
1945	34	376
1946	44	361
1947	50	468
1948	102	671
1949	164	921
1950	204	1171
1951	202	1257
1952	203	1241
1953	177	1182
1954	204	1236
1955	217	1226
1956	222	1211
1957	213	1221
1958	201	1148
1959	214	1241
1960	247	1309
1961	262	1407

Table A.6. (continued)

Year	Chemical-materials engineering[a] (1)	Chemistry-related fields[b] (2)
1962	315	1456
1963	347	1603
1964	387	1690
1965	545	1973
1966	517	2100
1967	504	2268
1968	559	2344
1969	627	2607
1970	686	2910
1971	658	2832

Note: Cf. Table 3.9.

a For the definition of this aggregate subject, see Table 3.9, note a.
b For the definition of this aggregate subject, see Table 3.9, note b.

Sources: *Column 1* – IB, Adkins, 1975, Table A–5.10, 290–293.
Column 2 – add column 1 and Table 3.3, column 4.

TABLE A.7.

Persons credentialled with a chemistry-related doctorate degree, 1930–1971

Year	Holders of terminal doctorate degrees			Persons who received a doctorate in fields indicated during preceding forty-one years	
	Chemistry[a] (1)	Chemical-materials engineering[b] (2)	Chemistry-related fields[c] (3)	Chemistry[a] (4)	Chemistry-related fields[c] (5)
1930	3 596	206	3 802	4 310	4 552
1931	3 899	240	4 139	4 650	4 930
1932	4 192	273	4 465	4 977	5 294
1933	4 538	318	4 856	5 362	5 729
1934	4 895	373	5 268	5 758	6 187
1935	5 216	424	5 640	6 109	6 594
1936	5 589	453	6 042	6 521	7 039
1937	6 005	491	6 496	6 983	7 545
1938	6 350	520	6 870	7 364	7 959
1939	6 751	544	7 295	7 809	8 433
1940	7 176	586	7 762	8 278	8 950
1941	7 719	633	8 352	8 867	9 593
1942	8 208	668	8 876	9 407	10 174
1943	8 609	684	9 293	9 874	10 663
1944	9 034	710	9 744	10 361	11 182
1945	9 257	735	9 992	10 643	11 496
1946	9 453	770	10 223	10 893	11 788
1947	9 743	809	10 552	11 242	12 184

Table A.7. (continued)

Year	Holders of terminal doctorate degrees			Persons who received a doctorate in fields indicated during preceding forty-one years	
	Chemistry[a]	Chemical-materials engineering[b]	Chemistry-related fields[c]	Chemistry[a]	Chemistry-related fields[c]
	(1)	(2)	(3)	(4)	(5)
1948	10 174	897	11 071	11 748	12 790
1949	10 780	1042	11 822	12 437	13 640
1950	11 581	1222	12 803	13 325	14 729
1951	12 460	1401	13 861	14 302	15 905
1952	13 317	1579	14 896	15 253	17 055
1953	14 133	1732	15 865	16 164	18 139
1954	14 966	1909	16 875	17 109	19 284
1955	15 769	2098	17 867	18 021	20 408
1956	16 543	2289	18 832	18 903	21 507
1957	17 327	2471	19 798	19 796	22 607
1958	18 042	2641	20 683	20 622	23 628
1959	18 822	2821	21 643	21 553	24 768
1960	19 627	3031	22 658	22 550	26 009
1961	20 481	3251	23 732	23 591	27 306
1962	21 318	3519	24 837	24 595	28 619
1963	22 253	3813	26 066	25 704	30 066
1964	23 221	4143	27 364	26 840	31 583
1965	24 293	4617	28 910	28 050	33 330
1966	25 496	5063	30 559	29 413	35 201

Table A.7. (continued)

Year	Holders of terminal doctorate degrees			Persons who received a doctorate in fields indicated during preceding forty-one years	
	Chemistry[a] (1)	Chemical-materials engineering[b] (2)	Chemistry-related fields[c] (3)	Chemistry[a] (4)	Chemistry-related fields[c] (5)
1967	26 851	5492	32 343	30 929	37 207
1968	28 208	5970	34 178	32 503	39 322
1969	29 731	6507	36 238	34 251	41 672
1970	31 465	7096	38 561	36 246	44 333
1971	33 134	7654	40 788	38 088	46 795

See Figures A.4–3 and A.4–4.

Note: Columns 3 and 5 provide the solid dots and triangles, respectively, for Table A.4–3. The open dots are from Table 2.6, column 1. For the dots and triangles in Figure A.4–4A, divide Table 2.6, column 1, by columns 4 and 5 above, respectively, then multiply by 100. For the dots and triangles in Part B, do the same thing, using columns 1 and 3 above as denominators.

[a] For the definition of "chemistry" as used here, see Table 3.1, note b.
[b] For the definition of this aggregate subject, see Table 3.9, note a.
[c] For the definition of this aggregate subject, see Table 3.9, note b.

Sources: *Columns 1 and 2* – IB, Adkins, 1975. Chemistry is from Table B–4.5, 474–475; chemical-materials engineering, Table B–4.10, 484–485.

Column 3 – add columns 1 and 2.

Column 4 – sum over the appropriate forty-one-year intervals of Table 3.3, column 4.

Column 5 – sum over the appropriate forty-one-year intervals of Table A.6, column 2.

TABLE B.1.
The chemical process industries

Standard industrial classification code: Major group	Industry group	Name
20		*Food and kindred products*
	201	Meat products
	202	Dairy products
	203	Canned and preserved fruits and vegetables
	204	Grain mill products
	205	Bakery products
	206	Sugar and confectionery products
	207	Fats and oils
	208	Beverages
	209	Miscellaneous food preparations and kindred products
26		*Paper and allied products*
	261	Pulp mills
	262	Paper mills, except building paper mills
	263	Paperboard mills
	264	Converted paper and paperboard products, except containers and boxes
	265	Paperboard containers and boxes
	266	Building paper and building board mills
28		*Chemicals and allied products*
	281	Industrial inorganic chemicals
	282	Plastics materials and synthetic resins, synthetic rubber, synthetic and other man-made fibers, except glass
	283	Drugs
	284	Soaps, detergents and cleaning preparations, perfumes, cosmetics and other toilet preparations
	285	Paints, varnishes, lacquers, enamels, and allied products
	286	Industrial organic chemicals
	287	Agricultural chemicals
	289	Miscellaneous chemical products
29		*Petroleum refining and related industries*
	291	Petroleum refining
	295	Paving and roofing materials
	299	Miscellaneous products of petroleum and coal

Table B.1. (continued)

Standard industrial classification code		
Major group	Industry group	Name
30		*Rubber and miscellaneous plastics and products*
	301	Tires and inner tubes
	302	Rubber and plastic footwear
	303	Reclaimed rubber
	304	Rubber and plastics hose and belting
	306	Fabricated rubber products, not elsewhere classified
	307	Miscellaneous plastics products
32		*Stone, clay, glass, and concrete products*
	321	Flat glass
	322	Glass and glassware, pressed or blown
	323	Glass products, made of purchased glass
	324	Cement, hydraulic
	325	Structural clay products
	326	Pottery and related products
	327	Concrete, gypsum and plaster products
	328	Cut stone and stone products
	329	Abrasive, asbestos, and miscellaneous non-metallic mineral products
33		*Primary metal industries*
	331	Blast furnaces, steel work, and rolling and finishing mills
	332	Iron and steel foundries
	333	Primary smelting and refining of nonferrous metals
	334	Secondary smelting and refining of nonferrous metals
	335	Rolling, drawing and extruding of nonferrous metals
	336	Nonferrous foundries (castings)
	339	Miscellaneous primary metal products

Source: II, OMB, 1972, 59–69, 100–105, 111–132, 136–152.

TABLE B.2.
The chemicals and allied products industry

Standard industrial classification code: Industry group	Industry	Name
281		*Industrial inorganic chemicals*
	2812	Alkalies and chlorine
	2813	Industrial gases
	2816	Inorganic pigments
	2819	Industrial organic chemicals, not elsewhere classified
282		*Plastics materials and synthetic resins, synthetic rubber, synthetic and other man-made fibers, except glass*
	2821	Plastics materials, synthetic resins, and nonvulcanizable elastomers
	2822	Synthetic rubber (vulcanizable elastomers)
	2823	Cellulosic man-made fibers
	2824	Synthetic organic fibers, except cellulosic
283		*Drugs*
	2831	Biological products
	2833	Medicinal chemicals and botanical products
	2834	Pharmaceutical preparations
284		*Soaps, detergents, and cleaning preparations, perfumes, cosmetics, and other toilet preparations*
	2841	Soaps and other detergents, except specialty cleansers
	2842	Specialty cleaning, polishing, and sanitation preparations
	2843	Surface active agents, finishing agents, sulfonated oils and assistants
	2844	Perfumes, cosmetics, and other toilet preparations
285	2851	*Paints, varnishes, lacquers, enamels, and allied products*
286		*Industrial organic chemicals*
	2861	Gum and wood chemicals
	2865	Cyclic (coal tar) crudes, and cyclic intermediates, dyes, and organic pigments (lakes and toners)
	2869	Industrial organic chemicals, not elsewhere classified
287		*Agricultural chemicals*
	2873	Nitrogenous fertilizers

Table B.2. (continued)

Standard industrial classification code		
Industry group	Industry	Name
	2874	Phosphatic fertilizers
	2875	Fertilizers, mixing only
	2879	Pesticides and agricultural chemicals, not elsewhere classified
289		*Miscellaneous chemical products*
	2891	Adhesives and sealants
	2892	Explosives
	2893	Printing ink
	2895	Carbon black
	2899	Chemicals and chemical preparations, not elsewhere classified

Note: The Chemicals and Allied Products Industry constitutes Major Group 28 of the Standard Industrial Classification of the Office of Management and Budget.

Source: II, OMB, 1972, 111–126.

TABLE C.1.

Accuracy of regression curve for American citations as percentage of non-German citations in *Chemische Berichte*, 1939–1969

Year	Observed by Computer Horizons (percentage) (1)	Logistic trend calculated from our data (percentage) (2)	Ratio of column 1 to column 2 (3)
1939	18.7	30.3	0.62
1949	36.2	36.1	1.00
1959	41.1	41.3	1.00
1969	54.3	45.6	1.19

Sources: *Column 1* – IB, Computer Horizons, 1972, Table 68, 71–72. For each year, divide citations in *Chemische Berichte* going to American papers by the difference between total citations and citations to German papers.

Column 2 – calculated by nonlinear regression on Table 6.7, column 5; see Appendix E.

TABLE E.1.

Estimated parameters from linear least-squares regression analyses (linear equation)

Observed values (Y)						Location of		
Location			Estimated parameters[a]		Pearson's r^2	Predicted values		
Table	Column	Number	K	M		Table	Column	Figure
			(1)	(2)	(3)	(4)	(5)	(6)
5.15	2	21	−1102.316 017	0.567 532 467 5	0.988	–	–	5.2–4
6.9	3	13	−1033.667 191	0.546 633 640 5	0.898	6.9	4	6.2–2 6.2–3
6.11	1 and 2	8	−1118.393 865	0.601 776 214 2	0.961	6.11	3	6.2–3
6.11	4	13	−899.332 591 7	0.468 535 646 1	0.944	6.11	5	6.2–3
6.12	1 and 2	8	1254.260 224	−0.633 234 903 8	0.972	6.12	3	6.2–3
6.12	4	16	1371.762 096	−0.670 600 309 7	0.852	6.12	5	6.2–3

Note: See Appendix E for explanation.

[a] Parameters are for the equation Y = K + MT, where Y represents the observed value and T the year. Since these parameters are intended for use as intermediates in calculating predicted values, all digits from the calculator display are provided, although they are significant to only a few places at best (see Appendix D).

Source: calculated using the linear regression subroutine on a Texas Instruments Programmable 58 calculator.

TABLE E.2.

Estimated parameters from log-linear least-squares regression analyses (exponential equation)

Observed values			Estimated parameters[a]				Location of		
Location							Predicted values		
Table	Column	Number	ln K (1)	ln B (2)	Annual growth rate R (%) (3)	Pearson's r^2 [b] (4)	Table (5)	Column (6)	Figure (7)
2.2	1	11	−89.709 405 34	0.048 217 631 6	4.9	0.936	2.2	2	2.1–2
2.4	1	104	−121.804 701 8	0.067 968 289	7.0	0.939	2.4	2	2.2–3
3.1	1	89	−95.843 072 8	0.055 503 545 1	5.7	0.984	–	–	3.1–4
									3.1–6
3.1	2	89	−72.655 530 17	0.042 561 130 1	4.3	0.978	–	–	3.1–6
3.1	4	89	−52.061 630 62	0.031 210 096 7	3.2	0.942	–	–	3.1–6
3.2	1	89	−68.011 847 17	0.040 647 356 5	4.1	0.916	–	–	3.1–4
									3.1–7
3.2	2	89	−87.581 810 38	0.049 259 662 9	5.0	0.959	–	–	3.1–7
3.2	4	89	−65.308 024 52	0.037 024 309 2	3.8	0.945	–	–	3.1–7
3.3	1	89	−112.419 715 5	0.062 117 281 9	6.4	0.974	–	–	3.1–4
									3.1–5
3.3	2	89	−102.009 08	0.056 240 28	5.8	0.974	–	–	3.1–5
3.3	4	89	−90.668 458 13	0.049 841 260 9	5.1	0.967	–	–	3.1–5
5.15	1	21	−85.528 960 11	0.044 913 636 8	4.6	0.987	–	–	5.2–4
5.16	1	28	−79.880 599 36	0.047 323 663 9	4.8	0.991	6.29	5	5.3–1
5.16	2	13	−73.655 971 69	0.041 885 124 7	4.3	0.974	5.17	2	5.3–1
							5.18	1	
							6.29	2	
5.16	3	10	−74.787 521 02	0.042 092 535 7	4.3	0.873	–	–	5.3–1

Table E.2. (continued)

Observed values			Estimated parameters[a]				Location of		
Location							Predicted values		
Table	Column	Number	ln K (1)	ln B (2)	Annual growth rate R (%) (3)	Pearson's r^2 [b] (4)	Table (5)	Column (6)	Figure (7)
5.16	4	2	110.287 767	0.060 888 556 4	6.3		–	–	5.3–1
5.16	5	21	–101.152 203 6	0.053 015 424 2	5.4	0.943			5.3–1
6.8[c]	1	14	–96.980 658 36	0.055 376 213 8	5.7	0.954	–	–	–
6.8[c]	4	14	–84.416 767 31	0.048 383 417 9	5.0	0.970	-	–	–
A.1	1	4	–100.151 680 3	0.053 879 295 2	5.5	0.998	A.1	2	A.2–2

Note: See Appendix E for explanation.

[a] Parameters are for the equation ln Y = ln K + (ln B)T, where ln Y represents the logarithm of the observed value and T the year. Since ln K and ln B are intended for use as intermediates in calculating predicted values, all digits from the calculator display are provided, although they are significant to only a few places at best (see Appendix D).
[b] This coefficient is not entirely appropriate for log-linear regressions; see Appendix E.
[c] Calculated in support of the assumption that the number of American and non-German chemical papers have been growing exponentially. For formulas used to calculate Table 6.8, columns 2, 3, 5, and 6, see Appendix C.

Source: calculated using the linear regression subroutine on a Texas Instruments Programmable 58 calculator.

TABLE E.3.

Estimated parameters from nonlinear least-squares regression analyses (Gompertz and logistic equations)

Observed values									Location of		
Location			Estimated parameters[a]			Asymptotic standard errors[b]			Predicted values		
Table	Column	Number	K (1)	A (2)	B (3)	K (4)	A (5)	B (6)	Table (7)	Column (8)	Figure (9)
A. Gompertz equation[c]											
2.2	1	11	255.687 01	0.002 933 11	0.980 952 33	24.0	0.005 33	0.002 48	2.2	3	2.1–3 2.1–4
B. Logistic equations[d]											
2.3	3	11	14.598 628	3.098 185 2	–0.059 329 31	0.349	0.0876	–	2.3	4	2.1–5
2.4	1	91	128 912.30	7.604 364 9	–0.093 627 56	1490	0.0381	–	2.4	3	2.2–4
2.4	1	47	18 972.708	7.552 141 8	–0.165	329	0.0650	–	2.4	4	2.2–7
2.4	1	45	122 008.04	9.042 052 7	–0.112 790 56	956	0.0305	–	2.4	5	2.2–7
2.5	3	9	1.392 005 4	3.407 149 4	–0.046 795 05	0.236	0.411	–	2.5	4	2.2–6
5.13[e]	1	37	9 941.294 5	7.339 704 6	–0.087 1	285	0.0953	–	5.13	3	5.2–1 5.2–2
6.7	5	16	57.163 064	2.763 644 1	–0.041 828 84	5.07	0.233	–	6.7 C.1	6 2	6.2–1

Table E.3. (continued)

Note: See Appendix E for explanation.

[a] Since these parameters are intended for use as intermediates in calculating predicted values, the first eight substantive digits from the computer printout are provided, although they are significant to only a few places at best (see Appendix D).

[b] Given to three significant digits.

[c] Parameters are for the equation $Y = KA^{BT}$, where Y represents the observed value and T = (year-1870).

[d] Parameters are for the equation $Y = K/(1 + e^{A+BT})$, where Y represents the observed value and T = (year-1870).

[e] Observed values shown in table multiplied by 1000 before regression.

Source: calculated on an IBM 370 computer using the NLIN procedure in SAS79. See II, Barr *et al*., 1976, 193–199, and II, SAS Institute Inc., 1979, 317–329.

BIBLIOGRAPHY

An Introductory Note

This bibliographic essay is intended to offer readers a guide to the esoteric and sometimes bewildering sources for the construction of chemical indicators. These range from annual surveys conducted by federal government agencies, through periodic accountings by professional organizations, to isolated studies of some small aspect of chemistry in American culture. The easy assumption of students of modern science that historical time series on money, manpower, and other resources are readily available contrasts sharply with the acute paucity of reliable long-run data.

Citations in this essay are in abbreviated form; all works are included in the full bibliography that follows, in section III unless indicated otherwise. Only the most general sources are noted here. Readers interested in the bibliography of a particular subject are directed to the footnote discussion of the appropriate tables and text sections above.

I. Historiography of American Chemistry

Chemistry and chemical engineering in the United States offer unparalleled opportunities for historical research on the interactions of pure and applied science. Chemistry has always been closely tied to agriculture, industry, and government, and the history of chemistry departments in American universities exemplifies trends in the establishment of new markets for scientific expertise. Despite these and other intriguing possibilities, the literature on the history of American chemistry is sparse. Edgar Fahs Smith's 1914 *Chemistry in America* is the only book-length attempt to survey the entire field, but it suffers from the author's proclivity for heroic biography. Despite his bias toward academic chemistry, Edward Beardsley's slim monograph on the *Rise of the American Chemistry Profession* (1964) remains the basic guide to nineteenth-century developments in chemical education, occupations, and institutional growth. Kenneth L. Taylor's insightful survey of 'Two Centuries of Chemistry' (1976) achieves a more reasonable balance between academic and industrial chemistry, and his bibliographic notes are a helpful guide to the relevant literature. The recent festschrift for Aaron Idhe (Parascandola and Whorton (eds.), 1983) offers several detailed essays.

The most comprehensive sources on the history of American chemistry are the series of commemorative publications sponsored by the American Chemical Society. *A Half-Century of Chemistry in America, 1876–1926* (1926), edited by C. A. Browne, contains authoritative surveys of various specialties. Browne and Mary E. Weeks published a useful *History of the American Chemical Society* in 1952, which can be supplemented with *A Century of Chemistry* (IB, 1976), edited by Herman Skolnik and Kenneth M. Reese. The 6 April 1976 issue of *Chemical and Engineering News* is a centennial review of the history of the American Chemical Society and research specialties in chemistry. Wyndham D. Miles's *American Chemists and Chemical Engineers* (IB, 1976), also prepared for the Society's hundredth anniversary, contains informative biographical sketches of 517 chemical worthies.

Although these publications present a wealth of information on diverse aspects of American chemistry and chemical technology, there is a noticeable lack of analytical studies dealing with the social and institutional history of the chemistry profession. Margaret Rossiter's *Emergence of Agricultural Science* (1975), a detailed study of mid-nineteenth century agricultural chemistry, is a notable exception, as is Robert Kohler's *From Medical Chemistry to Biochemistry* (1982). But we still lack studies analogous to E. Schmauderer (ed.), *Der Chemiker im Wandel der Zeiten* (1973) and C. Russell, N. G. Coley, and G. K. Roberts, *Chemists by Profession* (1977), on Germany and Britain, respectively.

On the history of the chemical industry in the United States, see Williams Haynes's encyclopedic *American Chemical Industry: A History* (6 vols., 1945–1954), an indispensable compendium of biographical vignettes, institutional and corporate histories, and information on the development of industrial chemical processes in America. L. F. Haber's *Chemical Industry during the Nineteenth Century* (1958) and *The Chemical Industry, 1900–1930* (1971) place American developments in a European context. Chemical engineering is also beginning to attract the attention of historians: see the recent volumes of essays edited by William Furter (1980 and 1982), and the AIChE Diamond Jubilee history (Reynolds, 1983).

II. Science Indicators and Social Indicators

Measurements of the "health" of science have a long history. In recent years, considerable attention has been given to the problems of collecting and refining time series on various aspects of the contemporary scientific and technological enterprise. The biennial series of *Science Indicators* reports (IA, NSB, 1973, 1975, 1977, 1979, 1981, and 1983) are the most prominent example of this activity. The array of indicators presented in these reports, revised and extended in response to both scholarly critiques and bureaucratic imperatives, have aroused increasing interest among policy analysts and other students of science at home and abroad. The science indicators movement, as pointed out in Chapter 1, is an important reference point for our inquiries into chemistry in America.

The best entry into the critical literature on science indicators is the volume *Toward a Metric of Science* (II, 1978), edited by Y. Elkana *et al.* Other assessments include David Drew, 'Science Indicators Redux' (II, 1979); U.S. General Accounting Office, 'Science Indicators: Improvements Needed in Design, Construction, and Interpretation' (II, 1979); and essays in the fall 1978 issue of the *4S Newsletter* and the October 1980 issue of *Scientometrics*. A convenient guide to this literature before 1979 is Roger Hahn's *Bibliography of Quantitative Studies on Science and Its History* (II, 1980).

The relations between science indicators and other social indicators are explored in O. D. Duncan, 'Science Indicators and Social Indicators" (II, 1978) and Harriet Zuckerman and Roberta B. Miller, 'Indicators of Science: Notes and Queries' (II, 1980). Robert J. Rossi and Kevin J. Gilmartin cover all aspects of research in the field in their *Handbook of Social Indicators* (II, 1980). Many of the issues discussed in their book, from the advantages and disadvantages of various data sources to guidelines for preparing social indicators reports, are equally applicable to the construction and analysis of chemical indicators. On the social indicators literature in general, see Leslie D. Wilcox *et al.*, *Social Indicators and Societal Monitoring: An Annotated Bibliography* (II, 1972) and Kevin J. Gilmartin *et al.*, *Social Indicators: An Annotated Bibliography of Current Literature* (II, 1979).

III. Statistical Sources for Chemical Indicators

Statistics on scientific and technological activities in the United States are now generated routinely by a panoply of public and private agencies. Foremost among them are the Bureau of the Census, the Bureau of Labor Statistics, the National Center for Education Statistics, the National Research Council, the National Science Foundation, and the American Chemical Society. These agencies and others provide information on resources available for science and technology, the distribution of money and manpower among government, industry, and academe, and the demographic, educational, and economic characteristics of the population of scientists and engineers in the United States. The plethora of reports on postwar American science and technology is matched by the paucity of statistical information on the period before 1950.

The principal source of statistical information on scientific and technological activity in the United States since the early 1950s is the National Science Foundation. The NSF's Division of Science Resources Studies conducts periodic surveys of the level of funding for research and development and the dynamics of the labor market in science and engineering occupations. The *Surveys of Science Resources Series* published by the Division is indispensable for anyone contemplating a foray into the field of chemical indicators. One way to become familiar with the range of data available from the NSF is to begin with two recent overviews of their studies of science and technology resources: *National Patterns of Science and Technology Resources* (IA, 1984) and *Science and Engineering Personnel: A National Overview* (IA, 1982). The first of these summarizes available data on research and development expenditures in government, industry, and academe, placing scientific activity in the context of national economic trends. The second concentrates on the supply and utilization of scientists and engineers, analyzing the labor market in terms of employment sectors, fields of specialization, work activity, and occupational mobility. A third summary report focuses on academic science alone: *Academic Science/Engineering, 1972–83* (IA, 1984).

Perhaps the best way to obtain a working familiarity with NSF data is to study carefully the chapters on resources for research and development, basic research, industrial R & D, and science and engineering personnel in the report *Science Indicators 1982* (IA, NSB, 1983). Like other volumes in this series, *SI-82* incorporates NSF findings on research and development in government industry, and academe. For more information on NSF statistics, technical details, and a description of the scope of each survey conducted by the Division of Science Resources Studies, see *A Guide to NSF Science Resources Data* (II, 1980), and *Project Summaries: FY 1983* (II, 1983).

Analysts seeking information specifically on chemists and chemical engineers may be frustrated on occasion to find that NSF analyses lump them with other physical scientists or engineers as a group. To remedy this situation, one turns naturally to the statistical compilations of the American Chemical Society. The ACS Office of Manpower Studies conducts annual surveys of the salaries and employment status of chemists and chemical engineers; the starting salaries and employment status of new graduates in chemistry and chemical engineering; and academic enrollments and degree conferrals in these two fields. (The academic survey is prepared in conjunction with the ACS Committee on Professional Training.) For some recent published reports of these surveys, see American Chemical Society, *Salaries 1984. Analysis of the American Chemical Society's 1984 Survey of Salaries and Employment* (IB, 1984); *idem, Starting Salaries 1983. Analysis of the American Chemical Society's Survey of Graduates in Chemistry and Chemical Engineering* (IB, 1984); and 'ACS Committee on Professional Training 1983 Annual Report' (IB, 1984). Both of the salary surveys are also summarized in *C & EN* annually.

In 1974, the ACS inaugurated a series of summary reports on the demography, education, employment, and remuneration of chemists and chemical engineers. Entitled *Professionals in Chemistry* (IB, 1975–), this series provides helpful analyses of the ACS surveys mentioned above, as well as annual surveys of the chemical industry conducted by *Chemical & Engineering News* (see below). In addition, there are helpful discussions of the comparability of ACS, NSF, and Bureau of Labor Statistics data on chemists and chemical engineers in earlier issues. After the 1979 edition, *Professionals in Chemistry* was placed on a biennial basis and restricted to statistical tables alone, with no figures or interpretive discussion. Still, the tables provide a wealth of information on the community of chemists and chemical engineers since World War II. (For the results of a survey of ACS members in the early 1940s, in some ways the institutional precursor of the activities of the Office of Manpower Studies, see IB, Fraser, 1942a–d.)

The time series on the aggregate number of chemists in the labor force can be obtained from the occupational breakdowns of the U.S. Census of Population. Some of the problems caused by changing occupational classifications over the last hundred years are discussed in Sections A.1 and A.2; these issues are analyzed (and variant measures proposed) in IA, Edwards, 1943, and IA, Kaplan and Casey, 1958. An alternate source for estimates of the population of employed chemists and chemical engineers is the Bureau of Labor Statistics. Bulletin 1781 of the BLS, *Employment of Scientists and Engineers*, 1950–70 (IA, 1973), was particularly helpful. See also Bureau of Labor Statistics, *Manpower Resources in Chemistry and Chemical Engineering*, Bulletin 1132 (IA, 1953). Readers particularly interested in the employment of chemists in academe, government, or industry are referred to the detailed discussion in Chapter 5.

Data on degree conferrals in chemistry have a checkered history. The first attempt to collect systematic figures was made by J. McKeen Cattell, who published his compilations in the columns of *Science* magazine. His data were superseded by the efforts of the U.S. Bureau (later Office) of Education, which published a biennial survey of data on degree conferrals and other aspects of American education. During the 1920s, these figures were supplemented for chemistry and the other sciences by the compilations of Callie Hull and C. J. West of the National Research Council's Research Information Service. Abbott Ferriss's *Indicators of Trends in American Education* (IB, 1969) was a pioneering attempt to create long-run series on degree conferrals by field. In *The Great American Degree Machine* (IB, 1975), Douglas Adkins makes a more sophisticated attempt to provide internally consistent figures on degree conferrals from 1890 to 1971 at the bachelors, masters, and doctorate levels.

Another source of information on doctorates in chemistry and chemical engineering deserves mention. Since 1946, the National Academy of Sciences has maintained a Doctorate Records File (DRF) containing information on all PhD recipients in the United States since 1920. Administered by the Commission on Human Resources of the National Research Council, the Doctorate Records File is complemented by the Comprehensive Roster of Doctorate Recipients, which includes information on over 400 000 PhDs. The Comprehensive Roster is composed of the DRF and the National Science Foundation's National Roster of Scientific and Technical Personnel. Since 1958, the DRF has been augmented annually by a Survey of Earned Doctorates, and the Comprehensive Roster is the basis for a biennial Survey of Doctorate Recipients. (For the results of the most recent published survey, see *Science, Engineering, and Humanities Doctorates in the United States: 1981 Profile* (IB, 1982)). Over the past 35 yr, the NAS has sponsored a number of analyses of the demography of the doctorate community in the United States. The 1978 NRC report on *A Century of Doctorates* (IB) is an extremely valuable summary of this work (and it contains an annotated bibliography of other studies based on the DRF and the Comprehensive Roster).

For information on the chemical industry, the basic statistical sources are the Censuses of Manufactures conducted by the Bureau of the Census. Begun in the early nineteenth century, these surveys provide information on employment, production, value added by manufacture, and other aspects of chemical manufacturing. These Census reports allow one to analyze the changing sectoral distribution of the chemical industry over the past hundred years, despite the usual difficulties caused by changing definitions (see Appendix B). Since 1949, the Census of Manufactures has been supplemented by sample surveys of U. S. manufacturing for intercensal years. For guides to the early materials, see Meyer H. Fishbein, *Censuses of Manufactures, 1810–1890* (II, 1973) and *Bureau of the Census Catalog of Publications* (II, 1974). For an indication of the scope of more recent manufacturing data, see the Census Bureau's *Guide to Industrial Statistics* (II, 1978).

The American Chemical Society is also a valuable source of information on the chemical industry. *Chemical and Engineering News* publishes annual surveys of 'Facts and Figures for the Chemical Industry' and 'Facts and Figures for Chemical R & D' which include both government and private data on the output, finances, performance, employment, and research expenses of the American chemical industry. (The latter survey also covers federal outlays and academic research.) The *C&EN* 'Facts and Figures' surveys begin with 1939 data, although publication commenced in the early 1950s. For the most recent of these surveys, see the 11 June and 23 July 1984 issues of *C&EN*. One particular feature of these surveys we used in Chapter 4 is the ranking of the top 50 chemical producers; cf. the annual Fortune 500 directory (usually published in early May).

For other measures of the economic impact of the chemical industry, we used national income data generated by the Bureau of Economic Analysis in the Department of Commerce; see, for example, *National Income and Product Accounts of the United States* (IA, 1981). Other more esoteric economic measures (such as total factor productivity) are discussed in Chapter 4. Readers interested in money and manpower invested in research and development by the chemical industry are referred to the specific sources mentioned in Section 5.1 and to the NSF series *Research and Development in Industry* produced by the Division of Science Resources Studies (e.g., IA, NSF, 1976c).

For a discussion of the statistical reliability and methodology of the principal federal government data bases, see Bureau of the Census, *Statistical Abstract of the United States: 1984* (IA, 1983), 905–923; this source also includes a subject guide to government statistical publications (pp. 924–967). A more comprehensive guide to government data is the Congressional Information Service's monthly *American Statistics Index* (1973–). The 1974 retrospective edition provides an informative discussion of federal statistical publications on an agency-by-agency basis, including most of the serials mentioned above. An analogous CIS monthly, the *Statistical Reference Index* (1980–), covers statistics produced by business, trade, and professional associations (including the American Chemical Society). Another helpful bibliography is Paul Wasserman *et al.* (eds.), *Statistics Sources*

(II, 1983), esp. pp. 292–294 (chemicals & allied products), 1408–1409 (R & D), and 1475 (scientists and engineers). *Statistics Sources*, however, lacks the analytical descriptions of each entry (down to the titles of tables) that make the *American Statistics Index* and the *Statistical Reference Index* so valuable. Theodore Peck's *Chemical Industries Information Sources* (II, 1979) is also useful.

Bibliographies like these are a boon to those seeking information on the variegated activities of American chemists and chemical engineers. However, we must close with a caveat. The profusion of statistical information on science and technology since 1950 is in stark contrast to the situation for earlier periods. The few studies that one unearths for those earlier periods suffer from suspect techniques. So the neophyte in chemical indicators with an interest in historical time series faces both a problem and an opportunity. The problem is obvious. But the opportunity is exciting: one can seek to invent the appropriate indicators and to discover the requisite data in the vast archive of chemical literature, constrained only by the limitations of one's energy and imagination.

BIBLIOGRAPHY

I. Data Sources

A. FEDERAL GOVERNMENT*

Andrews and Moylan, 1969. E. W. Andrews and M. Moylan. 'Scientific and Professional Employment by State Governments'. *Monthly Labor Review* **92** (August): 40–45.

BEA, 1977. Department of Commerce, Bureau of Economic Analysis. *The National Income and Product Accounts of the United States, 1929–74. Statistical Tables.*

BEA, 1981. Department of Commerce, Bureau of Economic Analysis. *The National Income and Product Accounts of the United States, 1929–76. Statistical Tables.* A Supplement to the *Survey of Current Business*, September.

BLS, 1951a. Bureau of Labor Statistics. *Federal White-Collar Workers. Their Occupations and Salaries. June 1951*. Bulletin No. 1117.

BLS, 1951b. Bureau of Labor Statistics. Division of Manpower and Employment Statistics. 'The Nation's Scientific and Technical Manpower'. Prepared by Helen Wood and Robert W. Cain. BLS Manpower Report No. 3. Mimeographed.

BLS, 1953. Bureau of Labor Statistics. *Manpower Resources in Chemistry and Chemical Engineering*. Bulletin 1132.

BLS, 1964. Bureau of Labor Statistics. *Employment of Scientific and Technical Personnel in State Government Agencies, 1962*. Bulletin 1412.

BLS, 1973. Bureau of Labor Statistics. *Employment of Scientists and Engineers, 1950–1970*. Bulletin 1781.

Chase, 1970. Office of Education. *A Directory of Graduate Deans at United States Universities, 1872–1970*. By John L. Chase.

Dunham *et al*., 1966. Office of Education, National Center for Educational Statistics. *Teaching Faculty in Universities and Four-Year Colleges. Spring 1963*. By Ralph E. Dunham, Patricia S. Wright, Marjorie O. Chandler. DHEW Publication No. OE–53022–63.

Edwards, 1943. Bureau of the Census, Sixteenth Census of the United States: 1940, Population. *Comparative Occupation Statistics for the United States, 1870 to 1940: A Comparison of the 1930 and the 1940 Census Occupation and Industry Classifications and Statistics; A Comparable Series of Occupation Statistics, 1870 to 1930; and a Social-Economic Grouping of the Labor Force, 1910 to 1940*. By Alba M. Edwards.

* Unless otherwise noted, all items in this section were published in Washington, D.C., by the U.S. Government Printing Office.

Fay, 1917. Albert H. Fay. *Preparedness Census of Mining Engineers, Metallurgists and Chemists*. U.S. Bureau of Mines Technical Paper No. 179.

FSA, 1952. Federal Security Agency, Office of Education, National Scientific Register. *Research and Development Personnel in Industrial Laboratories 1950*. Report of the National Academy of Sciences – National Research Council to National Scientific Register under Contract SAE-1219. Scientific Manpower Series No. 1.

Gertler, 1962. Diane B. Gertler. *Preliminary Report on Offerings and Enrollments in Grades 9–12 of Non-Public Secondary Schools, 1961–62*. USOE Publication No. OE–24008.

Gertler, 1963. Office of Education. *Statistics of Nonpublic Secondary Schools, 1960–61: Type of School, Enrollment, and Staff*. By Diane B. Gertler. USOE Circular No. 707, OE–200050.

Goldman *et al.*, 1964. Morris R. Goldman, Martin L. Marimont, and Beatrice N. Vaccara. 'The Interindustry Structure of the United States. A Report on the 1958 Input-Output Study'. *Survey of Current Business* **44** (November): 10–17.

Jessen *et al.*, 1937. Carl A. Jessen, Lester B. Herlihy, and Blanche K. Choate. *Subject Registrations in Private High Schools and Academies, 1932–33*. USOE Pamphlet No. 73.

Kaplan and Casey, 1958. David L. Kaplan and M. Claire Casey. *Occupational Trends in the United States 1900 to 1950*. Bureau of the Census Working Paper No. 5. Washington, D.C.: U.S. Department of Commerce.

Munroe, 1908. Charles E. Munroe. 'Chemicals and Allied Products'. In Department of Commerce and Labor, Bureau of the Census. *Manufactures: 1905. Part IV. Special Reports on Selected Industries*. 397–512. Bulletin 92.

NCES, 1968. National Center for Educational Statistics. *Earned Degrees Conferred, 1966–67: Part A – Summary Data*. By Marjorie O. Chandler. OE–54013–67, Part A.

NCES, 1969. National Center for Educational Statistics. *Earned Degrees Conferred, 1967–68: Part A – Summary Data*. By Mary Evans Harper and Marjorie O. Chandler. OE–54013–68–A.

NCES, 1971. National Center for Educational Statistics. *Earned Degrees Conferred, 1968–69: Part A – Summary Data*. By Mary Evans Harper and Marjorie O. Chandler. OE–54013–69–A.

NCES, 1972. National Center for Educational Statistics. *Earned Degrees Conferred, 1969–70: Summary Data*. By Mary Evans Harper. DHEW Publication No. (OE) 72–65.

NCES, 1973. National Center for Educational Statistics. *Earned Degrees Conferred, 1970–71*. By Mary Evans Harper. DHEW Publication No. (OE) 73–11412.

NCES, 1975a. National Center for Education Statistics. *Digest of Educational Statistics. 1974 Edition*. By W. Vance Grant and C. George Lind. NCES 75–210.

NCES, 1975b. National Center for Education Statistics. *Earned Degrees Conferred: 1971–72*. Prepared by Curtis O. Baker and Agnes Q. Wells. NCES 75–108.

NCES, 1975c. National Center for Education Statistics. *Summary of Offerings and Enrollments in Public Secondary Schools, 1972–73*. Prepared by Logan Osterndorf. NCES 76–150.

NCES, 1976. National Center for Education Statistics. *Earned Degrees Conferred, 1972–73 and 1973–74: Summary Data*. By Curtis O. Baker and Agnes Q. Wells. NCES 76–105.

NCES, 1977. National Center for Education Statistics. *Earned Degrees Conferred, 1974–75: Summary Data*. By Curtis O. Baker and Agnes Q. Wells. NCES 77–328.

NCES, 1978a. National Center for Education Statistics. *Digest of Education Statistics, 1977–78*. By W. Vance Grant and C. George Lind.

NCES, 1978b. National Center for Education Statistics. *Earned Degrees Conferred, 1975–76* [Summary Data]. By Stanley V. Smith and Agnes Q. Wells. NCES 78–300.

NCES, 1980a. National Center for Education Statistics. *Earned Degrees Conferred, 1976–77: Summary Data*. By Stanley V. Smith and Geneva Davis. NCES 79–309.

NCES, 1980b. National Center for Education Statistics. *Earned Degrees Conferred, 1977–78: Summary Data*. By Andrew J. Pepin and Agnes Q. Wells. NCES 80–346.

NSB, 1973. National Science Board. *Science Indicators 1972*. Report of the National Science Board, 1973. NSB 73–1.

NSB, 1975. National Science Board. *Science Indicators 1974*. Report of the National Science Board, 1975. NSB 75–1.

NSB, 1977. National Science Board. *Science Indicators 1976*. Report of the National Science Board, 1977. NSB 77–1.

NSB, 1979. National Science Board. *Science Indicators 1978*. Report of the National Science Board, 1979. NSB 79–1.

NSB, 1981. National Science Board. *Science Indicators 1980*. Report of the National Science Board, 1981. NSB 81–1.

NSB, 1983. National Science Board. *Science Indicators 1982*. Report of the National Science Board, 1983. NSB 83–1.

NSF, 1956. National Science Foundation. *Science and Engineering in American Industry. Final Report on a 1953–1954 Survey*. Prepared for the National Science Foundation by the U.S. Department of Labor, Bureau of Labor Statistics. NSF 56–16.

NSF, 1957a. National Science Foundation. 'Funds for Research and Development in Colleges and Universities, 1953–54'. *Reviews of Data on Research and Development* (March). NSF 57–9.

NSF, 1957b. National Science Foundation. *Scientific Manpower in the Federal Government, 1954*. NSF 57–32. Copy in Library of Congress.

NSF, 1959. National Science Foundation. *Science and Engineering in American Industry. Report on a 1956 Survey*. Prepared for the National Science Foundation by the U.S. Department of Labor. Bureau of Labor Statistics. NSF 59–50.

NSF, 1961a. National Science Foundation. *Employment of Scientific and Technical Personnel in State Government Agencies. Report on a 1959 Survey*. NSF 61–17.

NSF, 1961b. National Science Foundation. *Scientific and Technical Personnel in Industry 1960*. Prepared for the National Science Foundation by the U.S. Department of Labor. Bureau of Labor Statistics. NSF 61–75.

NSF, 1961c. National Science Foundation. *Scientists and Engineers in the Federal Government, October 1958*. NSF 61–43.

NSF, 1962. National Science Foundation. *Scientific and Technical Personnel in the Federal Government, 1959 and 1960*. NSF 62–26.

NSF, 1965. National Science Foundation. *Scientific and Technical Personnel in the Federal Government, 1961 and 1962*. Surveys of Science Resources Series. NSF 65–4.

NSF, 1967. National Science Foundation. *Scientific and Technical Personnel in the Federal Government, 1964*. Surveys of Science Resources Series. NSF 67–21.

NSF, 1968. National Science Foundation. Office of Economic and Manpower Studies. Scientific Manpower Studies Group. 'Scientific and Technical Personnel in the Federal Government, 1966'. *Reviews of Data on Science Resources* No. 14 (April). NSF 68–16.

NSF, 1969. National Science Foundation. *Scientific and Technical Personnel in the Federal Government, 1967*. Surveys of Science Resources Series. NSF 69–26.

NSF, 1970. National Science Foundation. *Scientific and Technical Personnel in the Federal Government 1968*. Surveys of Science Resources Series. NSF 70–24.

NSF, 1971a. National Science Foundation, Division of Science Resources Studies. 'Enrollment Increase in Science and Mathematics in Public Secondary Schools, 1948–49 to 1969–70'. *Science Resources Studies Highlights* (15 October). NSF 71–30.

NSF, 1971b. National Science Foundation. *Scientific, Technical, and Health Personnel in the Federal Government, 1969*. Surveys of Science Resources Series. NSF 70–44.

NSF, 1972. National Science Foundation. *Research and Development in Industry, 1970. Funds, 1970. Scientists and Engineers, January 1971*. Surveys of Science Resources Series. NSF 72–309.

NSF, 1973. National Science Foundation. Division of Science Resources Studies. Manpower Utilization Studies Group. 'Federal Scientific, Technical, and Health Personnel, 1971'. *Reviews of Data on Science Resources* No. 21 (September). NSF 73–309.

NSF, 1976a. National Science Foundation. Division of Science Resources Studies. Manpower Utilization Studies Group. 'Education and Work Activities of Federal Scientific and Technical Personnel, January, 1974'. *Reviews of Data on Science Resources* No. 27 (May). NSF 76–308.

NSF, 1976b. National Science Foundation. *Expenditures for Scientific and Engineering Activities at Universities and Colleges, Fiscal Year 1974*. Surveys of Science Resources Series. NSF 76–303.

NSF, 1976c. National Science Foundation. *Research and Development in Industry, 1974. Funds, 1974. Scientists and Engineers, January 1975*. Surveys of Science Resources Series. NSF 76–322.

NSF, 1982. National Science Foundation, Division of Science Resources Studies. *Science and Engineering Personnel: A National Overview*. NSF 82–318.

NSF, 1984a. National Science Foundation, Division of Science Resources Studies. *Academic Science/Engineering, 1972–83: R&D Funds, Scientists and Engineers, Graduate Enrollment and Support*. Surveys of Science Resources Series. NSF 84–322.

NSF, 1984b. National Science Foundation, Division of Science Resources Studies. *National Patterns of Science and Technology Resources, 1984*. NSF 84–311.

ORUS, 1879. Department of the Interior. *Official Register of the United States, Containing a List of Officers and Employees in the Civil, Military, and Naval Service on the Thirtieth of June, 1879; Together with a List of Ships and Vessels Belonging to the United States*. Volume I. Legislative, Executive, Judicial. Compiled and Printed under the Direction of the Secretary of the Interior.

ORUS, 1883. Department of the Interior. *Official Register of the United States, Containing a List of Officers and Employees in the Civil, Military, and Naval Service on the First of July, 1883; Together with a List of Ships and Vessels Belonging to the United States*. Volume I. Legislative, Executive, Judicial. Compiled under the Direction of the Secretary of the Interior, by J. G. Ames, Superintendent of Documents.

ORUS, 1887. Department of the Interior. *Official Register of the United States, Containing a List of Officers and Employees in the Civil, Military, and Naval Service on the First of July, 1887; Together with a List of Ships and Vessels Belonging to the United States*. Volume I. Legislative, Executive, Judicial. Compiled under the Direction of the Secretary of the Interior, by J. G. Ames, Superintendent of Documents.

ORUS, 1892. Department of the Interior. *Official Register of the United States, Containing a List of Officers and Employees in the Civil, Military, and Naval Service on the First of July, 1891; Together with a List of Vessels Belonging to the United States*. Volume I. Legislative, Executive, Judicial. Compiled under the Direction of the Secretary of the Interior, by J. G. Ames, Superintendent of Documents.

ORUS, 1897. Department of the Interior. *Official Register of the United States, Containing a List of Officers and Employees in the Civil, Military, and Naval Service, Together with a List of Vessels Belonging to the United States, July 1, 1897*. Volume I. Legislative, Executive, Judicial. Compiled under the Direction of the Secretary of the Interior, by Edward M. Dawson, Chief Clerk, Department of the Interior.

ORUS, 1901. Department of the Interior. *Official Register of the United States, Containing a List of Officers and Employees in the Civil, Military, and Naval Service, Together with a List of Vessels Belonging to the United States, July 1, 1901*. Volume I. Legislative, Executive, and Judicial. [By Authority of Congress.] Compiled under the Direction of the Secretary of the Interior, by Edward M. Dawson, Chief Clerk, Department of the Interior.

ORUS, 1905. Department of the Interior. *Official Register of the United States, Containing a List of Officers and Employees in the Civil, Military, and Naval Service, Together with a List of Vessels Belonging to the United States, July 1, 1905*. Volume I. Legislative, Executive, Judicial. [By Authority of Congress.] Compiled by the Chief Clerk of the Department of the Interior.

ORUS, 1911. Department of Commerce and Labor, Bureau of the Census. *Official Register. Persons in the Civil, Military, and Naval Service of the United States, and List of Vessels*. 1911. Volume I. Directory. Compiled by the Department of Commerce and Labor, Bureau of the Census.

Rowland, 1883. William L. Rowland. 'Report on the Manufacture of Chemical Products and Salt'. In Department of the Interior, Census Office. *Report on the Manufactures of the United States at the Tenth Census (June 1, 1880)*, Vol. II, 985–1028.

Smith, 1941. Malcolm L. Smith. 'Occupations and Salaries in Federal Employment'. *Monthly Labor Review* **52**: 66–85.

Swift, 1929. A. H. Swift. *The American Chemical Industry: Production and Foreign Trade in [the] First Quarter of [the] Twentieth Century*. U.S. Department of Commerce, Bureau of Foreign and Domestic Commerce, Trade Promotion Series No. 78.

USBC, 1913. Department of Commerce, Bureau of the Census. Thirteenth Census of the United States. *Manufactures, 1909*. Vol. X. *Reports for Principal Industries*. 'The Manufacture of Chemicals and Allied Industries', 529–592.

USBC, 1919. Department of Commerce, Bureau of the Census. *Census of Manufactures, 1914*. Vol. II. *Reports for Selected Industries and Detailed Statistics for Industries, by States*. 'Chemicals and Allied Industries', 452–533.

USBC, 1974. Bureau of the Census. *Characteristics of Persons in Engineering and Scientific Occupations: 1972*. Technical Paper No. 33.

USBC, 1975a. Bureau of the Census. *Current Population Reports*, Series P–25, No. 614, 'Estimates of the Population of the United States, by Age, Sex and Race: 1970 to 1975'.

USBC, 1975b. Bureau of the Census. *Historical Statistics of the United States, Colonial Times to 1970*. Bicentennial Edition, 2 vols.

USBC, 1977. Bureau of the Census. *Statistical Abstract of the United States: 1977*. 98th edition.

USBC, 1979. Bureau of the Census. *Statistical Abstract of the United States: 1979*. 100th edition.

USBC, 1983. Bureau of the Census. *Statistical Abstract of the United States: 1984*. 104th edition.

USBC-ASM, 1975. Bureau of the Census, Annual Survey of Manufactures: 1973. *General Statistics for Industry Groups and Industries*, M73(AS)-1.

USBC-ASM, 1978. Bureau of the Census, Annual Survey of Manufactures: 1976. *Industry Profiles*, M76(AS)-7.

USBC-COM, 1970. Bureau of the Census, Census of Manufactures: 1967. *Special Report Series: Concentration Ratios in Manufacturing*. MC67(S)-2.1.

USBC-COM, 1975. Bureau of the Census, Census of Manufactures: 1972. *Special Report Series: Concentration Ratios in Manufacturing*. MC72 (SR)-2.

USBC-COM, 1976. Bureau of the Census, Census of Manufactures: 1972. *Volume II. Industry Statistics. Part 2. SIC Major Groups 27–34*.

USBC-COM, 1979. Bureau of the Census, Census of Manufactures: 1977. *Summary Series: Selected Statistics for Industry Groups and Industries: 1977 and 1972*. Preliminary Report. MC77-S-1(P).

USBC-COP, 1964. Bureau of the Census, Census of Population: 1960. Vol. I, *Characteristics of the Population*, Part 1, United States Summary.

USBC-COP, 1973. Bureau of the Census, Census of Population: 1970. Vol. I, *Characteristics of the Population*, Part 1, United States Summary, Section 2.

USBE, 1911. Bureau of Education. *Report of the Commissioner of Education for the Year Ended June 30, 1910*. 2 vols.

USBE, 1912. Bureau of Education. *Report of the Commissioner of Education for the Fiscal Year Ended June 30, 1911*. 2 vols.

USBE, 1925. Bureau of Education. *Biennial Survey of Education, 1920–1922*. Vol. II. USBE *Bulletin*. 1924, No. 14.

USCSC, 1926. Civil Service Commission. *Opportunities for Chemists in the United States Civil Service. Including Table Showing Salaries and Distribution of Government Chemists*.

USCSC, 1929. Civil Service Commission. *Opportunities for Chemists in the United States Civil Service*.

USCSC, 1943. Civil Service Commission. *Opportunities for Chemists in Civilian War Service*.

USCSC, 1958. Civil Service Commission. Employment Statistics Office. *Occupations of Federal White-Collar Workers. February 28, 1957.* Pamphlet 56–1. June, 1958.

USCSC, 1972. Civil Service Commission. Manpower Statistics Division. *Occupations of Federal White-Collar Workers. October 31, 1969 and 1970.* Pamphlet SM 56–09.

USCSC, 1975. Civil Service Commission. Bureau of Manpower Information Systems. *Occupations of Federal White-Collar Workers. October 31, 1972 and 1973.* SM 56–10.

USCSC, 1977. Civil Service Commission. Bureau of Manpower Information Systems. Federal Civilian Manpower Statistics. *Occupations of Federal White-Collar Workers. October 31, 1974 and 1975.* SM 56–11.

USCSC, 1978. Civil Service Commission. Bureau of Personnel Management Information Systems. Workforce Analysis and Statistics Division. *Occupations of Federal White-Collar Workers. October 31, 1976.* SM 56–12.

USDI, 1883. Department of the Interior, Census Office. *Compendium of the Tenth Census (June 1, 1880).* . . . Part II, Table 103 – OCCUPATIONS, 1368–1377.

'U.S. National Income', 1976. 'U.S. National Income and Product Accounts: Revised Estimates. First Quarter of 1973 to First Quarter of 1976. . .'. *Survey of Current Business* **56** (July): 22–68.

'U.S. National Income', 1979. 'U.S. National Income and Product Accounts: Revised Estimates, 1976–1978'. *Survey of Current Business* **59** (July): 24–72.

USOE, 1930. Office of Education. *Biennial Survey of Education, 1926–1928.* USOE *Bulletin*, 1930, No. 16.

USOE, 1937. Office of Education. *Biennial Survey of Education, 1932–1934.* USOE *Bulletin*, 1935, No. 2.

USOE, 1954. Office of Education. *Biennial Survey of Education in the United States, 1948–50.*

USOE, 1963a. Office of Education. *Digest of Educational Statistics: 1963 Edition.* USOE *Bulletin*, 1963, No. 43. OE–10024–63.

USOE, 1963b. Office of Education. *Earned Degrees Conferred, 1961–1962: Bachelor's and Higher Degrees.* By Wayne E. Tolliver. OE–54013–62. Circular No. 719.

USOE, 1965. Office of Education. *Earned Degrees Conferred, 1962–1963: Bachelor's and Higher Degrees.* By Patricia Wright. OE–54013–63. Circular No. 777.

USOE, 1966. Office of Education. *Earned Degrees Conferred, 1963–1964: Bachelor's and Higher Degrees (Final Report).* By Patricia Wright. OE–54013–64. Misc. No. 54.

USOE, 1967. Office of Education. *Earned Degrees Conferred, 1964–1965: Bachelor's and Higher Degrees, Institutional Data.* By Paul L. Mason and Mabel C. Rice. OE–54013–65.

USOE, 1968. Office of Education. *Earned Degrees Conferred, 1965–1966.* By Marjorie O. Chandler and Mabel C. Rice. OE–54013–66.

USOE, 1972. Office of Education. *Digest of Educational Statistics, 1971 Edition.* By Kenneth A. Simon and W. Vance Grant. DHEW Publication No. (OE) 72–45.

USOE, 1974. Office of Education. *Digest of Educational Statistics: 1973 Edition.* By W. Vance Grant and C. George Lind. DHEW Publication No. (OE) 74–11103.

USOPM, 1979. Office of Personnel Management. *Occupations of Federal White-Collar Workers. October 31, 1977.* SM 56–13.

USOPM, 1980. Office of Personnel Management. *Occupations of Federal White-Collar Workers. October 31, 1978.* SM 56–14.

B. OTHER

AAAS, 1948. American Association for the Advancement of Science. *Summarized Proceedings for the Period from January, 1940, to January, 1948, and a Directory of Members as of December 31, 1947*. Washington, D.C.: American Association for the Advancement of Science.

'Academy of Sciences Picks', 1980. 'Academy of Sciences Picks 59 Members and 12 Associates'. *New York Times*, 27 April 1980.

ACS, 1973. American Chemical Society, Committee on Chemistry and Public Affairs. *Chemistry in the Economy*. Washington, D.C.: American Chemical Society.

ACS, 1974. American Chemical Society, Chemical Abstracts Service. *CAS Today: Facts and Figures about Chemical Abstracts Service*. Columbus, Ohio: American Chemical Society.

ACS, 1975. American Chemical Society, Office of Manpower Studies. *Professionals in Chemistry: 1974. A Comprehensive Report on: Growth and Characteristics, Work Activities and Employers, Salaries, Women Chemists, Supply/Demand*. By Panagis A. Benatatos and Maria D. Frizat. Washington, D.C.: American Chemical Society.

ACS, 1976. American Chemical Society, Office of Manpower Studies. *Professionals in Chemistry: 1975. A Comprehensive Report on: Characteristics, Remuneration, Employers, Minority Chemists, Economic Status of Women Chemists, Chemistry Postdoctoral Fellows, Supply/Demand: Projections*. Washington, D.C.: American Chemical Society.

ACS, 1977a. American Chemical Society. *Annual Report 1976*. In *Chemical and Engineering News* **55** (4 April): 37–60.

ACS, 1977b. American Chemical Society, Office of Manpower Studies. *Professionals in Chemistry: 1976. A Comprehensive Report on: Characteristics, Remuneration, Employers, Minority Chemists, Employment in the Chemical Industry, Chemistry Postdoctoral Fellows, Supply/Demand*. Washington, D.C.: American Chemical Society.

ACS, 1978a. 'ACS Committee on Professional Training Issues Its 1977 Annual Report'. *Chemical and Engineering News* **56** (10 April): 61–67.

ACS, 1978b. American Chemical Society. *Annual Report 1977*. In *Chemical and Engineering News* **56** (10 April): 31–54.

ACS, 1978c. '1978 Salary Survey: Little Change in Chemists' Job Status'. *Chemical and Engineering News* **56** (19 June): 31–37.

ACS, 1978d. American Chemical Society, Office of Manpower Studies. *Professionals in Chemistry: 1977. A Comprehensive Report on: Characteristics, Remuneration, Employers, Minority Chemists, Job Switching and Salaries, Chemistry Postdoctoral Fellows, Supply/Demand*. Washington, D.C.: American Chemical Society.

ACS, 1979a. American Chemical Society. *Annual Report 1978*. In *Chemical and Engineering News* **57** (9 April): 35–58.

ACS, 1979b. American Chemical Society, Office of Manpower Studies. *Professionals in Chemistry: 1978. A Comprehensive Report on: Characteristics, Remuneration, Employers, Minority Chemists, Supply/Demand*. Washington, D.C.: American Chemical Society.

ACS, 1980. American Chemical Society. *Annual Report 1979*. In *Chemical and Engineering News* **58** (14 April): 33–52.

ACS, 1981. American Chemical Society, Office of Manpower Studies. *Professionals in Chemistry: 1979–1980. A Digest of Information about the Chemical Profession.* Washington, D.C.: American Chemical Society.

ACS, 1983a. American Chemical Society. *Annual Report 1982.* In *Chemical and Engineering News* **61** (18 April): 37–56.

ACS, 1983b. American Chemical Society, Office of Manpower Studies. *Professionals in Chemistry: 1981–1982. A Digest of Information about the Chemical Profession.* Washington, D.C.: American Chemical Society.

ACS, 1984a. American Chemical Society, Office of Statistical Services. *Salaries 1984: Analysis of the American Chemical Society's 1984 Survey of Salaries and Employment.* Washington, D.C.: American Chemical Society.

ACS, 1984b. American Chemical Society, Office of Statistical Services. *Starting Salaries 1983: Analysis of the American Chemical Society's Survey of Graduates in Chemistry and Chemical Engineering.* Washington, D.C.: American Chemical Society.

ACS-CPT, 1955. American Chemical Society, Committee on Professional Training. *Directory of Graduate Research, 1955.* Washington, D.C.: American Chemical Society.

ACS-CPT, 1961. American Chemical Society, Committee on Professional Training. *Directory of Graduate Research, 1961.* Washington, D.C.: American Chemical Society.

ACS-CPT, 1963. American Chemical Society, Committee on Professional Training. *Directory of Graduate Research, 1963.* Washington, D.C.: American Chemical Society.

ACS-CPT, 1965. American Chemical Society, Committee on Professional Training. *Directory of Graduate Research, 1965.* Washington, D.C.: American Chemical Society.

ACS-CPT, 1967. American Chemical Society, Committee on Professional Training. *Directory of Graduate Research, 1967.* Washington, D.C.: American Chemical Society.

ACS-CPT, 1969. American Chemical Society, Committee on Professional Training. *Directory of Graduate Research, 1969.* Washington, D.C.: American Chemical Society.

ACS-CPT, 1974. American Chemical Society, Committee on Professional Training. *Directory of Graduate Research, 1973.* Washington, D.C.: American Chemical Society.

ACS-CPT, 1975. American Chemical Society, Committee on Professional Training. *Directory of Graduate Research, 1975.* Washington, D.C.: American Chemical Society.

ACS-CPT, 1977a. American Chemical Society, Committee on Professional Training. '1976 Annual Report on the ACS Committee on Professional Training'. *Chemical and Engineering News* **55** (21 March): 52–57.

ACS-CPT, 1977b. American Chemical Society, Committee on Professional Training. *Directory of Graduate Research, 1977.* Washington, D.C.: American Chemical Society.

ACS-CPT, 1979. American Chemical Society, Committee on Professional Training. *Directory of Graduate Research, 1979.* Washington, D.C.: American Chemical Society.

ACS-CPT, 1984. 'ACS Committee on Professional Training 1983 Annual Report'. *Chemical and Engineering News* **62** (7 May): 44–49.

Adkins, 1975. Douglas L. Adkins. *The Great American Degree Machine: An Economic Analysis of the Human Resource Output of Higher Education*. A Technical Report Sponsored by the Carnegie Commission on Higher Education. Berkeley, Calif.: Carnegie Commission on Higher Education.

ARPC, 1905. *Annual Reports on the Progress of Chemistry for 1904*. Vol. 1. London: Chemical Society.

ARPC, 1967. *Annual Reports on the Progress of Chemistry for 1966*. Vol. 63. London: Chemical Society.

Backman, 1970. Jules Backman. *The Economics of the Chemical Industry*. Washington, D.C.: Manufacturing Chemists' Association.

Baker, 1961. Dale B. Baker. 'Growth of Chemical Literature. Past, Present, and Future'. *Chemical and Engineering News* **39** (17 July): 78–81.

Baker, 1971. Dale B. Baker. 'World's Chemical Literature Continues to Expand'. *Chemical and Engineering News* **49** (12 July): 37–40.

Baker, 1976. Dale B. Baker. 'Recent Trends in Growth of Chemical Literature'. *Chemical and Engineering News* **54** (10 May): 23–27.

Baker, 1981. Dale B. Baker. 'Recent Trends in Chemical Literature Growth'. *Chemical and Engineering News* 59 (1 June): 29–34.

Baruch, 1932. Ismar Baruch. 'Positions in the Professional and Scientific Service of the United States Government'. In *University Training for the National Service. Proceedings of a Conference Held at the University of Minnesota, July 14 to 17, 1931*, 51–74. Minneapolis: University of Minnesota Press.

Baskerville *et al.*, 1902. Charles Baskerville, Louis Kahlenberg, Charles E. Munroe, William A. Noyes, and Edgar Fahs Smith. 'Report of the Census Committee'. In *Twenty-Fifth Anniversary of the American Chemical Society. New York City, April Twelfth and Thirteenth, 1901*, 99–137. Easton, Penna.: Chemical Publishing Company.

Benner, 1940. Gladys Vera Benner. *The Preparation in the Special Sciences in Institutions of Higher Learning of the Teachers of General Science in American Public Secondary Schools*. PhD dissertation, University of Pennsylvania.

Bergen, 1952. Catherine Bergen. 'Science Articles for the Layman'. *School Science and Mathematics* **52**: 687–693.

Bergen, 1955. Catherine Bergen. 'Science in the *New York Times*'. *School Science and Mathematics* **54**: 484–488.

Bobbitt, 1926. Franklin Bobbitt. *Curriculum Investigations*. Chicago: University of Chicago.

Breithut, 1917. Frederick E. Breithut. 'The Status and Compensation of the Chemist in Public Service'. *Journal of Industrial and Engineering Chemistry* **9**: 64–79.

Buckingham, 1926. B. R. Buckingham. *Supply and Demand in Teacher Training*. Bureau of Educational Research Monographs, No. 4. Columbus: Ohio State University.

Carter, 1966. Anne P. Carter. 'The Economics of Technological Change'. *Scientific American* **214** (April): 25–31.

Carter, 1970. Anne P. Carter. *Structural Change in the American Economy*. Cambridge, Mass.: Harvard University Press.

'Census', 1944. 'Census of Chemists and Chemical Engineers. Abstracted from National Roster of Scientific and Specialized Personnel as Reported April 1, 1944'. *Chemical and Engineering News* **22**: 1004–1005.

Clarke, 1881. Frank Wigglesworth Clarke. *A Report on the Teaching of Chemistry and Physics in the United States*. U.S. Bureau of Education, *Circulars of Information*, 1880, No. 6. Washington, D.C.: Government Printing Office.

Computer Horizons, 1972. Computer Horizons, Inc. *Annotated Bibliography of International Publication and Citation Comparisons*. Supplement to Second Annual Report, Contract No. NSF-C 627, 'Exploration of the Possibility of Generating Importance and Utilization Measures by Citation Indexing of Approximately 500 Journals in the Physical Sciences.' Chicago: Computer Horizons, Inc., August.

Cooper, 1941. Franklin S. Cooper. 'Locations and Extent of Industrial Research Activity in the United States'. In III, NRPB, 1941, 173–187.

Crane, 1944. E. J. Crane. 'Growth of Chemical Literature. Contributions of Certain Nations and the Effects of War'. *Chemical and Engineering News* **22**: 1478–1481, 1496.

DeLoach and Jeanes, 1963. Will S. DeLoach and Opey Dew Jeanes. 'Chemists as College Presidents'. *Science Education* **47**: 353–355.

'Du Pont Fellowships', 1936. 'Du Pont Fellowships for Research Chemists'. *Science* **83**: 178.

'Facts & Figures', 1978. 'Facts & Figures. The U.S. Chemical Industry'. *Chemical and Engineering News* **56** (12 June): 45–79.

Ferriss, 1969. Abbott L. Ferriss. *Indicators of Trends in American Education*. New York: Russell Sage Foundation.

'Finances', 1940. 'The Finances of the American Chemical Society'. *Science* **91**: 12.

Flinn and Cobb, 1921. Alfred D. Flinn and Ruth Cobb, comps. *Research Laboratories in Industrial Establishments of the United States Including Consulting Research Laboratories*. 2nd edition. Bulletin of the National Research Council, No. 16, December 1921. Washington, D.C.: National Research Council, National Academy of Sciences.

'Fortune', 1944. 'Fortune Management Poll'. *Fortune* **29** (May): 8ff.

'Fortune 500', 1976. *The Fortune Directory of the 500 Largest U.S. Industrial Corporations [1976]*. Supplement to *Fortune* **93** (May 1976).

Fraser, 1942a. Andrew Fraser, Jr. 'The Economic Status of the Members of the American Chemical Society. Approved Report of the Committee on Economic Status. [Part I]'. *Chemical and Engineering News* **20** (25 October): 1289–1293.

Fraser, 1942b. Andrew Fraser, Jr. 'The Economic Status of the Members of the American Chemical Society. Approved Report of the Committee on Economic Status. [Part II]'. *Chemical and Engineering News* **20** (25 November): 1497–1505.

Fraser, 1942c. Andrew Fraser, Jr. 'The Economic Status of the Members of the American Chemical Society. Approved Report of the Committee on Economic Status. [Part III]'. *Chemical and Engineering News* **20** (10 December): 1563–1574.

Fraser, 1942d. Andrew Fraser, Jr. 'The Economic Status of the Members of the American Chemical Society. Approved Report of the Committee on Economic Status. [Part IV]'. *Chemical and Engineering News* **20** (25 December): 1635–1643.

Fussler, 1949. Herman H. Fussler. 'Characteristics of the Research Literature Used by Chemists and Physicists in the United States'. *Library Quarterly* **19**: 19–35.

Gort, 1962. Michael Gort. *Diversification and Integration in American Industry*. National Bureau of Economic Research, No. 77, General Series. Princeton: Princeton University Press.

Guggenheim Foundation, 1975. John Simon Guggenheim Memorial Foundation. *Directory of Fellows, 1925–1974*. New York: John Simon Guggenheim Memorial Foundation.

Hamor, 1935. William A. Hamor. 'Living Chemists Now or Formerly in College Presidencies'. *Industrial and Engineering Chemistry, News Edition* **13** (20 November): 437, 481.

Harmon and Soldz, 1963. Lindsey R. Harmon and Herbert Soldz, comps. *Doctorate Production in United States Universities, 1920–1962. With Baccalaureate Origins of Doctorates in Sciences, Arts, and Professions*. Publication No. 1142. Washington, D.C.: National Academy of Sciences – National Research Council.

Hart, 1933. Hornell Hart. 'Changing Social Attitudes and Interests'. In President's Research Committee on Social Trends, *Recent Social Trends in the United States*, Vol. 1, 382–442. New York: McGraw-Hill Book Company.

Haynes (ed.), 1951. Williams Haynes (ed.). *Chemical Who's Who*. 3rd edition. New York: Lewis Historical Publishing Company.

Heineman, 1928. Ailsie M. Heineman. 'A Study of General Science Textbooks'. *General Science Quarterly* **13**: 11–23.

Hodge *et al.*, 1964. Robert W. Hodge, Paul M. Siegel, and Peter H. Rossi. 'Occupational Prestige in the United States, 1925–1963'. *American Journal of Sociology* **70**: 286–302.

Hopkins, 1925. L. Thomas Hopkins. 'A Study of Magazine and Newspaper Science Articles with Relation to Courses in Sciences for High Schools'. *School Science and Mathematics* **25**: 783–800.

Howe and Nelson, 1933. Harrison Howe and Edna Lucile Nelson. 'Presidents of the American Chemical Society'. *Industrial and Engineering Chemistry* **25**: 110–118.

Hull, 1938. Callie Hull, comp. *Industrial Research Laboratories of the United States Including Consulting Research Laboratories*. 6th edition. Bulletin of the National Research Council, No. 102, December 1938. Washington, D.C.: National Research Council, National Academy of Sciences.

Hull, 1940. Callie Hull, comp. *Industrial Research Laboratories of the United States Including Consulting Research Laboratories*. 7th edition. Bulletin of the National Research Council, No. 104, December 1940. Washington, D.C.: National Research Council, National Academy of Sciences.

Hull, 1946. Callie Hull, comp. *Industrial Research Laboratories of the United States Including Consulting Research Laboratories*. 8th edition. Bulletin of the National Research Council, No. 113, July 1946. Washington, D.C.: National Research Council, National Academy of Sciences.

Hull and West, 1932. Callie Hull and Clarence J. West. 'The Eighth Census of Graduate Research Students in Chemistry, 1931'. *Journal of Chemical Education* **9**: 1472–1473.

Hull and West, 1933. Callie Hull and Clarence J. West. 'The Ninth Census of Graduate Research Students in Chemistry, 1932'. *Journal of Chemical Education* **10**: 499–503.

Hull and West, 1934. Callie Hull and Clarence J. West. 'The Tenth Census of Graduate Research Students in Chemistry and Chemical Engineering, 1933'. *Journal of Chemical Education* **11**: 471–475, 542.

Hull and West, 1935. Callie Hull and Clarence J. West. 'The Eleventh Census of Graduate Research Students in Chemistry and Chemical Engineering, 1934'. *Journal of Chemical Education* **12**: 339–343.

Hull and West, 1936. Callie Hull and Clarence J. West. 'The Twelfth Census of Graduate Research Students in Chemistry and Chemical Engineering, 1935'. *Journal of Chemical Education* **13**: 339–343.

Hurd, 1970. Ethan A. Hurd. 'Patent Literature: Current Problems and Future Trends'. *Journal of Chemical Documentation* **10**: 167–173.

Hutson, 1923. P. W. Hutson. 'High-School Science Teachers: A Study of Their Training in Relation to the Subjects They Are Teaching'. *Educational Administration and Supervision* **9**: 423–438.

'Industrial Research', 1946. 'Industrial Research Personnel Doubled', *Chemical and Engineering News* **24**: 2905.

Kaplan, 1964. A. D. H. Kaplan. *Big Enterprise in a Competitive System*. Revised Edition. Washington, D.C.: Brookings Institution.

Kebler, 1920. Lyman F. Kebler. 'A Study of the Economic Status of Chemists in the Government Service, Ten Educational Institutions, and Thirty-Seven Firms'. *Journal of Industrial and Engineering Chemistry* **12**: 1206–1212.

Kendrick, 1961. John W. Kendrick. *Productivity Trends in the United States*. National Bureau of Economic Research, Number 71, General Series. Princeton: Princeton University Press.

Kevles and Harding, 1977. Daniel J. Kevles and Carolyn Harding. 'The Physics, Mathematics, and Chemical Communities in America, 1870–1915: A Statistical Survey'. Social Science Working Paper, Number 136. Pasadena: California Institute of Technology.

Kiefer, 1977. David M. Kiefer. 'Sales Climb for Most Big Chemical Producers'. *Chemical and Engineering News* **55** (2 May): 40–41.

Kirby, 1925. Thomas J. Kirby. *Subject Combinations in High School Teachers' Program*. University of Iowa Extension Bulletin No. 136. College of Education Series No. 14. Iowa City: University of Iowa.

Klopp, 1927. W. J. Klopp. 'A Study of the Offerings of General Science Texts'. *General Science Quarterly* **11**: 236–246.

Knode, 1944. Jay Carroll Knode. 'Presidents of State Universities'. *Scientific Monthly* **58**: 218–220.

Koelsche and Morgan, 1964. Charles L. Koelsche and Ashley G. Morgan, Jr. *Scientific Literacy in the Sixties*. Athens: University of Georgia.

Krieghbaum, 1967. Hillier Krieghbaum. *Science and the Mass Media*. New York: New York University Press.

'Membership Statistics', 1983. 'Membership Statistics'. *Journal of the Electrochemical Society* **130** (December): 475C.

Miles (ed.), 1976. Wyndham D. Miles (ed.). *American Chemists and Chemical Engineers*. Washington, D.C.: American Chemical Society.

Moreland, 1940. George B. Moreland, Jr. 'What Young People Want to Read About'. *Library Quarterly* **10**: 469–493.

Narin and Carpenter, 1975. Francis Narin and Mark P. Carpenter. 'National Publication and Citation Comparisons'. *Journal of the American Society for Information Science* **26**: 80–103.

NAS, 1948–1949. National Academy of Sciences – National Research Council. *Report*, Fiscal Year 1948–1949. Washington, D.C.: U.S. Government Printing Office, 1950.

NAS, 1953–1954. National Academy of Sciences – National Research Council. *Report*, Fiscal Year 1953–1954. Washington, D.C.: U.S. Government Printing Office, 1958.

NAS, 1965. National Academy of Sciences – National Research Council. Committee for the Survey of Chemistry. *Chemistry: Opportunities and Needs* [The 'Westheimer Report']. Washington, D.C.: National Academy of Sciences.

NAS, 1970–1971. National Academy of Sciences, National Academy of Engineering, and National Research Council. *Annual Report*, Fiscal Year 1970–1971. Washington, D.C.: U.S. Government Printing Office, 1974.

NAS, 1971. National Academy of Sciences. *Scientific, Technical, and Related Societies of the United States*. 9th edition. Washington, D.C.: National Academy of Sciences.

NAS, 1971–1972. National Academy of Sciences, National Academy of Engineering, Institute of Medicine, and National Research Council. *Annual Report*, Fiscal Year 1971–1972. Washington, D.C.: U.S. Government Printing Office, 1975.

NAS, 1974–1975. National Academy of Sciences, National Academy of Engineering, Institute of Medicine, and National Research Council. *Annual Report*, Fiscal Year 1974–1975. Senate Document No. 94–155. 94th Cong., 2d sess.

NAS, 1975–1976. National Academy of Sciences, National Academy of Engineering, Institute of Medicine, and National Research Council. *Annual Report*, Fiscal Year 1975–1976. Senate Document No. 94–258. 94th Cong., 2d sess.

'NAS Elects', 1980. 'NAS Elects New Members'. *Science* **208**: 580.

Navin, 1970. Thomas R. Navin. 'The 500 Largest American Industrials in 1917'. *Business History Review* **44**: 360–386.

NEA, 1950. National Education Association, National Commission on Teacher Education and Professional Standards. *Teacher Supply and Demand in the United States. Report of the 1950 National Teacher Supply and Demand Study*. By Ray C. Maul. Washington, D.C.: National Education Association.

NEA, 1954. National Education Association, National Commission on Teacher Education and Professional Standards. *The 1954 Teacher Supply and Demand Report*. Washington, D.C.: National Education Association.

NEA, 1958. National Education Association, Research Division. *Teacher Supply and Demand in Public Schools, 1958. Report of the Eleventh Annual National Teacher Supply and Demand Study*. Washington, D.C.: National Education Association.

NEA, 1962. National Education Association, Research Division. *Teacher Supply and Demand in Public Schools, 1962. Fifteenth Annual Survey*. Research Report 1962–R8. Washington, D.C.: National Education Association.

NEA, 1965. National Education Association, Research Division. *Teacher Supply and Demand in Public Schools, 1965*. Research Report 1965–R10. Washington, D.C.: National Education Association.

NEA, 1967. National Education Association, Research Division. *The American Public-School Teacher, 1965–66*. Research Report 1967–R4. Washington, D.C.: National Education Association.

NEA, 1970. National Education Association, Research Division. *Teacher Supply and Demand in Public Schools, 1969*. Research Report 1969–R14. Washington, D.C.: National Education Association.

Nobel Foundation, 1972a. Nobel Foundation. *Nobel: The Man and His Prizes*. 3rd edition. New York: American Elsevier Publishing Company.

Nobel Foundation, 1972b–. Nobel Foundation. *Les Prix Nobel* . . . . Stockholm: P. A. Norstedt.

Norris, 1925. James F. Norris. 'Second Census of Graduate Research Students in Chemistry'. *Journal of Industrial and Engineering Chemistry* **17**: 755–756.

Novak, 1942. Benjamin J. Novak. *An Analysis of the Science Content of the New York Times and of Selected General Science Textbooks*. EdD thesis, Temple University.

NRC, 1938. National Research Council. *National Research Fellowships, 1919–1938.* Washington, D.C.: National Research Council.

NRC, 1978. National Research Council. Commission on Human Resources, Board on Human-Resource Data and Analyses. *A Century of Doctorates: Data Analyses of Growth and Change. U.S. PhDs – Their Numbers, Origins, Characteristics, and the Institutions from Which They Come.* Lindsey R. Harmon, Project Director. Report to the National Science Foundation, the National Endowment for the Humanities, and the U.S. Office of Education. Washington, D.C.: National Academy of Sciences.

NRC, 1982. National Research Council, Office of Scientific and Engineering Personnel, Commission on Human Resources. *Science, Engineering, and Humanities Doctorates in the United States: 1981 Profile.* Based on the 1981 Survey of Science, Engineering, and Humanities Doctorates. Betty D. Maxfield, Project Director. Washington, D.C.: National Academy Press.

'Number of Grads', 1980. 'Number of Grads Changes Little'. *Chemical and Engineering News* **58** (21 April): 58–61 and 66–68.

Pair *et al.*, 1978. Mary Wilson Pair, Denise Akey, Nancy Yakes, *et al.* (ed.), *Encyclopedia of Associations.* 12th edition, 3 vols. Detroit: Gale Research Company.

Parsons, 1933. Charles L. Parsons. 'Statement to the Membership: Your Society's Problem, Past and Future'. *Industrial and Engineering Chemistry, News Edition* **11**: 120–123.

Perazich and Field, 1940. George Perazich and Philip M. Field. *Industrial Research and Changing Technology.* Report No. M–4 of the National Research Project of the Work Projects Administration. Philadelphia: Work Projects Administration.

Porter and Livesay, 1969. P. Glenn Porter and Harold C. Livesay. 'Oligopolists in American Manufacturing and Their Products, 1909–1963'. *Business History Review* **43**: 282–298.

Rand, 1950. Myron J. Rand, comp. *Industrial Research Laboratories of the United States Including Consulting Research Laboratories.* 9th edition. Bulletin of the National Research Council, No. 120, November 1950. Washington, D.C.: National Research Council, National Academy of Sciences.

Rand, 1951a. Myron J. Rand. 'The National Research Fellowships'. *Scientific Monthly* **73**: 71–80.

Rand, 1951b. Myron J. Rand. 'Research and Development Personnel in Industrial Laboratories'. *Chemical and Engineering News* **29**: 3604–3608.

Searle and Ruch, 1926. A. H. Searle and G. M. Ruch. 'Science in Magazines. A Study of Science Articles in Magazines'. *School Science and Mathematics* **26**: 389–396.

Siegfried, 1952. Robert Siegfried. *A Study of Chemical Research Publications from the United States before 1880.* PhD dissertation, University of Wisconsin.

Skolnik and Reese, 1976. Herman Skolnik and Kenneth M. Reese (eds.). *A Century of Chemistry: The Role of Chemists and the American Chemical Society.* Washington, D.C.: American Chemical Society.

Sorenson and Sorenson, 1973. J. S. Sorenson and D. D. Sorenson. 'A Comparison of Science Content in Magazines in 1964–1965 and 1969–1970'. *Journalism Quarterly* **50**: 97–101.

Terleckyj, 1963. Nestor E. Terleckyj. *Research and Development: Its Growth and Composition.* National Industrial Conference Board, Studies in Business Economics, No. 82. New York: National Industrial Conference Board.

'Top 100', 1977. 'The Top 100 Industrials . . . Sixty Years of Corporate Ups, Downs and Outs'. *Forbes* **120** (September 15): 127 ff.

UN, 1952. United Nations, Department of Economic Affairs. *Demographic Yearbook. Annuaire Demographique. 1952.* 4th issue. New York: United Nations.

UN, 1971. United Nations, Department of Economic and Social Affairs, Statistical Office. *Demographic Yearbook. Annuaire Demographique. 1970.* 22nd issue. New York: United Nations.

UN, 1976. United Nations, Department of Economic and Social Affairs, Statistical Office. *Demographic Yearbook. Annuaire Demographique. 1975.* 27th issue. New York: United Nations.

'Utilization', 1947. 'Utilization of Chemists and Chemical Engineers in World War II'. *Chemical and Engineering News* **25**: 2206–2208.

Van Doren, 1948. Lloyd Van Doren. 'Early History of the American Institute of Chemists.' *The Chemist* **25**: 183–191.

Waples and Tyler, 1931. Douglas Waples and Ralph W. Tyler. *What People Want to Read About.* Chicago: American Library Association and University of Chicago Press.

West and Hull, 1927a. Clarence J. West and Callie Hull, comps. *Handbook of Scientific and Technical Societies and Institutes of the United States and Canada*. Bulletin of the National Research Council, No. 58 (May 1927). Washington: D.C.: National Research Council, National Academy of Sciences.

West and Hull, 1927b. Clarence J. West and Callie Hull. 'Third Census of Graduate Research Students in Chemistry'. *Journal of Chemical Education* **4**: 909–911. (Also in *Reprint and Circular Series of the National Research Council*, No. 79.)

West and Hull, 1928. Clarence J. West and Callie Hull. 'The Fourth Census of Graduate Research Students in Chemistry, 1927'. *Journal of Chemical Education* **5**: 882–884. (Also in *Reprint and Circular Series of the National Research Council*, No. 84.)

West and Hull, 1929. Clarence J. West and Callie Hull. 'The Fifth Census of Graduate Research Students in Chemistry, 1928'. *Journal of Chemical Education* **6**: 1338–1340.

West and Hull, 1930. Clarence J. West and Callie Hull. 'The Sixth Census of Graduate Research Students in Chemistry, 1929'. *Journal of Chemical Education* 7: 1674–1675.

West and Hull, 1931a. Clarence J. West and Callie Hull, comps. *Industrial Research Laboratories of the United States Including Consulting Research Laboratories*. 4th edition. Bulletin of the National Research Council, No. 81, January 1931. Washington, D.C.: National Research Council, National Academy of Sciences.

West and Hull, 1931b. Clarence J. West and Callie Hull. 'The Seventh Census of Graduate Research Students in Chemistry, 1930'. *Journal of Chemical Education* 8: 1374–1375.

West and Hull, 1933. Clarence J. West and Callie Hull, comps. *Industrial Research Laboratories of the United States Including Consulting Research Laboratories*. 5th edition. Bulletin of the National Research Council, No. 91, August 1933. Washington, D.C.: National Research Council, National Academy of Sciences.

West and Risher, 1927. Clarence J. West and Ervye L. Risher. *Industrial Research Laboratories of the United States Including Consulting Research Laboratories*. 3rd edition. Bulletin of the National Research Council, No. 60, January 1927. Washington, D.C.: National Research Council, National Academy of Sciences.

Wiley, 1917. Harvey W. Wiley. 'The Chemist in the Public Service'. *Journal of Industrial and Engineering Chemistry* 9: 81–84.

Xerox University Microfilms, 1973+. Xerox University Microfilms. *Comprehensive Dissertation Index, 1961–1972.* 37 vols. Ann Arbor, Mich.: Xerox University Microfilms, 1973. An annual supplement to this series in book form began with a 1973 issue. The machine-readable data base version of this source is updated monthly.

Yakes and Akey (eds.), 1980. Nancy Yakes and Denise Akey (eds.), *Encyclopedia of Associations*. 14th edition, 3 vols. Detroit: Gale Research Company.

Zanetti, 1924. J. E. Zanetti. 'Census of Graduate Research Students in Chemistry'. *Journal of Industrial and Engineering Chemistry* **16**: 402–403.

II. Bibliography, Historiography, and Methodology

Aymard, 1972. Maurice Aymard. 'The Annales and French Historiography (1929–72)'. *Journal of European Economic History* **1**: 491–511.

Bard, 1974. Yonathan Bard. *Nonlinear Parameter Estimation*. New York: Academic Press.

Barr *et al.*, 1976. Anthony J. Barr, James H. Goodnight, John P. Sall, and Jane T. Helwig. *A User's Guide to SAS 76*. Raleigh, N.C.: SAS Institute, Inc.

Beer and Lewis, 1965. John J. Beer and W. David Lewis. 'Aspects of the Professionalization of Science'. In *The Professions in America*, edited by Kenneth S. Lynn, 110–130. Boston: Beacon Press.

Blau and Duncan, 1967. Peter M. Blau and Otis Dudley Duncan. *The American Occupational Structure*. New York: John Wiley.

Bloch, 1949. Herbert A. Bloch. 'An Analysis of National Publication Trends and Publishers' Best Sellers as an Index of Cultural Transition'. *Journal of Educational Sociology* **22**: 287–303.

BOB, 1957. U.S. Bureau of the Budget, Office of Statistical Standards, Technical Committee on Industrial Classification. *Standard Industrial Classification Manual*. Washington, D.C.: U.S. Government Printing Office.

BOB, 1967. Executive Office of the President, Bureau of the Budget, Office of Statistical Standards. *Standard Industrial Classification Manual 1967*. Washington, D.C.: U.S. Government Printing Office.

Bogue, 1981. Allan G. Bogue. 'Numerical and Formal Analysis in United States History'. *Journal of Interdisciplinary History* **12**: 137–175.

Bowman, 1964. Raymond T. Bowman. 'Comments on "*Qui Numerare Incipit Errare Incipit*" by Oskar Morgenstern'. *American Statistician* **18** (June): 10–20.

Carroll, Sturchio, and Bud, 1976. P. Thomas Carroll, Jeffrey L. Sturchio, and Robert F. Bud. '[Report on the] Symposium on Quantitative Methods in History of Science'. *4S Newsletter* **1**, No. 4: 10–12.

Cattell, 1898. J. McKeen Cattell. 'Doctorates Conferred by American Universities for Scientific Research'. *Science* **8**: 197–201.

Cattell, 1903. J. McKeen Cattell. 'Doctorates Conferred by American Universities'. *Science* **18**: 257–262.

Cattell, 1911. J. McKeen Cattell. 'Doctorates Conferred by American Universities'. *Science* **34**: 193–202.

Chaunu, 1970. Pierre Chaunu. 'L'histoire sérielle. Bilan et perspectives'. *Revue Historique* **243**: 297–320.

Chaunu, 1973. Pierre Chaunu. 'Un nouveau champ pour l'histoire sérielle: le quantitatif au troisième niveau'. *Mélanges en l'Honneur de Fernand Braudel: Méthodologie de l'Histoire et des Sciences Humaines*, 105–125. Paris: Privat, Editeur.

Chubin, 1983. Daryl E. Chubin. *Sociology of Sciences: An Annotated Bibliography on Invisible Colleges, 1972–1981*. Garland Bibliographies in Sociology, 2. Garland Reference Library of Social Science, 127. New York: Garland Publishing, Inc.

Cole and Eales, 1917. F. J. Cole and Nellie B. Eales. 'The History of Comparative Anatomy: A Statistical Analysis of the Literature'. *Science Progress* **11**: 578–596.

Conk, 1978a. Margo Anderson Conk. 'Occupational Classifications in the United States Census: 1870–1940'. *Journal of Interdisciplinary History* **9**: 111–130.

Conk, 1978b. Margo A. Conk. 'Social Mobility in Historical Perspective'. *Marxist Perspectives* **1**, No. 3: 52–69.

Counts, 1925. G. S. Counts. 'The Social Status of Occupations'. *School Review* **33**: 16–27.

Cozzens *et al.*, 1978. Susan Cozzens. Ming Ivory, Janice LaPorte, Henry Small, and Janet Stanley. *Citation Analysis: An Annotated Bibliography*. 2nd edition. Philadelphia: Institute for Scientific Information.

Croxton *et al.*, 1967. Frederick E. Croxton, Dudley J. Cowden, and Sidney Klein. *Applied General Statistics*. 3rd edition. Englewood Cliffs, N.J.: Prentice-Hall, Inc.

Daniels, 1967. George H. Daniels. 'The Process of Professionalization in American Science: The Emergent Period, 1820–1860'. *Isis* **58**: 151–166.

Daniels, 1970. George H. Daniels. 'The Big Questions in the History of American Technology'. *Technology and Culture* **11**: 1–21.

Daniels (ed.), 1972. George H. Daniels (ed.). *Nineteenth-Century American Science: A Reappraisal*. Evanston, Ill.: Northwestern University Press.

Drew, 1979. David Drew. 'Science Indicators Redux'. *4S Newsletter* **4**, No. 1 (winter): 24–34.

Duncan, 1974. Otis Dudley Duncan. 'Developing Social Indicators'. *Proceedings of the National Academy of Sciences* **71**: 5096–5102.

Duncan, 1978. Otis Dudley Duncan. 'Science Indicators and Social Indicators'. In II, Elkana *et al.* (eds.), 1978, 31–38.

Dupree, 1966. A. Hunter Dupree. 'The History of American Science – A Field Finds Itself'. *American Historical Review* **71**: 863–874.

Edge, 1977. David O. Edge. 'Why I Am Not a Co-Citationist'. *4S Newsletter* **2**, No. 3: 13–19.

Edge, 1979. David Edge. 'Quantitative Measures of Communication in Science: A Critical Review'. *History of Science* **17**: 102–134.

Edge and Mulkay, 1976. David O. Edge and Michael J. Mulkay. *Astronomy Transformed: The Emergence of Radio Astronomy in Britain*. New York: Wiley-Interscience.

Ehrenberg, 1978. A. S. C. Ehrenberg. 'Graphs or Tables?' *The Statistician* **27**: 87–96.

Elkana *et al.* (eds.), 1978. Yehuda Elkana, Joshua Lederberg, Robert K. Merton, Arnold Thackray, and Harriet Zuckerman (eds.). *Toward a Metric of Science: The Advent of Science Indicators*. New York: John Wiley and Sons.

Erickson, 1975. Charlotte Erickson. 'Quantitative History [review article]'. *American Historical Review* **80**: 351–365.

Ferriss, 1979. Abbott L. Ferriss. 'The U.S. Federal Effort in Developing Social Indicators'. *Social Indicators Research* **6**: 129–152.

Fienberg and Goodman, 1974. Stephen E. Fienberg and Leo A. Goodman. '*Social Indicators, 1973*: Statistical Considerations'. In *Social Indicators, 1973: A Review Symposium*, edited by Roxann A. Van Dusen, 63–85. Washington, D.C.: Social Science Research Council, Center for Coordination of Research on Social Indicators.

Fishbein, 1973. Meyer H. Fishbein. *The Censuses of Manufactures, 1810–1890*. Reference Information Paper No. 50. Washington, D.C.: U.S. National Archives and Records Service.

Floud, 1973. Roderick Floud. *An Introduction to Quantitative Methods for Historians*. Princeton: Princeton University Press.

Fogel, 1975. Robert W. Fogel. 'The Limits of Quantitative Methods in History'. *American Historical Review* **80**: 329–350.

Forster, 1978. Robert Forster. 'Achievements of the Annales School'. *Journal of Economic History* **38**: 58–76.

Freidson, 1970. Eliot Freidson. *Profession of Medicine: A Study of the Sociology of Applied Knowledge*. New York: Dodd, Mead and Company.

Freidson, 1977. Eliot Freidson. 'The Future of Professionalisation'. In *Health and the Division of Labor*, edited by M. Stacey *et al.*, 14–38. London: Croom Helm.

Furet, 1971. François Furet. 'Quantitative History'. *Daedalus* **100**: 151–167. (Reprinted in *Historical Studies Today*, edited by Felix Gilbert and Stephen R. Graubard, 45–61. New York: W. W. Norton, 1971.)

Garfield, 1979. Eugene Garfield. *Citation Indexing – Its Theory and Application in Science, Technology, and Humanities*. New York: Wiley-Interscience.

Garfield, Malin, and Small, 1978. Eugene Garfield, Morton V. Malin, and Henry Small. 'Citation Data as Science Indicators'. In II, Elkana *et al.* (eds.), 1978, 179–207.

Garfield, Sher, and Torpie, 1964. Eugene Garfield, Irving H. Sher, and Richard J. Torpie. *The Use of Citation Data in Writing the History of Science*. Philadelphia: Institute for Scientific Information.

Gilbert and Woolgar, 1974. G. Nigel Gilbert and Steve Woolgar. 'The Quantitative Study of Science: An Examination of the Literature'. *Science Studies* **4**: 279–294.

Gilmartin *et al.*, 1979. Kevin J. Gilmartin, Robert J. Rossi, Leonard S. Lutomski, and Donald F. B. Reed. *Social Indicators: An Annotated Bibliography of Current Literature*. Garland Reference Library of Social Science, 62. New York: Garland Publishing, Inc.

Goldstein, 1959. Harold T. Goldstein. *Historical Comparability of Census of Manufactures Industries, 1929–1958*. Bureau of the Census Working Paper No. 9. Washington, D.C.: U.S. Department of Commerce.

Gonzalez *et al.*, 1975. Maria E. Gonzalez, Jack L. Ogus, Gary Shapiro, and Benjamin J. Tepping. 'Standards for Discussion and Presentation of Errors in Survey and Census Data'. *Journal of the American Statistical Association* **70**, No. 351, Pt. II.

Graham *et al.* (eds.), 1983. Loren Graham, Wolf Lepenies, and Peter Weingart. *Functions and Uses of Disciplinary Histories*. Sociology of the Sciences Yearbook, 7. Dordrecht and Boston: D. Reidel.

Griffith, Drott, and Small, 1977. Belver C. Griffith, M. Carl Drott, and Henry G. Small. 'On the Use of Citations in Studying Scientific Achievements and Communication'. *4S Newsletter* **2**, No. 3: 9–13.

Hahn, 1980. Roger Hahn. *A Bibliography of Quantitative Studies on Science and Its History*. Berkeley Papers in History of Science, III. Berkeley: University of California.

Hamblin *et al.*, 1973. Robert L. Hamblin *et al. A Mathematical Theory of Social Change*. New York: John Wiley and Sons.

Harnett, 1975. Donald L. Harnett. *Introduction to Statistical Methods*. 2nd edition. Reading, Mass.: Addison-Wesley Publishing Company.

Hershberg and Dockhorn, 1976. Theodore Hershberg and Robert Dockhorn. 'Occupational Classification'. *Historical Methods Newsletter* 9: 59–77.

Hexter, 1972. J. H. Hexter. 'Fernand Braudel and the *Monde Braudellien* . . .'. *Journal of Modern History* **44**: 480–539.

Holton, 1978. Gerald Holton. 'Can Science Be Measured?' In II, Elkana *et al.* (eds.), 1978, 39–68.

Hughes, 1965. Everett C. Hughes. 'Professions'. In *The Professions in America*, edited by Kenneth S. Lynn, 1–14. Boston: Beacon Press.

Hughes, 1978. Thomas Parke Hughes. 'Neue Themen und Ergebnisse der Technikgeschichtsschreibung in den USA'. *Geschichte und Gesellschaft* **4**: 258–271. Reprinted in *Technology and Culture* **20** (1979): 697–711.

Katz, 1972. Michael B. Katz. 'Occupational Classification in History'. *Journal of Interdisciplinary History* **3**: 63–88.

Keyfitz, 1977. Nathan Keyfitz. *Applied Mathematical Demography*. New York: Wiley-Interscience.

King *et al.*, 1976. D. W. King, D. D. McDonald, N. K. Roderer, and B. L. Wood. *Statistical Indicators of Scientific and Technical Communication, 1960–1980*. 2 vols. A Report to the National Science Foundation, Division of Science Information, under Contract NSF C-878. Washington, D.C.: U.S. Government Printing Office.

Kohlstedt, 1977. Sally Gregory Kohlstedt. 'Reassessing Science in Antebellum America'. *American Quarterly* **29**: 444–453.

Kousser, 1980. J. Morgan Kousser. 'Quantitative Social-Scientific History'. In *The Past before Us: Contemporary Historical Writing in the United States*, edited by Michael Kammen, 433–456. Ithaca, N.Y.: Cornell University Press.

Kruskal, 1978. William Kruskal. 'Taking Data Seriously'. In II, Elkana *et al.* (eds.), 1978, 139–169.

Kuznets, 1951. Simon Kuznets. 'Statistical Trends and Historical Changes'. *Economic History Review*, 2d ser., **3**: 265–278.

Lee *et al.*, 1957. Everett S. Lee, Ann Ratner Miller, Carol P. Brainerd, and Richard A. Easterlin. *Population Redistribution and Economic Growth: United States, 1870–1950, I: Methodological Considerations and Reference Tables*. Prepared under the direction of Simon Kuznets and Dorothy Swaine Thomas. American Philosophical Society *Memoirs*, Vol. 45. Philadelphia: American Philosophical Society.

Lemaine *et al.* (eds.), 1976. Gerard Lemaine, Roy MacLeod, Michael Mulkay, and Peter Weingart (eds.). *Perspectives on the Emergence of Scientific Disciplines*. Maison des Sciences de l'Homme, Paris, Publications, 4. The Hague: Mouton, and Chicago: Aldine.

Leupold *et al.*, 1982. Andrea Leupold, Peter Weingart, and Matthias Winterhager. *Wissenschaftsindikatoren und Quantitative Wissenschaftsforschung. Eine annotierte Bibliographie*. Report Wissenschaftsforschung, 18. Bielefeld: B. Kleine Verlag.

Lurie, 1972. Edward Lurie. 'The History of Science in America: Development and New Directions'. In II, Daniels (ed.), 1972, 3–21.

McCloskey, 1976. Donald N. McCloskey. 'Does the Past Have Useful Economics?' *Journal of Economic Literature* **14**: 434–461.

McCormmach, 1971. Russell McCormmach. 'Editor's Foreword'. *Historical Studies in the Physical Sciences* **3**: ix–xxiv.

Machlup, 1962. Fritz Machlup. *The Production and Distribution of Knowledge in the United States*. Princeton: Princeton University Press.

Merton, 1977. Robert K. Merton. 'The Sociology of Science: An Episodic Memoir'. In *The Sociology of Science in Europe*, edited by Robert K. Merton and Jerry Gaston, 3–141. Carbondale: Southern Illinois University Press.

Millerson, 1964. Geoffrey Millerson. *The Qualifying Associations: A Study in Professionalization*. London: Routledge and Kegan Paul.

Morgenstern, 1963a. Oskar Morgenstern. *On the Accuracy of Economic Observations*. 2nd edition, completely revised. Princeton: Princeton University Press.

Morgenstern, 1963b. Oskar Morgenstern. '*Qui Numerare Incipit Errare Incipit*'. *Fortune* **68** (October): 142–144, 173–174, 178, 180.

Morgenstern, 1964. Oskar Morgenstern. '*Fide Sed Ante Vide*: Remarks to Mr. R. T. Bowman's "Comments" '. *American Statistician* **18** (October): 15–16, 25.

Narin and Moll, 1977. Francis Narin and Joy K. Moll. 'Bibliometrics'. *Annual Review of Information Science and Technology* **12**: 35–58.

NORC, 1947. National Opinion Research Center. 'Jobs and Occupations: A Popular Evaluation'. *Opinion News* **9** (1 September): 3–13.

NSF, 1980. National Science Foundation, Division of Science Resources Studies. *A Guide to NSF Science Resources Data*. Washington, D.C.: U.S. National Science Foundation.

NSF, 1983. National Science Foundation, Division of Science Resources Studies. *Project Summaries: FY1983*. NSF 83–326. Washington, D.C.: U.S. National Science Foundation.

Olinick, 1978. Michael Olinick. *An Introduction to Mathematical Models in the Social and Life Sciences*. Reading, Mass.: Addison-Wesley Publishing Company.

OMB, 1972. Executive Office of the President, Office of Management and Budget, Statistical Policy Division. *Standard Industrial Classification Manual 1972*. Washington, D.C.: U.S. Government Printing Office.

Peck, 1979. Theodore P. Peck (ed.). *Chemical Industries Information Sources*. Management Information Guide Series, 29. Detroit: Gale Research Company.

Price, 1961. Derek J. de Solla Price. *Science since Babylon*. New Haven: Yale University Press.

Price, 1963. Derek J. de Solla Price. *Little Science, Big Science*. New York: Columbia University Press.

Price, 1965. Derek J. de Solla Price. 'Networks of Scientific Papers'. *Science* **149**: 510–515.

Price, 1970. Derek J. de Solla Price. 'Citation Measures of Hard Science, Soft Science, Technology and NonScience'. In *Communication among Scientists and Engineers*, edited by C. E. Nelson and D. K. Pollock, 3–22. Lexington, Mass.: D.C. Heath and Company.

Price, 1976. Derek J. de Solla Price. 'A General Theory of Bibliometric and Other Cumulative Advantage Processes'. *Journal of the American Society for Information Science* **27**: 292–306.

Pyenson, 1977. Lewis Pyenson. ' "Who the Guys Were": Prosopography in the History of Science'. *History of Science* **15**: 155–188.

Reingold, 1976. Nathan Reingold. 'Definitions and Speculations: The Professionalization of Science in America in the Nineteenth Century'. In *The Pursuit of Knowledge in the Early American Republic*, edited by Alexandra Oleson and Sanborn C. Brown, 33–69. Baltimore: Johns Hopkins University Press.

Reingold (ed.), 1976. Nathan Reingold (ed.). *Science in America since 1920*. History of Science Selections from *Isis*, edited by Robert P. Multhauf. New York: Neale Watson Academic Publications.

Reingold (ed.), 1979. Nathan Reingold (ed.). *The Sciences in the American Context: New Perspectives*. Washington, D.C.: Smithsonian Institution Press.

Rose, 1967. Steven Rose. 'The S Curve Considered'. *Technology and Society*. **4**: 33–39.

Rosenberg, 1964. Charles E. Rosenberg. 'On the Study of American Biology and Medicine: Some Justifications'. *Bulletin of the History of Medicine* **38**: 364–374.

Rosenberg, 1970. Charles E. Rosenberg. 'On Writing the History of American Science'. In *The State of American History*, edited by Herbert J. Bass, 183–196. Chicago: Quadrangle.

Rosenberg, 1979. Charles E. Rosenberg. 'Toward an Ecology of Knowledge: On Discipline, Context, and History'. In III, Oleson and Voss (eds.), 1979, 440–455.

Rosenberg, 1983. Charles E. Rosenberg. 'Science in American Society: A Generation of Historical Debate'. *Isis* **74**: 356–367.

Rossi and Gilmartin, 1980. Robert J. Rossi and Kevin J. Gilmartin. *The Handbook of Social Indicators: Sources, Characteristics, and Analysis*. New York: Garland STPM Press.

Rothenberg, 1982. Marc Rothenberg. *The History of Science and Technology in the United States: A Selective and Critical Bibliography*. New York: Garland Publishing, Inc.

SAS Institute Inc., 1979. *SAS User's Guide, 1979 Edition*. Raleigh, N.C.: SAS Institute Inc.

Shapin and Thackray, 1974. Steven Shapin and Arnold Thackray. 'Prosopography as a Research Tool in History of Science: The British Scientific Community, 1700–1900'. *History of Science* **12**: 1–28.

Sheldon and Parke, 1975. Eleanor B. Sheldon and Robert Parke. 'Social Indicators'. *Science* **188**: 693–699.

Siegel, 1971. Paul M. Siegel. *Prestige in the American Occupational Structure*. PhD dissertation, University of Chicago.

Small, 1977. Henry Small. 'A Co-Citation Model of a Scientific Specialty: A Longitudinal Study of Collagen Research'. *Social Studies of Science* **7**: 139–166.

M. Smith, 1943. Maphaeus Smith. 'An Empirical Scale of Prestige Status of Occupations'. *American Sociological Review* **8**: 185–192.

T. Smith, 1975. Thomas Smith. 'Reconstructing Occupational Structures: The Case of the Ambiguous Artisans'. *Historical Methods Newsletter* **8**: 134–146.

Spear, 1969. Mary E. Spear. *Practical Charting Techniques*. New York: McGraw-Hill.

Sprague, 1978. D. N. Sprague. 'A Quantitative Assessment of the Quantification Revolution'. *Canadian Journal of History* **13**: 177–192.

Stoianovich, 1976. Traian Stoianovich. *French Historical Method: The Annales Paradigm*. Ithaca, N.Y.: Cornell University Press.

Stone, 1971. Lawrence Stone. 'Prosopography'. *Daedalus* **100**: 46–79.
Sullivan, White, and Barboni, 1977. Daniel Sullivan, D. Hywel White, and Edward J. Barboni. 'Co-Citation Analyses of Science: An Evaluation'. *Social Studies of Science* 7: 223–240.
Thackray, 1978. Arnold Thackray. 'Measurement in the Historiography of Science'. In II, Elkana *et al.* (eds.), 1978, 11–30.
Treiman, 1977. Donald J. Treiman. *Occupational Prestige in Comparative Perspective*. Quantitative Studies in Social Relations, edited by Peter H. Rossi. New York: Academic Press.
Tufte, 1983. Edward R. Tufte. *The Visual Display of Quantitative Information*. Cheshire, Conn.: Graphics Press.
USBC, 1880. 'Alphabetical List of Occupations [1880?]'. Series 75, Records of the Tenth Census. Records of the Bureau of the Census. Record Group 29. Washington, D.C.: U.S. National Archives and Records Service.
USBC, 1915. Bureau of the Census. *Index to Occupations. Alphabetical and Classified*. Washington, D.C.: U.S. Government Printing Office.
USBC, 1921. Bureau of the Census. *Classified Index to Occupations*. Washington, D.C.: U.S. Government Printing Office.
USBC, 1930. Bureau of the Census. *Classified Index to Occupations, Fifteenth Census of the United States*. Washington, D.C.: U.S. Government Printing Office.
USBC, 1940. Bureau of the Census. *Sixteenth Census of the United States (1940), Classified Index of Occupations*. Washington, D.C.: U.S. Government Printing Office.
USBC, 1950. Bureau of the Census. *1950 Census of Population, Classified Index of Occupations and Industries*. Washington, D.C.: U.S. Government Printing Office.
USBC, 1960a. Bureau of the Census. *The Accuracy of Census Statistics with and without Sampling*. Technical Paper No. 2. Washington, D.C.: U.S. Government Printing Office.
USBC, 1960b. Bureau of the Census. *1960 Census of Population, Classified Index of Occupations and Industries*. Washington, D.C.: U.S. Government Printing Office.
USBC, 1971. Bureau of the Census. *1970 Census of Population, Classified Index of Industries and Occupations*. Washington, D.C.: U.S. Government Printing Office.
USBC, 1972. Bureau of the Census. *1970 Occupation and Industry Classification Systems in Terms of Their 1960 Occupation and Industry Elements*. By John A. Priebe, Joan Heinkel, and Stanley Greene. Technical Paper No. 26. Washington, D.C.: U.S. Government Printing Office.
USBC, 1974. U.S. Bureau of the Census. *Bureau of the Census Catalog of Publications, 1790–1972*. Washington, D.C.: U.S. Government Printing Office.
USBC, 1978. U.S. Bureau of the Census. *Guide to Industrial Statistics*. 1976 edition. Washington, D.C.: U.S. Government Printing Office.
USCSC, 1950–. Civil Service Commission, Bureau of Programs and Standards. *Handbook of Occupational Groups and Series of Classes Established under the Federal Position-Classification Plan*. Washington, D.C.: U.S. Government Printing Office.
USCSC, 1951–. Civil Service Commission. *Qualification Standards for Classification Act Positions*. Civil Service Handbook X118. Washington, D.C.: U.S. Government Printing Office.
USCSC, 1966. Civil Service Commission. *Basic Training Course in Position Classification*. Personnel Methods Series No. 11. Part 1, *Fundamentals and the Federal Plan*. Washington, D.C.: U.S. Government Printing Office.

USGAO, 1979. U.S. General Accounting Office. *Science Indicators: Improvements Needed in Design, Construction, and Interpretation*. Report by the Comptroller General of the United States. PAD–79–35. September 25.

Van Tassel and Hall (eds.), 1966. David D. Van Tassel and Michael G. Hall (eds.). *Science and Society in the United States*. Homewood, Ill.: Dorsey Press.

Wasserman *et al.* (eds.), 1983. Paul Wasserman, Jacqueline O'Brien, Daphne A. Grace, and Kenneth Clansky (eds.). *Statistics Sources: A Subject Guide to Data on Industrial, Business, Social, Educational, Financial, and Other Topics for the United States and Internationally*. 8th edition. 2 vols. Detroit: Gale Research Company.

Weisz and Kruytbosch, 1982. Diane B. Weisz and Carlos E. Kruytbosch. *Studies of Scientific Disciplines: An Annotated Bibliography*. NSF 83–7. Washington, D.C.: U.S. National Science Foundation.

West, 1920. Clarence Jay West. *A Reading List on Scientific and Industrial Research and the Service of the Chemist to Industry*. Reprint and Circular Series of the National Research Council, No. 9. Washington, D.C.: National Research Council.

Whitrow, 1976. Magda Whitrow (ed.). *ISIS Cumulative Bibliography: A Bibliography of the History of Science Formed from ISIS Critical Bibliographies 1–90, 1913–1965*. Vol. 3: *Subjects*. London: Mansell, in conjunction with the History of Science Society.

Wilcox *et al.*, 1972. Leslie D. Wilcox, Ralph M. Brooks, George M. Beal, and Gerald E. Klonglan. *Social Indicators and Societal Monitoring: An Annotated Bibliography*. New York: Elsevier Scientific Publishing Company.

Wrigley, 1969. Edward Wrigley. *Population and History*. New York: McGraw-Hill.

Ziman, 1978. John Ziman. 'From Parameters to Portents – and Back'. In II, Elkana *et al.* (eds.), 1978, 261–284.

Zuckerman and Miller, 1980. Harriet Zuckerman and Roberta B. Miller. 'Indicators of Science: Notes and Queries'. *Scientometrics* **2**: 347–353.

III. Other Books and Articles

AAAS, 1941. American Association for the Advancement of Science, Committee on the Improvement of Science Instruction for Purposes of General Education. 'Chemistry Instruction for Purposes of General Education'. *Journal of Chemical Education* **18**: 10–14.

AACC, 1965. 'American Association of Cereal Chemists: The First Fifty Years'. *Cereal Science Today* **10** (April): 121–132, 173–184.

Abelson, 1965. Philip H. Abelson. 'Chemistry: Opportunities and Needs'. *Science* **150**: 1247.

ACS, 1924. American Chemical Society, Committee on Chemical Education. 'Correlation of High School and College Chemistry'. *Journal of Chemical Education* **1**: 87–99.

ACS, 1925. American Chemical Society, Committee on Chemical Education. 'Report of the Committee on the Relationship Between High-School and College Chemistry'. *Journal of Chemical Education* **2**: 269–275.

ACS, 1927. American Chemical Society, Committee on Chemical Education. 'Correlation of High School and College Chemistry'. *Journal of Chemical Education* **4**: 640–656.

ACS, 1944. American Chemical Society. *Vocational Guidance in Chemistry and Chemical Engineering*. Revised edition. Washington, D.C.: American Chemical Society.

ACS, 1951a. American Chemical Society. *Chemistry . . . Key to Better Living*. Diamond Jubilee Volume. Washington, D.C.: American Chemical Society.

ACS, 1951b. American Chemical Society, Chemical and Engineering News. *Careers in Chemistry and Chemical Engineering*. Washington, D.C.: American Chemical Society.

ACS, 1979. American Chemical Society, Department of Public and Member Relations. *The Chemist and the Public*. 2nd edition. Washington, D.C.: American Chemical Society.

Adams, 1935. Roger Adams. 'The Relation of the University Scientists to the Chemical Industries'. *Industrial and Engineering Chemistry, News Edition* **13**: 365–367.

Anderson, 1958. Oscar E. Anderson, Jr. *The Health of a Nation: Harvey W. Wiley and the Fight for Pure Food*. Chicago: University of Chicago Press.

AOAC, 1934. Association of Official Agricultural Chemists. *Golden Anniversary of the Association of Official Agricultural Chemists, 1884–1934*. Washington, D.C.: Association of Official Agricultural Chemists.

AOCS, 1947. American Oil Chemists' Society. *Presidents and Committees of the American Oil Chemists' Society, 1909–1947*. Chicago: American Oil Chemists' Society.

Ash *et al.*, 1915. Charles S. Ash *et al.* 'Symposium on the Contributions of the Chemist to American Industries'. *Journal of Industrial and Engineering Chemistry* 7: 273–304.

Ashford, 1942. T. A. Ashford. 'The Problem of Chemistry in General Education'. *Journal of Chemical Education* **19**: 260–263.

Ayre and Lemaire, 1977. Pamela Ayre and Patricia M. Lemaire (eds.). *Careers Nontraditional*. Washington, D.C.: American Chemical Society.

Bacon, 1915. Raymond F. Bacon. 'The Object and Work of the Mellon Institute'. *Journal of Industrial and Engineering Chemistry* 7: 343–347.

Bacon, 1916. Raymond F. Bacon. 'The Industrial Fellowships of the Mellon Institute'. *Science* **43**: 453–456.

Bacon and Hamor, 1916. Raymond F. Bacon and William A. Hamor. *The American Petroleum Industry*. 2 vols. New York: McGraw-Hill.

Badger, 1929. W. L. Badger. 'The Demand for Chemical Engineers'. *Industrial and Engineering Chemistry, News Edition* 7 (20 March): 3.

Ball, 1938. C. R. Ball. *Federal, State, and Local Administrative Relationships in Agriculture*. 2 vols. Berkeley: University of California Press.

Bartlett, 1941. Howard R. Bartlett. 'The Development of Industrial Research in the United States'. In III, NRPB, 1941, 19–77.

Bartow, 1939. Virginia Bartow. 'Chemical Genealogy'. *Journal of Chemical Education* **16**: 236–238.

Bartow, 1941. Virginia Bartow. 'Developments in the Chemistry Department, University of Illinois, 1926–1941'. In University of Illinois, Department of Chemistry, *Developments during the Period 1927–1941, Publications of the Department, Courses of Study, Faculty, Doctor of Philosophy Degrees in Chemistry*, 27–45. Urbana: University of Illinois.

Baskerville (ed.), 1911. Charles Baskerville (ed.). *Municipal Chemistry. A Series of Thirty Lectures by Experts on the Application of the Principles of Chemistry to the City....* New York: McGraw-Hill.

Bates, 1965. Ralph S. Bates. *Scientific Societies in the United States*. 3rd edition. Cambridge, Mass.: MIT Press.

Battle, 1912. Kemp P. Battle. *History of the University of North Carolina*. 2 vols. Raleigh: Edwards and Broughton.

Beal, 1927. George D. Beal. 'Ten Years of Chemistry at Illinois (1916–1927)'. In *Special Circular of the Department of Chemistry, 1916–1927*, University of Illinois Bulletin, Vol. 24, No. 52: 9–23.

Beardsley, 1964. Edward H. Beardsley. *The Rise of the American Chemistry Profession, 1850–1900*. University of Florida Monographs, Social Sciences, No. 23. Gainesville: University of Florida Press.

Beaton, 1957. Kendall Beaton. *Enterprise in Oil: A History of Shell in the United States*. New York: Appleton-Century-Crofts.

Beck, 1956. Alfred D. Beck. 'Progress Report on the Development of General Science Curriculum Program in the Public Schools of New York City'. *Science Education* **40**: 134–136.

Ben-David, 1971. Joseph Ben-David. *The Scientist's Role in Society: A Comparative Study*. Englewood Cliffs, N.J.: Prentice-Hall.

Berenson, 1963. Conrad Berenson (ed.). *The Chemical Industry: Viewpoints and Perspectives*. New York: Interscience.

Bernhard *et al.* (eds.), 1982. Carl Gustaf Bernhard, Elisabeth Crawford, and Per Sörbom (eds.). *Science, Technology, and Society in the Time of Alfred Nobel*. Nobel Symposium, 52. New York: Pergamon Press.

Bernstein, 1978. Peter W. Bernstein. 'Peter Grace's Long Search for Security'. *Fortune* **97** (8 May): 116–133.

Bigelow, 1908. W. D. Bigelow. 'Chemical Positions in the Government Service'. *Science* **27**: 481–488.

Billinger, 1930. R. D. Billinger. 'Thomas Messinger Drown'. *Journal of Chemical Education* 7: 2875–2886.

Billinger, 1939. R. D. Billinger. 'The Chandler Influence in American Chemistry'. *Journal of Chemical Education* **16**: 253–257.

Billmeyer and Kelley, 1975. Fred W. Billmeyer, Jr., and Richard N. Kelley. *Entering Industry: A Guide for Young Professionals*. New York: John Wiley and Sons.

Birr, 1966. Kendall Birr. 'Science in Industry'. In II, Van Tassel and Hall (eds.), 1966, 35–80.

Birr, 1979. Kendall Birr. 'Industrial Research Laboratories'. In II, Reingold (ed.), 1979, 193–207.

Blank and Stigler, 1957. David M. Blank and George J. Stigler. *The Demand and Supply of Scientific Personnel*. National Bureau of Economic Research, General Series, No. 62. New York: National Bureau of Economic Research.

Bledstein, 1976. Burton J. Bledstein. *The Culture of Professionalism: The Middle Class and the Development of Higher Education in America*. New York: W. W. Norton.

Bogert, 1908. Marston Taylor Bogert. 'American Chemical Societies'. *Journal of the American Chemical Society* **30**: 163–182.

Bogert *et al.*, 1919. Marston T. Bogert *et al.* 'Report of the Committee on War Service for Chemists'. *Journal of Industrial and Engineering Chemistry* **11**: 413–415.

Bogert, 1931. Marston T. Bogert. 'Biographical Memoir of Charles Frederick Chandler, 1836–1925'. *Biographical Memoirs of the National Academy of Sciences* **14**: 125–181.

Boig and Howerton, 1952a. Fletcher S. Boig and Paul W. Howerton. 'History and Development of Chemical Periodicals in the Field of Organic Chemistry: 1877–1949'. *Science* **115**: 25–31.

Boig and Howerton, 1952b. F. S. Boig and P. W. Howerton. 'History and Development of Chemical Periodicals in the Field of Analytical Chemistry, 1877–1950'. *Science* **115**: 555–560.

Boorstin, 1965. Daniel J. Boorstin. *The Americans: The National Experience*. New York: Random House.

Boorstin, 1973. Daniel J. Boorstin. *The Americans: The Democratic Experience*. New York: Random House.

Borscheid, 1976. Peter Borscheid. *Naturwissenschaft, Staat und Industrie in Baden (1848–1914)*. Industrielle Welt, Bd. 17. Stuttgart: Ernst Klett Verlag.

Borth, 1942. Christy Borth. *Pioneers of Plenty: Modern Chemists and Their Work*. Enlarged edition. New York: New Home Library.

Bowen, 1924. Catherine Drinker Bowen. *A History of Lehigh University*. Bethlehem, Pa.: Lehigh Alumni Bulletin.

Bowles and Gintis, 1976. Samuel Bowles and Herbert Gintis. *Schooling in Capitalist America: Educational Reform and the Contradictions of Economic Life*. New York: Basic Books.

Brode, 1971. Wallace Reed Brode. 'Manpower in Science and Engineering, Based on a Saturation Model'. *Science* **173**: 206–213.

Brooks, 1958. Benjamin T. Brooks. 'The Current Research Effort in the United States, Russia, and Other Countries'. *Journal of Chemical Education* **35**: 468–469.

Brooks and Smythe, 1975. R. R. Brooks and L. E. Smythe. 'The Progress of Analytical Chemistry, 1910–1970'. *Talanta* **22**: 495–504.

Browne (ed.), 1926. Charles A. Browne (ed.). *A Half-Century of Chemistry in America, 1876–1926*. An Historical Review Commemorating the Fiftieth Anniversary of the American Chemical Society. Philadelphia, September 6–11, 1926. Golden Jubilee Number of the *Journal of the American Chemical Society* **48**, No. 8A (20 August).

Browne, 1938a. C. A. Browne. 'Dr. Thomas Antisell and His Associates in the Founding of the Chemical Society of Washington'. *Journal of the Washington Academy of Sciences* **28**: 213–232.

Browne, 1938b. C. A. Browne. 'The Chemical Society of Washington and Its Part in the Reorganization of the American Chemical Society'. *Journal of the Washington Academy of Sciences* **28**: 233–246.

Browne, 1939. C. A. Browne. 'Charles Edward Munroe, 1849–1938'. *Journal of the American Chemical Society* **61**: 1301–1316.

Browne and Weeks, 1952. C. A. Browne and Mary Elvira Weeks. *A History of the American Chemical Society: Seventy-Five Eventful Years*. Washington, D.C.: American Chemical Society.

Bruchey, 1975. Stuart Bruchey. *Growth of the Modern American Economy*. New York: Dodd, Mead and Company.

Burns and Enck, 1977. Robert M. Burns, with Ernest G. Enck. *A History of the Electrochemical Society, 1902–1976*. Princeton, N.J.: Electrochemical Society.

BVI, 1922. Bureau of Vocational Information. *Women in Chemistry: A Study of Professional Opportunities*. Studies in Occupations, No. 4. New York: Bureau of Vocational Information.

Caldwell, 1892. George C. Caldwell. 'The American Chemist'. *Journal of the American Chemical Society* **14**: 331–349.

Calhoun, 1960. Daniel H. Calhoun. *The American Civil Engineer: Origin and Conflicts*. Cambridge, Mass.: Harvard University Press.

Calhoun, 1965. Daniel H. Calhoun. *Professional Lives in America: Structure and Aspiration, 1750–1850*. Cambridge, Mass.: Harvard University Press.

Calvert, 1967. Monte A. Calvert. *The Mechanical Engineer in America, 1830–1910: Professional Cultures in Conflict*. Baltimore: Johns Hopkins Press.

Campbell, 1964. J. Arthur Campbell. 'CHEM Study – An Approach to Chemistry Based on Experiments'. In *New Curricula*, edited by Robert W. Heath, 82–93. New York: Harper and Row.

Carlisle, 1943. Norman V. Carlisle. *Your Career in Chemistry*. With a preface by C. M. A. Stine. New York: E. P. Dutton.

Carmichael, 1974. Emmett B. Carmichael. 'Fifty Golden Years: History and Accomplishments of the American Institute of Chemists'. *The Chemist* **51** (September): 22–23; (October): 28–29; (November): 27–28; (December): 23–25.

Carroll, 1982. P. Thomas Carroll. *Academic Chemistry in America, 1876–1976: Diversification, Growth, and Change*. PhD dissertation, University of Pennsylvania.

Chamberlain and Browne, 1926. Joseph S. Chamberlain and C. A. Browne (eds.). *Chemistry in Agriculture: A Cooperative Work Intended to Give Examples of the Contributions Made to Agriculture by Chemistry*. New York: Chemical Foundation.

Chandler, 1962. Alfred D. Chandler, Jr. *Strategy and Structure: Chapters in the History of the Industrial Enterprise*. Cambridge, Mass.: MIT Press.

Chandler, 1969. Alfred D. Chandler, Jr. 'The Structure of American Industry in the Twentieth Century: A Historical Overview'. *Business History Review* **43**: 255–281.

Chandler, 1977. Alfred D. Chandler, Jr. *The Visible Hand: The Managerial Revolution in American Business*. Cambridge, Mass.: Harvard University Press, Belknap Press.

C&EN, 1976. 'Centennial: American Chemical Society, 1876 to 1976'. Special issue of *Chemical and Engineering News* **54** (6 April): 1–206.

'Chemical', 1950. 'The Chemical Century'. *Fortune* **41** (March): 69–76, 114 ff.

Chemical Foundation, 1923. [Chemical Foundation]. *The Future Independence and Progress of American Medicine in the Age of Chemistry*. New York: Chemical Foundation.

'Chemical Industry', 1937. 'Chemical Industry: I'. *Fortune* **16** (December): 83–90, 157–162.

'Chemical Warfare Service', 1918. 'Chemical Warfare Service'. *Journal of Industrial and Engineering Chemistry* **10**: 675–684.

'Chemicals', 1957. 'Chemicals Lose Old-Time Fizz'. *Business Week*, 20 April: 89–94.

Cheyney, 1940. Edward Potts Cheyney. *History of the University of Pennsylvania, 1740–1940*. Philadelphia: University of Pennsylvania Press.

Chittenden, 1945. R. H. Chittenden. *The First Twenty-Five Years of the American Society of Biological Chemists*. New Haven, Conn.: American Society of Biological Chemists.

B. Clarke, 1937. Beverly L. Clarke. 'The Role of Analytical Chemistry in Industrial Research. III. The Training of Analysts'. *Journal of Chemical Education* **14**: 561–563.

F. W. Clarke, 1885. F. W. Clarke. 'The Relations of the Government to Chemistry'. *Bulletin of the Chemical Society of Washington* **1**: 9–22.

F. W. Clarke, 1909. F. W. Clarke. 'The Chemical Work of the United States Geological Survey'. *Science* **30**: 161–171.

M. Clarke, 1977. Margaret Jackson Clarke. *The Federal Government and the Fixed Nitrogen Industry, 1915–1926*. PhD dissertation, Oregon State University.

Clarkson, 1923. Grosvenor B. Clarkson. *Industrial America in the World War: The Strategy behind the Line, 1917–1918*. Noston and New York: Houghton Mifflin Company.

Cleaveland, 1959. Frederic N. Cleaveland. *Science and State Government*. Chapel Hill: University of North Carolina Press.

Clement, 1911. J. K. Clement. 'The Work of the Chemical Laboratories of the Bureau of Mines'. *Journal of Industrial and Engineering Chemistry* **3**: 96–99.

Coben, 1976. Stanley Coben. 'Foundation Officials and Fellowships: Innovation in the Patronage of Science'. *Minerva* **14**: 225–240.

Cochran, 1972. Thomas C. Cochran. *American Business in the Twentieth Century*. Cambridge, Mass.: Harvard University Press.

Cochrane, 1966. Rexmond C. Cochrane. *Measures for Progress: A History of the National Bureau of Standards*. National Bureau of Standards, Miscellaneous Publication No. 275. Washington, D.C.: U.S. Government Printing Office.

Cochrane, 1978. Rexmond C. Cochrane. *The National Academy of Sciences: The First Hundred Years, 1863–1963*. Washington, D.C.: National Academy of Sciences.

Coith, 1943. Herbert Coith. *So You Want to Be a Chemist?* New York: McGraw-Hill.

Cole and Cole, 1973. Jonathan and Stephen Cole. *Social Stratification in Science*. Chicago: University of Chicago Press.

CRS, 1975. U.S. Library of Congress, Congressional Research Service, Science Policy Research Division. *The National Science Foundation and Pre-College Science Education: 1950–1975*. Report Prepared for the Subcommittee on Science, Research, and Technology of the Committee on Science and Technology, U.S. House of Representatives. Serial T, January 1976. 94th Cong., 2d sess. Washington, D.C.: U.S. Government Printing Office.

CRSE, 1918. Commission on the Reorganization of Secondary Education. *Cardinal Principles of Secondary Education*. U.S. Bureau of Education *Bulletin*, 1918, No. 35. Washington, D.C.: U.S. Government Printing Office.

CRSE, 1920. Commission on the Reorganization of Secondary Education. *Reorganization of Science in Secondary Schools*. U.S. Bureau of Education *Bulletin*, 1920, No. 26. Washington, D.C.: U.S. Government Printing Office.

Cushman, 1920. Allerton S. Cushman. *Chemistry and Civilization*. Boston: R. G. Badger. Second edition in 1925.

Davis *et al.*, 1972. Lance E. Davis *et al. American Economic Growth: An Economist's History of the United States*. New York: Harper and Row.

Dede and Hardin, 1973. C. J. Dede and J. Hardin. 'Reform, Revisions, Reexaminations: Secondary Science Education since World War II'. *Science Education* **57**: 485–491.

'D'Ianni Wins', 1978. 'D'Ianni Wins Close ACS Election'. *Chemical and Engineering News* **56** (20 November): 6.

Dietz, 1943. David Dietz. *The Goodyear Research Laboratory*. Akron, Ohio: Goodyear Tire and Rubber Company.

Dietz, 1955. David Dietz. *Harvest of Research: The Story of the Goodyear Chemical Division*. Akron, Ohio: Goodyear Tire and Rubber Company.

Dolby, 1977. R. G. A. Dolby. 'The Transmission of Two New Scientific Disciplines from Europe to North America in the Late Nineteenth Century'. *Annals of Science* **34**: 287–310.

Dudley, 1898. C. B. Dudley. 'The Dignity of Analytical Work'. *Journal of the American Chemical Society* **20**: 81–96.

Duncan, 1907. Robert Kennedy Duncan. *The Chemistry of Commerce: A Simple Interpretation of Some New Chemistry in Its Relation to Modern Industry*. New York: Harper and Brothers.

Duncan, 1913. Robert K. Duncan. 'Industrial Fellowships: Five Years of an Educational Industrial Experiment'. *Journal of the Franklin Institute* **175**: 43–57.

Dupree, 1957. A. Hunter Dupree. *Science in the Federal Government: A History of Policies and Activities to 1940*. Cambridge, Mass.: Harvard University Press, Belknap Press.

Dupree, 1963. A. Hunter Dupree. 'Central Scientific Organisation in the United States Government'. *Minerva* **1**: 453–469.

Dutton, 1942. William S. Dutton. *Du Pont: One Hundred and Forty Years*. New York: Charles Scribner's Sons.

E. I. du Pont, 1927. E. I. du Pont de Nemours and Company. *Annual Report 1927*. Wilmington, Del.: E. I. du Pont de Numours and Company.

E. I. du Pont, 1935. E. I. du Pont de Nemours and Company. *Annual Report 1935*. Wilmington, Del.: E. I. du Pont de Nemours and Company.

E. I. du Pont, 1936. E. I. du Pont de Nemours and Company. *Annual Report 1936*. Wilmington, Del.: E. I. du Pont de Nemours and Company.

E. I. du Pont, 1937. E. I. du Pont de Nemours and Company. *Annual Report 1937*. Wilmington, Del.: E. I. du Pont de Nemours and Company.

Enos, 1962. John L. Enos. *Petroleum Progress and Profits: A History of Process Innovation*. Cambridge, Mass.: MIT Press.

Ewing, 1976. Galen W. Ewing. 'Analytical Chemistry: The Past 100 Years'. In III, C&EN, 1976, 128–142.

Farnham *et al.*, 1925. Dwight T. Farnham, James A. Hall, R. W. King, and H. E. Howe. *Profitable Science in Industry*. New York: Macmillan Company.

Farrell, 1926. Hugh Farrell. *What Prive Progress? The Stake of the Investor in the Discoveries of Science*. New York: G. P. Putnam's Sons.

Fay, 1931. Paul J. Fay. 'The History of Chemistry Teaching in American High Schools'. *Journal of Chemical Education* **8**: 1533–1562.

Ford, 1976a. Gerald Ford. 'National Medal of Science'. The President's Remarks at the Presentation Ceremony in the East Room, October 18, 1976. *Weekly Compilation of Presidential Documents* **12**, No. 43: 1535–1536.

Ford, 1976b. Gerald Ford. 'Nobel Prize Winners'. Statement by the President on Americans Winning Every Nobel Prize for 1976, October 21, 1976. *Weekly Compilation of Presidential Documents* **12**, No. 43: 1551–1552.

Forman, Heilbron, and Weart, 1975. Paul Forman, John Heilbron, and Spencer Weart. *Physics circa 1900: Personnel, Funding, and Productivity of the Academic Establishments. Historical Studies in the Physical Sciences*, **5**. Princeton: Princeton University Press.

Frazer, 1930. [J. C. W. Frazer]. 'The Francis P. Garvan Chair of Chemical Education'. *Journal of Chemical Education* 7: 277–282.

French, 1946. John C. French. *A History of the University Founded by Johns Hopkins.* Baltimore: Johns Hopkins Press.

Friedel, 1983. Robert D. Friedel. *Pioneer Plastic: The Making and Selling of Celluloid.* Madison: University of Wisconsin Press.

Friedman, 1973. Lawrence Friedman. *A History of American Law.* New York: Simon and Schuster.

Furter (ed.), 1980. William F. Furter (ed.). *History of Chemical Engineering.* Advances in Chemistry Series, 190. Washington, D.C.: American Chemical Society.

Furter (ed.), 1982. William F. Furter (ed.). *A Century of Chemical Engineering.* New York: Plenum.

Galambos, 1979. Louis Galambos. 'The American Economy and the Reorganization of the Sources of Knowledge'. In III, Oleson and Voss (eds.), 1979, 269–282.

Garvan, 1919. Francis P. Garvan. *Address . . . at the Annual Dinner of the National Cotton Manufacturers' Association.* Hotel Biltmore, New York, Friday evening, April 25, 1919. n.p., n.d. Copy in Widener Library, Harvard University (Econ7540. 09.19bx).

Garvan, 1921. Francis P. Garvan. *Address . . . Delivered before the Joint Session of the Society of Chemical Industry and the American Chemical Society at Columbia University.* September 7, 1921. n.p., n.d. Copy in Widener Library, Harvard University (Econ7540.09.21bx).

Garvan, 1922. Francis P. Garvan. *The Chemical Question. An Open Letter to Warren G. Harding, President of the United States.* New York: Chemical Foundation. Copy in Widener Library, Harvard University (Econ7540.09.22bx).

Garvan, 1929. Francis P. Garvan. 'Acceptance Address [for Medal of the American Institute of Chemists]'. *Journal of Chemical Education* **6**: 1072–1075.

Gellner, 1964. Ernest Gellner. *Thought and Change.* Chicago: University of Chicago Press.

Getman, 1940. Frederick H. Getman. *The Life of Ira Remsen.* Easton, Pa.: Journal of Chemical Education.

Ginnings, 1925. P. M. Ginnings. 'Supply and Demand as Related to Chemistry Employment'. *Industrial and Engineering Chemistry, News Edition* **3** (20 September): 9.

Ginnings, 1929. P. M. Ginnings. 'Supply and Demand as Related to Chemistry Employment – II'. *Industrial and Engineering Chemistry, News Edition* **7** (20 January): 1.

Ginnings, 1930. P. M. Ginnings. 'Supply and Demand as Related to Chemical Employment – III'. *Industrial and Engineering Chemistry, News Edition* **8** (10 March): 8.

Ginnings, 1934. P. M. Ginnings. 'Supply and Demand as Related to Chemical Employment – IV'. *Industrial and Engineering Chemistry, News Edition* **12** (20 February): 69.

Ginzberg, 1979. Eli Ginzberg. 'The Professionalization of the U.S. Labor Force'. *Scientific American* **240** (March): 48–53.

Gordon, 1943. Neil E. Gordon. 'The Section, Division, and Journal of Chemical Education: A Brief Historical Retrospect'. *Journal of Chemical Education* **20**: 369–372, 405.

Gordon *et al.*, 1931. Neil E. Gordon, Lyman C. Newell, Charles H. Herty, and Paul Smith. 'Report of the Committee on Prize Essays at the End of the Contests (1923–1931) Conducted under the Auspices of the American Chemical Society with Funds Provided by Mr. and Mrs. Francis P. Garvan of New York City'. *Journal of Chemical Education* 8: 2031–2039.

Grady and Chittum, 1940. Roy I. Grady and John W. Chittum *et al. The Chemist at Work*. Easton, Pa.: Journal of Chemical Education.

Graham, 1941. Charles C. Graham. 'Some Data Pertinent to Textbooks of General Science'. *Science Education* **25**: 35–41.

Greenberg, 1965. Daniel S. Greenberg. 'Chemistry: A "Little Science" Would Like a Little More Money'. *Science* **150**: 1267–1270.

Grier, 1926. N. M. Grier. 'A Preliminary Report on the Progress and Encouragement of Science Instruction in American Colleges and Universities, 1912–1922'. *School Science and Mathematics* **26**: 753–764, 872–881, 931–940.

Griffith, 1924. Ivor Griffith. 'Chemistry as a Source of New Wealth'. *Annals of the American Academy of Political and Social Science* **115**: 38–46.

Guralnick, 1979. Stanley M. Guralnick. 'The American Scientist in Higher Education, 1820–1910'. In II, Reingold (ed.), 1979, 99–141.

L. F. Haber, 1958. L. F. Haber. *The Chemical Industry during the Nineteenth Century: A Study of the Economic Aspect of Applied Chemistry in Europe and North America*. Oxford: Clarendon Press.

L. F. Haber, 1971. L. F. Haber. *The Chemical Industry, 1900–1930: International Growth and Technological Change*. Oxford: Clarendon Press.

S. Haber, 1974. Samuel Haber. 'The Professions and Higher Education in America: A Historical View'. In *Higher Education and the Labor Market*, edited by Margaret S. Gordon, 237–280. New York: McGraw-Hill.

H. Hale, 1921. Harrison Hale. *American Chemistry: A Record of Achievement. The Basis for Future Progress*. New York: D. Van Nostrand. Second edition in 1928.

H. Hale, 1932. Harrison Hale. 'The History of Chemical Education in the United States from 1870 to 1914'. *Journal of Chemical Education* **9**: 729–744.

W. J. Hale, 1932. William J. Hale. *Chemistry Triumphant. The Rise and Reign of Chemistry in a Chemical World*. Baltimore: Williams and Wilkins.

W. J. Hale, 1934. William J. Hale. *The Farm Chemurgic: Farmward the Star of Destiny Lights Our Way*. Boston: Stratford.

W. J. Hale, 1949. William J. Hale. *Farmer Victorious: Money, Mart, and Mother Earth*. New York: Coward-McCann.

Hall, 1939. Carrol C. Hall. 'Trends in the Organization of High-School Chemistry since 1920'. *Journal of Chemical Education* **16**: 116–120.

Hammond, 1929. John Hays Hammond. 'The Chemical Engineer'. In *The Profession of Engineering*, edited by Dugald C. Jackson and W. Paul Jones, 107–116. New York: John Wiley and Sons.

Hannaway, 1976. Owen Hannaway. 'The German Model of Chemical Education in America: Ira Remsen at Johns Hopkins (1876–1913)'. *Ambix* **23**: 145–164.

Harding, 1947. T. Swann Harding. *Two Blades of Grass: A History of Scientific Development in the United States Department of Agriculture*. Norman: University of Oklahoma Press.

Harkins and Wallace 1959. Malcolm J. Harkins and Charles E. Wallace. 'The Chemical Industry'. In *The Development of American Industries: Their Economic Significance*, 4th edition, edited by John G. Glover and Rudolph L. Lagai, 297–331. New York: Simmons-Boardman Publishing Corporation.

'Harrison Wins', 1976. 'Harrison Wins Handily in ACS Election'. *Chemical and Engineering News* **54** (22 November): 6.

Hart, 1893. Edward Hart. 'Twenty-Five Years' Progress in Analytical Chemistry'. *Proceedings of the American Association for the Advancement of Science* **42**: 87–97.

Hart, 1922. Edward Hart. 'II. Random Recollections of an Old Professor'. *Chemical Age* **30**: 446–450.

Haskell, 1977. Thomas L. Haskell. 'Power to the Experts'. *New York Review of Books* **24** (13 October): 28–33.

Hawkins, 1960. Hugh Hawkins. *Pioneer: A History of the Johns Hopkins University*. Ithaca, N.Y.: Cornell University Press.

Hawkins, 1972. Hugh Hawkins. *Between Harvard and America: The Educational Leadership of Charles W. Eliot*. New York: Oxford University Press.

Hayes, 1911. C. W. Hayes, comp. *The State Geological Surveys of the United States*. U.S. Geological Survey Bulletin, No. 465. Washington: U.S. Government Printing Office.

Haynes, 1933. Williams Haynes. *Chemical Economics*. New York: D. Van Nostrand Company.

Haynes, 1936. Williams Haynes. *Men, Money and Molecules*. Garden City, N.Y.: Doubleday, Doran and Company.

Haynes, 1943. Williams Haynes. *This Chemical Age: The Miracle of Man-Made Materials*. 2nd edition. New York: A. A. Knopf.

Haynes, 1945a. Williams Haynes. *American Chemical Industry: The World War I Period, 1912–1922*. Vol. II of *American Chemical Industry: A History*. New York: D. Van Nostrand.

Haynes, 1945b. Williams Haynes. *American Chemical Industry: The World War I Period, 1912–1922*. Vol. III of *American Chemical Industry: A History*. New York: D. Van Nostrand.

Haynes, 1948. Williams Haynes. *American Chemical Industry: The Merger Era, 1923–1929*. Vol. IV of *American Chemical Industry: A History*. New York: D. Van Nostrand.

Haynes, 1949. Williams Haynes (ed.). *American Chemical Industry: The Chemical Companies*. Vol. VI of *American Chemical Industry: A History*. New York: D. Van Nostrand.

Haynes, 1954a. Williams Haynes. *American Chemical Industry: Background and Beginnings*. Vol. I of *American Chemical Industry: A History*. New York: D. Van Nostrand.

Haynes, 1954b. Williams Haynes. *American Chemical Industry: Decade of New Products, 1930–1939*. Vol. V of *American Chemical Industry: A History*. New York: D. Van Nostrand.

Heath *et al*., 1944. Roy E. Heath *et al*. 'Symposium on Industrial Demands for Non-laboratory Chemists'. *Journal of Chemical Education* **21**: 269–288, 293.

Hendrick, 1919. Ellwood Hendrick. 'The Chemist of the Future'. In his *Percolator Papers*, 199–228. New York: Harper and Brothers.

Hendrickson, 1961. Walter B. Hendrickson. 'Nineteenth-Century State Geological Surveys: Early Government Support of Science'. *Isis* **52**: 357–371.

Henry, 1947. N. B. Henry (ed.). *Science Education in American Schools*. 46th Yearbook of the National Society for the Study of Education. Part I. Chicago: National Society for the Study of Education.

Henry, 1960. N. B. Henry (ed.). *Rethinking Science Education*. 59th Yearbook of the

National Society for the Study of Education. Part I. Chicago: National Society for the Study of Education.

Herty, 1924. Charles H. Herty. 'American Progress in Dye Manufactures'. *Industrial and Engineering Chemistry* **16**: 1021–1023.

Herty, 1925. Charles H. Herty. 'The Future of the Synthetic Organic Chemical Industry in America'. *Journal of Chemical Education* **2**: 519–532.

Herty, 1929. Charles H. Herty. 'Advance of the Nation through the Science of Chemistry'. *Journal of Chemical Education* **6**: 1037–1044.

Hillebrand, 1910. W. F. Hillebrand. 'Chemistry in the Bureau of Standards'. *Journal of Industrial and Engineering Chemistry* **2**: 423–426.

Hirt, 1954. Francis L. Hirt. 'Growth Characteristics of the Economy Illustrated by the Chemical Industry'. *Survey of Current Business* **34** (September): 10–14, 22.

Hixson, 1937. Arthur W. Hixson. 'Francis P. Garvan, 1875–1937'. *Industrial and Engineering Chemistry, News Edition* **15**: 539–540.

Holland and Pringle, 1928. Maurice Holland and Henry F. Pringle. *Industrial Explorers*. New York: Harper and Brothers.

Holland and Spraragen, 1932. Maurice Holland and W. Spraragen. 'Chemical Research in the Depression'. *Industrial and Engineering Chemistry* **24**: 956–960.

Hopkins, 1935. B. S. Hopkins. 'The Cultural Value of Chemistry in General Education'. *Journal of Chemical Education* **12**: 418–428.

Hopkins *et al.*, 1936. B. S. Hopkins, L. W. Mattern, Wilhelm Segerblom, and Neil E. Gordon. 'An Outline of Essentials for a Year of High School Chemistry'. *Journal of Chemical Education* **13**: 175–179.

Hougen, 1977. Olaf A. Hougen. 'Seven Decades of Chemical Engineering'. *Chemical Engineering Progress* **73**: 89–104.

Howe, 1925a. Harrison E. Howe. 'Earning Power of Chemical Research'. In III, Farnham *et al.*, 1925, 70–99.

Howe, 1925b. Harrison E. Howe. 'Profit-Earning Research'. In III, Farnham *et al.*, 1925, 100–124.

Howe, 1925c. Harrison E. Howe. 'The Waste Problem. Methods of Procedure toward Research'. In III, Farnham *et al.*, 1925, 125–142.

Howe, 1926. Harrison E. Howe. *Chemistry in the World's Work*. New York: D. Van Nostrand.

Howe, 1937. Harrison E. Howe. 'The Chemical Industries'. In U.S. National Resources Committee, *Technological Trends and National Policy Including the Social Implications of New Inventions*, 289–314. Washington, D.C.: U.S. Government Printing Office.

Howe, 1941. Harrison E. Howe. 'Chemistry in Industrial Research'. In III, NRPB, 1941, 223–235.

Howe (ed.), 1924–1925. Harrison E. Howe (ed.). *Chemistry in Industry: A Cooperative Work Intended to Give Examples of the Contributions Made to Industry by Chemistry*. 2 vols. New York: Chemical Foundation.

Hunter, 1925. George W. Hunter. 'The Place of Science in the Secondary School'. *School Review* **33**: 370–381, 453–466.

Hunter, 1932. George W. Hunter. 'The Sequence of Science in the Junior and Senior High School'. *Science Education* **16**: 103–115.

Hunter, 1933. George W. Hunter. 'Science Sequence in the Junior and Senior High Schools'. *School Science and Mathematics* **33**: 214–223.

Hunter and Parker, 1942. George W. Hunter and Alice L. Parker. 'The Subject Matter of General Science'. *School Science and Mathematics* **42**: 869–877.

Hurd, 1949. Paul deHart Hurd. *A Critical Analysis of the Trends in Secondary School Science Teaching from 1895 to 1948*. PhD dissertation, Stanford University.

Hurd and Rowe, 1964. Paul deHart Hurd and M. B. Rowe. 'Science in the Secondary School'. *Review of Educational Research* **34**: 286–297.

Ihde, 1964. Aaron J. Ihde. *The Development of Modern Chemistry*. New York: Harper and Row.

Jackson, 1970. Charles O. Jackson. *Food and Drug Legislation during the New Deal*. Princeton: Princeton University Press.

Jenkins, 1975. Reese Jenkins, *Images and Enterprise: Technology and the American Photographic Industry, 1839–1925*. Baltimore: Johns Hopkins University Press.

H. Johnson, 1968. Howard C. E. Johnson. 'The Rise of the United States Chemical Industry, 1918–1968'. *Chemical Week* **103** (16 November): 105–145.

P. Johnson, 1950. P. G. Johnson. *The Teaching of Science in Public High Schools*. U.S. Office of Education *Bulletin*, 1950, No. 9. Washington, D.C.: U.S. Government Printing Office.

Jones, 1969. Daniel P. Jones. *The Role of Chemists in Research on War Gases in the United States during World War I*. PhD dissertation, University of Wisconsin.

Judd, 1933. Charles H. Judd. *Problems of Education in the United States*. Recent Social Trends Monographs. New York: McGraw-Hill.

'Judging', 1928. 'Judging an Industry'. *Industrial and Engineering Chemistry* **20**: 881–882.

Kahn, 1961. Alfred E. Kahn. 'The Chemical Industry'. In *The Structure of American Industry: Some Case Studies*, 3rd edition, edited by Walter Adams, 233–276. New York: Macmillan.

Kargon, 1977. Robert H. Kargon. *Science in Victorian Manchester: Enterprise and Expertise*. Baltimore: Johns Hopkins University Press.

Kaufman, 1927. G. Kaufman. 'A Chemistry Curriculum for Every Man's Son and Daughter'. *Journal of Chemical Education* **4**: 976–978.

Kessel, 1973. William G. Kessel. 'The Beginning of the Journal and the Division of Chemical Education'. *Journal of Chemical Education* **50**: 803–807.

Kett, 1968. Joseph F. Kett. *The Formation of the American Medical Profession: The Role of Institutions, 1780–1860*. New Haven: Yale University Press.

Kevles, 1978. Daniel J. Kevles. *The Physicists: The History of a Scientific Community in Modern America*. New York: A. A. Knopf.

Kevles, 1979. Daniel Kevles. 'The Physics, Mathematics, and Chemistry Communities: A Comparative Analysis'. In III, Oleson and Voss (eds.), 1979, 139–172.

Kidd, 1959. Charles V. Kidd. *American Universities and Federal Research*. Cambridge, Mass.: Harvard University Press, Belknap Press.

Kirkpatrick (ed.), 1933. Sidney D. Kirkpatrick (ed.). *Twenty-Five Years of Chemical Engineering Progress*. Silver Anniversary Volume, American Institute of Chemical Engineers. New York: American Institute of Chemical Engineers. (Reprinted Freeport, N.Y.: Books for Libraries Press, 1968.)

Kirkpatrick, 1941. Sidney D. Kirkpatrick. 'The Chemical Engineer in Industrial Research'. In III, NRPB, 1941, 306–315.

Klein, 1917. Otto H. Klein. 'The Chemist in the Service of the City of New York'. *Journal of Industrial and Engineering Chemistry* **9**: 79–81.

Knapp and Goodrich, 1952. R. H. Knapp and H. B. Goodrich. *Origins of American Scientists: A Study Made under the Direction of a Committee of the Faculty of Wesleyan University*. Chicago: University of Chicago Press, for Wesleyan University.

Knoblauch *et al.*, 1962. H. C. Knoblauch, E. M. Law, and W. P. Meyer. *State Agricultural Experiment Stations, A History of Research Policy and Procedure*. U.S. Department of Agriculture, Miscellaneous Publication No. 904. Washington, D.C.: U.S. Government Printing Office.

Kohler, 1982. Robert E. Kohler. *From Medical Chemistry to Biochemisty: The Making of a Biomedical Discipline*. Cambridge: Cambridge University Press.

Kohlstedt, 1976. Sally Gregory Kohlstedt. *The Formation of the American Scientific Community: The American Association for the Advancement of Science, 1848–1860*. Urbana: University of Illinois Press.

Kolesnik, 1958. Walter B. Kolesnik. *Mental Discipline in Modern Education*. Madison: University of Wisconsin Press.

Kopperl, 1976. Sheldon J. Kopperl. 'T. W. Richards' Role in American Graduate Education in Chemistry'. *Ambix* **23**: 165–174.

Krug, 1960. Edward A. Krug. *The Secondary School Curriculum*. New York: Harper and Brothers.

Krug, 1964. Edward A. Krug. *The Shaping of the American High School, 1880–1920*. Madison: University of Wisconsin Press.

Krug, 1972. Edward A. Krug. *The Shaping of the American High School, 1920–1941*. Madison: University of Wisconsin Press.

Küppers *et al.*, 1982. Günter Küppers, Peter Weingart, and Norbert Ulitzka. *Die Nobelpreise in Physik und Chemie, 1901–1929. Materialen zum Nominierungsprozess*. Report Wissenschaftsforschung, 23. Bielefeld: B. Kleine Verlag.

Kuhn, 1971. Thomas S. Kuhn. 'The Relations between History and the History of Science'. *Daedalus* **100**: 271–304.

Laitinen and Ewing, 1977. Herbert A. Laitinen and Galen W. Ewing (eds.). *A History of Analytical Chemistry*. Washington, D.C.: Division of Analytical Chemistry, American Chemical Society.

Landis, 1944. Walter S. Landis. *Your Servant the Molecule*. New York: Macmillan.

Larson, 1950. Robert Lourie Larson. *Charles Frederick Chandler: His Life and Work*. PhD dissertation, Columbia University.

Larson, 1977. Magali Sarfatti Larson. *The Rise of Professionalism: A Sociological Analysis*. Berkeley: University of California Press.

Larson *et al.*, 1971. Henrietta M. Larson, Evelyn H. Knowlton, and Charles S. Popple. *New Horizons, 1927–1950*. History of Standard Oil Company (New Jersey), Vol. III. New York: Harper and Row.

Layton, 1971. Edwin T. Layton, Jr. *The Revolt of the Engineers: Social Responsibility and the American Engineering Profession*. Cleveland: Case Western Reserve University Press.

Lehman, 1955. Harvey C. Lehman. 'Ages at Time of First Election of Presidents of Professional Organizations'. *Scientific Monthly* **80**: 293–298.

Lewis, 1967. W. David Lewis. 'Industrial Research and Development'. In *Technology in Western Civilization*, edited by Melvin Kranzberg and Carroll W. Pursell, Jr., Vol. 2, 615–634. New York: Oxford University Press.

Little, 1928. Arthur D. Little. *The Handwriting on the Wall: A Chemist's Interpretation*. Boston: Little, Brown and Company.

Lorant, 1967. John H. Lorant. 'Technological Change in American Manufacturing during the 1920's'. *Journal of Economic History* **27**: 243–246.

Lorant, 1975. John Herman Lorant. *The Role of Capital-Improving Innovations in American Manufacturing during the 1920's*. Dissertations in American Economic History, edited by Stuart Bruchey. New York: Arno Press.

Lowrance, 1977. William W. Lowrance. 'The NAS Surveys of Fundamental Research, 1962–1974, in Retrospect'. *Science* **197**: 1254–1260.

McKenna, 1908. Charles F. McKenna. 'The Justification of the American Institute of Chemical Engineers'. *Transactions of the American Institute of Chemical Engineers* **1**: 8–20.

MacLeod and Andrews, 1967. R. M. MacLeod and E. K. Andrews. *Selected Science Statistics Relating to Research Endowment and Higher Education, 1850–1914*. Science Policy Research Unit, University of Sussex.

McMillen, 1946. Wheeler McMillen. *New Riches from the Soil: The Progress of Chemurgy*. New York: D. Van Nostrand Company.

Mahr, 1914. Hermann W. Mahr. 'The Broader Applications of Chemistry by the Municipality'. *Journal of Industrial and Engineering Chemistry* **6**: 1030–1032.

Manning, 1967. Thomas G. Manning. *Government in Science: The U.S. Geological Survey*. Lexington: University of Kentucky Press.

Mansfield, 1968. Edwin Mansfield. *The Economics of Technological Change*. New York: W. W. Norton.

Marshall, 1938–1941. Louis Marshall. 'The Young Chemist and the Government Service.' *The Chemist* **15** (1938): 367–376; **16** (1939): 124–134, 161–165, 195–202, 240–249, 300–304, 344–349, 365–374, and 397–405; **17** (1940): 15–25, 41–47, 80–89, 250–253, 299–307, 334–344, 365–371, and 394–398; and **18** (1941): 70–80 and 101–106.

Mason, 1911. William P. Mason. 'Contribution of Chemistry to Sanitation'. *Journal of Industrial and Engineering Chemistry* **3**: 292–295.

Mattern, 1928. Louis W. Mattern. 'The Correlation of High School Chemistry with First Year College Chemistry'. *Journal of Chemical Education* **5**: 1627–1633.

Mees, 1920. C. E. K. Mees. *The Organization of Industrial Scientific Research*. New York: McGraw-Hill.

Mellon Institute, 1924. Mellon Institute of Industrial Research of the University of Pittsburgh. *Industrial Fellowships: The System of Practical Cooperation between Science and Industry as Formulated by Robert Kennedy Duncan*. Pittsburgh: Mellon Institute of Industrial Research.

Merrill and Ridgway, 1969. Richard J. Merrill and David W. Ridgway. *The CHEM Study Story*. San Francisco: W. H. Freeman and Company.

Merritt, 1969. Raymond H. Merritt. *Engineering in American Society, 1850–1875*. Lexington: University Press of Kentucky.

Metzger, 1961. Walter P. Metzger. *Academic Freedom in the Age of the University*. New York: Columbia University Press.

Miles and Kuslan, 1969. Wyndham D. Miles and Louis Kuslan. 'Washington's First Consulting Chemist, Henri Erni'. *Records of the Columbia Historical Society of Washington, D.C.* **46**: 154–166.

Morrell, 1972. J. B. Morrell. 'The Chemist Breeders: The Research Schools of Liebig and Thomas Thomson'. *Ambix* **19**: 1–46.

Morrison, 1937. A. Cressy Morrison. *Man in a Chemical World: The Service of Chemical Industry*. New York: Charles Scribner's Sons.

Mowery, 1983a. David C. Mowery. 'Industrial Research and Firm Size, Survival, and Growth in American Manufacturing, 1921–1946: An Assessment'. *Journal of Economic History* **43** (1983): 953–980.

Mowery, 1983b. David C. Mowery. 'The Relationship between Intrafirm and Contractual Forms of Industrial Research in American Manufacturing, 1900–1940'. *Explorations in Economic History* **20** (1983): 351–374.

Mueller, 1962. Willard F. Mueller. 'The Origins of the Basic Inventions Underlying Du Pont's Major Product and Process Innovations, 1920 to 1950'. In National Bureau of Economic Research, *The Rate and Direction of Inventive Activity: Economic and Social Factors* (Special Conference Series No. 12), 323–346. Princeton: Princeton University Press.

Muh, 1954. Edward Muh. 'The Chemical Foundation, Inc.' 2 vol. typescript history. Francis P. Garvan Estate Collection, Small File No. 13. Laramie, Wyo.: American Heritage Center, University of Wyoming.

Mulkay, 1976. Michael J. Mulkay. 'The Mediating Role of the Scientific Elite'. *Social Studies of Science* **6**: 445–470.

Naess, 1948. Ragnar D. Naess. 'Commercial Chemical Development: An Investment Manager Looks at the Chemical Industry'. *Chemical and Engineering News* **26**: 841–847.

Nash, 1963. Gerald D. Nash. 'The Conflict between Pure and Applied Science in Nineteenth-Century Public Policy: The California State Geological Survey, 1860–1874'. *Isis* **54**: 217–228.

Nelson, 1959. Ralph L. Nelson. *Merger Movements in American Industry, 1895–1956*. National Bureau of Economic Research, No. 66, General Series. Princeton: Princeton University Press.

Newman, 1938. Albert B. Newman. 'Development of Chemical Engineering Education in the United States'. Supplement to *Transactions of the American Institute of Chemical Engineers*, Vol. 34, No. 3A.

Nobel Foundation, 1964. Nobel Foundation. *Nobel Lectures Including Presentation Speeches and Laureates' Biographies. Chemistry, 1942–1962*. New York: Elsevier Publishing Company.

Nobel Foundation, 1966a. Nobel Foundation. *Nobel Lectures Including Presentation Speeches and Laureates' Biographies. Chemistry, 1901–1921*. New York: Elsevier Publishing Company.

Nobel Foundation, 1966b. Nobel Foundation. *Nobel Lectures Including Presentation Speeches and Laureates' Biographies. Chemistry, 1922–1941*. New York: Elsevier Publishing Company.

Nobel Foundation, 1972. Nobel Foundation. *Nobel Lectures Including Presentation Speeches and Laureates' Biographies. Chemistry, 1963–1970*. New York: Elsevier Publishing Company.

'Nobel Sweep', 1976. 'Nobel Sweep'. *New York Times*, October 20, 1976, 44, cols. 1–2.

Noble, 1977. David F. Noble. *America by Design: Science, Technology and the Rise of Corporate Capitalism*. New York: A. A. Knopf.

Norris, 1925. James F. Norris. 'Academic Research and Industry'. *Industrial and Engineering Chemistry* **17**: 1088–1089.

Norris, 1932. James F. Norris. 'Research and Industrial Organic Chemistry'. *Science* **75**: 5–10.

Noyes, 1908. W. A. Noyes. 'Chemical Publications in America in Relation to Chemical Industry'. *Science* **28**: 225–227.

NRPB, 1941. U.S. National Resources Planning Board. *Research – A National Resource*, Vol. 2, *Industrial Research*. Washington, D.C.: U.S. Government Printing Office.

NSB, 1976. U.S. National Science Board. *Science at the Bicentennial: A Report from the Research Community*. NSB 76–1. Washington, D.C.: U.S. Government Printing Office.

NSB, 1978. U.S. National Science Board. *Basic Research in the Mission Agencies: Agency Perspectives on the Condition and Support of Basic Research. Report of the National Science Board, 1978*. NSB 78–1. Washington, D.C.: U.S. Government Printing Office.

NSSE, 1932. National Society for the Study of Education. *A Program for Teaching Science*, edited by G. M. Whipple. 31st Yearbook, Part I. Bloomington, Ill.: Public School Publishing Company.

Oden, 1977. Jack P. Oden. 'Charles Holmes Herty and the Birth of the Southern Newsprint Paper Industry, 1927–1940'. *Journal of Forest History* **21** (April 1977): 77–89.

Ogden, 1974. William R. Ogden. 'An Analysis of Published Research Pertaining to Objectives for the Teaching of Secondary School Chemistry as Reflected in Selected Professional Periodicals, 1919–1972'. *School Science and Mathematics* **74**: 120–128.

Ogden, 1976. William R. Ogden. 'Contributions of Major Committee Reports to the Teaching of High School Chemistry: 1893–1975'. *School Science and Mathematics* **76**: 461–465, 599–604, 639–646.

Oleson and Voss (eds.), 1979. Alexandra Oleson and John Voss (eds.). *The Organization of Knowledge in Modern America, 1860–1920*. Baltimore: Johns Hopkins University Press.

O'Rourke, 1940. L. J. O'Rourke. *Opportunities in Government Employment. Getting a Job in Federal, State or Municipal Government*. New York: Garden City Publishing Company.

Parascandola. 1983. John Parascandola. 'Charles Holmes Herty and the Effort to Establish an Institute for Drug Research in Post World War I America'. In III, Parascandola and Whorton (eds.), 1983, 85–103.

Parascandola and Whorton (eds.), 1983. John Parascandola and James C. Whorton (eds.). *Chemistry and Modern Society: Historical Essays in Honor of Aaron J. Ihde*. ACS Symposium Series, 228. Washington, D.C.: American Chemical Society.

Parker, 1909. P. J. Parker. 'The Industrial Chemist and His Journal'. *Journal of Industrial and Engineering Chemistry* **1**: 1–2.

Parmelee, 1920. H. C. Parmelee. 'Need of Professional Solidarity among Chemists'. *Chemical and Metallurgical Engineering* **23**: 1113–1114.

Peaslee, 1929. W. D. A. Peaslee. 'Supply and Demand as Related to Chemistry Employment'. *Industrial and Engineering Chemistry, News Edition* 7 (10 March): 14.

Penick, *et al.*, 1972. James L. Penick, Jr., Carroll W. Pursell, Jr., Morgan B. Sherwood, and Donald C. Swain (eds.). *The Politics of American Science: 1939 to the Present*. Revised edition. Cambridge, Mass.: MIT Press.

Perazich, 1951. George Perazich. 'Research: Who, Where, How Much'. *Chemical Week* **69** (27 October): 17–24.

Perloff *et al.*, 1960. Harvey S. Perloff, Edgar S. Dunn, Jr., Eric E. Lampard, and Richard F. Muth. *Regions, Resources, and Economic Growth*. Baltimore: Johns Hopkins Press.

'Petrochemicals', 1955. 'Petrochemicals: Where the Growth Shows'. *Business Week*, 15 October: 78–82.

Pfetsch, 1974. Frank R. Pfetsch. *Zur Entwicklung der Wissenschaftspolitik in Deutschland, 1750–1914*. Berlin: Duncker und Humblot.

Porter, 1973. Glenn Porter. *The Rise of Big Business, 1860–1910*. New York: Thomas Y. Crowell Company.

Pound, 1936. Arthur Pound. *Industrial America: Its Way of Work and Thought*. Boston: Little, Brown, and Company.

Powell, 1922. Fred Wilbur Powell. *The Bureau of Mines: Its History, Activities and Organization*. Institute for Government Research, Service Monographs of the United States Government, No. 3. New York: D. Appleton and Company.

Prescott, 1954. Samuel C. Prescott. *When M.I.T. Was "Boston Tech", 1861–1916*. Cambridge, Mass.: Technology Press.

Price, 1970. Derek J. de Solla Price. 'Measuring the Size of Science'. *Proceedings of the Israel Academy of Sciences and Humanities* **4**: 98–111.

Pursell, 1968. Carroll W. Pursell, Jr. 'The Administration of Science in the Department of Agriculture, 1933–1940'. *Agricultural History* **42**: 231–240.

Pursell, 1969. Carroll W. Pursell, Jr. 'The Farm Chemurgic Council and the United States Department of Agriculture, 1935–1939'. *Isis* **60**: 307–317.

Pursell, 1972. Carroll Pursell. 'Science and Industry'. In II, Daniels (ed.), 1972, 231–248.

Pursell, 1979. Carroll Pursell. 'Science Agencies in World War II: The OSRD and Its Challengers'. In II, Reingold (ed.), 1979, 359–378.

Rae, 1979. John Rae. 'The Application of Science to Industry'. In III, Oleson and Voss (eds.), 1979, 249–268.

Reader, 1966. W. J. Reader. *Professional Men: The Rise of the Professional Classes in Nineteenth-Century England*. London: Weidenfeld and Nicholson.

Reed, 1983. Germaine M. Reed. 'Charles Holmes Herty and the Promotion of Southern Economic Development'. *South Atlantic Quarterly* **82** (1983): 424–436.

Reich, 1977. Leonard S. Reich. 'Research, Patents, and the Struggle to Control Radio: A Study of Big Business and the Uses of Industrial Research'. *Business History Review* **51**: 208–235.

Reich, 1980. Leonard S. Reich. 'Science'. In *Encyclopedia of American Economic History*, edited by Glenn Porter, 3 vols., I, 281–293. New York: Charles Scribner's Sons.

Reich, 1983. Leonard S. Reich. 'Irving Langmuir and the Pursuit of Science and Technology in the Corporate Environment'. *Technology and Culture* **24** (1983): 199–221.

Reid, 1972. E. Emmet Reid. *My First One Hundred Years*. New York: Chemical Publishing Company.

Reingold, 1977. Nathan Reingold. 'The Case of the Disappearing Laboratory'. *American Quarterly* **29**: 79–101.

Reitz, 1973. Jeffrey G. Reitz. 'The Flight from Science Revisited: Career Choice of Science and Engineering in the 1950s and 1960s'. *Science Education* **57**: 121–134.

'Report', 1925. 'Report of the Committee on Prize Essays of the American Chemical Society, 1923–1924'. *Journal of Chemical Education* **2**: 3–11.

Reynolds, 1983. Terry S. Reynolds. *75 Years of Progress – A History of the American Institute of Chemical Engineers, 1908–1983*. New York: American Institute of Chemical Engineers.

Rhodes, 1977. Edward L. Rhodes. 'Indicators of Trends in the Learning of Science in High School'. In III, Terleckyj (ed.), 1977, 187–201.

Richards, 1902. Joseph W. Richards. 'The American Electrochemical Society'. *Transactions of the American Electrochemical Society* **1**: 1–2.

Richardson, 1908. W. D. Richardson. 'The Usefulness of Chemistry in the Industries'. *Science* **27**: 801–812.

Riese, 1977. Reinhard Riese. *Die Hochschule auf dem Wege zum wissenschaftlichen Grossbetrieb. Die Universität Heidelberg und das badische Hochschulwesen, 1860–1914*. Industrielle Welt, Bd. 19. Stuttgart: Ernst Klett Verlag.

Rocke and Ihde, 1979. Alan J. Rocke and Aaron J. Ihde. 'A Badger Chemist Genealogy'. *Journal of Chemical Education* **56**: 93–95.

[Roeber], 1902. [E. F. Roeber]. 'Greeting'. *Electrochemical Industry* **1**: 1.

Rogers, 1961. Jerome S. Rogers. 'Brief History of the American Leather Chemists Association'. *Journal of the American Leather Chemists Association* **56**: 404–416.

Roscoe, 1887. Henry Enfield Roscoe. *Record of Work Done in the Chemical Department of the Owens College, 1857–1887*. London: Macmillan.

Rose, 1920. Robert E. Rose. *Molecules and Man*. Wilmington, Del.: E. I. du Pont de Nemours and Company.

Rosen, 1956. Sidney Rosen. 'The Rise of High School Chemistry in America (to 1920)'. *Journal of Chemical Education* **33**: 627–633.

C. Rosenberg, 1972. Charles E. Rosenberg. 'Science, Technology, and Economic Growth: The Case of the Agricultural Experiment Station Scientist, 1875–1914'. In II, Daniels (ed.), 1972, 181–209.

C. Rosenberg, 1976. Charles E. Rosenberg. *No Other Gods: On Science and American Social Thought*. Baltimore: Johns Hopkins University Press.

C. Rosenberg, 1977. Charles E. Rosenberg. 'Rationalization and Reality in the Shaping of American Agricultural Research, 1875–1914'. *Social Studies of Science* 7: 401–422.

H. Rosenberg, 1960. H. H. Rosenberg. 'Research Planning and Program Development in the National Institutes of Health: The Experience of a Relatively New and Growing Agency'. *Annals of the American Academy of Political and Social Science* **327**: 103–113.

N. Rosenberg, 1972. Nathan Rosenberg. *Technology and American Economic Growth*. New York: Harper and Row.

Rossiter, 1975. Margaret W. Rossiter. *The Emergence of Agricultural Science: Justus Liebig and the Americans, 1840–1880*. New Haven: Yale University Press.

Rossiter, 1977. Margaret W. Rossiter. 'The Charles F. Chandler Collection'. *Technology and Culture* **18**: 222–230.

Rossiter, 1979. Margaret W. Rossiter. 'The Organization of the Agricultural Sciences'. In III, Oleson and Voss (eds.), 1979, 211–248.

Rossiter, 1982. Margaret W. Rossiter. *Women Scientists in America: Struggles and Strategies to 1940*. Baltimore: Johns Hopkins University Press.

Rothstein, 1972. William G. Rothstein. *American Physicians in the Nineteenth Century: From Sects to Science*. Baltimore: Johns Hopkins University Press.

Rudolph, 1962. Frederick Rudolph. *The American College and University: A History*. New York: Vintage.

Russell, Coley, and Roberts, 1977. Colin A. Russell, with Noel G. Coley, and Gerrylynn K. Roberts. *Chemists by Profession: The Origin and Rise of the Royal Institute of Chemistry*. Milton Keynes: Open University Press.

Sanders, 1977. Howard J. Sanders. 'Supply and Demand for Chemists – Looking to 1985'. *Chemical and Engineering News* **55** (4 July): 18–30.

Sanders and Seltzer, 1977. Howard J. Sanders and Richard J. Seltzer. 'Employment Outlook 1978: Career Opportunities for Chemical Professionals'. *Chemical and Engineering News* **55** (24 October): 21–49.

Schmauderer (ed.), 1973. Eberhard Schmauderer (ed.). *Der Chemiker im Wandel der Zeiten: Skizzen zur geschichtlichen Entwicklung des Berufsbildes*. Weinheim: Verlag Chemie.

SCI, 1931. Society of Chemical Industry. *Journal of the Society of Chemical Industry, Special Jubilee Number, July 1931*. Supplement to the Journal of the Society of Chemical Industry, 17 July 1931. London: Society of Chemical Industry.

Semple, 1963. Robert B. Semple. 'What Is So Different about the Chemical Industry?' In III, Berenson, 1963, 7–12.

Servos, 1979. *Physical Chemistry in America, 1890–1933: Origins, Growth, and Definition*. PhD dissertation, Johns Hopkins University.

Servos, 1982. 'A Disciplinary Program That Failed: Wilder D. Bancroft and the *Journal of Physical Chemistry*, 1896–1933'. *Isis* **73** (1982): 207–232.

Sharp, 1940. C. H. Sharp. 'The Trends in the Subject Matter of High School Chemistry'. *Science Education* **24**: 383–386.

Shaw, 1960. B. T. Shaw. 'Research Planning and Control in the United States Department of Agriculture: The Experience of an Old and Well-Established Research Agency'. *Annals of the American Academy of Political and Social Science* **327**: 95–102.

Sherman, 1951. Joseph V. Sherman. 'Chemical Research Serves as a Barometer of Future Growth Possibilities'. *Barron's* **31** (16 April): 11–12.

Shils, 1979. Edward Shils. 'The Order of Learning in the United States: The Ascendancy of the University'. In III, Oleson and Voss (eds.), 1979, 19–47.

Shreve, 1967. R. Norris Shreve. *Chemical Process Industries*. 3rd edition. New York: McGraw-Hill Book Company.

Siebring, 1954. B. R. Siebring. 'Institutional Influences in the Undergraduate Training of Ph.D. Chemists'. *Journal of Chemical Education* **31**: 195–200.

Siebring, 1961. B. R. Siebring. 'Institutional Influences in the Graduate Training of Eminent Chemists'. *Journal of Chemical Education* **38**: 630–632.

Slosson, 1919. Edwin E. Slosson. *Creative Chemistry. Descriptive of Recent Achievements in the Chemical Industries*. New York: Century.

Slosson, 1925. Edwin E. Slosson. *Sermons of a Chemist*. New York: Harcourt, Brace and Company.

B. Smith and Karlesky, 1977. Bruce L. R. Smith and Joseph J. Karlesky. *The State of Academic Science: The Universities in the Nation's Research Effort*. 2 vols. New York: Change Magazine Press.

E. F. Smith, 1914. Edgar Fahs Smith. *Chemistry in America: Chapters from the History of the Science in the United States*. New York: D. Appleton and Company.

E. F. Smith, 1926. Edgar Fahs Smith. 'Mineral Chemistry'. In *A Half Century of Chemistry in America*, special supplement to the *Journal of the American Chemical Society* 48, No. 8A: 69–88.

E. F. Smith, 1927. Edgar Fahs Smith. *Old Chemistries*. New York: McGraw-Hill Book Company.

E. F. Smith, 1929. Edgar Fahs Smith. *Charles Mayer Wetherill, 1825–1871*. Easton, Pa.: Journal of Chemical Education.

E. L. Smith, 1934. Earl L. Smith. 'Chemical Securities'. *Industrial and Engineering Chemistry* **26**: 608–611.

Soule, 1963. Roland P. Soule. 'Chemical Industry's Problem: Slowing Profits'. In III, Berenson, 1963, 13–23.

Spiegel-Rösing and Price, 1977. Ina Spiegel-Rösing and Derek J. de Solla Price (eds.). *Science, Technology, and Society: A Cross-Disciplinary Perspective*. Beverly Hills, Calif.: Sage Publications.

Spring, 1972. Joel H. Spring. *Education and the Rise of the Corporate State*. Boston: Beacon Press.

'Stacy Tops Crawford', 1977. 'Stacy Tops Crawford in ACS Election'. *Chemical and Engineering News* **55** (28 November): 4.

Stevens, 1971. Rosemary Stevens. *American Medicine and the Public Interest*. New Haven: Yale University Press.

Stewart, 1945. Irwin Stewart. *Organizing Scientific Research for War: The Administrative History of the Office for Scientific Research and Development*. Boston: Little, Brown.

Stieglitz, 1919. Julius Stieglitz. Introduction to III, Slosson, 1919, [xv–xx].

Stieglitz, 1928. Julius Stieglitz (ed.). *Chemistry in Medicine: A Cooperative Treatise Intended to Give Examples of Progress Made in Medicine with the Aid of Chemistry*. New York: Chemical Foundation.

Stieglitz *et al.*, 1919. Julius Stieglitz *et al.* 'Report by the Committee on Publication of Compendia of Chemical Literature, Etc.' *Journal of Industrial and Engineering Chemistry* **11**: 415–417.

Stine, 1925. C. M. A. Stine. 'The Kinship of Du Pont Products'. *Du Pont Magazine* **19** (March): 1–2, 15–16.

Stine, 1928. Charles M. A. Stine. 'Chemical Engineering in Modern Industry'. *Transactions of the American Institute of Chemical Engineers* **21**: 45–54.

Stine, 1940. C. M. A. Stine. 'The Rise of the Organic Chemical Industry in the United States'. In *Smithsonian Institution Annual Report, 1940*, 177–192. Smithsonian Publication No. 3606. Washington, D.C.: Smithsonian Institution.

Stollberg *et al.*, 1960. R. Stollberg *et al.* 'The Status of Science Teaching in Elementary and Secondary Schools'. In III, Henry, 1960, 82–96.

Strauss and Rainwater, 1962. Anselm L. Strauss and Lee Rainwater, with Marc J. Swartz and Barbara G. Berger, and a contribution by W. Lloyd Warner. *The Professional Scientist: A Study of American Chemists*. Chicago: Aldine Publishing Company.

Stryker, 1961. Perrin Stryker. 'Chemicals: The Ball is Over'. *Fortune* **64** (October): 125–127, 207–208ff.

Sturchio, 1981. Jeffrey L. Sturchio. *Chemists and Industry in Modern America: Studies in the Historical Application of Science Indicators*. PhD dissertation, University of Pennsylvania.

Sulzberger, 1976. C. L. Sulzberger. 'And Helped by the Brain Drain'. *New York Times*, October 24, Section IV, 15, cols. 1–3.

Summerlin and Craig, 1966. Leo Summerlin and Sara P. Craig. 'Evolution of the High School Chemistry Text and Present Implications'. *Science Education* **50**: 223–233.

Szabadváry, 1966. Ferenc Szabadváry. *History of Analytical Chemistry*. New York: Pergamon Press.

Tarbell and Tarbell, 1979. D. Stanley and Ann Tracy Tarbell. 'The Role of Roger Adams in American Science'. *Journal of Chemical Education* **56**: 163–165.

Tarbell and Tarbell, 1981. D. Stanley and Ann Tracy Tarbell. *Roger Adams: Scientist and Statesman*. Washington, D.C.: American Chemical Society.

Tarbell, Tarbell, and Joyce, 1980. D. Stanley Tarbell, Ann Tracy Tarbell, and R. M. Joyce. 'The Students of Ira Remsen and Roger Adams'. *Isis* **71**: 620–626.

Taylor, 1976. Kenneth L. Taylor. 'Two Centuries of Chemistry'. In *Issues and Ideas in America*, edited by Benjamin J. Taylor and Thurman J. White, 267–284. Norman: University of Oklahoma Press.

Terleckyj (ed.), 1977. Nestor E. Terleckyj (ed.). *The State of Science and Research: Some New Indicators*. Westview Special Studies in Science and Technology. Boulder, Col.: Westview Press, for the National Planning Association.

Thackray, 1977. Arnold W. Thackray. 'The University, The Past: The United States'. In *The University and Medicine: The Past, the Present, and Tomorrow. Report of an Anglo-American Bicentennial Conference*, edited by John Z. Bowers and Elizabeth F. Purcell, 21–40. New York: Josiah Macy, Jr., Foundation.

Thompson *et al.*, 1920. G. W. Thompson *et al.* 'Symposium on Maintenance and Preservation of Our Chemical Industries'. *Transactions of the American Institute of Chemical Engineers* **11**: 249–382.

Thorpe, 1916. Edward Thorpe. *The Right Honourable Sir Henry Enfield Roscoe . . . A Biographical Sketch*. London: Longmans, Green and Company.

Tobey, 1971. Ronald C. Tobey. *The American Ideology of National Science, 1919–1930*. Pittsburgh: University of Pittsburgh Press.

True, 1937. A. C. True. *A History of Agricultural Experimentation and Research in the United States, 1607–1925*. U.S. Department of Agriculture, Miscellaneous Publication No. 251. Washington, D.C.: U.S. Government Printing Office.

Turley, 1953. H. G. Turley. 'The American Leather Chemists Association'. *Journal of the American Leather Chemists Association* 48: 594–600.

U.S. Congress, 1945. U.S. Congress, Senate. *Government's Wartime Research and Development, 1940–1944*. Report of the Subcommittee on War Mobilization to the Committee on Military Affairs. Parts I and II. 79th Cong., 1st sess.

Vagtborg, 1976. Harold Vagtborg. *Research and American Industrial Development: A Bicentennial Look at the Contributions of Applied Research*. New York: Pergamon Press.

Van Antwerpen and Fourdrinier, 1958. F. J. van Antwerpen and Sylvia Fourdriner. *Highlights: The First Fifty Years of the American Institute of Chemical Engineers*. New York: American Institute of Chemical Engineers.

Vanderbilt, 1976. Byron M. Vanderbilt. 'America's First R & D Center'. *Industrial Research* **15** (November): 27–31.

Vanderbilt, 1979. Byron M. Vanderbilt. 'Who Was First in R & D?' *Journal of Chemical Education* **56**: 319–320.

Vatter, 1963. Harold G. Vatter. *The U.S. Economy in the 1950s: An Economic History*. New York: W. W. Norton.

Veysey, 1965. Laurence R. Versey. *The Emergence of the American University*. Chicago: University of Chicago Press.

Wagstaff, 1950. Henry McGilbert Wagstaff. *Impressions of Men and Movements at the University of North Carolina*. Edited with a prefatory note by Louis R. Wilson. Chapel Hill: University of North Carolina Press.

Weart, 1976. Spencer Weart. 'The Rise of "Prostituted" Physics'. *Nature* **262**: 13–17.

Weart, 1979. Spencer R. Weart. 'The Physics Business in America, 1919–1940: A Statistical Reconnaissance'. In II, Reingold (ed.), 1979, 295–358.

Webb, 1959. Hanor A. Webb. 'How General Science Began'. *School Science and Mathematics* **59**: 421–430.

Weber, 1925. Gustavus A. Weber. *The Bureau of Standards: Its History, Activities and Organization*. Institute for Government Research, Service Monographs of the United States Government, No. 35. Baltimore: Johns Hopkins Press.

Weber, 1928. Gustavus A. Weber. *The Bureau of Chemistry and Soils: Its History, Activities and Organization*. Institute for Government Research, Service Monographs of the United States Government, No. 52. Baltimore: Johns Hopkins Press.

Weinberg, 1967. Alvin M. Weinberg. *Reflections on Big Science*. Cambridge, Mass.: MIT Press.

White, 1928. Alfred H. White. 'Chemical Engineering Education in the United States'. *Transactions of the American Institute of Chemical Engineers* **21**: 55–85.

White, 1931. Alfred H. White. 'Occupations and Earnings of Chemical Engineering Graduates'. *Transactions of the American Institute of Chemical Engineers* **27**: 221–252.

Whittemore, 1975. Gilbert F. Whittemore, Jr. 'World War I, Poison Gas Research, and the Ideals of American Chemists'. *Social Studies of Science* **5**: 135–163.

Wiebe, 1967. Robert H. Wiebe. *The Search for Order, 1877–1920*. New York: Hill and Wang.

Wigglesworth, 1928. Henry Wigglesworth. 'Chemical Industries'. In *Representative Industries of the United States*, edited by H. T. Warshow, 130–181. New York: Henry Holt and Company.

Wilensky, 1964. Harold L. Wilensky. 'Mass Society and Mass Culture'. *American Sociological Review* **29**: 173–197.

Wiley, 1899. H. W. Wiley. 'The Relation of Chemistry to the Progress of Agriculture'. In *Yearbook of the United States Department of Agriculture, 1899*, 201–258. Washington: Government Printing Office, 1900.

Wilford, 1976a. John Noble Wilford. 'Ford Accused by 10 Nobel Winners of Inaccurate Claims on Science Aid'. *New York Times*, October 27, 30, cols. 5–6.

Wilford, 1976b. John Noble Wilford. 'U.S. Science Is Fine, But Could Be Better', *New York Times*, October 24, Section IV, 16, cols. 1–6.

Williamson *et al.*, 1963. Harold F. Williamson, Ralph L. Andreano, Arnold F. Daum, and Gilbert C. Klose. *The American Petroleum Industry: The Age of Energy, 1899–1959*. Evanston, Ill.: Northwestern University Press.

L. R. Wilson, 1957. Louis R. Wilson. *The University of North Carolina, 1900–1930. The Making of a Modern University*. Chapel Hill: University of North Carolina Press.

M. Wilson, 1975. M. Kent Wilson. 'The Top Twenty and the Rest: Big Chemistry and Little Funding'. *Annual Review of Physical Chemistry* **26**: 1–16.

Wise, 1980. George Wise. 'A New Role for Professional Scientists in Industry: Industrial Research at General Electric, 1900–1916'. *Technology and Culture* **21** (1980): 408–429.

Wise, 1983. George Wise. 'Ionists in Industry: Physical Chemistry at General Electric, 1900–1915'. *Isis* **74** (1983): 7–21.

Yates, 1929. Raymond Francis Yates. 'Wall Street and the Research Laboratory'. *Scientific American* **141**: 382–384.

Yerkes, 1920. Robert M. Yerkes (ed.). *The New World of Science: Its Development during the War*. New York: The Century Company.

'Zettlemoyer Wins', 1979. 'Zettlemoyer Wins ACS Election'. *Chemical and Engineering News* **57** (19 November): 7–8.

Zuckerman, 1970. Harriet Zuckerman. 'Stratification in American Science'. *Sociological Inquiry* **40**: 235–257.

Zuckerman, 1977. Harriet Zuckerman. *Scientific Elite: Nobel Laureates in the United States*. New York: Free Press.

Zuckerman and Merton, 1972. Harriet Zuckerman and Robert K. Merton. 'Age, Aging, and Age Structure in Science'. In *A Theory of Age Stratification*, Vol. 3 of *Aging and Society*, edited by M. W. Riley, Marylin Johnson, and Ann Foner, 292–356. New York: Russell Sage Foundation.

INDEX

Numbers in boldface type are page references for the illustrations and the text tables; those in italics for the reference tables following the main text. A lowercase n follows page references for footnotes. Numbers in parentheses are years.

GPSR Compliance

The European Union's (EU) General Product Safety Regulation (GPSR) is a set of rules that requires consumer products to be safe and our obligations to ensure this.

If you have any concerns about our products, you can contact us on ProductSafety@springernature.com

In case Publisher is established outside the EU, the EU authorized representative is:

Springer Nature Customer Service Center GmbH
Europaplatz 3
69115 Heidelberg, Germany

Batch number: 10823811

Printed by Printforce, the Netherlands